The Handbook of Environmental Chemistry

Volume 5 Water Pollution
Part E

O. Hutzinger
Editor-in-Chief

Springer

Berlin
Heidelberg
New York
Barcelona
Hong Kong
London
Milan
Paris
Tokyo

Pollutant-Solid Phase Interactions Mechanisms, Chemistry and Modeling

By
T. A. T. Aboul-Kassim · B. R. T. Simoneit

Springer

Environmental chemistry is a rather young and interdisciplinary field of science. Its aim is a complete description of the environment and of transformations occurring on a local or global scale. Environmental chemistry also gives an account of the impact of man's activities on the natural environment by describing observed changes.

"The Handbook of Environmental Chemistry" provides the compilation of today's knowledge. Contributions are written by leading experts with practical experience in their fields. The Handbook will grow with the increase in our scientific understanding and should provide a valuable source not only for scientists, but also for environmental managers and decision makers.

ISSN 1433-6863
ISBN 3-540-41650-1
Springer-Verlag Berlin Heidelberg New York

Library of Congress Cataloging-in-Publication Data

The Natural environment and the biogeochemical cycles / with
contributions by P. Craig ... [et al.].
v. <A–F > : ill. ; 25 cm. – (The Handbook of environmental chemistry :
v. 1) Includes bibliographical references and indexes.
ISBN 0-387-09688-4 (U.S). – ISBN 3-540-55255-3 (pt. F : Berlin). –
ISBN 0-387-55255-3 (pt. F : New York)
1. Biogeochemical cycles. 2. Environmental chemistry.
I. Craig, P. J., 1944– . II. Series.
QD31. H335 vol. 1 [QH344] 628.5 s

Springer-Verlag Berlin Heidelberg New York
a member of BertelsmannSpringer Science+Business Media GmbH
http//www.springer.de

Printed in Germany

Production Editor: Christiane Messerschmidt, Rheinau
Cover Design: E. Kirchner, Springer-Verlag
Typesetting: Fotosatz-Service Köhler GmbH, Würzburg
Printed on acid-free paper SPIN: 106 383 18 52/3020 – 5 4 3 2 1 0

The Handbook of Environmental Chemistry

Preface

Environmental Chemistry is a relatively young science. Interest in this subject, however, is growing very rapidly and, although no agreement has been reached as yet about the exact content and limits of this interdisciplinary discipline, there appears to be increasing interest in seeing environmental topics which are based on chemistry embodied in this subject. One of the first objectives of Environmental Chemistry must be the study of the environment and of natural chemical processes which occur in the environment. A major purpose of this series on Environmental Chemistry, therefore, is to present a reasonably uniform view of various aspects of the chemistry of the environment and chemical reactions occurring in the environment.

The industrial activities of man have given a new dimension to Environmental Chemistry. We have now synthesized and described over five million chemical compounds and chemical industry produces about hundred and fifty million tons of synthetic chemicals annually. We ship billions of tons of oil per year and through mining operations and other geophysical modifications, large quantities of inorganic and organic materials are released from their natural deposits. Cities and metropolitan areas of up to 15 million inhabitants produce large quantities of waste in relatively small and confined areas. Much of the chemical products and waste products of modern society are released into the environment either during production, storage, transport, use or ultimate disposal. These released materials participate in natural cycles and reactions and frequently lead to interference and disturbance of natural systems.

Environmental Chemistry is concerned with reactions in the environment. It is about distribution and equilibria between environmental compartments. It is about reactions, pathways, thermodynamics and kinetics. An important purpose of this Handbook, is to aid understanding of the basic distribution and chemical reaction processes which occur in the environment.

Laws regulating toxic substances in various countries are designed to assess and control risk of chemicals to man and his environment. Science can contribute in two areas to this assessment; firstly in the area of toxicology and secondly in the area of chemical exposure. The available concentration ("environmental exposure concentration") depends on the fate of chemical compounds in the environment and thus their distribution and reaction behaviour in the environment. One very important contribution of Environmental Chemistry to the above mentioned toxic substances laws is to develop laboratory test methods, or mathematical correlations and models that predict the environmental fate of new chemical compounds. The third purpose of this Handbook is to help in the basic understanding and development of such test methods and models.

The last explicit purpose of the Handbook is to present, in concise form, the most important properties relating to environmental chemistry and hazard assessment for the most important series of chemical compounds.

At the moment three volumes of the Handbook are planned. Volume 1 deals with the natural environment and the biogeochemical cycles therein, including some background information such as energetics and ecology. Volume 2 is concerned with reactions and processes in the environment and deals with physical factors such as transport and adsorption, and chemical, photochemical and biochemical reactions in the environment, as well as some aspects of pharmacokinetics and metabolism within organisms. Volume 3 deals with anthropogenic compounds, their chemical backgrounds, production methods and information about their use, their environmental behaviour, analytical methodology and some important aspects of their toxic effects. The material for volume 1, 2 and 3 was each more than could easily be fitted into a single volume, and for this reason, as well as for the purpose of rapid publication of available manuscripts, all three volumes were divided in the parts A and B. Part A of all three volumes is now being published and the second part of each of these volumes should appear about six months thereafter. Publisher and editor hope to keep materials of the volumes one to three up to date and to extend coverage in the subject areas by publishing further parts in the future. Plans also exist for volumes dealing with different subject matter such as analysis, chemical technology and toxicology, and readers are encouraged to offer suggestions and advice as to future editions of "The Handbook of Environmental Chemistry".

Most chapters in the Handbook are written to a fairly advanced level and should be of interest to the graduate student and practising scientist. I also hope that the subject matter treated will be of interest to people outside chemistry and to scientists in industry as well as government and regulatory bodies. It would be very satisfying for me to see the books used as a basis for developing graduate courses in Environmental Chemistry.

Due to the breadth of the subject matter, it was not easy to edit this Handbook. Specialists had to be found in quite different areas of science who were willing to contribute a chapter within the prescribed schedule. It is with great satisfaction that I thank all 52 authors from 8 countries for their understanding and for devoting their time to this effort. Special thanks are due to Dr. F. Boschke of Springer for his advice and discussions throughout all stages of preparation of the Handbook. Mrs. A. Heinrich of Springer has significantly contributed to the technical development of the book through her conscientious and efficient work. Finally I like to thank my family, students and colleagues for being so patient with me during several critical phases of preparation for the Handbook, and to some colleagues and the secretaries for technical help.

I consider it a privilege to see my chosen subject grow. My interest in Environmental Chemistry dates back to my early college days in Vienna. I received significant impulses during my postdoctoral period at the University of California and my interest slowly developed during my time with the National Research

Council of Canada, before I could devote my full time of Environmental Chemistry, here in Amsterdam. I hope this Handbook may help deepen the interest of other scientists in this subject.

Amsterdam, May 1980 *O. Hutzinger*

Twentyone years have now passed since the appearance of the first volumes of the Handbook. Although the basic concept has remained the same changes and adjustments were necessary.

Some years ago publishers and editor agreed to expand the Handbook by two new open-end volume series: Air Pollution and Water Pollution. These broad topics could not be fitted easily into the headings of the first three volumes. All five volumes series are integrated through the choice of topics and by a system of cross referencing.

The outline of the Handbook is thus as follows:

1. The Natural Environment and the Biochemical Cycles,
2. Reaction and Processes,
3. Anthropogenic Compounds,
4. Air Pollution,
5. Water Pollution.

Rapid developments in Environmental Chemistry and the increasing breadth of the subject matter covered made it necessary to establish volume-editors. Each subject is now supervised by specialists in their respective fields.

A recent developments is the accessibility of all new volumes of the Handbook from 1990 onwards, via the Springer Homepage http:// www.springer.de or http://Link.springer.de/series/hec/ or http://Link.springer-ny.com/series/hec/.

During the last 5 to 10 years there was a growing tendency to include subject matters of societal relevance into a broad view of Environmental Chemistry. Topcis include LCA (Live Cycle Analysis), Environmental Management, Sustainable Development and others. Whilst these topics are of great importance for the development and acceptance of Environmental Chemistry Publishers and Editors have decided to keep the Handbook essentially a source of information on "hard sciences".

With books in press and in preparation we have now well over 40 volumes available. Authors, volume-editors and editor-in-chief are rewarded by the broad acceptance of the "Handbook" in the scientific community.

Bayreuth, June 2001 *Otto Hutzinger*

Contents

Foreword

Organic compounds, both natural and synthetic, are a vital part of everyday life. They come in various forms such as foods, fuels, antibiotics, plastic containers, rubber tires, agricultural fertilizers, photocopying compounds, etc. Society cannot survive in its present form without these materials from the chemical industry. Growth in the numbers of organic chemicals during recent decades has been extraordinary. Presently, more than 70,000 organic compounds are in commercial production with approximately 1000 added each year. Most are complex compounds that can be released directly and/or indirectly to the surrounding environment. Of these, more than 1000 compounds are of environmental concern because of their production quantities, toxicity, persistence, and tendency to bioaccumulate. A view is emerging in relation to environmental protection and hazardous substance management that: (1) some organic chemicals and/or organic leachates from solid waste materials (SWMs) and contaminated sites are of such extreme environmental concern that all use should be highly controlled including isolation for disposal, and (2) most hazardous substances are of sufficient social value that their continual use, production and disposal are justified. For these chemicals their sources, fate, behavior and effects must be fully assessed and understood. Assessment and understanding are essential for society to accept risks of adverse ecological or human health effects.

Concern regarding the adverse effects of organic compounds has resulted in the worldwide initiation of plans for the registration of new chemicals before commercial use has commenced. Examples of these are the US Toxic Substances Control Act, the Environmental Contaminants Act in Canada, and the Scheme for the Hazard Assessment of Chemicals (OECD) Guidelines in the European Community.

In order to have a better and healthy environment, it is important that the environmental chemodynamics of such complex organic mixtures (COMs) and/or leachates from SWMs and contaminated sites be predicted accurately. It is imperative that the molecular organic composition of such pollutant mixtures, their transport processes and migration in and between the various multimedia environments be fully understood. In addition, their chemical and biochemical transformation processes (e.g., sorption, desorption, photolysis, volatilization and biodegradation) as well as their effects on the interacting organisms, that occupy these multimedia, should be completely delineated.

Answers to some of these questions can be found in the present book which addresses those properties of an organic compound present in solution and/or leachate that determine its tendency to: (1) adsorb on solid phase systems (i.e.

soil, sediment, suspended matter, colloids and biocolloids/biosolids), (2) leach through the subsurface environment, (3) volatilize/evaporate into the atmosphere, or (4) be absorbed across a biological membrane and bioconcentrated/ biomagnified in the aquatic environment. In addition, after being released into the environment, an organic pollutant leached from complex mixtures may be photochemically degraded, oxidized/reduced, hydrolyzed or metabolized by microorganisms. The important consideration in environmental chemodynamics of organic pollutants and leachates of COMs is which pollutants react in a given transformation process, what product will be formed and at what rate these changes occur. An understanding of these ideas is important in many aspects such as defining exposure and predicting hazard from such complex mixtures, and providing a basis for developing strategies for preventing or minimizing exposure.

Accordingly, the present book, entitled *"Pollutant-Solid Phase Interactions: Mechanisms, Chemistry and Modeling"*, is considered an essential step toward a comprehensive understanding of an important chemodynamic mechanism such as the interactions between organic pollutants (i.e., present in aqueous systems or as SWM leachates) and various solid phases. The present volume is divided into five interrelated chapters. Each chapter has its own objectives.

The first chapter, entitled *"Organic Pollutants in Aqueous-Solid Phase Environments: Types, Analyses and Characterizations"*, has three main goals. The first is to present a complete review of the most toxic organic pollutant types which are present in both aqueous and solid phase environments. These pollutants include petroleum hydrocarbons, pesticides, phthalates, phenols, PCBs, organotin compounds and surfactants. The second goal is to provide a complete and comprehensive critical review of the different analytical techniques used for the determination of these organic compounds. The third objective is to discuss and review the up-to-date instrumental developments and advancements for the identification and characterization of such organic compounds.

The second chapter, entitled *"Interaction Mechanisms between Organic Pollutants and Solid Phase Systems"*, discusses the different solid phase compositions, the interaction mechanisms between these phases and the organic pollutants, and factors affecting sorption mechanisms with the role-played by humic substances in such interactions.

Chapter three, entitled *"Sorption/Desorption of Organic Pollutants from Complex Mixtures: Modeling, Kinetics, Experimental Techniques and Transport Parameters"*, reviews the most widely used modeling techniques in the field in order to analyze sorption/desorption data generated from environmental systems. Some important aspects of the kinetics, chemistry and modeling approaches of sorption/desorption processes of solid phase systems are discussed. In addition, a background theory and experimental techniques for the different sorption/ desorption processes are considered, while the estimation of transport parameters of organic pollutants from such laboratory studies are presented and evaluated.

Chapter four, entitled *"QSAR/QSPA and Multicomponent Joint Toxic Effect Modeling of Organic Pollutants at Aqueous-Solid Phase Interfaces"*, reviews several interdisciplinary approaches, such as: (1) some physical and chemical properties of organic pollutants in complex mixtures which can affect their sorption/ desorption chemodynamics, (2) the fundamentals of both quantitative-structure activity and structure-property relationships, with special emphasis on using molecular connectivity indices as useful properties for pollutant mobility and bioavailability prediction, and (3) the multicomponent joint toxic/gentoxic effect models to predict the bioavailable fraction and action of organic pollutants at aqueous-solid phase interfaces. Then, it discusses and evaluates how these interdisciplinary approaches can be applied and integrated using a group of toxic and carcinogenic pollutants such as PCBs and PAHs.

The last chapter, entitled *"Microbial Transformations at Aqueous-Solid Phase Interfaces: A Bioremediation Approach"*, presents the basic principles of microbial associations at aqueous-solid phase interfaces, and the types and mechanisms of biodegradation and biotransformation, showing how those principles relate to bioremediation engineering technologies. It considers some of the microbiological, chemical, environmental, engineering and technological aspects of biodegradation/biotransformation.

It is important to mention that information about chemodynamics of organic pollutants at aqueous-solid phase interfaces is too large, diverse and multidisciplinary, and its knowledge base is expanding too rapidly to be fully covered in a single book. Nevertheless, we have attempted to present, as much as we can, the most important and valid key principles that underlie the science and engineering for possible chemistry, interactions, and modeling at these interfaces. Thus, knowledge from one or two disciplines was not sufficient, information from many disciplines was needed. For instance, the characterization of organic pollutants requires knowledge of organic chemistry/geochemistry; their fate and behavior follow chemical and physical principles; changes in toxicity, hazard and exposure represent topics of concern in environmental toxicology; the areas (media) containing the organic pollutants represent interfacial environments with unique properties; and the modeling techniques are based on approaches common in environmental engineering. Thus, this book is addressed to chemists, ecologists, oceanographers, environmental scientists and environmental engineers and should prove to be of use.

We hope that the present information helps to continue the search for creative and economical ways of limiting the release of contaminants into the environment, developing highly sensitive techniques for tracking organic pollutants once released, and to find effective methods to remediate our spoiled resources and environment.

Corvallis, Oregon, USA, March 2001

Tarek A. T. Aboul-Kassim
Bernd R. T. Simoneit

CHAPTER 1

Organic Pollutants in Aqueous-Solid Phase Environments: Types, Analyses and Characterizations

Tarek A.T. Aboul-Kassim[1], Bernd R.T. Simoneit[2]

[1] Department of Civil, Construction and Environmental Engineering, College of Engineering, Oregon State University, 202 Apperson Hall, Corvallis, OR 97331, USA
e-mail: Tarek.Kassim@orst.edu

[2] Environmental and Petroleum Geochemistry Group, College of Oceanic and Atmospheric Sciences, Oregon State University, Corvallis, OR 97331, USA
e-mail: simoneit@oce.orst.edu

In order to study the chemodynamic behavior (i.e., fate and transport) of organic pollutants in the environment and their interactions with various solid phase systems, our goals in this chapter are to address these aspects. The first is to present a review of the most toxic organic pollutant types which are present in both aqueous and solid phase environments. These pollutants include petroleum hydrocarbons, pesticides, phthalates, phenols, PCBs, organotin compounds, and surfactants as well as complex organic mixtures (COMs) of pollutants leached from solid waste materials (SWMs) in landfills and disposal sites. The term *solid phase* system is used here to indicate soil-particulate matter, sediment, suspended, and biological materials. The second goal is to provide a comprehensive review of the different analytical techniques used for the determination of these organic compounds. The third objective is to discuss and evaluate the current instrumental developments and advances for the identification and characterization of these organic compounds. This chapter serves as the backbone for the subsequent chapters in the present volume, and aids in understanding the various interaction mechanisms between organic pollutants and diverse solid phase surfaces, their chemistry, and applicable modeling techniques.

Keywords. Organic pollutants, Hydrocarbons, Pesticides, Phthalates, Phenols, PCBs, Surfactants, Instrumentation, Identification, Characterization, Aqueous-solid phase systems

The Handbook of Environmental Chemistry Vol. 5 Part E
Pollutant-Solid Phase Interactions: Mechanism, Chemistry and Modeling
(by T.A.T. Aboul-Kassim, B.R.T. Simoneit)

List of Abbreviations

BSTFA	Bis(trimethylsilyl)trifluoroacetamide
CI	Chemical ionization
COMs	Complex organic mixtures
CSIA	Compound specific isotope analysis
DEHP	Diethyl phthalate
DOP	Dioctyl phthalate
ECD	Electron capture detector
EI	Electron impact
EPA	Environmental Protection Agency
ESI	Electrospray ionization
FAB	Fast-atom bombardment
FI	Field ionization
GC	Gas chromatography
GC-AED	Gas chromatography with atomic emission detection

GC-FPD	Gas chromatograph with flame photometric detection
GC-MS	Gas chromatography-mass spectrometry
GPC	Gel permeation chromatography
HCs	Hydrocarbons
HPLC	High performance liquid chromatography
HTGC-MS	High temperature gas chromatography-mass spectrometry
IDMS	Isotope dilution mass spectrometry
IRMS	Isotope ratio mass spectrometry
ITD	Ion trap detector
LC	Liquid chromatography
LIMS	Laser ionization mass spectrometry
LLE	Liquid-liquid extraction
MALDI	Matrix-assisted laser desorption ionization
MS	Mass spectrometry
OCPs	Organochlorine pesticides
PAEs	Phthalic acid esters
PAHs	Polycyclic aromatic hydrocarbons
PCBs	Polychlorinated biphenyls
PD	Plasma desorption
PGD	Plasma and glow discharge
RIMS	Resonance ionization mass spectrometry
SFC	Supercritical fluid chromatography
SFE	Supercritical fluid extraction
SIMS	Secondary ionization mass spectrometry
SPE	Solid phase extraction
SPME	Solid phase microextraction
SSJ/LIF	Supersonic jet laser-induced fluorescence
SWMs	Solid waste materials
TOC	Total organic carbon
TOF-MS	Time of flight-mass spectrometry
TPs	Transformation products

1 Introduction

The twenty-first century can properly be called the age of organic chemistry due to the huge worldwide increase in organic chemical production (more than 70,000 compounds) and utilization. Many of these organic compounds have proven to be toxic, carcinogenic, and mutagenic to various aquatic organisms and, directly and/or indirectly, to humans [1]. The dramatic increase in the production of organic chemicals has completely altered our immediate human environment and provided a wealth of new compounds which, in many cases, were more toxic and carcinogenic than the parent compounds.

With environmental protection high on the agenda of many industrial countries, new rules and regulations are currently being set up for monitoring greater numbers of hazardous organic pollutants. Organic pollutants present in the various environmental multimedia may occur naturally [2] and/or derive from anthropogenic sources [3–13]. Anthropogenic input may derive from industrial sources [14–20], urban wastes [21–35], agricultural activity [36–44], and from degradation products [45–52]. Organic pollutants have different polarities and chemical properties; hence, low detection limits are necessary for studying the fate and transport of these organic compounds in and/or within the different environmental multimedia, as well as their interactive behavior with other solid phase surfaces.

Accordingly, environmental organic analysis has expanded dramatically in the last 25 years. With the development of commercially available gas chromatography-mass spectrometer (GC-MS) systems, there has been a significant increase in the number of organic pollutant fingerprints that have been discovered and identified [53–73]. Identities of individual compounds or compositional fingerprints can be determined by highly sophisticated and advanced instruments [5, 64, 74–88] and are used to provide information about the type [62, 64, 82, 89–92], amount [89, 93–96], and source confirmation [1, 53–55, 97] of these pollutants.

Different terms have been used in the literature to describe various environmental organic pollutants/contaminants that are characterized in terms of their molecular structures [1, 53–55]. The term chemical fossil was first used by Eglinton and Calvin [98] to describe organic compounds in the geosphere whose carbon skeleton suggested an unambiguous link with a known natural product. In addition, other terms such as biological markers, organic tracers, biomarkers, or molecular fossils, have also been used to describe such organic compounds [1, 53–56, 60, 61, 63, 66, 68–73]. In line with the current trends in environmental organic chemistry and for the sake of consistency, the term molecular marker (MM) suggested by Aboul-Kassim [1] will be used in this book to describe both naturally occurring (i.e., biological and hence biomarker) and/or anthropogenically derived organic (i.e., non-biomarker) compounds that are present in both aqueous and solid phase environments.

The main objectives of this chapter are: (1) to review the different toxic organic pollutants present in both liquid and solid (i.e., sediment, soil, suspended matter and biosolids as bacteria, plankton, etc.) phase environments as well as complex organic mixture (COM) leachates from solid waste materials of landfills and disposal sites; (2) to summarize the most recent analyses of these MM pollutants; and (3) to discuss the optimum instrumental analytical methods for organic pollutant characterization.

It is intended that the review of the different aspects and goals in this chapter provides an up-to-date background for the succeeding chapters in this volume. This will clarify the discussions about the different interaction mechanisms between organic pollutants and various solid phases, their chemistry, and applicable modeling techniques that are presented in the subsequent chapters.

2
Types of Organic Pollutants

Approximately one-half of the industrially produced organic chemicals reach the global environment via direct and/or indirect routes, for example agricultural practices, municipal and industrial wastes, and landfill effluents. These products include a variety of pesticides and their metabolites, aliphatic and aromatic organic derivatives of petroleum hydrocarbons and plastics, organic solvents and detergents, phenols, PCBs, and organotin compounds. When these substances reach the natural environment, various degradation and transfer processes are initiated. The chemical properties of each organic compound (such as molecular structure, volatility, ionic charge and ionizability, polarizability, and water-solubility) determine which processes predominate. Currently the prevalent opinion is that interaction processes, leading to activation inactivation, physical sorption, and/or chemical binding or partitioning are among the most widespread and important phenomena affecting toxic organic pollutants in the global environment. Some general considerations and properties of major organic pollutant groups, of relevance to the environment and of importance to human health, will be summarized briefly in the following subsections.

2.1
Petroleum Hydrocarbons

Hydrocarbons (HCs) of petroleum origin are widespread organic pollutants that are found in both aquatic and solid phase environments [1, 53–56, 99, 100]. The most common groups of compounds are aliphatic and polycyclic aromatic hydrocarbons (PAHs). Of these the PAHs are toxic, carcinogenic, and sometimes mutagenic to both aquatic organisms and ultimately humans [1]. The following is a brief description of each group.

2.1.1
Aliphatic Compounds

Aliphatic hydrocarbons, a diverse suite of compounds, are an important lipid fraction which is either natural (i.e., from photosynthesis by marine biota inhabiting the surface waters or by terrestrial vascular plants) or anthropogenic (i.e., of petroleum origin from land runoff, and/or industrial inputs). Aliphatic hydrocarbons have been studied and characterized from various environmental multimedia [1, 53–56, 99–109].

Aliphatic hydrocarbons of petroleum origin (Fig. 1) (also coal) in the environment are usually composed of:

1. Homologous long chain *n*-alkane series ranging from $<C_{15}$ to $>C_{38}$ with no carbon number predominance [1, 53–55, 73, 109–114]
2. Unresolved complex mixture (UCM) of branched and cyclic hydrocarbons [1, 53–56, 68, 70, 113, 115–119]

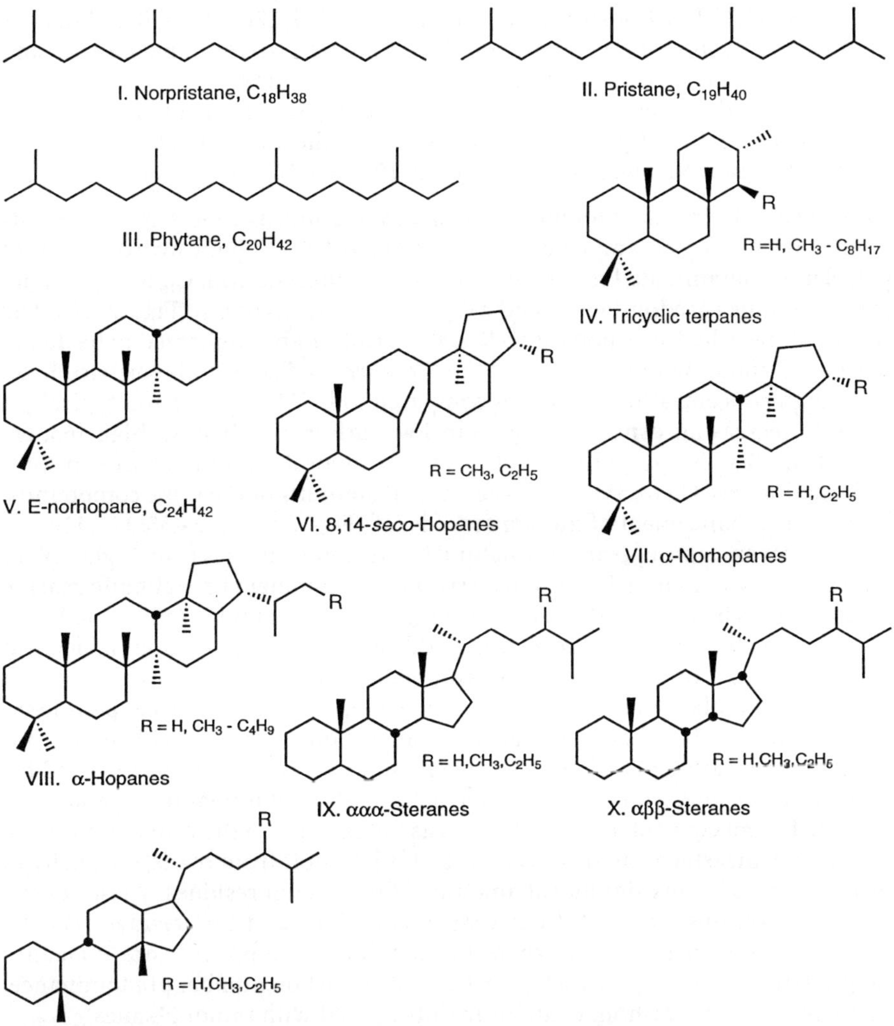

Fig. 1. Chemical structures of some aliphatic hydrocarbon molecular markers as cited in the text

3. Isoprenoid hydrocarbons such as norpristane (2,6,10-trimethylpentadecane), pristane (2,6,10,14-tetramethylpentadecane), and phytane (2,6,10,14-tetramethylhexadecane) (Structures I–III, Fig. 1) [1, 53–56, 68, 70, 120–123]
4. Tricyclic terpanes (Structure IV, Fig. 1), usually ranging from $C_{19}H_{34}$ to $C_{30}H_{56}$, and in some cases to $C_{45}H_{86}$ [68, 124–126]
5. Tetracyclic terpanes such as 17,21- and 8,14-*seco*-hopanes (Structures V–VI, Fig. 1) [125–127]
6. Pentacyclic triterpanes, such as the 17α(H),21β(H)-hopane series (Structures VII–VIII, Fig. 1), consisting of 17α(H)-22,29,30-trisnorhopane (T_m),

17α(H),21β(H)-29-norhopane, and the extended 17α(H),21β(H)-hopanes (>C_{31}) with subordinate amounts of the 17β(H),21α(H)-hopane series and 18α(H)-22,29,30-trisnorneohopane (T_s), [1, 53–55, 114]
7. Steranes and diasteranes with the 5α(H),14α(H),17α(H)-configuration (IX), 5α(H),14β(H),17β(H)-configuration (X), and the 13α(H),17β(H)-diasteranes (Structure XI, Fig. 1) (e.g., [1, 53–55, 101, 103, 105–107, 117]).

Typical GC-MS traces of aliphatic hydrocarbon patterns representative of different environmental samples are shown in Fig. 2. The aliphatic hydrocarbons of petroleum contaminated sediment and water are present from C_{16} to C_{38} with no carbon number predominance and a C_{max} at C_{21} and C_{30} or C_{32} (Figs. 2a, b). The source of these hydrocarbons as well as the UCM can be confirmed to be due to petroleum input by the presence of the biomarkers discussed below. Crude oil has a high concentration of alkanes compared to UCM (Fig. 2c) and typically a smooth decreasing concentration from low carbon numbers to high [63, 66, 111]. The alkanes <C_{20} are initially lost by evaporation and subsequent biodegradation (see Chap. 5) removes additional amounts of the same compounds, leaving an enhancement of the isoprenoids (cf., Figs. 2a, b) [53–55, 111, 116]. In contrast, an example of primarily natural background alkanes from higher plant waxes is shown in Fig. 2d. This is a terrigenous component brought into marine environments by river washout and atmospheric fallout and is sedimented with minerals. Such *n*-alkanes have a strong odd carbon number predominance and a C_{max} at C_{27}, C_{29}, or C_{31} [56]. A minor component from petroleum is also present as UCM and *n*-alkanes from C_{20} to C_{26}. An example of hydrothermal petroleum is shown in Fig. 2e, where the *n*-alkanes range from C_{13} to C_{25} with significant amounts of isoprenoids. There have been numerous compositions reported for petroleums formed from the hydrothermal alteration of immature organic matter in sediment covered marine rift areas as for example in the Gulf of California and the northeastern Pacific Ocean [128–130]. Runoff from roads, especially in urban areas, contains significant amounts of petroleum residues. These consist of lubrication oils, particles from vehicle emissions and fuel residues [1]. An example is shown in Fig. 2f, where the dominant components are *n*-alkanes ranging from C_{22} to C_{38}, with a C_{max} at C_{29} and no carbon number predominance. In other cases, the washout contains mainly a UCM with minor alkanes.

Characteristic examples of biomarker distributions typical for petroleum consisting of tricyclic terpanes (key ion *m/z* 191), hopanes (key ion *m/z* 191), and steranes/diasteranes (key ions 217, 218, 259) are shown in Fig. 3. The tertacyclic terpanes are not major components in the *m/z* 191 plots, because their key ion is at *m/z* 123. The tricyclic terpanes range from C_{21} to C_{29}, with a C_{max} at C_{23} and no C_{22} and C_{27}. The mature hopanes range from C_{27} to C_{35}, with the 17α(H),21β(H) configuration and the homologs >C_{31} are resolved into the C-22*S* and *R* diastereomers [68, 73, 68, 114]. The steranes range from C_{27} to C_{29} and are generally less concentrated than the hopanes. The mature sterane series have the 5α(H),14α(H),17α(H)- and 5α(H),14β(H),17β(H)-configurations with all homologs also resolved into the respective C-21 *S* and *R* diastereomers (Figs. 3b, c). The diastereomers also range from C_{27} to C_{29} and in part coelute with the steranes (Fig. 3b). A summary of the identifications of the various aliphatic hydrocarbons just discussed is given in Table 1.

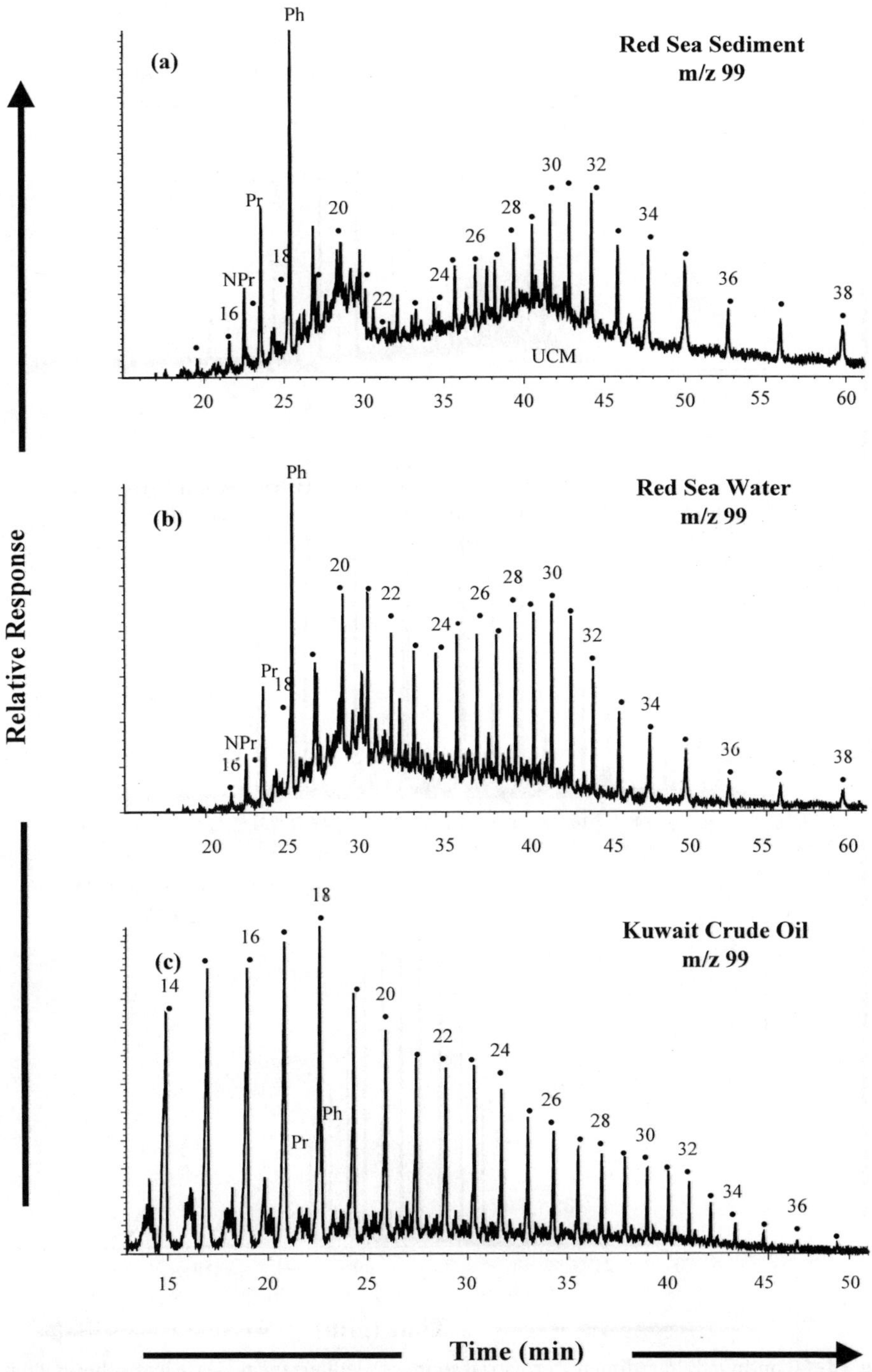

Fig. 2a–c. GC-MS traces (m/z 99 key ion) of various aliphatic hydrocarbon fractions from different environmental matrices: **a** sediment – Red Sea; **b** water – Red Sea; **c** Kuwait crude oil spill

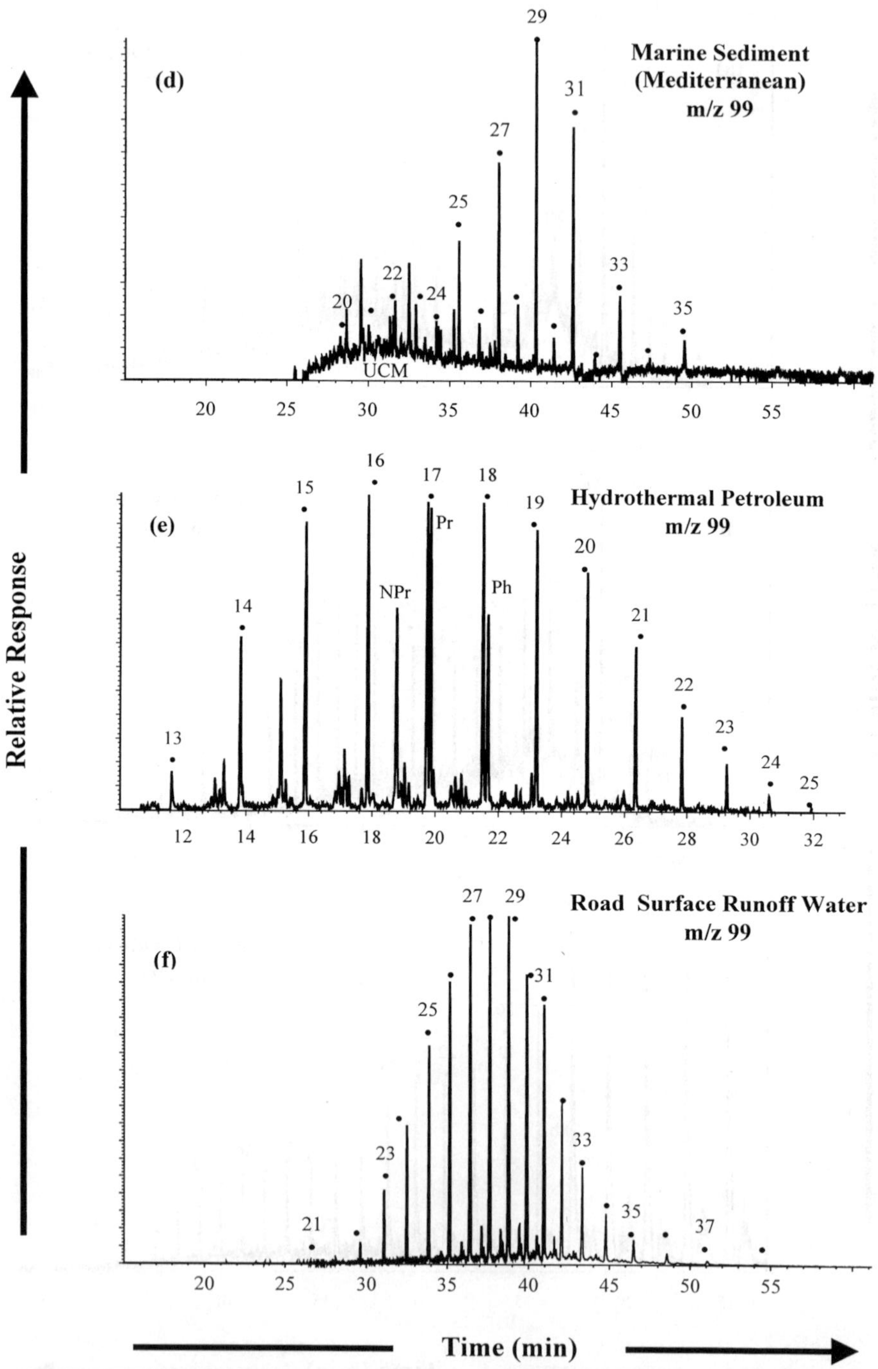

Fig. 2d–f (continued) **d** sediment, terrestrial source – Mediterranean Sea; **e** hydrothermal petroleum – Guaymas basin, Gulf of California; **f** road surface runoff water

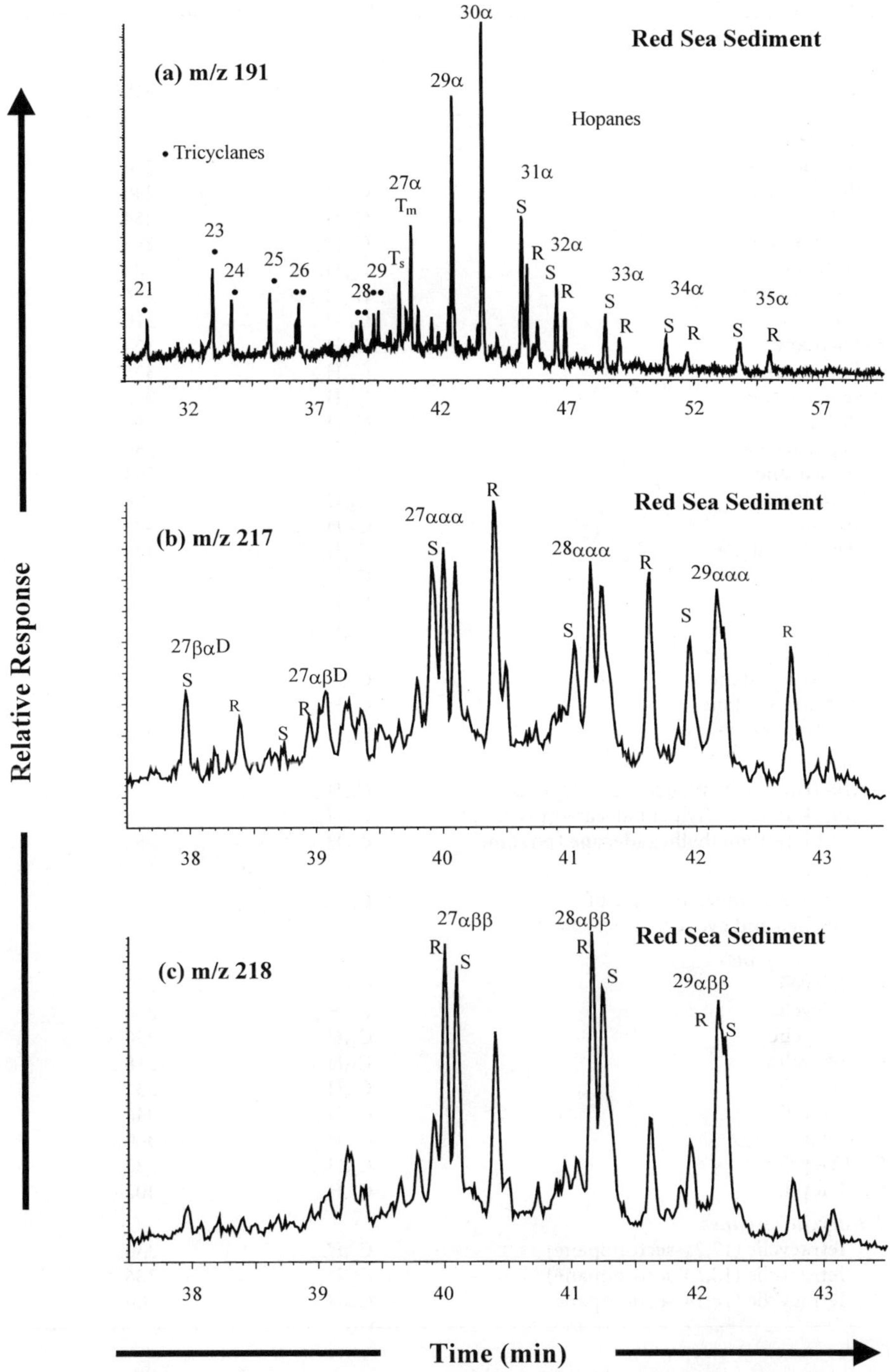

Fig. 3 a–c. GC-MS key ion traces representing the: **a** m/z 191 tricyclanes and $\alpha\beta$ hopane series; **b** m/z 217 $\alpha\alpha\alpha$-steranes and diasteranes; **c** m/z 218 $\alpha\beta\beta$-steranes (Red Sea sediment)

Table 1. Typical hydrocarbon identifications and chemical compositions (representative structures are shown in Fig. 1)

Compound Name	Composition	MW
n-Alkanes		
n-Hexadecane	$C_{16}H_{34}$	226
n-Heptadecane	$C_{17}H_{36}$	240
n-Octadecane	$C_{18}H_{38}$	254
n-Nonadecane	$C_{19}H_{40}$	268
n-Eicosane	$C_{20}H_{42}$	282
n-Heneicosane	$C_{21}H_{44}$	296
n-Docosane	$C_{22}H_{46}$	310
n-Tricosane	$C_{23}H_{48}$	324
n-Tetracosane	$C_{24}H_{50}$	338
n-Pentacosane	$C_{25}H_{52}$	352
n-Hexacosane	$C_{26}H_{54}$	366
n-Heptacosane	$C_{27}H_{56}$	380
n-Octacosane	$C_{28}H_{58}$	394
n-Nonacosane	$C_{29}H_{60}$	408
n-Triacontane	$C_{30}H_{62}$	422
n-Hentriacontane	$C_{31}H_{64}$	436
n-Dotriacontane	$C_{32}H_{66}$	450
n-Tritriacontane	$C_{33}H_{68}$	464
n-Tetratriacontane	$C_{34}H_{70}$	478
n-Pentatriacontane	$C_{35}H_{72}$	492
n-Hexatriacontane	$C_{36}H_{74}$	506
n-Heptatriacontane	$C_{37}H_{76}$	520
n-Octatriacontane	$C_{38}H_{78}$	534
Isoprenoids		
2,6,10-Trimethylpentadecane (norpristane)	$C_{18}H_{38}$	254
2,6,10,14-Tetramethylpentadecane (pristane)	$C_{19}H_{40}$	268
2,6,10,14-Tetramethylhexadecane (phytane)	$C_{20}H_{42}$	282
UCM		
Unresolved complex mixture of branched and cyclic hydrocarbons	C_{12}–C_{27}	
Tricyclic Terpanes		
C_{19}-Tricyclic	$C_{19}H_{34}$	262
C_{20}-Tricyclic	$C_{20}H_{36}$	276
C_{21}-Tricyclic	$C_{21}H_{38}$	290
C_{23}-Tricyclic	$C_{23}H_{42}$	318
C_{24}-Tricyclic	$C_{24}H_{44}$	332
C_{25}-Tricyclic	$C_{25}H_{46}$	346
C_{26}-Tricyclic	$C_{26}H_{48}$	360
C_{28}-Tricyclic	$C_{28}H_{52}$	388
C_{29}-Tricyclic	$C_{29}H_{54}$	402
Tetracyclic terpanes		
C_{24}-Tetracyclic (17,21-seco-hopane)	$C_{24}H_{42}$	330
C_{28}-Tetracyclic (18,14-seco-hopane)	$C_{28}H_{50}$	386
C_{29}-Tetracyclic (18,14-seco-hopane)	$C_{29}H_{52}$	400

Table 1 (continued)

Compound name	Composition	MW
Pentacyclic triterpanes		
18α(H)-22,29,30-Trisnorneohopane (Ts)	$C_{27}H_{46}$	370
17α(H)-22,29,30-Trisnorhopane (Tm)	$C_{27}H_{46}$	370
17α(H),21β(H)-29-Norhopane	$C_{29}H_{50}$	398
17α(H),21β(H)-Hopane	$C_{30}H_{52}$	412
17α(H),21β(H)-Homohopane (22S)	$C_{31}H_{54}$	426
17α(H),21β(H)-Homohopane (22R)	$C_{31}H_{54}$	426
17α(H),21β(H)-Bishomohopane (22S)	$C_{32}H_{56}$	440
17α(H),21β(H)-Bishomohopane (22R)	$C_{32}H_{56}$	440
17α(H),21β(H)-Trishomohopane (22S)	$C_{33}H_{58}$	454
17α(H),21β(H)-Trishomohopane (22R)	$C_{33}H_{58}$	454
17α(H),21β(H)-Tetrakishomohopane (22S)	$C_{34}H_{60}$	468
17α(H),21β(H)-Tetrakishomohopane (22R)	$C_{34}H_{60}$	468
17α(H),21β(H)-Pentakishomohopane (22S)	$C_{35}H_{62}$	482
17α(H),21β(H)-Pentakishomohopane (22R)	$C_{35}H_{62}$	482
Diasteranes		
13α(H),17β(H)-Diacholestane (20S)	$C_{27}H_{48}$	372
13α(H),17β(H)-Diacholestane (20R)	$C_{27}H_{48}$	372
Steranes		
5α(H),14α(H),17α(H)-Cholestane (20S)	$C_{27}H_{48}$	372
5α(H),14β(H),17β(H)-Cholestane (20R)	$C_{27}H_{48}$	372
5α(H),14β(H),17β(H)-Cholestane (20S)	$C_{27}H_{48}$	372
5α(H),14α(H),17α(H)-Cholestane (20R)	$C_{27}H_{48}$	372
5α(H),14α(H),17α(H)-Ergostane (20S)	$C_{28}H_{50}$	386
5α(H),14β(H),17β(H)-Ergostane (20R)	$C_{28}H_{50}$	386
5α(H),14β(H),17β(H)-Ergostane (20S)	$C_{28}H_{50}$	386
5α(H),14α(H),17α(H)-Ergostane (20R)	$C_{28}H_{50}$	386
5α(H),14α(H),17α(H)-Sitostane (20S)	$C_{29}H_{52}$	400
5α(H),14β(H),17β(H)-Sitostane (20R)	$C_{29}H_{52}$	400
5α(H),14β(H),17β(H)-Sitostane (20S)	$C_{29}H_{52}$	400
5α(H),14α(H),17α(H)-Sitostane (20R)	$C_{29}H_{52}$	400

2.1.2
Polycyclic Aromatic Compounds

Polycyclic aromatic hydrocarbons (PAHs, sometimes also called polynuclear aromatics, PNA) are a hazardous class of widespread pollutants. The parent structures of the common PAHs are shown in Fig. 4 and the alkylated homologs are generally minor in combustion emissions. PAHs are produced by all natural combustion processes (e.g., wild fires) and from anthropogenic activity such as fossil fuels combustion, biomass burning, chemical manufacturing, petroleum refining, metallurgical processes, coal utilization, tar production, etc. [6, 9, 15, 18, 20, 24, 131–139].

PAHs are neutral, nonpolar organic molecules consisting of two or more fused benzene rings arranged in various configurations with hydrophobicity increasing with molecular weight (Fig. 4). Many members of this class of

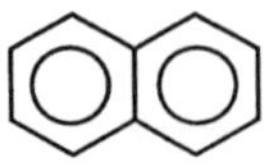
I. Naphthalene

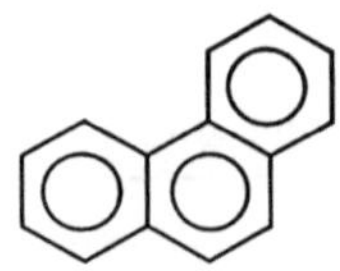
II. Phenanthrene

III. Anthracene

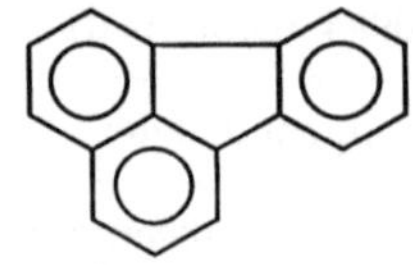
IV. Fluoranthene

V. Pyrene

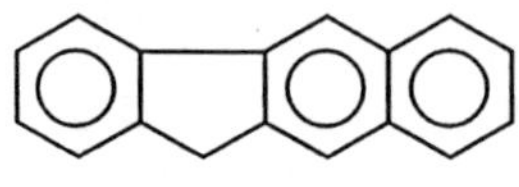
VI. Benzo[b]fluorene

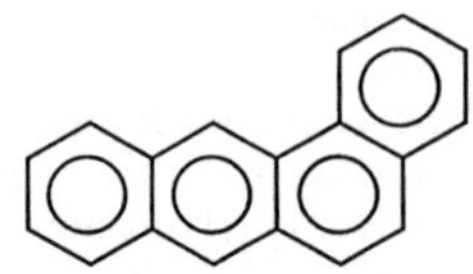
VII. Benz[a]anthracene

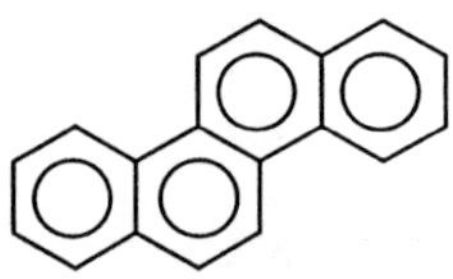
VIII. Chrysene

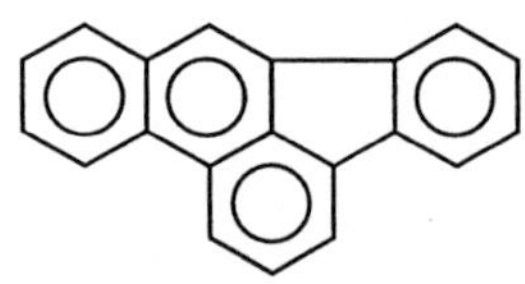
IX. Benzo[b]fluoranthene

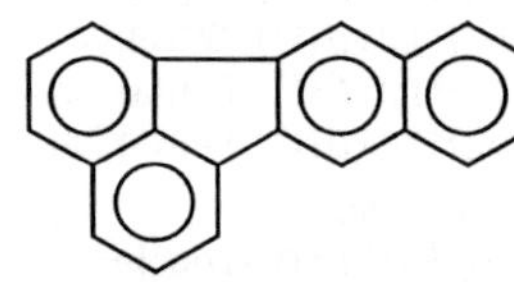
X. Benzo[k]fluoranthene

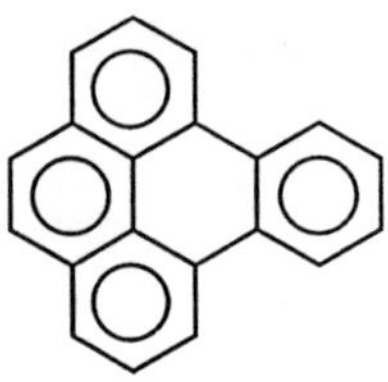
XI. Benzo[e]pyrene

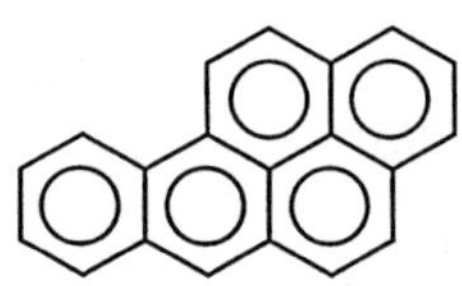
XII. Benzo[a]pyrene

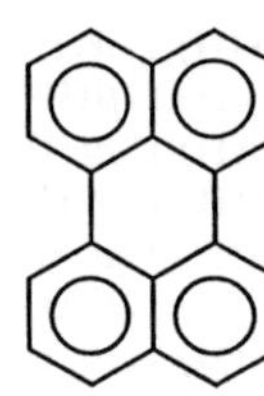
XIII. Perylene

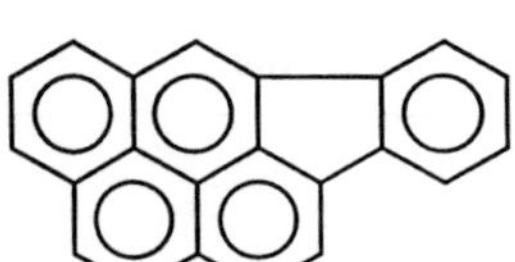
XIV. Indeno[1,2,3-cd]pyrene

XV. Benzo[ghi]perylene

XVI. Anthanthrene

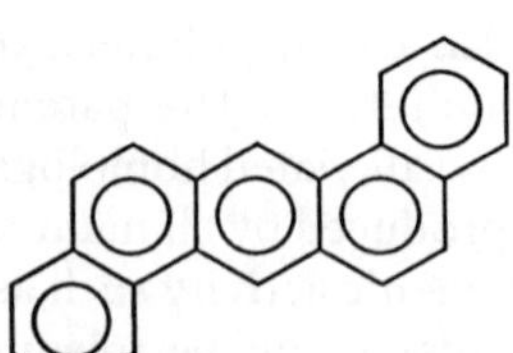
XVII. Dibenz[a,h]anthracene

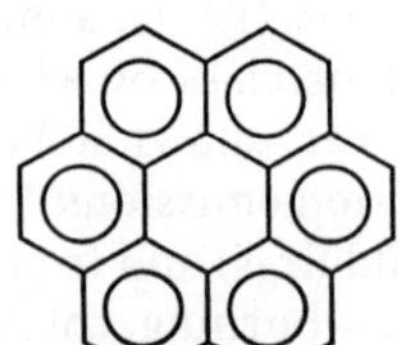
XVIII. Coronene

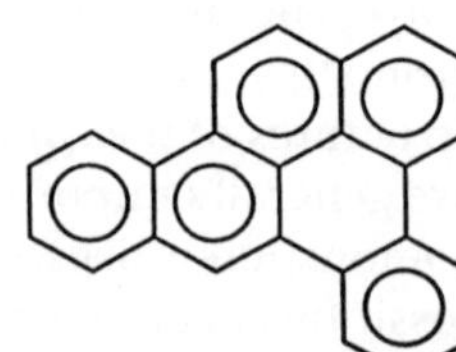
XIX. Dibenzo[a,e]pyrene

Fig. 4. Chemical structures of some examples of polycyclic aromatic hydrocarbons

compounds have been identified to exhibit toxic and mutagenic properties [140–142]. The World Health Organization has, therefore, recommended limits for certain PAHs in drinking water and the US-EPA has included 16 PAHs in its list of priority pollutants to be monitored in industrial effluents.

Although there is evidence that the environmental sources of PAHs also include natural inputs such as combustion (e.g., forest fires [139]), sediment diagenesis [56, 139], geological phenomena (e.g., tar pits, seepage from rock formations, and biological conversion of natural precursors [139]), most of the PAHs contamination of aquifers, soils, sediments, and water bodies comes from anthropogenic sources [9, 15, 18, 20, 24, 131–137]. Hence, the occurrence of PAHs in both aquatic and solid phase environments is generally recognized as contamination from anthropogenic sources. This is a cause for environmental concern because PAHs can be hazardous at very low concentrations and some PAHs are degraded relatively slowly. Because PAHs are hydrophobic, adsorption is very important in determining their fate in surface and subsurface water-soil/sediment systems.

Characteristic examples of typical distributions of PAHs in various environmental samples (GC-MS analysis) are shown in Fig. 5. The PAH distribution in a fallout sample from Alexandria shows a wide range of compounds with a predominance of high molecular weight PAHs such as pyrene, benzo[*a*]pyrene, anthanthrene and benzo[*g,h,i*]perylene (Fig. 5a). This represents a thermogenic/pyrolytic origin for these PAHs in the atmospheric organic matter at Alexandria City. Similarly, a leachate from municipal solid waste (MSW) bottom incineration ash, currently generated in large quantities in the United States and used as a highway construction and repair material, shows the presence of several high molecular weight PAH compounds such as fluoranthene, pyrene, benz[*a*]anthracene, benzo[*b*+*k*]fluoranthenes, benzo[*e*]pyrene, benzo[*a*]pyrene, indenopyrene, benzoperylene, dibenzanthracene, anthanthrene, dibenzoperylene, and coronene (Fig. 5b). This confirms the high temperature pyrolytic source for these compounds which can present a serious health and ecosystem hazard due to their toxic and genotoxic characters (see Chap. 4). On the other hand, a hydrothermal petroleum sample from Escanaba Trough, Northeast Pacific Ocean [143] shows an abundance of low molecular weight PAHs such as naphthalene, phenanthrene, etc., with some of their alkylated C_1- and C_2-homologs (Fig. 5c), indicating a single petroleum end member source for this sample.

The alkyl-substituent pattern for some PAHs series (e.g., alkylnaphthalenes, phenanthrene/anthracene, pyrene/fluoranthene, *m/z* 228 and *m/z* 252) are shown in Figs. 6–9, respectively. The parent PAHs and their alkylated homologs are determined in GC-MS data by monitoring their corresponding molecular weights. For example, for the naphthalene series the ions at *m/z* 128, 142 methylnaphthalenes, 156 C_2-naphthalenes, 170 C_3-naphthalenes, and 184 C_4-naphthalenes are monitored (Fig. 6). The GC elution orders of the C_2-naphthalene and C_3-naphthalene isomers have been reported [144, 145].

The phenanthrene/anthracene series is shown in Fig. 7 and the major peak in the *m/z* 234 trace has the retention index of retene which is generally derived from conifer wood burning. Sometimes there is a triplet of peaks in the same C_4 plot due to benzonaphthothiophenes ($C_{10}H_{16}S$) which are components of some

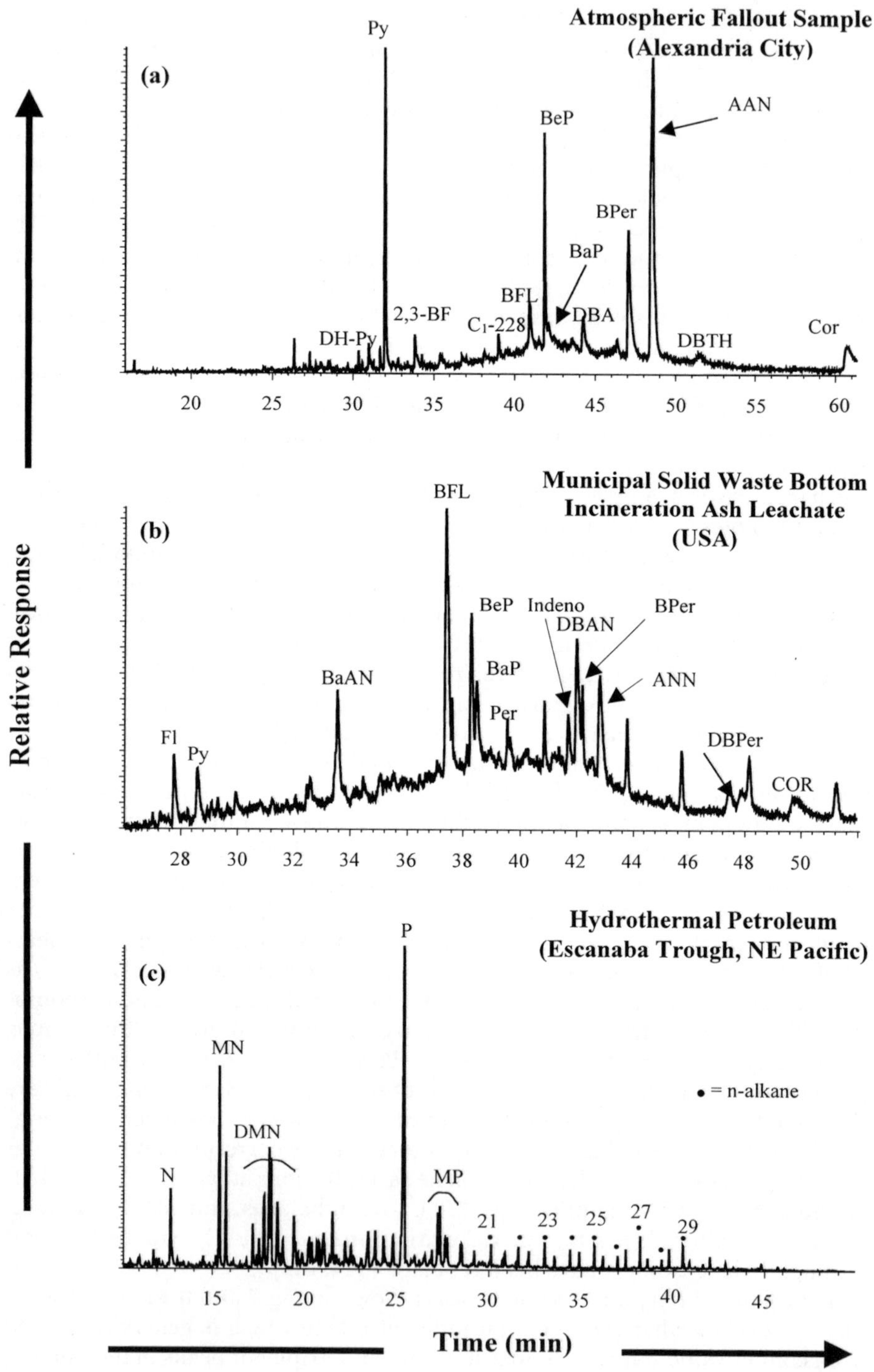
Atmospheric Fallout Sample
(Alexandria City)
(a)
Py
BeP
AAN
BPer
BaP
BFL
2,3-BF
C1-228
DBA
DH-Py
DBTH
Cor
20 25 30 35 40 45 50 55 60
Municipal Solid Waste Bottom
Incineration Ash Leachate
(USA)
(b)
BFL
BeP
Indeno
DBAN
BPer
BaP
BaAN
ANN
Per
Fl
Py
DBPer
COR
28 30 32 34 36 38 40 42 44 46 48 50
Hydrothermal Petroleum
(Escanaba Trough, NE Pacific)
(c)
P
MN
• = n-alkane
DMN
N
MP
21
23
25
27
29
15 20 25 30 35 40 45
Relative Response
Time (min)

Fig. 5 a–c. A typical distribution of polycyclic aromatic hydrocarbons in: **a** atmospheric fallout sample, Alexandria City – Egypt; **b** bottom incineration ash leachate of municipal solid waste – USA; **c** hydrothermal petroleum, Escanaba Trough, NE Pacific Ocean. PAH Compound identifications: N = naphthalene, MN = methylnaphthalene, DMN = dimethylnaphthalenes, P = phenanthrene, MP = methylphenanthrene, Fl = fluoranthene, Py = pyrene, BaAN = benz-[*a*]anthracene, DH-Py = dihydropyrene, 2,3-BF = 2,3-benzofluorene, BFL = benzo[*b,k*]fluoranthene, BeP = benzo[*e*]pyrene, BaP = benzo[*a*]pyrene, Per = perylene, C_1-228 = methyl-228 series, Indeno = indeno[1,2,3-*c,d*]pyrene, DBAN = dibenz[*a,h*]anthracene, BPer = benzo[*g,h,i*] perylene, AAN = anthanthrene, DBTH = dibenzothiophene, Cor = coronene, DBP = dibenzo [*a,e*]pyrene, DBPer = dibenzo[*g,h,i*]perylene

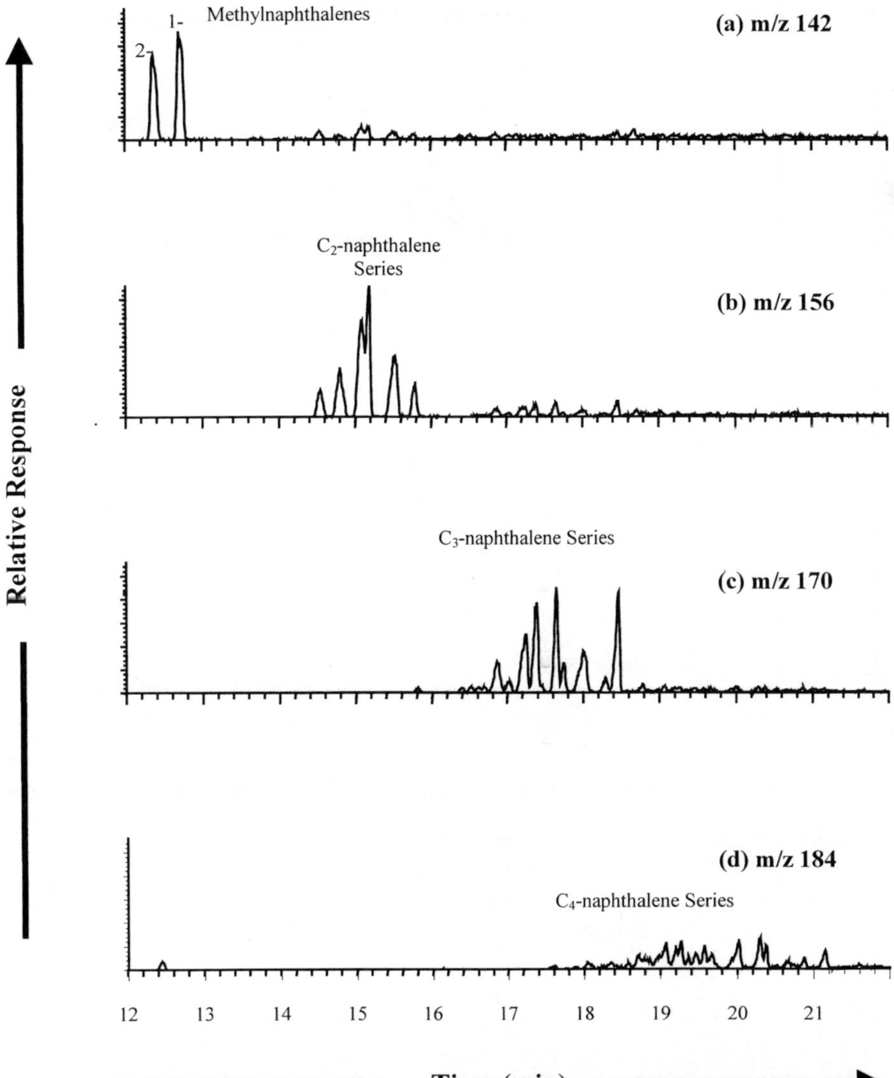

Fig. 6 a–d. Alkyl-substituted naphthalene series (GC-MS key ions: *m/z* 142, 156, 170, and 184) from a Red Sea sediment sample

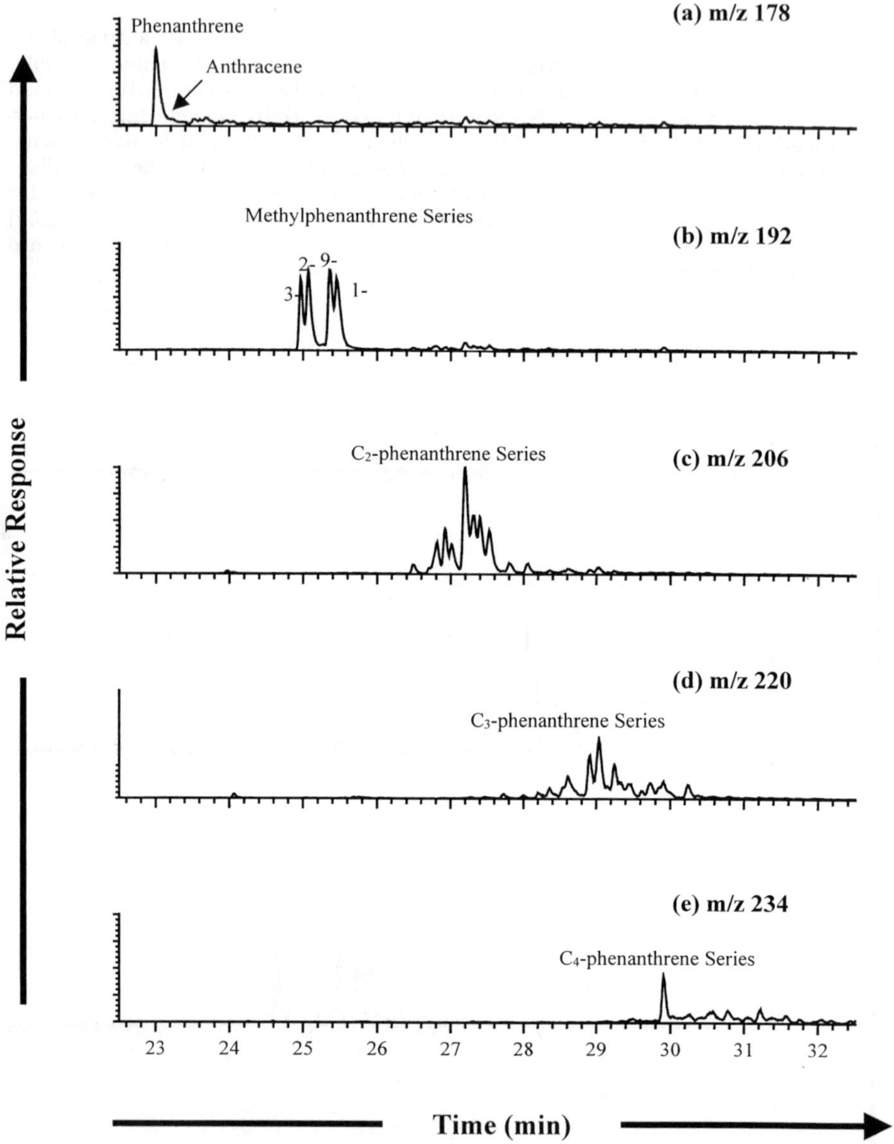

Fig. 7a–e. Alkyl-substituted phenanthrene series (GC-MS key ions: *m/z* 178, 192, 206, 220, and 234) from a bottom ash sample from a coal fired power plant

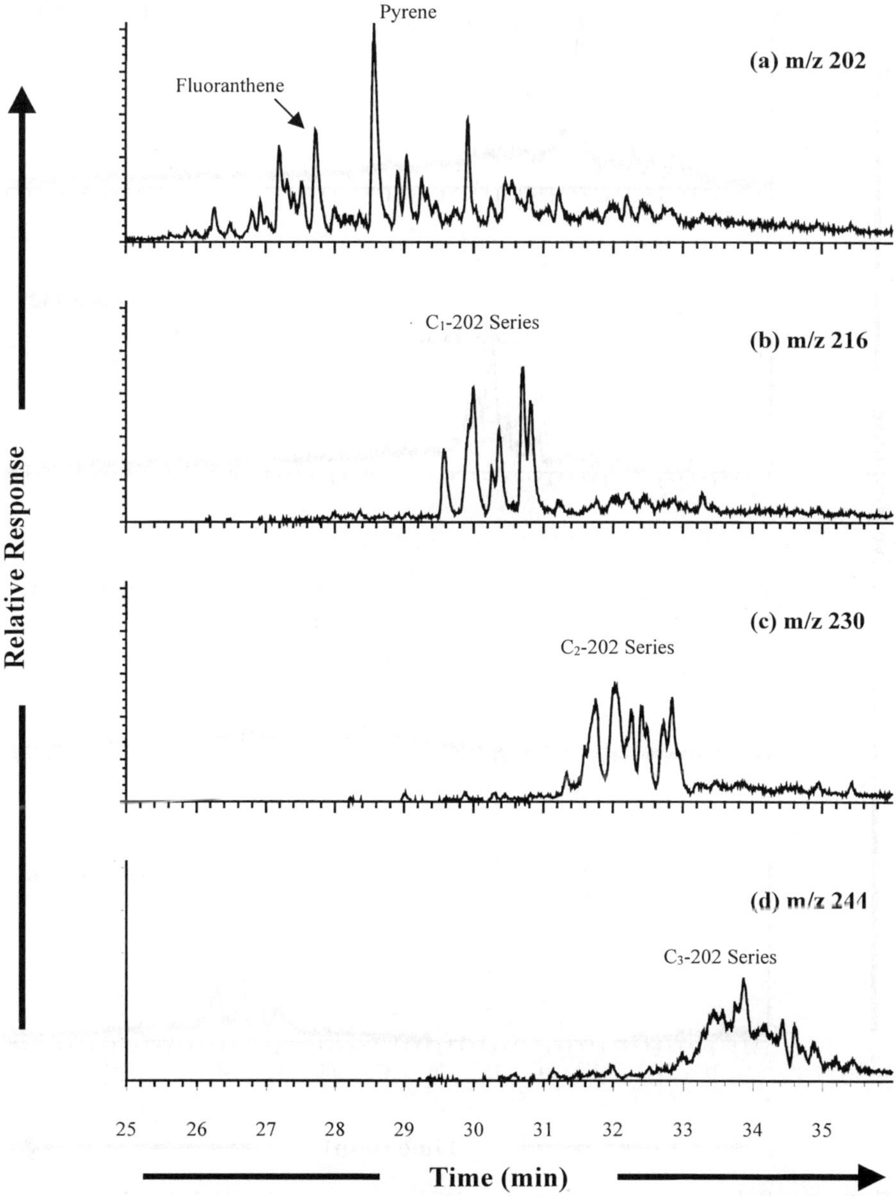

Fig. 8a–d. Alkyl-substituted pyrene/fluoranthene series (key ions: *m/z* 202, 216, 230, and 244) from a bottom ash sample from a coal fired power plant

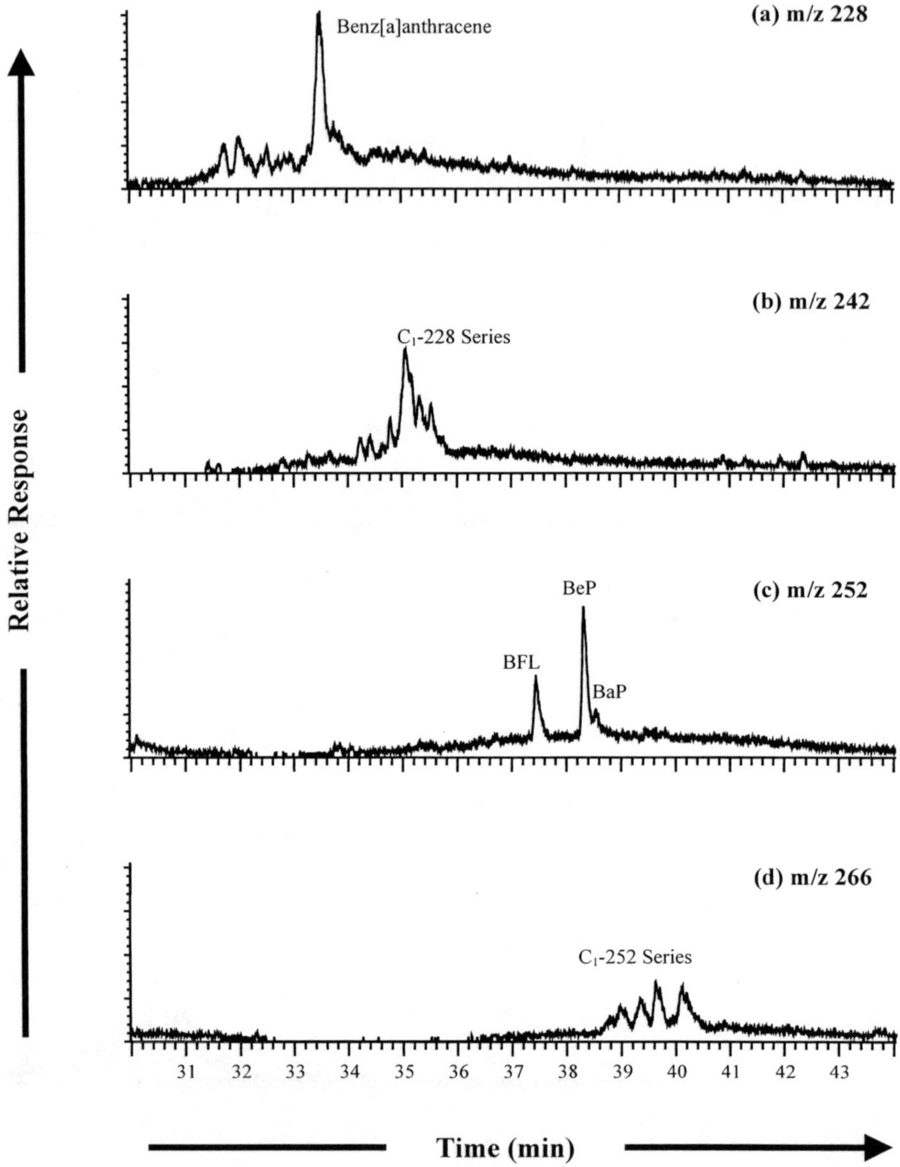

Fig. 9a–d. Alkyl-substituted 228 series (GC-MS key ions: *m/z* 228, 242, 252, and 266, respectively) from a bottom ash sample from a coal fired power plant

Table 2. Typical polycyclic aromatic hydrocarbon identifications and chemical compositions (representative structures are shown in Fig. 4)

Compound Name	Composition	MW
PAHs		
Naphthalene	$C_{10}H_8$	128
Phenanthrene	$C_{14}H_{10}$	178
Anthracene	$C_{14}H_{10}$	178
Fluoranthene	$C_{16}H_{10}$	202
Pyrene	$C_{16}H_{10}$	202
2,3-Benzofluorene	$C_{17}H_{12}$	216
Benz[*a*]anthracene	$C_{18}H_{12}$	228
Chrysene	$C_{18}H_{12}$	228
Benzo[*b*]fluoranthene	$C_{20}H_{12}$	252
Benzo[*k*]fluoranthene	$C_{20}H_{12}$	252
Benzo[*e*]pyrene	$C_{20}H_{12}$	252
Benzo[*a*]pyrene	$C_{20}H_{12}$	252
Perylene	$C_{20}H_{12}$	252
Indeno[1,2,3-*c,d*]pyrene	$C_{22}H_{12}$	276
Dibenz[*a,h*]anthracene	$C_{22}H_{14}$	278
Benzo[*g,h,i*]perylene	$C_{22}H_{12}$	276
Anthanthrene	$C_{22}H_{12}$	276
Coronene	$C_{24}H_{12}$	300
Dibenzo[*a,e*]pyrene	$C_{24}H_{14}$	302
Alkyl-substituted PAHs		
2-Methylnaphthalene (2MN)	$C_{11}H_{10}$	142
1-Methylnaphthalene (1MN)	$C_{11}H_{10}$	142
Dimethylnaphthalenes	$C_{12}H_{12}$	156
Trimethylnaphthalenes	$C_{13}H_{14}$	170
Tetramethylnaphthalenes	$C_{14}H_{16}$	184
3-Methylphenanthrene (3MP)	$C_{15}H_{12}$	192
2-Methylphenanthrene (2MP)	$C_{15}H_{12}$	192
9-Methylphenanthrene (9MP)	$C_{15}H_{12}$	192
1-Methylphenanthrene (1MP)	$C_{15}H_{12}$	192
Dimethylphenanthrenes	$C_{16}H_{14}$	206
Trimethylphenanthrenes	$C_{17}H_{16}$	220
Tetramethylphenanthrenes	$C_{18}H_{18}$	234
Methylpyrenes/fluoranthenes	$C_{17}H_{12}$	216
Dimethylpyrenes/fluoranthenes	$C_{18}H_{14}$	230
Trimethylpyrenes/fluoranthenes	$C_{20}H_{16}$	244
Methyl-228	$C_{19}H_{14}$	242
C_2-288	$C_{20}H_{16}$	256
C_3-228	$C_{21}H_{18}$	270
Methyl-252	$C_{21}H_{14}$	266
C_2-252	$C_{22}H_{16}$	280
C_3-252	$C_{23}H_{18}$	294
C_4-252	$C_{24}H_{20}$	308

high sulfur crude oils. The GC elution orders of the C_2- and C_3-phenanthrenes/anthracenes have been reported [146, 147].

Alkyl fluoranthenes/pyrenes (Fig. 8) and the alkylated *m/z* 228 and 252 series (Fig. 9) are observed mainly from incomplete combustion processes of petroleum and coal. Compound identifications on the figures are summarized in Table 2 with names, compositions, and molecular weights.

2.2 Pesticides

Several hundred-pesticide compounds of diverse chemical structures are widely used in the United States and Europe for agricultural and non-agricultural purposes (Fig. 10). Some are substitutes for organochlorines, which were banned due to their toxicity, persistence, and bioaccumulation in environmental matrices. According to a report published by the US-EPA, a total of 500,000 tons of pesticides was used in 1985 [144, 145, 148]. As far as specific pesticides are concerned, worldwide consumption of Malathion and Atrazine in 1980 amounted to 24,000 and 90,000 tons, respectively [149, 150]. In the Mediterranean countries, 2100 tons of Malathion (active ingredient) were sprayed during the same period compared to 9700 tons in Asia [150].

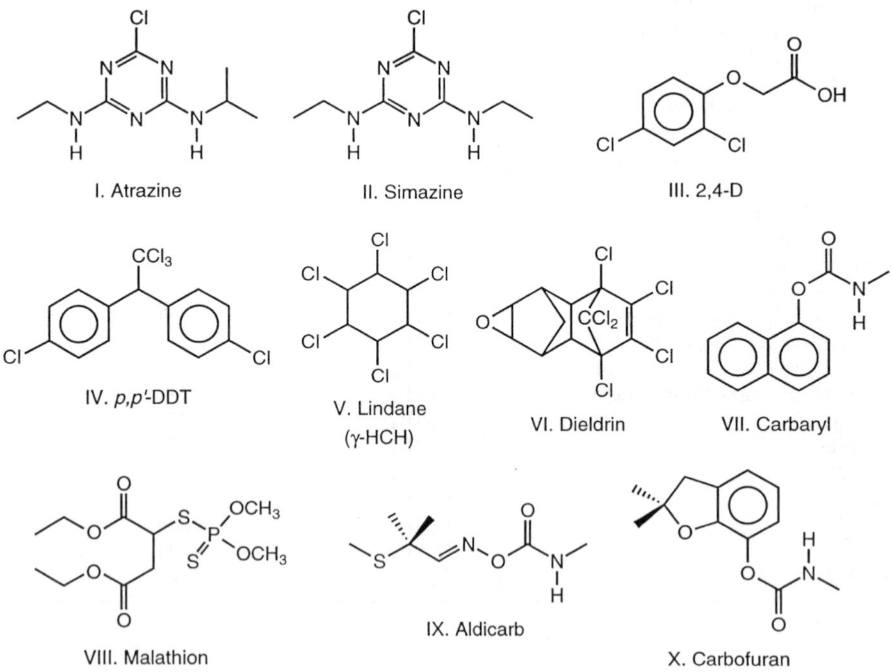

Fig. 10. Chemical structures of various pesticides

2.2.1 *Pesticide Groups*

Organic pesticides which have been and are still being used belong to numerous different families of organic chemicals and may be grouped in various ways. In the present chapter, the classification used is based on the interactive properties toward humic substances (HS) covering solid phases as will be discussed later in the next chapter. The following pesticide groups will be considered: cationic, basic, acidic, and non-ionic. Selected pesticides for various applications such as herbicides, insecticides, fungicides, and germicides will be discussed and are listed in Table 3.

2.2.1.1 *Cationic Compounds*

Bipyridilium herbicides such as Diquat and Paraquat (Structures I, II, Fig. 10, Table 3) are the only important compounds of this group that have been thoroughly investigated in relation to interactions with aquatic and soil HS [151, 152]. They are available as dibromide and dichloride salts, respectively, and are used as herbicides and desiccants. These compounds were shown to be toxic to humans [153, 154]. The solubility of cationic pesticides is generally high in aqueous solutions, where they dissociate readily to form divalent cations. Diquat and Paraquat are nonvolatile compounds and do not escape as vapors from aquatic or soil systems. They are known to photodecompose readily when exposed to sun or UV light, but are not photodecomposed when adsorbed onto particulate matter, and are able to form well-defined charge-transfer complexes with phenols and many other donor molecules [152].

2.2.1.2 *Basic Compounds*

The most important and extensively studied pesticides of this group (Fig. 10, Table 3) are Amitrole and several members of the family of *s*-triazines [89, 151, 153, 155, 156]. Amitrole had been widely used as a herbicide, but its uses as a registered product for application on food crops were canceled starting in 1971 because it was suspected of inducing thyroid tumors in rats [157–162]. Amitrole is soluble in water, with a weak basic character ($PK_b = 10$) and behaves chemically as a typical aromatic amine.

s-Triazines (Fig. 10, Table 3) which are currently used as herbicides are substituted diamino-*s*-triazines which have a chlorine, methoxy, methylthio, or azido group attached to the C-3 ring atom. The presence of electron-rich nitrogen atoms confers to *s*-triazines the well-known electron-donor ability, i.e., weak basicity and the capacity to interact with electron acceptor molecules, giving rise to electron-donor acceptor (charge-transfer) complexes.

Atrazine, one of the herbicides most widely used in the United States and European countries over the last 30 years, is employed for pre- and post-emergence weed control on corn, wheat, barley, and sorghum fields, and on railway

Table 3. Some common pesticides and related compounds with chemical names given in the text

Type	Common name	Chemical class	Use[a]	CAS #	Chemical Name
Cationic	Diquat dibromide	Nitrogen-containing compound	H	85–00–7	1,1′-Ethylene-2,2′-bipyridylium dibromide, monohydrate
	Paraquat	Nitrogen-containing compound	H	1910–42–5	1,1′-Dimethyl-4,4′-bipyridylium, dichloride
Basic	Amitrole	Triazole	H	61–82–5	3-Amino-1,2,4-triazole
	Atrazine	Triazine	H	1912–24–9	2-Chloro-4-(ethylamino)-6-(isopropylamino)-*s*-triazine
	Simazine	Triazine	H	122–34–9	2-Chloro-4,6-bis(ethylamino)-*s*-triazine
Acidic	2,4-Dinitrophenol	Nitrophenol	I; F; AC; AD	51–28–5	2,4-Dinitrophenol
	Pentachlorophenol	Organochlorine	F; M; AD	87–86–5	Pentachlorophenol
	Picloram	Amine	H	1918–02–1	4-Amino-3,5,6-trichloropicolinic acid
	2,4-D	Chlorophenoxy acid	H	94–75–7	(2,4-Dichlorophenoxy)acetic acid
	2,4,5-T	Chlorophenoxy acid	H	93–72–1	(±)-2-(2,4,5-Trichlorophenoxy) propanoic acid
Non-ionic	*o,p*′-DDT	Organochlorine	I	789–02–6	1,1,1-Trichloro-2-(*p*-chlorophenyl)-2-(*o*-chlorophenyl)ethane
	p,p′-DDT	Organochlorine	I	50–29–3	1,1,1-Trichloro-2,2-bis(*p*-chlorophenyl) ethane
	p,p′-DDE	*p,p*′-DDT degradate	I	72–55–9	1,1-Dichloro-2,2-bis(*p*-chlorophenyl) ethane
	p,p′-DDD	Organochlorine *p,p*′-DDT degradate	I	72–54–8	1,1-Dichloro-2,2-bis(*p*-chlorophenyl) ethane
	Toxaphene	Organochlorine	I	8001–35–2	Polychlorinated camphene
	Lindane (γ-HCH)	Organochlorine	I	58–89–9	1α,2α,3β,4α,5α,6β-Hexachlorocyclohexane
	Chlordane	Organochlorine	I	57–74–9	1,2,4,5,6,7,8,8-Octachloro-3a,4,7,7a-tetrahydro-4,7-methanoindan
	Heptachlor	Organochlorine	I	76–44–8	1,4,5,6,7,8,8-Heptachloro-3a,4,7,7a-tetrahydro-4,7-methano-1*H*-indene
	Aldrin	Organochlorine	I	309–00–2	(1α,4α,4aβ,5α,8α,8aβ)-1,2,3,4,10,10-Hexachloro-1,4,4a,5,8,8a-hexahydro-1,4:5,8-dimethanonaphthalene
	Dieldrin	Organochlorine	I	60–57–1	1,2,3,4,10,10-Hexachloro-6,7-epoxy-1,4,4a,5,6,7,8,8a-octahydro-(endo,exo)1,4:5,8-dimethanonaphthalene
	Endrin	Organochlorine	I	72–20–8	1,2,3,4,10,10-Hexachloro-6,7-epoxy-1,4,4a,5,6,7,8,8a-octahydro-(endo,endo)1,4:5,8-dimethanonaphthalene

Table 3 (continued)

Type	Common name	Chemical class	Use[a]	CAS #	Chemical Name
Non-ionic	Malathion	Organophosphorus	I	121–75–5	*O,O*-Dimethyl-*S*-[1,2-bis(ethoxycarbonyl)ethyl]dithiophosphate
	Parathion	Organophosphorus	I	56–38–2	*O,O*-Diethyl-*O*-4-nitrophenyl)phosphorothioate
	Propham	Carbamate	H; PGR	122–42–9	1-Methylethylphenyl carbamate
	Carbaryl	Carbamate	I	63–25–2	1-Naphthalenyl-*N*-methyl carbamate
	Methiocarb	Carbamate	I; M; AC	2032–65–7	3,5-Dimethyl-4-(methylthio)phenylmethyl carbamate
	Aldicarb	Carbamate	I; N; AC	116–06–3	2-Methyl-2-(methylthio)propionaldehyde *O*-(methyl-carbamoyl)oxime
	Carbofuran	Carbamate	I; N	1563–66–2	2,3-Dihydro-2,2-dimethyl-7-benzofuranyl methyl carbamate
	Fenuron	Urea	H	101–42–8	1,1-Dimethyl-3-phenyl urea
	Diuron	Urea	H	330–54–1	3-(3,4-Dichlorophenyl)-1,1-dimethyl urea
	Fluometuron	Urea	H	2164–17–2	1,1-Dimethyl-3-(α,α,α-trifluoro-*m*-tolyl) urea
	Propanil	Amide	H	709–98–8	*N*-(3,4-Dichlorophenyl)propanamide
	Propachlor	Acetanilide	H	1918–16–7	2-Chloro-*N*-(1-methylethyl)-*N*-phenyl acetanilide
	Alachlor	Acetanilide	H	15972–60–8	2-Chloro-*N*-(2,6-diethylphenyl)-*N*-(methoxymethyl)acetamide
	Trifluralin	Dinitroaniline	H	1582–09–8	2,6-Dinitro-*N,N*-dipropyl-4-(trifluoromethyl)benzamine
	Nitralin	Dinitroaniline	H	4726–14–1	4-Methylsulfonyl-2,6-dinitro-*N,N*-dipropylaniline
	Benfluralin	Dinitroaniline	H	1861–40–1	*N*-Butyl-*N*-ethyl-α,α,α-trifluoro-2,6-dinitro-*p*-tolidine
	Profluralin	Dinitroaniline	H	26399–36–0	2,6-Dinitro-*N*-cyclopropylmethyl-*N*-propyl-4-(trifluoromethyl) benzenamide
	Diphenamid	Amide	H	957–51–7	*N,N*-Dimethyl-2,2-diphenylacetamide
	Thiobencarb	Thiocarbamate	H	28249–77–6	*S*-4-Chlorobenzyl diethylthiocarbamate
	Dichlorobenil	Organochlorine	H	1194–65–6	2,6-Dichlorobenzonitrile

[a] H = Herbicide; I = Insecticide; M = Molluscicide; N = Nematocide; F = Fungicide, P = plant growth regulator; AD = Adjuvant; AC = Acaricide.

and roadside verges [157–159]. In this regard, in England and Wales alone, the non-agricultural use of this herbicide represented 140 tons of active ingredients whereas France accounted for 43 tons during 1989 [163]. Not surprisingly, it has been detected in ground- and surface-waters throughout the world [144, 145, 148, 163–166].

Symmetric-triazines have low solubilities in water, with the 2-chloro-*s*-triazines being less soluble than the 2-methylthio and 2-methoxy analogues. Water solubility increases at pH values where strong protonation occurs, e.g., between pH 5.0 and 3.0 for 2-methoxy- and 2-methylthio-*s*-triazines, and at pH $\leq$2.0 for 2-chloro-*s*-triazines. Structural modifications of the substituents significantly affect solubility at all pH levels. Increasing solubility is associated with increasing electron-donating capability of the substituents at C-2 and increasing size and branching of the N-alkyl groups at the C-4 and C-6 positions. The *s*-triazines and especially the chloro-*s*-triazines are hydrolyzed in aqueous systems [153]. Chloro- and methylthio-*s*-triazines are also partly photodecomposed in aqueous systems by UV and IR radiation, while methoxy-substituted compounds are not photodegradable [167]. Most *s*-triazines are relatively volatile, so they can be lost from aquatic and soil systems by evaporative processes [157–159, 161, 162].

2.2.1.3
Acidic Compounds

This group of pesticides comprises different families of chemicals with herbicidal action including substituted phenols, chlorinated aliphatic acids, chlorophenoxy alkanoic acids, and substituted benzoic acids, which possess carboxyl or phenolic functional groups capable of ionization in aqueous media to yield anionic species [47, 151, 168–170].

Chlorinated aliphatic acids have the highest water solubility and the strongest acidity among this group of compounds due to the strong electronegative inductive effect of the chlorine atoms replacing the hydrogens in the aliphatic chain of these acids. The water solubilities of the phenoxy alkanoic acids are low as they have a considerable lipophilic component. Most commercial formulations of these herbicides, however, contain the compound in the soluble salt form; thus the anionic species predominate in neutral aqueous systems, while at low pH levels they are present in the molecular rather than the anionic form. Dinitrophenols and pentachlorophenol (Fig. 10, Table 3) are generally of intermediate solubility in water, while they are highly water-soluble as alkali salts which represent most of their common commercial formulations.

With the exception of picloram and phenols (Fig. 10, Table 3), acidic pesticides are considered nonvolatile from aqueous and soil systems [153]. Some ester formulations of these compounds also behave as herbicides. They do not ionize in solution and are less water-soluble than the acid or salt forms. They are eventually hydrolyzed to acid anions in aqueous and soil systems, but in the ester form are non-ionic and relatively volatile.

2,4-D and 2,4,5-T (Fig. 10, Table 3) are among the most widely known and used phenoxy alkanoic acids. These two herbicides were used as defoliants in Vietnam.

Teratogenic (fetus deforming) effects on rats and mice were reported for 2,4,5-T and the isooctyl ester of 2,4-D, while mortality and physical abnormalities were shown to increase in chick embryos of gamebird eggs sprayed with 2,4-D at rates commonly used in field applications [153, 166]. The most extensively used halogenated benzoic acid herbicides are Chloramben and Dicamba.

2.2.1.4
Nonionic Compounds

Pesticides of this category (Fig. 10, Table 3) do not ionize significantly in aqueous systems and vary widely in their chemical composition and properties (i.e., water solubility, polarity, molecular volume, and tendency to volatilization).

Chlorinated hydrocarbon insecticides are among the most widely known and studied group of nonionic pesticides [151]. DDT, in particular, has been studied more than any other pesticide (Fig. 10, Table 3). It has been implicated as detrimental to numerous wildlife species and to accumulate in the food chain [171]. Several chlorinated hydrocarbons have been detected in various marine and terrestrial organisms, food crops, surface waters, and soils. Toxaphene, Lindane, Chlordane, and Heptachlor (Fig. 10, Table 3) have been found in the biosphere in much smaller levels than DDT, Aldrin, and Dieldrin [153, 172]. The DDT content of phytoplankton in the sea has been shown to increase since 1955 even though the amount used has been declining since 1965 [153]. With the exception of Lindane, all these compounds are insoluble in water. DDT is about ten times more insoluble than the other compounds of this family, and thus it is considered to be immobile in soil solid systems. Endrin, Dieldrin, and Aldrin show higher water solubility and are, therefore, slightly mobile in soils. The vapor pressure of chlorinated hydrocarbons (Fig. 10, Table 3) varies widely from low (e.g., DDT, Endrin, and Dieldrin [171]) to moderate (e.g., Toxaphene and Aldrin [172]) to high (e.g., Chlordane and Lindane) and very high (e.g., Heptachlor). Volatilization of DDT from soils and other surfaces is, therefore, almost insignificant; however, it converts to DDE which is more volatile.

DDT converts in part to *p,p′*-DDE over time in the environment, especially in sediments [151, 171]. An example of the total aliphatic extract of a sediment from the Los Angeles Bight contaminated with *p,p′*-DDE is shown in Fig. 11. The TIC trace shows a major UCM and the minor resolved peaks are normal alkanes (primarily from higher plant wax), with mature 17α(H),21β(H)-hopanes (from petroleum residues as is the UCM). The mass spectrum of *p,p′*-DDE is shown in Fig. 12a, registering the molecular ion cluster at *m/z* 316–320. DDE is detected in the *m/z* 246 fragmentogram (Fig. 11d), appearing as a small peak in the TIC trace and DDT is not detectable in this sample.

Organophosphates (Fig. 10, Table 3) are more toxic than chlorinated hydrocarbons, in particular to humans, but they exhibit lower persistence in soils and do not seem to accumulate in soil fauna or concentrate in birds and fish [74]. This behavior is also related to an enhanced water solubility and lower vapor pressure of organophosphates. Malathion and Parathion (Fig. 10, Table 3) insecticides are known to be chemically hydrolyzed and biodegraded by micro-

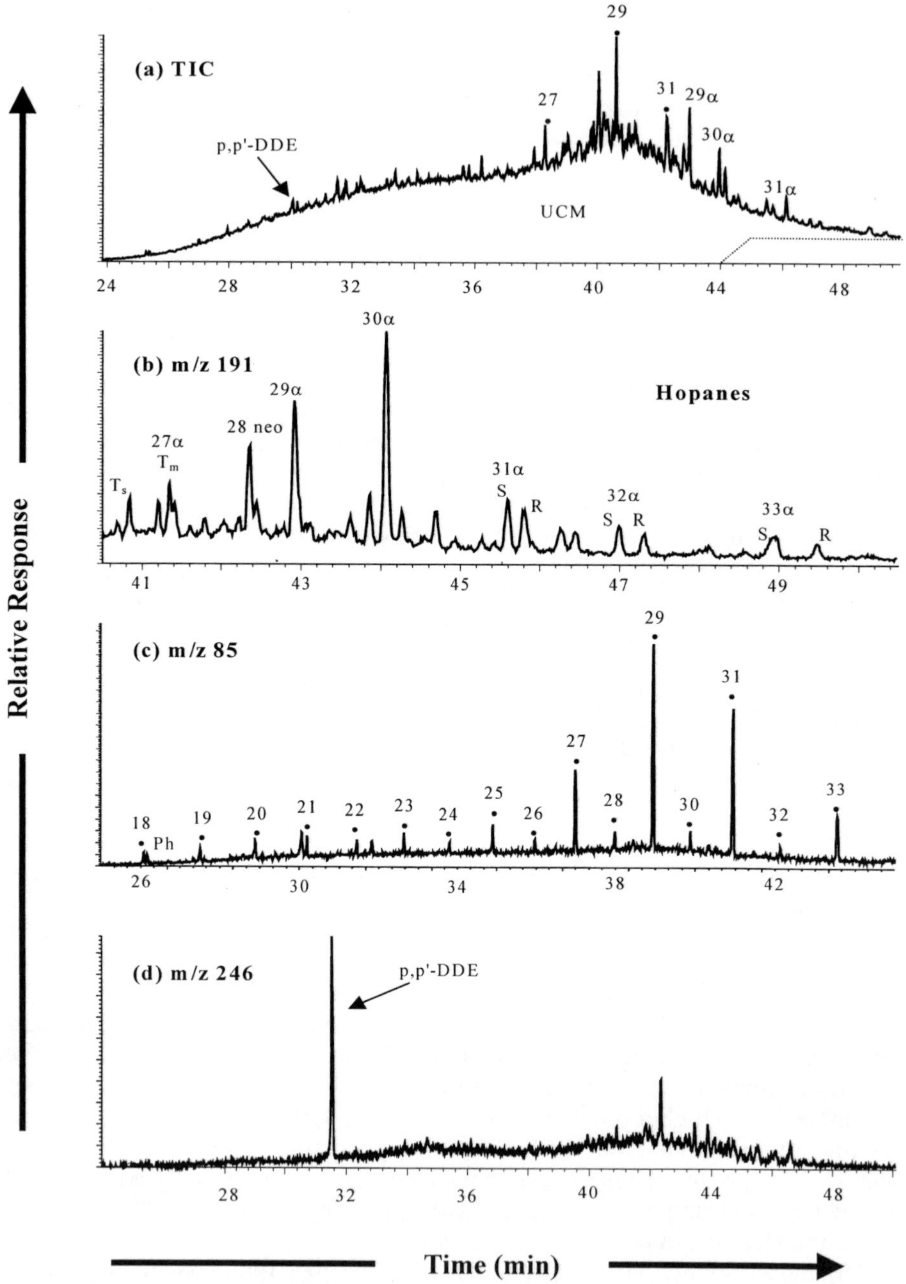

Fig. 11a–d. A GC-MS trace showing a typical distribution of a pesticide polluted sample from the Los Angeles Bight

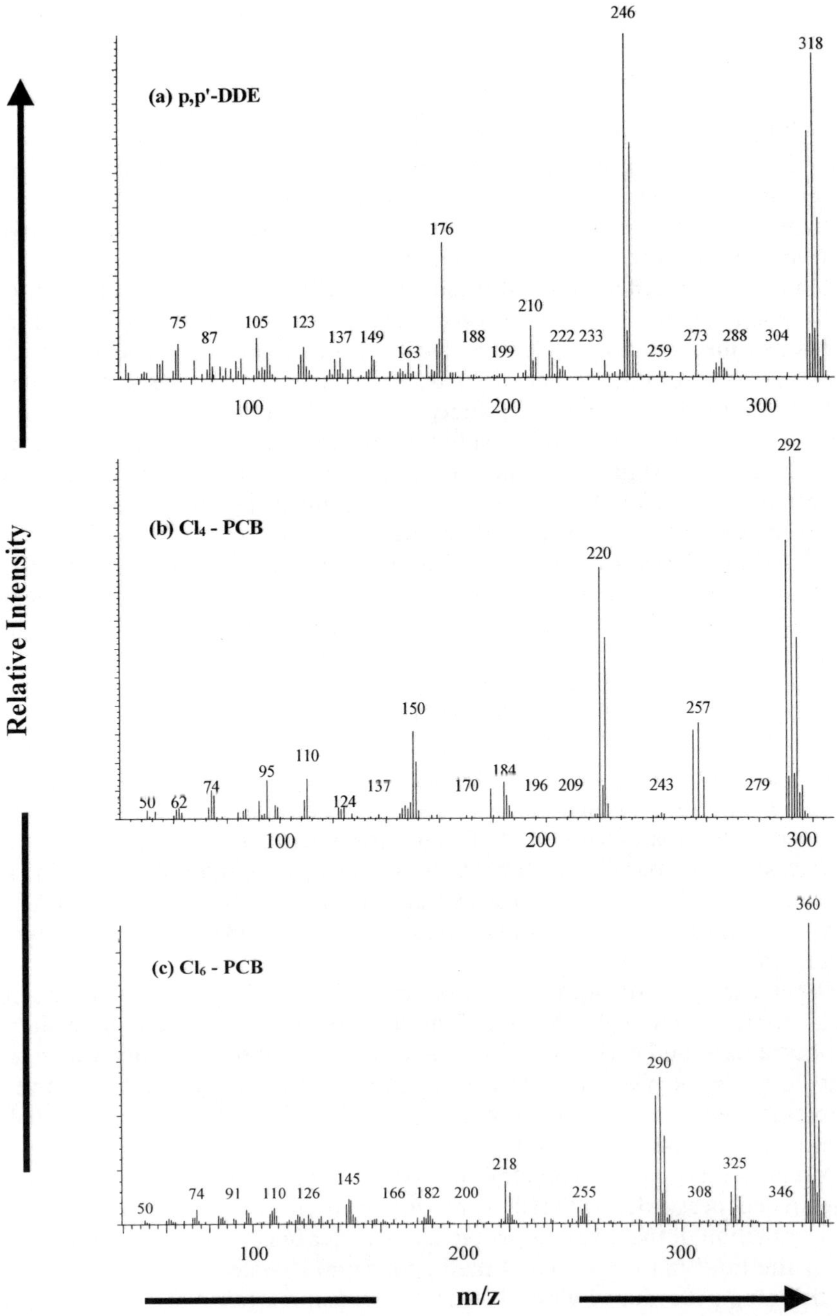

Fig. 12 a–c. Mass spectra of some halogenated compounds: **a** p,p′-DDE; **b** Cl_4-PCB; **c** Cl_6-PCB

organisms in soil systems. The most important organophosphate herbicide is Glyphosate.

Phenylcarbamates, or carbanilates, generally exhibit low water solubilities, and thus they are almost immobile in soil systems. Chlorpropham and Propham are readily volatilized from soil systems, but Terbutol and Carbaryl (Fig. 10, Table 3) are not. Ester- and amide-hydrolysis, N-dealkylation and hydroxylation are among the chemical reactions that carbamates undergo. The *N*-methylcarbamate insecticides (Fig. 10, Table 3) commonly used in soils are Carbaryl, Methiocarb, Aldicarb, and Carbofuran [74, 173].

More than 25 different substituted urea herbicides are currently commercially available [30, 173]. The most important are phenylureas and Cycluron, which has the aromatic nucleus replaced by a saturated hydrocarbon moiety. Benzthiazuron and Methabenzthiazuron are more recent selective herbicides of the class, with the aromatic moiety replaced by a heterocyclic ring system. With the exception of Fenuron, substituted ureas (i.e., Diuron, Fluometuron, Fig. 10, Table 3) exhibit low water solubilities, which decrease with increasing molecular volume of the compound. The majority of the phenylureas have relatively low vapor pressures and are, therefore, not very volatile. These compounds show electron-donor properties and thus they are able to form charge transfer complexes by interaction with suitable electron acceptor molecules. Hydrolysis, acylation, and alkylation reactions are also possible with these compounds.

The most important substituted anilide herbicides (Fig. 10, Table 3) are Propanil, Propachlor, and Alachlor [43, 151, 175–178].

Substituted dinitroanilines (Fig. 10, Table 3) are an important series of selective herbicides commercially introduced in agriculture in the 1960s. Trifluralin is the most prominent member of this series. Nitralin and Benfluralin have also received widespread usage, while Profluralin is a relatively recent herbicide of this class. Dinitroanilines show very low water solubilities. Nitralin and Benfluralin have low vapor pressures and are nonvolatile, while Trifluralin is relatively volatile. All these compounds have been shown to be relatively immobile in soil systems.

Other examples of nonionic compounds (Fig. 10, Table 3) are the phenylamide herbicides (e.g., Diphenamid, moderately water soluble and nonvolatile), thiocarbamate, and carbothioate herbicides (e.g., Thiobencarb, low water solubility, high vapor pressure, relative mobility in soil systems) and benzonitrile herbicides (e.g., Dichlobenil, low water solubility, low vapor pressure, relative immobility in most soils) [151].

A representative gas chromatogram with ECD of the analysis of various polar chlorinated pesticides isolated from cod liver oil [179] is shown in Fig. 13. Determination of the polar chlorinated pesticides in cod liver oil required clean up of the lipid matrix with a dimethylformamide/water/hexane liquid-liquid partitioning procedure followed by isolation using a normal-phase LC procedures, and final analysis by GC-ECD [179].

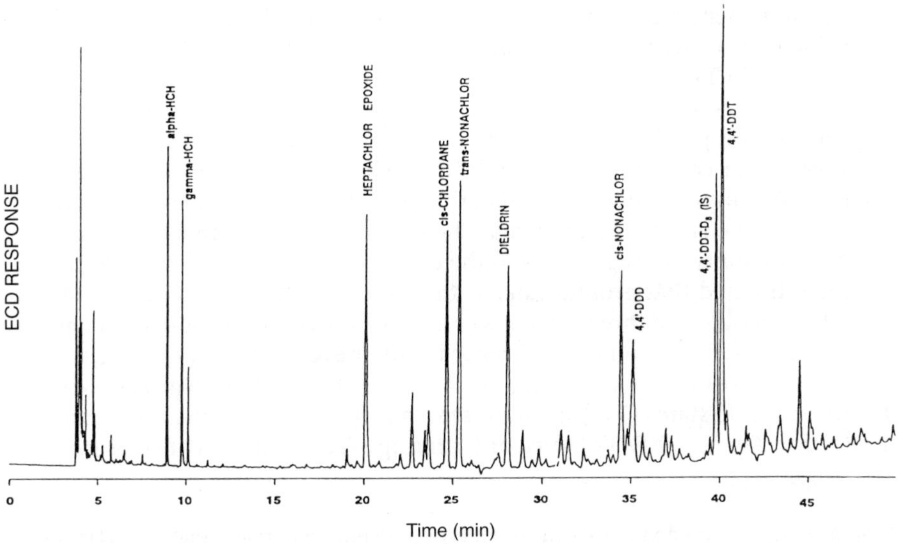

Fig. 13. A GC-ECD chromatogram of polar pesticide fraction analyzed in cod liver oil. Column: 60-m capillary column with 5% phenyl-substituted methylpolysiloxane phase (after [179] with permission)

2.2.2
Priority Lists

Due to the environmental impact of pesticides, several priority lists have been published to help protect the quality of drinking and surface waters. Table 4 lists the different pesticides from the 76/464/EC Directive (i.e., the so-called black list [168, 171–174]). Following the three general parameters (toxicity, persistence, and input) for selecting the priority list of pollutants in the United Kingdom, a "red-list" of substances that include several pesticides, most of them common to the EC list, was established.

A priority list for preventing the contamination of ground- and drinking waters by pesticides in Europe, which considers pesticides used in quantities

Table 4. Pesticides listed in the 76/464/EC Council Directive on pollution caused by dangerous substances discharged into the aquatic environment of the community (Black List)

2,4-D	Dichlorprop	Hexachlorbenzene	Parathion-ethyl
2,4,5-T	Dichlorvos	Linuron	Parathion-methyl
Aldrin	Dieldrin	Malathion	Phoxim
Atrazine	Dimethoate	MCPA	Propanil
Azinphos-ethyl	Disulfoton	Mecoprop	Pyrazon
Azinphos-methyl	Endosulfan	Metamidophos	Simazine
Chlordane	Endrin	Mevinphos	Triazophos
Coumaphos	Fenitrothion	Monolinuron	Trichlorfon
DDT	Fenthion	Omethoate	Trifuralin
Demeton	Heptachlor	Oxydemeton-methyl	

over 50 tons per annum (and over 500 are underlined) and their capacities as probable or transient leachable substances, was published [171, 177, 180, 181] and is listed in Table 5.

Following considerations based on usage information, physico-chemical properties, and persistence, a priority list of herbicides was established for the Mediterranean countries, i.e., France, Italy, Greece, and Spain ([168, 182, 183] Table 6). This list considers selected herbicides which can cause contamination of estuarine and coastal environments. The selection of pollutants has been based on the availability of usage data and the consideration of half-lives [182, 183].

It is estimated that groundwater is the source of drinking water for 90% of rural households and three-quarters of all US cities. In total, more than one-half of the US citizens rely on ground water for their everyday needs. Because of the amount of information indicating the presence of pesticides in ground-water in the different US states [148], a joint research project between the Environmental Protection Agency (EPA)'s Office of Drinking Water and the Office of Pesticide

Table 5. Pesticides used in Europe in amounts over 50 tons per annum that were classified as probable or transient leachers

2,4-D	Cyanazine	Isoproturon	Prometryn
Alachlor	Dalapon	Linuron	Propham
Aldicarb	Diazinon	Maneb	Propiconazole
Amitrole	Dichlobenil	MCPA	Propyzamide
Atrazine	Dimethoate	MCPP	Pyrethrin
Benazoline	Dinoseb	Metamitron	Simazine
Bentazone	Diuron	Metazachlor	Terbutryn
Bromofenoxim	DNOC	Methabenzthiazuron	Terbutylazine
Carbaryl	EPTC	Metham-sodium	Triademinol
Carbendazim	Ethofumesate	Methiocarb	Trichlorfon
Carbetamide	Ethoprophos	Metochlor	Trichloroacetic acid
Chloridazon	Fenamiphos	Oxydemeton methyl	Vinclozolin
Chlorpyrifos	Fluroxypyr	Phenmedipham	Ziram
Chlortoluron	Iprodione	Prochloraz	

Table 6. Herbicides of potential concern in the Mediterranean region

Alachlor	Dinoterb	Mecoporp	Pendimethalin
Amitrole	Diquat	Metamitron	Phenmedipham
Atrazine	Diuron	Metazachlor	Prometryn
Bentazone	DNOC	Methabenzthiazuron	Simazine
Bromoxynil	EPTC	Metobromuron	Trichloroacetic acid
Butylate	Ethalfuralin	Metochlor	Terbumeton
Carbetamide	Ethofumesate	Metoxuron	Terbutylazine
Chlortoluron	Flamprop-M-isopropyl	Mertribuzin	Terbutryn
2,4-D	Glyphosphate	Molinate	Tri-allate
Di-allate	Isoproturon	Napropamide	Trifluralin
Dichlobenil	Linuron	Neburon	
Dichlofop-methyl	MCPA	Paraquat	

Table 7. Pesticides and transformation products (TPs) included in the US National Pesticide Survey

EPA method #	Pesticides and transformation products		
504	For the determination of 1,2-dibromoethane (EDB) and 1,2-dibromo-3-chloropropane (DBCP) in water by hexane microextraction and GC		
	EDB	1,2-Dichloropropane	*trans*-1,3-Dichloropropene
	DBCB	*cis*-1,3-Dichloropropene	
507	For the determination of nitrogen- and phosphorus-containing pesticides in water by extraction with dichloromethane and detection by GC-NPD		
	Alachlor	Ethoprop	Prometryn
	Ametraton	Fenamiphos	Pronamide
	Ametryn	Fenamirol	Propazine
	Atrazine	Fluridone	Simazine
	Bromacil	Hexazinone	Simetryn
	Butachlor	Merphos	Stirofos
	Butylate	Metachlor	Tebuthiuron
	Carboxin	Methyl paraoxon	Terbacil
	Chloropham	Metribuzin	Terbufos
	Cycloate	Mevinphos	Terbutryn
	Diazinon	MGK 264	Tetrachlorvinphos
	Dichlorvos Diphenamid	Molinate	Triademefon
	Disulfoton	Napropamide	Tricyclazole
	Disulfoton sulfone	Norflurazon	Vernolate
	Disulfoton sulfoxide	Perbulate	
	EPTC	Prometon	
508	For the determination of chlorinated pesticides in ground water by extraction with dichloromethane and detection by GC-ECD		
	Aldrin	Dieldrin	γ-HCH
	α-Chlordane	Endosulfan I	Heptachlor
	γ-Chlordane	Endosulfan II	Heptachlor-epoxide
	Chlorneb	Endosulfan sulfate	Hexachlorbenzene
	Chlorobenzilate	Endrin	Metoxychlor
	Chlorothalonil	Endrin aldehydes	*cis*-Permethrin
	DCPA	Etridiazole	*trans*-Permethrin
	4,4′-DDD	α-HCH	Propachlor
	4,4′-DDE	β-HCH	Trifluralin
	4,4′-DDT	δ-HCH	
515.1	For the determination of chlorinated acids in ground water by adjusting the samples' pH to 12, shaking for 1 h to hydrolyze derivatives, removing the extraneous inorganic material by a solvent wash, and sample acidification. The chlorinated acids are extracted with diethyl ether; the acids are converted to their methyl esters using diazomethane as derivatizing agent; excess derivatizing agent is removed and the esters are determined by GC-ECD		
	Acifluorfen	Dicamba	4-Nitrophenol
	2,4-DB	3,5-Dichlorobenzoic acid	PCP
	Bentazone	Dalapon	Picloram
	Chloramben	Dichlorprop	2,4,5-T
	2,4-D	Dinoseb	2,4,5-TP
	DCPA acid metabolites	5-Hydroxydicamba	

Programs was conducted based on a statistically survey of pesticide contamination of drinking water wells. During this National Pesticide Survey, 1349 drinking water wells were sampled and analyzed for 127 pesticides [149, 150]. Pesticides and pesticide degradation products previously detected in ground water and pesticides regulated under the Safe Drinking Water Act, were automatically included in this priority list [184]. The compounds were grouped according to their method of analysis and thus seven methods were used which covered all the 127 analytes. These are indicated in Table 7 [185].

Some general comments can be made about the different priority lists presented in Tables 4–7 as follows:

- Although in some cases there is an agreement on which priority pesticides to monitor, such as Atrazine, 2,4-D, Linuron, and Dimethoate, which represent different chemical groups, in other cases there is complete disagreement. That is the case, for example, with the carbamates, which have a relatively high importance in US monitoring programs (Table 7). The EPA has developed an excellent method for analysis of these pesticides in water to very low limits of detection. In contrast, in Europe, in the first black list of pesticides there were no carbamates at all (Table 4). As they were not included in the first list of hazardous substances in Europe, no tradition of monitoring carbamates was established, although its use has been reported in several countries, such as The Netherlands, Spain, United Kingdom, and Italy.
- The official EPA method for monitoring carbamate pesticides (Method 531.1) has seldom been used in Europe, although it is a highly sensitive and robust method.
- The leachability of carbamates through ground and well waters has been studied as part of the National Pesticide Survey in the USA. In Europe, where the same sources are also important for drinking water, no planning has been undertaken in this regard. The percentage of ground water used for drinking purposes in Europe is close to 100% for Denmark, and 85% for Italy, Germany, France, and the United Kingdom, whereas in Spain it is in the region of 30%.
- The National Pesticide Survey list (Table 7) is the only one that specifically considers the transformation products (TPs) of pesticides. This is remarkable because in the European Community regulations the importance of TPs of pesticides is indicated [165], and there is no mention of specific TPs. This specification in the European Community list is vague, thus making it difficult for laboratories currently involved in monitoring programs to select and assess the TPs of importance.

2.3 PCBs

Since Jensen's initial detection of polychlorinated biphenyls (PCBs) in biological tissue during the 1970s [186, 187] and the subsequent realization that these compounds (Fig. 14) were potentially harmful to wildlife and man, there has been a continuous development in both the analytical techniques to determine these

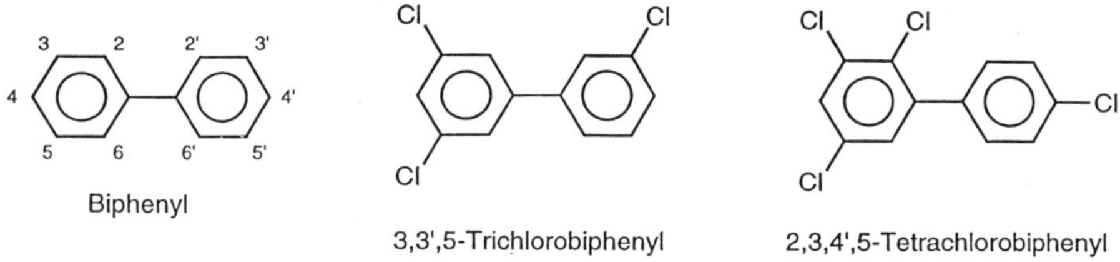

Fig. 14. Examples of chemical structures of PCBs as cited in the text

compounds [36, 62, 86, 188–192] and in the assessment of their biological effects [21–28, 193]. PCBs have been manufactured in substantial amounts since the 1920s [194]. Their use in the electrical, paint, pigments, paper, and cardboard industries and subsequent disposal into the environment [21, 31, 138, 195–201] during the intervening years has allowed sufficient time for them to spread to the remotest areas of the world before any control on use or disposal was implemented. Their high hydrophobicity, lipid solubility, and persistence have resulted in widespread contamination of biota to the extent that all environmental compartments that have been analyzed contain measurable levels of these pollutants [31, 138, 195–197, 199–204].

The early analyses of PCBs were made with packed gas chromatographic columns with electron capture detection and industrial formulations to quantify a total value for PCBs [205]. This early technology did not have the resolution to separate individual PCB congeners and the most appropriate method to estimate these pollutants at that time was unquestionably by the summation of the peak heights or areas of the low-resolution chromatogram. Some workers recognized the potential errors in such estimates and attempted to obtain a single response by perchlorination to the decachlorobiphenyl (CB 209) [205– 207]. The need to improve the separation, identification, and quantification of the individual PCB isomers has been reinforced by measurement of the toxic and biological effects of specific congeners [22, 25–28; 208–210]. With the present methodology and instrumental detection limits for low concentrations [211–213], it is now possible to measure individual PCBs routinely at levels of pg/kg, and with care at fg/kg.

Various PCB congeners and lower polarity pesticide fractions analyzed from cod liver oil is shown in Fig. 15 [179]. Measurement of the PCB congeners and pesticides in the cod liver oil required clean-up of the lipid matrix with a dimethylformamide/water/hexane liquid-liquid partitioning procedure followed by isolation of the PCBs and pesticides using a normal-phase LC procedures. The normal-phase LC procedures separate the analytes into two fractions, one containing the PCBs and the lower polarity chlorinated pesticides (HCB, 2,4′-DDE, and 4,4′-DDE) (Fig. 15) and the second containing the more polar chlorinated pesticides. The separation of PCBs and pesticides reduces the possible coelution of many of the pesticides with PCB congeners of interest. These two fractions were then analyzed by GC-ECD.

The salient features of the GC-MS data for the neutral extract components separated from PCB contaminated sediment in New Bedford harbor, Massachusetts are given in Fig. 16. The TIC trace indicates a major UCM with super-

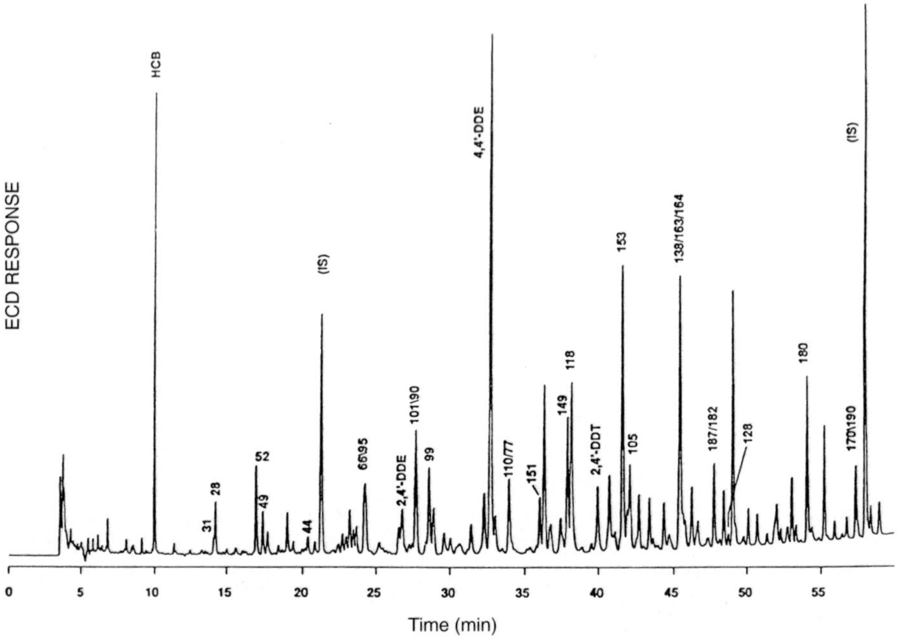

Fig. 15. A GC-ECD chromatogram of the PCB and lower polarity pesticide fraction analyzed from cod liver oil. Column: 60-m capillary column with 5% phenyl-substituted methylpolysiloxane phase (after [179] with permission). PCB compound identifications: (31) 2,4′,5-Trichlorobiphenyl, (28) 2,4,4′-Trichlorobiphenyl, (52) 2,2′,5,5′-Tetrachlorobiphenyl, (49) 2,2′,4,5′-Tetrachlorobiphenyl, (44) 2,2′,3,5′-Tetrachlorobiphenyl, (66/95) mixture of 2,3′,4,4′-Tetrachlorobiphenyl (major component) and 2,2′,3,5′,6-Pentachlorobiphenyl (minor component), (101/90) mixture of 2,2′,4,5,5′-Pentachlorobiphenyl (major component) and 2,2′,3,4′,5-Pentachlorobiphenyl (minor component), (99) 2,2′,4,4′,5-Pentachlorobiphenyl, (110/77) 2,3,3′,4′,6-Pentachlorobiphenyl, (151) 2,2′,3,5,5′,6-Hexachlorobiphenyl, (149) 2,2′,3,4′,5′,6-Hexachlorobiphenyl, (118) 2,3′,4,4′,5-Pentachlorobiphenyl, (153) 2,2′,4,4′,5,5′-Hexachlorobiphenyl, (105) 2,3,3′,4,4′-Pentachlorobiphenyl, (138/163/164) mixture of 2,2′,3,4,4′,5′-Hexachlorobiphenyl (major component), 2,3,3′,4′,5,6-Hexachlorobiphenyl and 2,3,3′,4′,5′,6-Hexachlorobiphenyl (minor component), (187/182) mixture of 2,2′,3,4′,5,5′,6-Heptachlorobiphenyl (major component) and 2,3,3′,4,4′,5,6-Heptachlorobiphenyl (minor component), (128) 2,2′,3,3′,4,4′-Hexachlorobiphenyl, (180) 2,2′,3,4,4′,5,5′-Heptachlorobiphenyl, (170/190) mixture of 2,2′,3,3′,4,4′,5-Heptachlorobiphenyl (major component) and 2,3,3′,4,4′,5,6-Heptachlorobiphenyl (minor component), and (IS) internal standard

imposed peaks due to elemental sulfur (S_6, S_7, and S_8), PCBs, and the mature 17α(H),21β(H)-hopanes. The latter are fingerprinted in the *m*/*z* 191 plot and confirm that they and the UCM are derived from petroleum residues (lubricating oils). The PCBs can be identified by GC-MS from their mass spectra as for example those shown in Fig. 12b, c. They can also be detected by the key ions as for example *m*/*z* 292, 326, and 360 (Fig. 16c, d). However, ECD-GC (e.g., Fig. 15) is considered more sensitive if the PCBs are present as trace constituents.

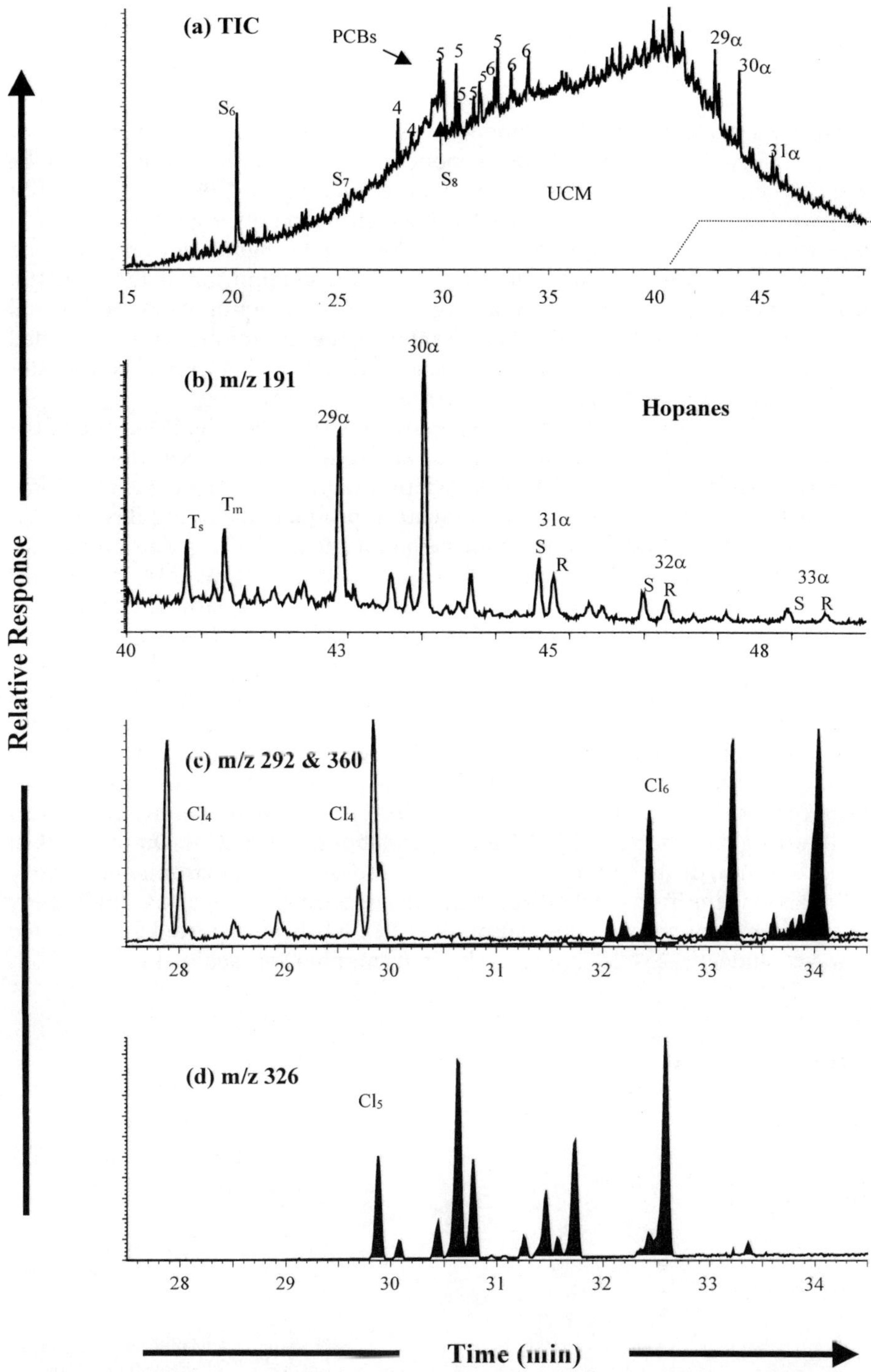

Fig. 16 a–d. GC-MS traces representing: **a** TIC; **b** *m/z* 191 hopane series; **c** *m/z* 292 and 360 series; **d** *m/z* 326 series of a PCB contaminated sediment sample (New Bedford harbor, MA)

2.4 Phthalates

Esters of 1,2-benzenedicarboxylic acid (phthalic acid esters, PAEs, phthalates) comprise a group of organic compounds used in large quantities by present day society (Fig. 17). The worldwide production of PAEs was estimated to be 4.2×10^9 kg during 1994 and has increased by roughly 50% during the last 20 years [214]. PAEs are mainly used as plasticizers in polyvinyl chloride (PVC) plastics and may constitute up to 67% of their total weight. They are also used in a variety of other products such as cosmetics, ammunition, inks, etc. [215]. Due to their broad range of applications, PAEs are ubiquitous environmental contaminants. In 1975, the rate of PAEs entering the environment was estimated at approximately 2.3×10^7 kg annually as a result of leaching from plastic wastes and the direct application of various formulations [216].

The phthalate ester di-(2-ethylhexyl)phthalate (DEHP) (Fig. 17) is one of the most abundant organic xenobiotics in the environment, accounting for approximately 40–50% of the global annual PAE production [217]. DEHP is an important and popular additive in many industrial products including flexible PVC materials and household products such as paint and glues [215]. The annual global production of DEHP has been estimated to $1-20 \times 10^6$ tons [218, 219]. DEHP is now considered a ubiquitous contaminant in many aquatic and terrestrial environments [215, 220]. The main sources of DEHP in the environment are incineration, direct evaporation, and sewage treatment plants (where DEHP is often found in elevated concentrations in the dewatered sewage sludge). There has been a growing concern regarding the potential health risks associated with DEHP. Although DEHP is considered relatively nontoxic, carcinogenic and mutagenic effects of DEHP on aquatic organisms and laboratory animals have been reported [218, 221, 222]. There has also been an increased focus on likely xeno-estrogenic effects of DEHP and its metabolites [218, 223]. On the basis of these findings, the need for a better understanding of the environmental fate of DEHP is evident. Transport of DEHP in soil has been examined in a single study [224] whereas microbial degradation of DEHP has been reported for activated sewage sludge [225–227] and a limited number of sediments and soils [228–231].

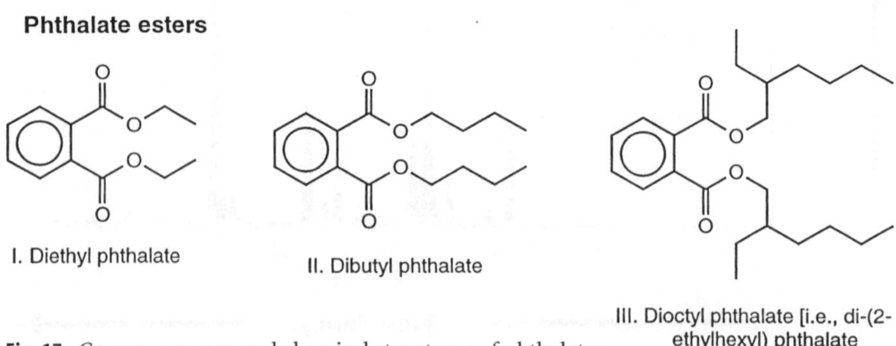

Fig. 17. Common names and chemical structures of phthalates

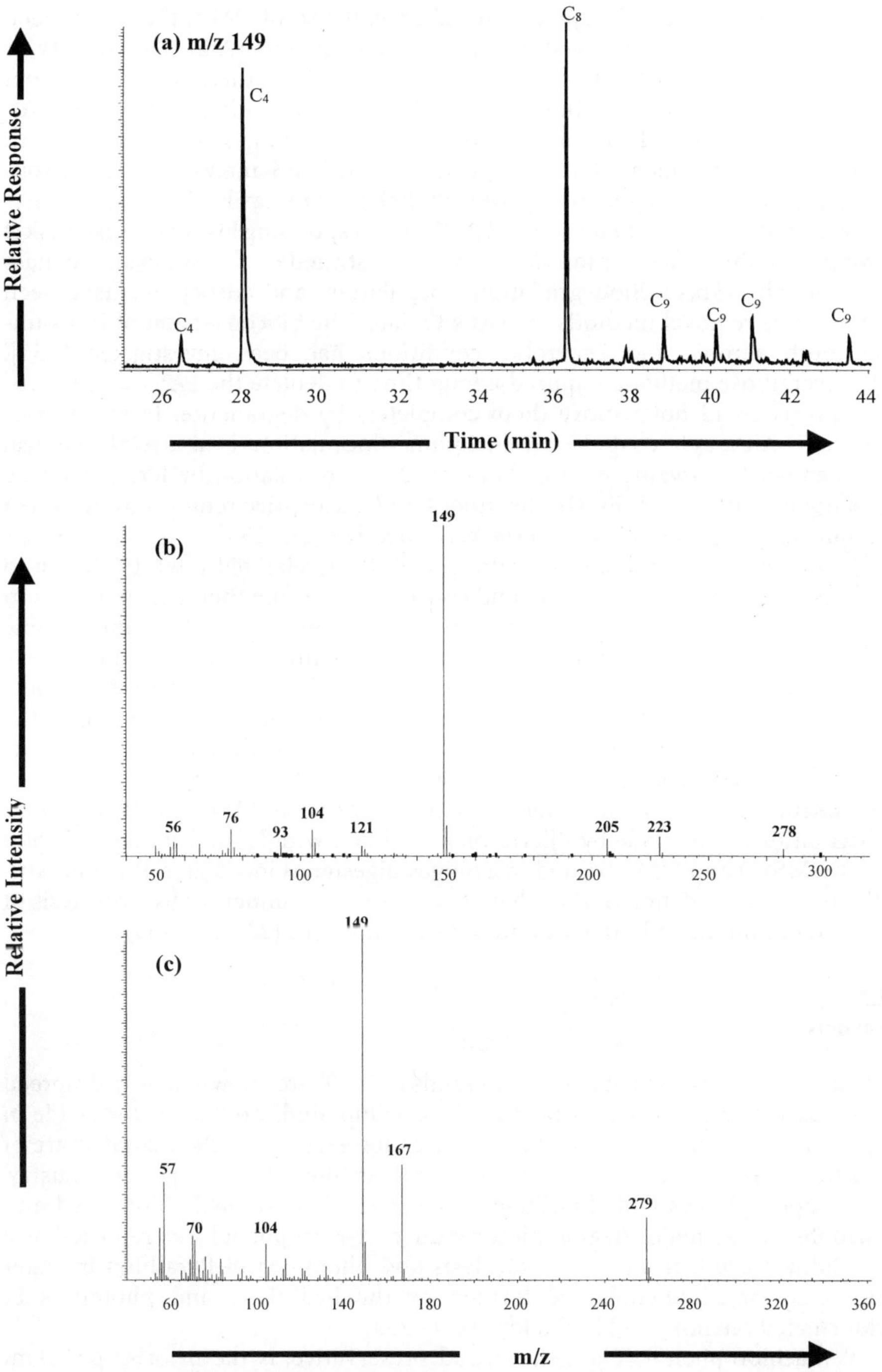

Fig. 18 a–c. An example of: **a** a phthalate ester GC MS fingerprint of an environmental sample; **b** *m/z* 149 C_4-phthalate; **c** *m/z* 149 C_8-phthalate

As a result of the widespread and abundant use of PAEs, they have been widely dispersed and detected in waters and sediments [232]. The toxicity or biological effects of PAEs have been reported [233, 234]; therefore, it is prudent to establish a method for precise analysis, characterization, removal, and/or bioremediation of PAEs from both aqueous and solid phase environments. An example of a phthalate ester fingerprint in the GC-MS analysis of an environmental sample is shown in Fig. 18a. Phthalates are easily detected by their characteristic key ion at *m/z* 149 and by the corresponding loss of one ester alkyl group from the molecular ion (M^+). This is illustrated on the two example mass spectra (Fig. 18b, c). Biodegradation, coagulation, and adsorption have been reported as removal methods for PAEs to date. The bioconversion of PAEs under both aerobic and anaerobic conditions has been investigated [235]. However, those methods required a long time to deplete the PAEs, and microorganisms could not remove them completely by degradation from aqueous solution. Although coagulation including flocculation is a useful removal mechanism for organic micropollutants [236], coagulation by ferric chloride was not effective for PAEs. On the other hand, adsorptive removal by activated carbon and biosorption by bacteria were effective [237, 238].

Studies on the aerobic degradation of PAEs accelerated after 1972, due to doubts about their degradability and concerns regarding their accumulation in the environment. In 1973, Saeger and Tucker [239] reported on the aerobic degradation of PAEs in activated sludge, and since then, numerous studies have shown that PAEs can be transformed by inoculates from various aerobic environments [230, 240–247]. Under anaerobic methanogenic conditions, the capacity for PAE transformation appears to vary among the habitats investigated and the PAEs studied. Some PAEs were shown to be degraded by sewage sludge inoculates, whereas others were more persistent [214, 248, 249]. Similar observations were made by Ejlertsson et al. [250] with landfill municipal solid waste (MSW) and MSW treated in a biogas digester as inoculates. Previous studies on the degradation of PAEs have shown that it commences by hydrolysis of the ester bond under both oxic and anoxic conditions [226, 241, 249].

2.5
Phenols

Phenol and substituted phenol compounds (Fig. 19) are known to be widespread as components of industrial wastes. These compounds are made worldwide in the course of many industrial processes, as for example in the manufacture of plastics, dyes, drugs, and antioxidants, and in the pulp and paper industry. Organophosphorus and chlorinated phenoxyacids also yield chlorinated and nitrophenols as major degradation products. 4-Nitrophenol was reported as a breakdown product after the hydrolysis and photolysis of Parathion in water and chlorinated phenols are formed by the hydrolysis and photolysis of chlorinated phenoxyacid herbicides [251–253].

Pentachlorophenol (Fig. 19), a wood preservative, is the priority pollutant within the group of chlorophenols that has been most released into the environment. Phenols are also breakdown products from natural organic com-

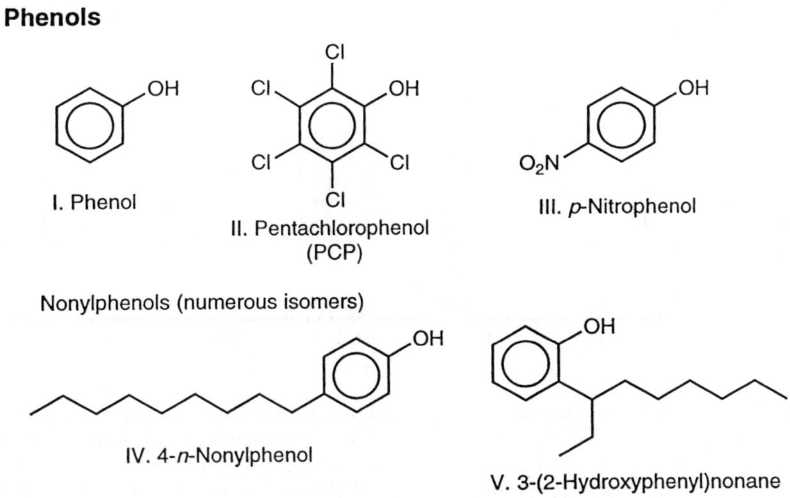

Fig. 19. Names and structures of phenol and substituted phenols

pounds such as humic substances, lignins, and tannins which are widely distributed throughout the environment. Figure 20 shows a typical GC-MS trace of a phenol-contaminated soil sample collected in the Bitterfeld region, Germany [254]. The GC-MS trace shows various chlorophenols (e.g., 2-chlorophenol, 2,4-dichlorophenol, 4-chlorophenol, 4-chloro-3-methylphenol, 2,3,5-trichlorophenol, 2,4,6-trichlorophenol, 2,3,4-trichlorophenol, 2,3,4,6-tetrachlorophenol, pentachlorophenol). Wennrich et al. [254] determined chlorophenols in contaminated soils using accelerated solvent extraction (ASE) with water as the solvent combined with solid-phase microextraction (SPME) and GC-MS analysis. Two different extraction procedures with respect to extraction temperature, extraction time and the effect of small amounts of organic modifiers (5% acetonitrile) on the extraction yields is represented by both upper and lower GC-MS traces in Fig. 20.

A hydrolysis step is involved in the pulp industry in order to concentrate the cellulose from wood. This uses large-scale processes whereby a liquid fraction, the lignocellulose, is formed as a by-product in the process, and contains high levels of phenolic components and their derivatives. These compounds also constitute an environmental problem due to their possible introduction into rivers, lakes, and/or seas. Chlorophenols from the cellulose bleaching process have traditionally attracted most of the interest in the analysis of industrial waste because of their high toxicity.

Phenols and related compounds are highly toxic to humans and aquatic organisms, thus becoming a cause for serious concern in the environment when they enter the food chain as water pollutants. Even at very low levels (i.e., <1 ppb), phenols affect the taste and odor of water and fish [253]. New environmental regulations introduced throughout the world place greater emphasis on treatment of this industrial waste. This fact has been realized by the pulp in-

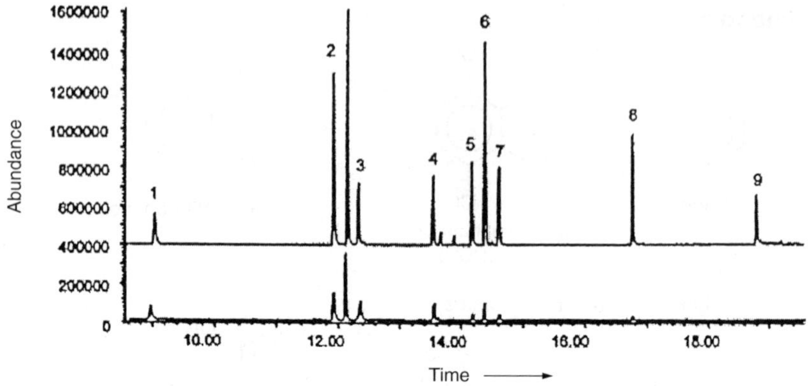

Fig. 20. A typical GC-MS trace of a phenol contaminated soil sample, Bitterfeld, Germany (after [254] with permission). Chlorophenols were extracted using ASE-SPME: upper chromatogram, procedure B; lower chromatogram, ASE conditions of water, 150 °C, 15 min. Peak identifications: (1) 2-chlorophenol, (2) 2,4-dichlorophenol, (3) 4-chlorophenol, (4) 4-chloro-3-methylphenol, (5) 2,3,5-trichlorophenol, (6) 2,4,6-trichlorophenol, (7) 2,3,4-trichlorophenol, (8) 2,3,4,6-tetrachlorophenol, (9) pentachlorophenol

dustries in both Europe and North America. Their waste waters as well as other industrial contaminated waste waters can in principle be treated by municipal sewage treatment plants.

2.6 Organotin Compounds

Organotin compounds (Fig. 21) are used worldwide as insecticides, fungicides, bactericides, acaricides, wood preservatives, plastic stabilizers, and antifouling agents [75, 59, 255, 256] and are therefore found in numerous environmental compartments as for example water, sediments, biological tissue, sewage sludge, etc. [257]. Due to their high toxicity for aquatic organisms, the application of tributyltin (TBT) and triphenyltin (TPT) (Fig. 21) as marine antifouling agents

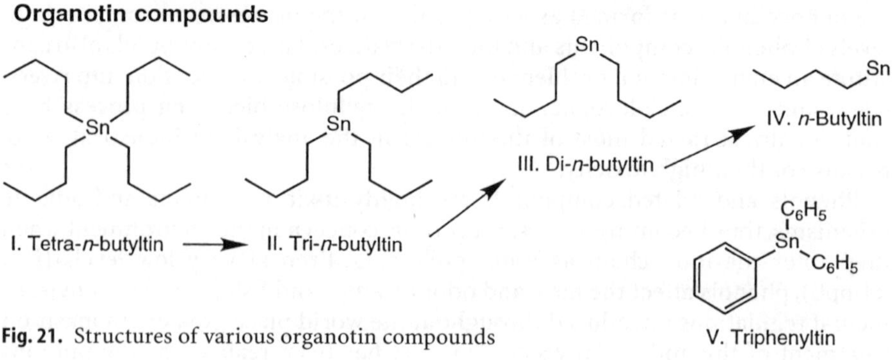

Fig. 21. Structures of various organotin compounds

has been restricted [17, 258–263]. Despite these restrictions, TBT and TPT, as well as their major metabolites dibutyltin (DBT), monobutyltin (MBT), diphenyltin (DPT), and monophenyltin (MPT) are still found in natural waters and sediments at concentration levels which may be critical for the most sensitive organisms [257, 264].

Despite the partial restrictions imposed on TBT by most countries, it is estimated that around 1200 tons per year of TBT is used for the protection of ship hulls [265]. High contamination of port waters has often been reported and waters near ports have also been affected to a lesser degree [264, 266–276]. Residual contamination in the open sea has been studied less, particularly since the detection limits of available analytical methods were inadequate. For instance in the northeastern Mediterranean, the contamination level was below the analytical threshold of 0.1 ng/l water at two reference stations in the open sea [277]. Notable concentrations have been measured in Tokyo Bay and in the Strait of Malacca where ship traffic is heavy, whereas concentrations elsewhere in the open sea have remained below the analytical threshold [278]. However, indirect measurements have suggested the presence of TBT at trace amounts in oceanic waters. Analysis of squid livers and the use of bioconcentration factors have indicated that TBT contamination could reach 0.8 ng/l in waters of the Northern Hemisphere and 0.4 ng/l in those of the Southern Hemisphere [279]. Moreover, the contamination of marine mammals constitutes an indication of TBT presence in Atlantic and Pacific waters [17, 280–282]. In the North Sea, a correlation has been found between physiological abnormalities in whelks and the intensity of shipping traffic [283]. Recently, analysis of TBT in deep-sea organisms collected from Suruga Bay, Japan, suggested that butyltin pollution has reached deep waters [284]. In the Mediterranean, total butyltin concentrations have ranged from 1200 ng/g to 2200 ng/g in dolphin liver [17]. All these studies tend to show the presence of trace amounts of TBT in the open sea. The coastal waters of the northwestern Mediterranean Sea are known to be contaminated by TBT [267, 268]. The density of marinas along the Italian, French, and Spanish coasts accounts in part for this contamination, and there is also considerable commercial and naval ship traffic.

Exposure of humans to butyltin compounds used as stabilizers or as biocides in household articles has been regarded as a source in addition to the ingestion of contaminated foodstuffs. Residues of butyltin compounds, including mono- (MBT), di- (DBT), and tributyltins (TBT), measured in human blood collected from central Michigan, USA are shown in Fig. 22a [285]. Acidified blood samples (2–3 ml) were homogenized with 70 ml of 0.1% tropolone in acetone, and the solvent was transferred to 100 ml of 0.1% tropolone in benzene. Moisture in the organic extract was removed using 35 g of anhydrous sodium sulfate. The sample was concentrated to 5 ml using a rotary evaporator at 40 °C. The concentrated extract was propylated by a Grignard reaction with *n*-propylmagnesium bromide, about 2 mol/l in tetrahydrofuran. The derivatized extract was purified by passing it through a column packed with 6 g of wet Florisil and the eluant from the Florisil column was rotary evaporated to 0.5–3 ml. Butyltin compounds were quantified by capillary gas chromatography with flame photometric detection (GC-FPD).

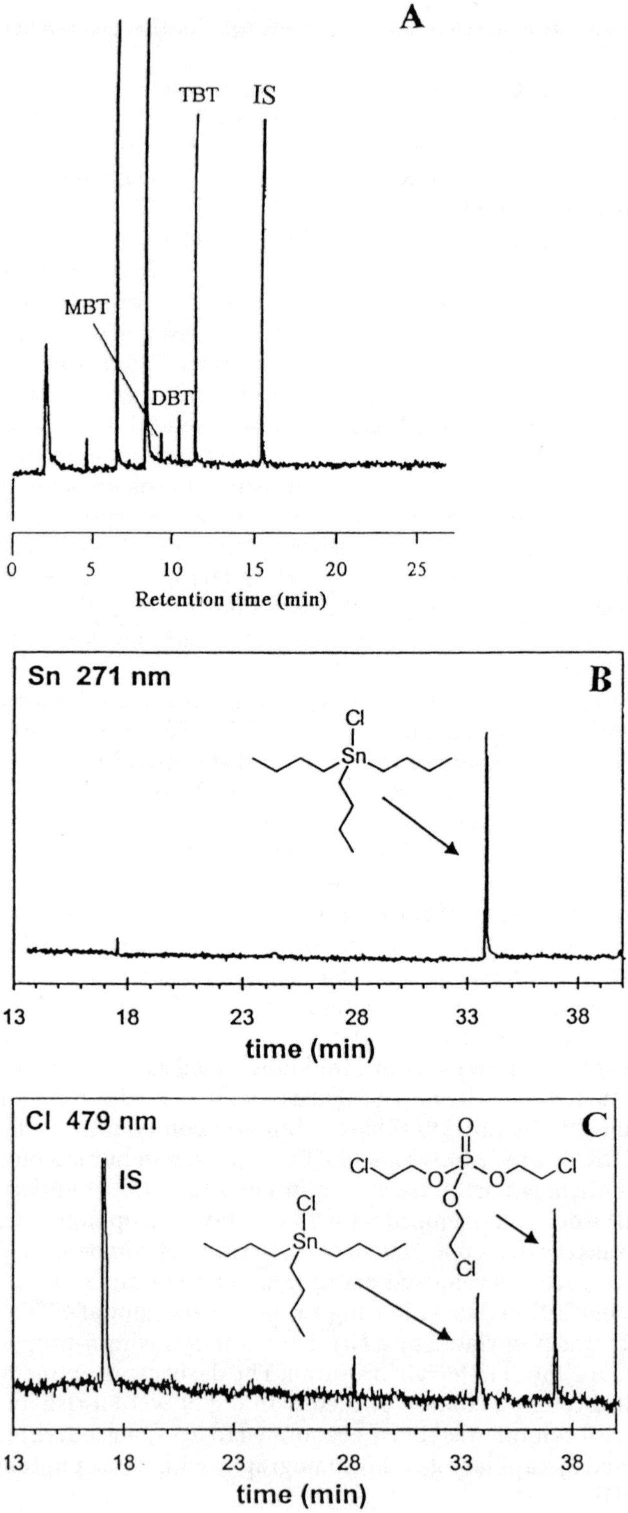
A
TBT
IS
MBT
DBT
0 5 10 15 20 25
Retention time (min)
Sn 271 nm
B
Cl
Sn
13 18 23 28 33 38
time (min)
Cl 479 nm
C
IS
Cl
Sn
O
P
13 18 23 28 33 38
time (min)

Organotin compounds enriched from a diethylether extract of a snow sample collected from the city of Gdansk, Poland and analyzed are shown in Fig. 22b, c [286]. Gas chromatography with atomic emission detection (GC-AED) run in the chlorine and tin channels, respectively, revealed the presence of tributyltin chloride and this was subsequently confirmed by GC-MS and GC-AED analyses of an internal standard solution (e.g., 1-chlorooctane) of that compound. Quantification was based on the response to chlorine (wavelength 479 nm) in the AED system, and a detection limit of 0.5–1 ng/l was achieved for all the reference substances.

Widespread usage of the organotin compounds motivated numerous studies in order to elucidate environmental contamination and impacts [12, 280, 286–289]. The following is a summary of the impact of such usage:

- Physiological abnormalities such as growth reduction in marine microalgae [290], shell thickening, and spat failure in oysters [80, 292] and physiological changes in gastropods [293] and whelks [294, 295] were reported due to organotin compound usage.
- Environmental monitoring and toxicological studies dealing with water [266, 296–298], sediment [299–301], mussels [300], and fish [296, 297, 302] imply that these compounds continue to pose a major ecotoxicological threat in the aquatic environment.
- Significant bioaccumulation of butyltins in higher trophic organisms and their appropriateness as bioindicators of aquatic organotin pollution was reported [281, 282, 303–305].
- Moreover, butyltin accumulation in other marine vertebrates indicated greater accumulation in various organs [280, 282, 295, 306].

Desorption of organotins from harbor sediments has been suggested as source of contamination of the aquatic environment [289, 307]. To study the occurrence and fate of organotins in the environment, in particular their transport and degradation in sediments and at the sediment-water interface, as well as their interactive characteristics with the different solid phase systems, precise and sensitive analytical methods are needed for the aqueous and solid phases, respectively. The challenges for such analytical techniques are: (1) only small samples of sediment pore water (30–60 ml) can be collected, (2) organotin compounds cover a broad range of polarity and hydrophobicity, and (3) organotins, in particular triphenyltin compounds, are unstable under drastic extraction conditions [308–310]. In addition, various binding interactions, such as ion exchange [311], hydrophobic partitioning [312], or surface complexation [313] must be overcome to desorb organotins from sediments.

Fig. 22a–c. Typical examples of organotin contaminated samples: **a** GC-FPD chromatograms of butyltin in human blood extracts (concentrations are <17 ng/ml, 16 ng/ml, and 85 ng/ml of MBT, DBT, and TBT, respectively, after [285] with permission; **b** GC-AED analysis of a diethylether extract of a snow sample collected in the tin channel (Sn = 271 nm) – Gdansk, Poland, after [286] with permission; **c** GC-AED analysis of a diethylether extract of a snow sample collected in the chlorine channel (Cl = 479 nm) – Gdansk, Poland, after [286] with permission

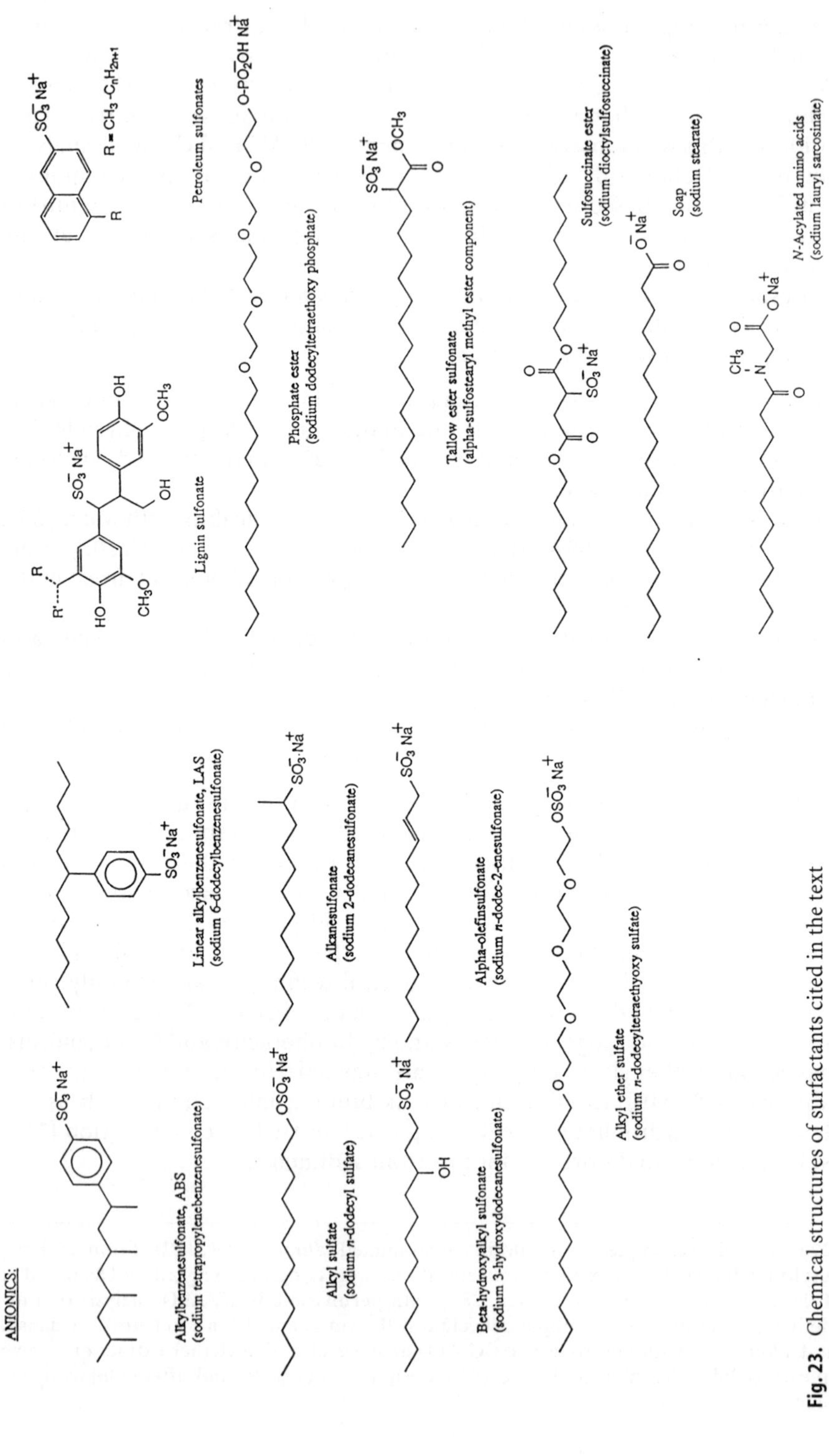

Fig. 23. Chemical structures of surfactants cited in the text

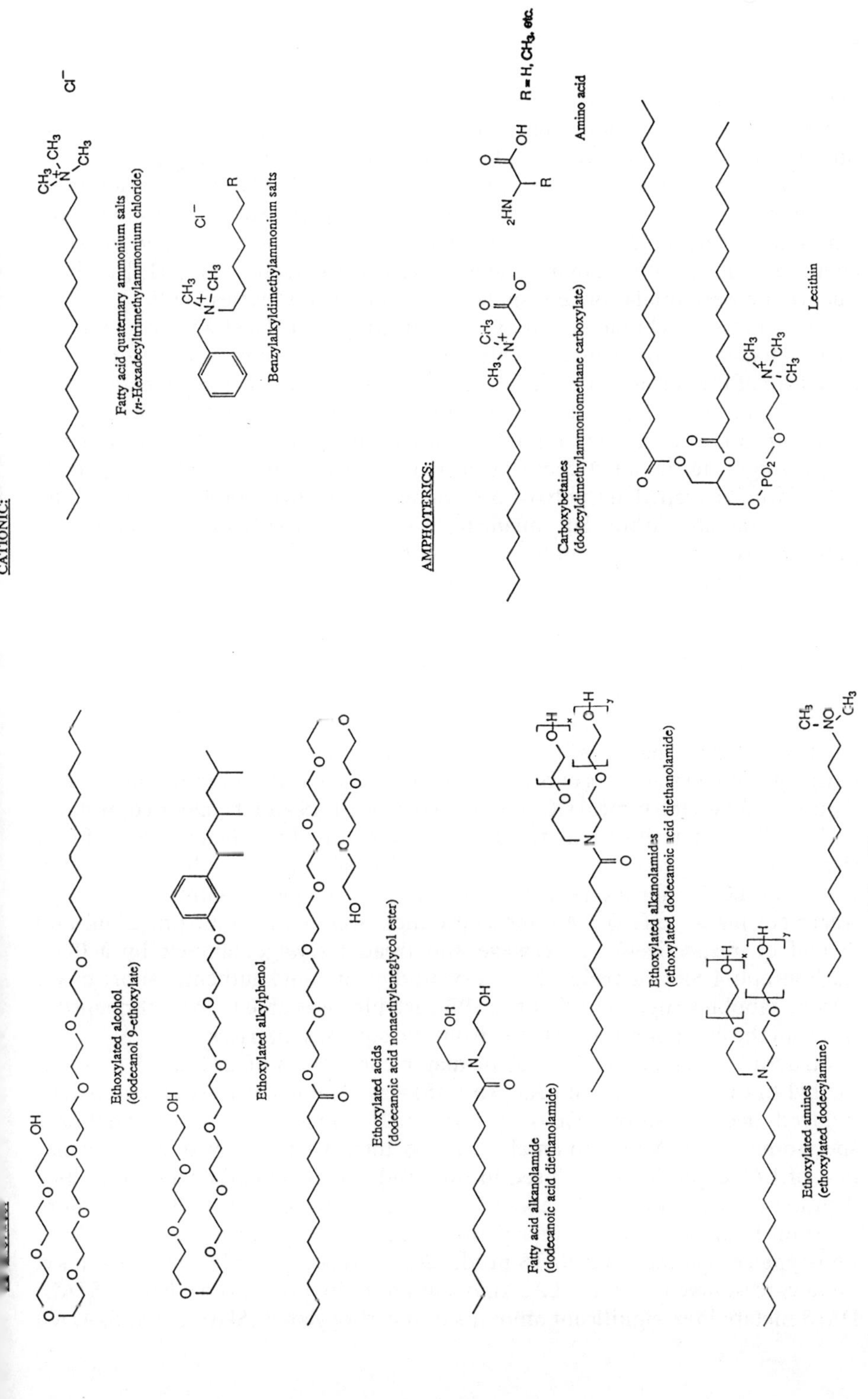

Fig. 23 (continued)

2.7 Surfactants

Surfactants (Fig. 23) represent one of the major and most versatile groups of organic compounds produced around the world [314]. Their main uses are industrial, 54% (cleaning products, food, and industrial processing), household, 29% (laundry, dishwashing, etc.) and personal care, 17% (soaps, shampoos, cosmetics). The worldwide production in 1988 [315] was 2.8 million tons. Surfactants, natural [316, 317] or synthetic, change the solubility and physicochemical state of other environmental micro-constituents [318, 319] and influence their accumulation and spreading at phase boundaries [320].

Surfactants are characterized by concentrating at surfaces and reducing the surface tension [316]. A prerequisite for this surface activity is an asymmetric structure of the surfactant molecule which consists of a water-repellent (hydrophobic) and a water-attracting (hydrophilic) part. In surfactants, the hydrophobic group is a relatively long aliphatic hydrocarbon chain (10–20 carbon atoms), which might be the alkyl chain of fatty acids, alkylbenzenes, alcohols, alkylphenols, polyoxyethylene, polyoxypropylene, etc. The hydrophilic groups can be sulfonate, sulfate, carboxylate, quaternary ammonium, sucrose, polypeptide, or polyoxyethylene [321]. In addition to the hydrocarbon chains, many surface-active agents have been synthesized with fluorocarbon moieties as the hydrophobic, and in this case oleophobic, portion of the molecule. The hydrophobic and hydrophilic parts of the surfactant molecule are in a balanced mutual relationship. Depending on the molecular structure, surfactants can be subdivided into groups discussed in the following sections.

While much attention has been paid to assess contamination levels of linear alkyl benzenesulfonate (LAS) surfactants in the environment, only a few papers have reported on the levels of breakdown products of LAS and coproducts [322]. Figure 24 shows the typical GC-MS (TIC) data for LAS and its major coproducts such as dialkyl tetralinsulfonates (DATS), and methyl-branched isomers of LAS (iso-LAS). These compounds were analyzed based on solid-phase extraction (SPE) and LC-MS for monitoring these analytes in aqueous samples of sewage treatment plants (STPs). LAS and coproducts were extracted from 25 ml and 200 ml of, respectively, raw sewage and treated sewage samples by a 0.5-g Carbograph 4 SPE cartridge. Recovery studies of some authentic short-chain LAS metabolites suggested that the SPE cartridge was able to extract quantitatively all the compounds of interest from the aqueous matrices.

Structure elucidation (Fig. 25) of major coproducts of LAS are DATS and methyl-branched isomers of LAS (iso-LAS) was obtained by in-source collision-induced decomposition (CID) spectra [323]. Liquid chromatography/mass spectrometry (LC/MS) with an electrospray interface to follow biotransformation of LAS coproducts was used in this study. In general, the laboratory biodegradation experiment of LAS and coproducts showed that DATS were more resistant than iso-LAS to primary biodegradation. Biotransformation of both LAS-type compounds and DATS produced, besides expected sulfophenyl alkyl monocarboxylated (SPAC) LAS and sulfotetralin alkylcarboxylated (STAC) DATS metabolites, significant amounts of dicarboxylated (SPADC and STADC)

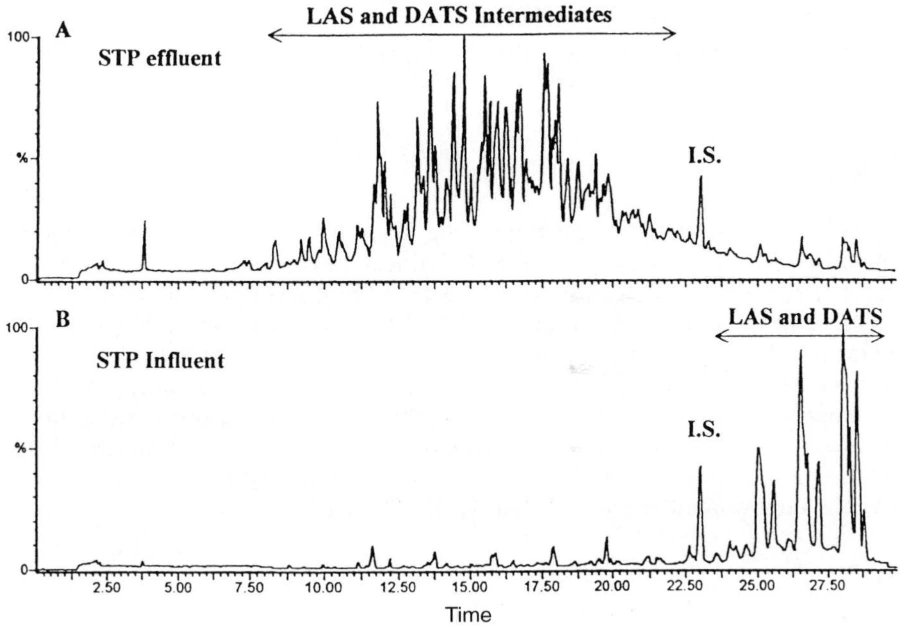

Fig. 24 a, b. Typical total ion current mass chromatograms for LAS, coproducts, and their intermediates in: **a** effluent; **b** influent samples of an activated sludge sewage treatment plant (after [322] with permission)

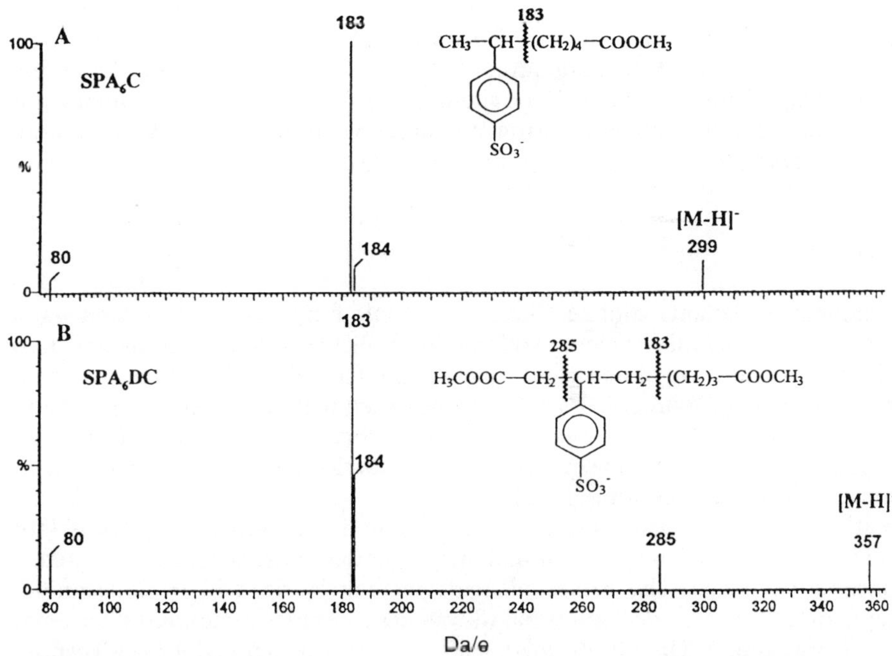

Fig. 25 a, b. CID spectra of two metabolites of LAS-type molecules (after [323] with permission)

species. SPADCs were less persistent than STADCs. After more than 5 months from the beginning of the experiment, 40% and 35% of the initial amounts of DATS and iso-LAS, respectively, were not mineralized.

2.7.1 *Anionic*

A wide range of anionic surfactants (Fig. 23) has been classified into groups, including alkyl benzene sulfonates (ABS), linear alkyl benzene sulfonates (LAS), alcohol sulfates (AS), alcohol ether sulfates (AES), alkyl phenol ether sulfates (APES), fatty acid amide ether sulfates (FAES), alpha-olefin sulfates (AOS), paraffin sulfonates, alpha sulfonated fatty acids and esters, sulfonated fatty acids and esters, mono- and di-ester sulfosuccinates, sulfosuccinamates, petroleum sulfonates, phosphate esters, and ligno-sulfonates. Of the anionic surfactants, ABS and LAS continue to be the major products of anionic surfactants [314, 324]. Anionic surfactants have been extensively monitored and characterized in various environmental matrices [34, 35, 45, 325–329].

2.7.2 *Cationic*

The only cationic surfactant (Fig. 23) found in any quantity in the environment is ditallow dimethylammonium chloride (DTDMAC), which is mainly the quaternary ammonium salt distearyldimethylammonium chloride (DSDMAC). The organic chemistry and characterization of cationic surfactants has been reported and reviewed [330–332]. The different types of cationic surfactants are fatty acid amides [333], amidoamine [334], imidazoline [335], petroleum feed stock derived surfactants [336], nitrile-derived surfactants [337], aromatic and cyclic surfactants [338], non-nitrogen containing compounds [339], polymeric cationic surfactants [340], and amine oxides [341].

2.7.3 *Nonionic*

Nonionic surfactants contain (Fig. 23) no ionic functionalities, as their name implies, and include ethylene oxide adducts (EOA) of alkylphenols and fatty alcohols. Production of detergent chain-length fatty alcohols from both natural and petrochemical precursors has now increased with the usage of alkylphenol ethoxylates (APEO) for some applications. This is environmentally less acceptable because of the slower rate of biodegradation and concern regarding the toxicity of phenolic residues [342].

The traditional major source for the nonionic surfactant industry is fatty acid triglycerides from both animal and vegetable sources as the saturated or unsaturated acids. The saturated acids include lauric acid (*n*-dodecanoic), myristic acid (*n*-tetradecanoic), palmitic acid (*n*-hexadecanoic), and stearic acid (*n*-octadecanoic). The unsaturated acids include oleic acid (*Z*-9-octadecenoic) and linoleic acid (*Z*,*Z*-9,12-octadecadienoic). Of the 200 non-ionic surfactants produced worldwide [330, 333], the distribution of product types is in the following order: oxyalkylated linear alcohols (OALA) (43%), oxyalkylated alkylphenols (OAAP) (25%), oxyalkylated fatty acids (OAFA) (18%), fatty acid amides (FAA) (5%), other oxyalkylated (OAs) compounds (6%), and miscellaneous (3%). Various nonionic surfactants have been studied and characterized in different environmental multimedia [17, 343–351].

2.7.4
Amphoteric (Zwitterionic)

Amphoteric surfactants (Fig. 23) are surface-active agents containing both anionic and cationic functional groups or moieties capable of carrying both ionic charges [314]. However, the term *amphoteric surfactants* or *amphoterics* is used generally to refer to materials that show amphoteric properties. The term *ampholytes* or *ampholytic surfactants*, though synonymous with amphoterics, is used to refer more specifically to surfactants which can accept or donate a proton, such as amino acids. A simple example of this type is 3-dimethyldodecylaminepropane sulfonate (DMDAPS). Within this group are also a number of important natural triglycerides (e.g., lecithin) and alkylbetaines. The latter are obtained by reacting an alkyldimethylamine with sodium chloroacetate and, because they are compatible with skin, they are used in the cosmetics field [352].

Although these surfactants represent less than 1% of the U.S. production of surfactants, the market use is increasing dramatically because of their unique properties [353]. Of particular importance is the synergistic effect that amphoteric surfactants have when used in conjunction with other types of surfactants. The non-eye-stinging characteristic of these compounds has been responsible for the upsurge in the baby shampoo market over time [354, 355].

In general, the main pollution problems associated with surfactants can be summarized as (1) foaming in river and wastewater treatment plants [314, 326, 344, 348, 349, 356, 357], (2) transformation to bioactive metabolites (i.e., polyethoxylated alkylphenols, estrogenic compounds) under aerobic and anaerobic conditions [315, 356], and (3) formation of certain cationics which are toxic to microorganisms at high concentrations [356, 357].

The presence of some surfactants or their by-products in the aquatic environment has been considered as a potential marker of pollution [45, 325]. Thus, the presence of alkylbenzene sulfonates in groundwater has been used as an indicator of the age of the groundwater [358]. Linear alkylbenzenes can act as tracers of domestic waste in the marine environment [34, 35, 359, 360] and trialkylamines as indicators of urban sewage in sludge, coastal waters, and sediments [17, 33, 45, 325, 327, 346, 361]. Analysis, identification, and characterization of surfactants are extensively reviewed and discussed by Aboul-Kassim and Simoneit [314], while pollution problems associated with these compounds are reviewed by Aboul-Kassim and Simoneit [356].

3
Analysis of Environmental Organic Pollutants

The power of analytical instrumentation currently available makes it possible to detect organic pollutants at extremely low concentrations in various environmental samples [64, 362–365]. Such low detection limits are essential if pollutants are to be measured with the accuracy and precision required for modeling their chemodynamic behavior. Most of the work on organic analysis and characterization has resulted from the use of GC and GC-MS.

The isolation of the analyte (i.e., the pollutant of interest) from both the matrix (i.e., the extraction process) and other bulk and trace organics (i.e., the clean-up process) must be fully optimized and highly efficient. Apart from instrumental calibration, the analytical variability of any GC or GC-MS determination of trace organics is primarily caused by interference from non-target compounds, which have not been removed from the extract. Increasing the specificity of the detectors does not necessarily remove the problem, but merely serves to hide the direct evidence of the interference. Varying amounts of extractants, which co-elute with the analyte, will affect the detector signal, giving rise to a reduced or even negative response [362, 363]. Improved reliability and robustness of a method is more likely achieved by efficient sample preparation than by some form of screening by a selective detector [366].

3.1
Recovery Measurements

Recovery measurements are one of the most difficult aspects in organic analysis. These measurements are often completed, with the minimum number of replicate determinations over a limited concentration range, to justify optimistically the use of a method. Experiments designed to obtain the efficiency of the analytical method often implicitly assume that this also includes the efficiency of extraction from the matrix [366].

The basic requirement is to estimate how much of the analyte has been removed from the natural matrix by a given extraction technique. However, the widespread practice of simply adding a known amount of the analyte to the matrix, usually in an organic solvent, prior to extraction and subsequent analysis, does not answer this question. This type of spiked sample analysis determines the accuracy and precision of the subsequent analytical steps, but does not necessarily measure the efficiency of extraction.

To determine the efficiency of extraction, it is imperative that the pollutant is bound to the matrix in a similar configuration to that which exists in the environment. The extraction efficiency can then be measured for that analyte in a specific matrix configuration. At present, water is the only matrix where this can be achieved in a relatively straightforward way. The analytes are added below the surface of the sample in a small volume of water miscible solvent. The water must be completely mixed and allowed to stand at least overnight prior to extraction to allow the pollutants to come into equilibrium with the other organic materials, particularly humic matter. The spiked water sample must be analyzed in its entirety, including the inner surfaces of the container, either separately or as a single determination.

Solid phases (such as sediment, soil, suspended matter, and biosolids) can be doped with known amounts of the analyte by adding the pollutants in a small volume of water-miscible solvent such as acetone, to the sample and the interstitial water. The sediment-solid phase and pore water are mixed thoroughly in a closed container for not less than 24 h and then allowed to settle for a similar period prior to a final mixing. The sediment solids can subsequently be freeze dried if nonvolatile analytes are being determined, but for more volatile compounds (e.g., chlorobenzenes), the sediment solids should be drained of any excess water and extracted as a wet sample. The filtered pore water should also be analyzed. If the organics are mixed completely with the sediment and are given sufficient time to adsorb and diffuse into the sediment surface, then most lipophilic, hydrophobic compounds will be associated almost completely with the organic fraction in the sediment [367, 368]. The sample should be analyzed in its entirety to reduce errors associated with any sample heterogeneity.

The extraction efficiency of organic compounds from solid matrices using the established techniques can be compared with an in situ measurement. For example, Lai et al. [369] used supersonic jet laser-induced fluorescence spectroscopy (SSJ/LIF) to determine PAHs in sediment-solids. The essential elements of the SSJ/LIF are: (1) a pressurized sample chamber where the solid sample is heated; (2) a nozzle connecting the sample chamber to the fluorescence cell; and (3) an evacuated fluorescence chamber through which the laser beam is passed. The samples were heated to 200 °C to produce a vapor of PAHs and other compounds. The LIF signal appeared within 20 s and persisted for 5–30 min. By selecting the correct monitoring wavelength (i.e., 386.74 nm for benzo[*a*]pyrene and 367.44 nm for pyrene), it was possible to distinguish between the two PAHs. Quantitative analyses were carried out by alternating the standards and unknown samples and comparing the integrals of the LIF signal. This technique is both precise and accurate with the limit of determination of 900 ng/g for benzo[*a*]pyrene and 200 ng/g for pyrene in the sediment solids.

Regular, routine sample recovery measurements can be made by using the method of standard addition. The matrix is spiked with the analytes in a small volume of solvent at a level which is 50%, 100%, 150%, and 200% above the estimated level in the sample. A number of independent replicates should be made at each level. Provided that sufficient material is available the sample can be analyzed prior to spiking. In case of limited size (e.g., small tissue samples) a number of samples may be pooled and homogenized for such recovery experiments.

Standard addition to wet sediment should be made in a water-miscible solvent (e.g., acetone or methanol). Any convenient solvent can be used to spike dry sediment. Standard addition to tissue samples can be made by first spiking a small amount of silica and allowing the solvent to evaporate. The silica is then ground with the tissue prior to extraction.

Following the analysis of the spiked samples, the data are plotted and modeled to determine the average recovery and the confidence interval of the method by generating a regression equation model. Once this recovery is established then a single or duplicate recovery sample can be analyzed at periodic intervals to check the validity of the regression equation. In this way a series of data are obtained over a period of time to give a long-term estimate for the method efficiency.

Isotope dilution mass spectrometry (IDMS) is another method to overcome the problem of sample recovery [370–372]. The ^{13}C-labeled isotope of the analyte is added to the sample at the commencement of the analysis and the ratio of the labeled and unlabeled compound is measured by MS. This technique eliminates the need for recovery measurements and automatically accounts for any losses in the determination [373]. The two major limitations of this method are the cost and availability of the labeled compounds and the need to use the MS as a detector.

3.2
Pre-Extraction and Preservation Treatments

Solid samples collected in the field are usually preserved by freezing immediately, either on board ship, in the field, or at the laboratory [374]. Rapid preservation is vital if the integrity of the sample is to be maintained. Sediment cores should be sectioned and each sub-sample frozen individually. Some core samplers allow the whole core to be frozen in situ prior to sectioning. This technique is preferable, if these facilities are available, since it allows the unconsolidated top sections to be handled more easily [375].

Sediment-solid and soil-solid samples can be treated in different ways prior to extraction depending on the purpose of the research program. Sediments or soils are stored more conveniently as dried powders. However, this technique is not appropriate if relatively volatile pollutants such as l-ring aryl hydrocarbons (e.g., alkylbenzenes, chlorohydrocarbons, chlorobenzenes), PAH (e.g., naphthalene) are to be determined. In such cases, the sediment or soil should remain frozen prior to analysis and extracted wet.

Most trace organic pollutants are associated with the organic fraction of sediments or soils, since they partition into the lipids and waxes on particle surfaces. A large proportion of the total organic carbon (TOC) is usually associated with finer particles and an arbitrary value of <63 μm has been selected to isolate most of the organic fraction of sediment-solids [376]. When this fraction is required for a separate analysis, it is advisable to wet sieve the sample, since dried sediments must be re-ground to break up agglomerates. It should be noted that re-grinding does not produce the original particle size distribution of the sediment-solids or soil-solids. The sieved samples which are to be analyzed for the less volatile component can be freeze-dried or air-dried. The resultant sediment brick will require gentle grinding to obtain a free flowing powder [374].

3.3
Extraction Techniques

Although selective extraction of organic compounds appears to be an attractive option, the different types of adsorption sites on solid phases require an exhaustive technique to recover the maximum amount of the analyte from the substrate. This is particularly true where the compounds to be extracted cover a broad range of polarity, reactivity, and molecular size. Therefore, extraction is primarily a process of separating all of the analytes as completely as possible from the bulk of the matrix. This process will inevitably carry along unwanted co-extracted materials. Selective extraction only possible when a small number of chemically similar compounds is to be isolated [377, 378]. The following is an overview of the most commonly used extraction techniques.

3.3.1
Supercritical Fluid Extraction

The first use of supercritical fluid extraction (SFE) as an extraction technique was reported by Zosel [379]. Since then there have been many reports on the use of SFE to extract PCBs, phenols, PAHs, and other organic compounds from particulate matter, soils and sediments [362, 363, 380–389]. The attraction of SFE as an extraction technique is directly related to the unique properties of the supercritical fluid [390]. Supercritical fluids, which have been used, have low viscosities, high diffusion coefficients, and low flammabilities, which are all clearly superior to the organic solvents normally used. Carbon dioxide (CO_2, [362, 363]) is the most common supercritical fluid used for SFE, since it is inexpensive and has a low critical temperature (31.3 °C) and pressure (72.2 bar). Other less commonly used fluids include nitrous oxide (N_2O), ammonia, fluoroform, methane, pentane, methanol, ethanol, sulfur hexafluoride (SF_6), and dichlorofluoromethane [362, 363, 391]. Most of these fluids are clearly less attractive as solvents in terms of toxicity or as environmentally benign chemicals. Commercial SFE systems are available, but some workers have also made inexpensive modular systems [390].

Levy et al. [392] briefly investigated alternative fluids for on-line SFE-capillary GC with CO_2, N_2O, and SF_6 for the extraction of PAHs and alkanes from solid waste, sediment, and shale rock. They initially compared the extraction efficiency of pure fluids and then some fluid mixtures. They found that 20% SF_6 in CO_2 was more effective at 375 bar, and 50 °C for 30 min than each pure fluid for removing both PAHs and alkanes.

McNally and Wheeler [393, 394] applied SFE to the analysis of sulfonylurea herbicides and their metabolites in soil-solids. Engelhardt and Gross [395] analyzed Aldrin, Lindane, and 4,4′-DDT in spiked soil samples using SFE followed by supercritical fluid chromatography (SFC). Lopez-Avila et al. [396] used SFE to extract a series of organochlorine and organophosphorus pesticides from sand using CO_2 and CO_2 modified with acetone. They also examined the extraction and recoveries of these pesticides from sand over a range of temperatures and pressures. Most recoveries were <50% and the recovery of four PAHs was <20%. The efficiency of extraction increased to between 28% and 93% with the addition of 10% methanol as a modifier. Only four of 16 PAHs tested had recoveries <50%. The recoveries for another matrix ranged between 23% and 107%. Of the 15 PAHs tested, 8 were <50% [396]. The most influential parameters were the extraction time and pressure, followed by moisture content and sample size.

Young and Weber [397] presented an equilibrium and rate study of analyte-matrix interactions in SFE in aqueous matrices, while correlation of SFE with supercritical fluid chromatography (SFC) in aqueous media has been reported by Yu et al. [398]. Tena et al. [399] screened PAHs in soil by on-line fiber-optic-interfaced SFE spectrometry.

3.3.2
Soxhlet Extraction

Soxhlet extraction is commonly used for the extraction of non-polar and semi-polar trace organics from a wide variety of solid phases (i.e., sediments, soils, etc.) [192, 366, 380, 400–404]. The size of the systems can vary, but the more common configurations use between 100 ml and 200 ml of solvent to extract 20–200 g of sample. Larger systems can be used, but require proportionally more solvent. It is essential to match the solvent polarity to the solute solubility and to wet the matrix thoroughly with the solvent when extraction commences.

Sediments and soils need to be thoroughly wetted with solvent to obtain an efficient extraction. Surface tension of the solvent across the pores of dry sediment are sufficient to prevent complete diffusion of the liquid into the micro-cavities of the sediment. Non-polar solvents do not readily wet the surface of dry sediments and are too immiscible with water to be able to penetrate water-wet material. This problem can largely be overcome by dampening the sediment with an electrolyte (e.g., 1% ammonium chloride, overnight) or by using an azeotrope or a binary mixture such as acetone with hexane or dichloromethane which has sufficient polarity and water solubility to wet the particle surfaces. If there is a need to remove waxes and lipids of a sample, it can be saponified prior to extraction [1, 53–55, 366]. In some cases, this technique can result in an even higher recovery [405]. On the other hand, Garcia-Ayuso et al. [406] introduced the microwave-assisted Soxhlet extraction technique and reported its advantages over other regular Soxhlet and/or different extraction procedures.

3.3.3
Blending and Ultrasonic Extraction

The simplest extraction technique is to blend or ultrasonically agitate a sample with an appropriate organic solvent at room temperature. Apart from the polarity of the solvent, the efficiency of the extraction is dependent upon the homogeneity of the sample and the mixing/ultrasonication/blending/soaking time. The mixture of sample and organic solvent are separated from each other by centrifugation or filtration and washing with solvent. Blending has been used for solid phase and other environmental samples [189, 366, 407–410].

The extraction of aromatic chlorophenols (e.g., chloroguaiacols, chlorocatechols) is complicated by the different sorption processes that control their binding within the soil-sediment structure [411–413]. The free, physically adsorbed chlorophenolics can be extracted with solvent, but this may only account for 1–5% of the total concentration of these pollutants in the sediment. Martinsen et al. [414] found that *n*-hexane or cyclohexane and iso-propanol recovered <1% of the tri- and tetra-catechols in sediment solids. Remberger et al. [368] attempted to extract both the "free" and the "bound" fractions with an acetonitrile/hexane/methyl *tert*-butyl ether solvent mixture. However, a higher recovery (25–100%) was obtained by using methanolic potassium hydroxide. Wells et al. [405] reported the same improvement with saponification for the recovery of some PCBs from sewage sludge during an intercomparison exercise. Brezny and Joyce [411] made a comparative study of the recovery of ten chlorophenols from soils using conventional solvent extraction and in situ acetylation. Four different extraction methods were compared: (1) sonication of sodium sulfate dried soil with dichloromethane; (2) similar sonication with ethyl acetate; (3) soaking with acetonitrile and ascorbic acid, leaving overnight, treating with dilute sulfuric acid and then extracting with methyl *tert*-butyl ether; and (4) acetylation with acetic anhydride in pyridine followed by sonication with ethyl acetate. Some of the chlorophenols gave good recoveries for all methods, e.g., 4,5-dichloroguaiacol, 3,4,5-trichloroveratrole, tetrachloroguaiacol, and tetrachloroveratrole (80–99%) while recoveries of others like 4,5-dichlorocatechol (1.3–59%) and 4,6-dichloroguaiacol (49–74%) were much improved by acetylation. This direct derivatization and extraction also acetylated other compounds in the matrix which made the subsequent determination more difficult. Therefore, Brezny and Joyce [411] recommended that the chlorophenols were extractively acetylated rather than directly.

3.3.4 *Liquid-Liquid Extraction*

Liquid-liquid extraction (LLE) is based on the partition of organic compounds between the aqueous sample and an immiscible organic solvent. The efficiency of an extracting solvent depends on the affinity of the compound for this solvent, as measured by the partition coefficient (i.e., on the ratio of volumes of each phase and on the number of extraction steps). Solvent selection for the extraction of environmental samples is described and reported in many reviews and recent articles [364–366, 415–420] and is related to the nature of the analyte. Non-polar or slightly polar solvents are generally chosen. Hexane and cyclohexane are typical solvents for extracting aliphatic hydrocarbons [421] and other non-polar pollutants such as organochlorine or organophosphorus pesticides [422]. Dichloromethane and chloroform are certainly the most common solvents for extracting non-polar to medium polarity organic pollutants [1, 53–55, 423]. The large selection of available pure solvents, providing a wide range of solubility and selective properties, is often claimed as an inherent advantage of LLE techniques. In fact, each solvent is seldom specific toward a class of compounds and LLE is mainly used for the wide spectrum of compounds extracted. The so-called lipid fraction is obtained by extraction with chloroform or dichloromethane and contains many organic compounds such as aliphatic and aromatic hydrocarbons, ketones, alcohols, fatty acids, sterols, etc. [1, 53–55, 424, 425].

LLE can be performed simply using separatory funnels. The partition coefficient should therefore be large because there is a practical limit to the phase-volume ratio and the number of extractions. When the partition coefficient is small and the sample very dilute, a large volume must be handled and continuous liquid-liquid extractors should be used. The extractions then take several hours. Such extractors have been described in the literature [364, 426–429]. The partition coefficient may be increased by adjusting the pH to prevent ionization of acids or bases or by forming ion pairs or hydrophobic complexes with metal ions, for example. The solubility of analytes in the aqueous phase can be reduced by adding salts. Fractionation of samples into acidic, basic, and neutral fractions can be attained by successive extractions at different pH [430–432].

It is difficult to compare recoveries obtained by different laboratories because their extraction conditions (pH, phase ratio, number and time-length of extractions, salinity) are generally different. Sample volumes can be very high, up to 200 l [433], and 50 l of surface water [434] or 20 l of sea water allow the extraction of 5 ng/l of alkanes. When using a specific detection method, the sample volume can be lower: 2 ng/l of PAH was determined from 1 l of river water using liquid chromatography and fluorescence detection [435]. Chlorophenols below the 10 ng/l level were determined from 100 ml of sea water with electron capture detection (ECD) GC [436].

The LLE of relatively polar and water-soluble organic compounds is, in general, difficult. The recovery obtained from 1 l of water with dichloromethane is 90% for Atrazine but lower for its more polar, degradation products, i.e., di-isopropyl- (16%), di-ethyl- (46%), and hydroxy-atrazine (46%). By carrying out LLE with a mixture of dichloromethane and ethyl acetate with 0.2 mol/l ammonium formate, the extraction recoveries for the three degradation products were increased to 62%, 87%, and 65%, respectively [437].

3.3.4.1
Concentration Procedures

LLE results in the extraction of the analyte into a relatively large volume of solvent which can be concentrated using a rotary evaporator to a few milliliters. Further concentration to a few hundred microliters can be carried out by passing a gentle stream of pure gas (usually dry N_2) over the surface of the extract contained in a small conical vial. The solvent-evaporation method is slow and has a risk of contamination. Micro-extractors have been described, and have the advantage of avoiding the further concentration of organic solvents [417–438]. One of them allows the handling of an aqueous volume up to 980 ml, extracted with 200 μl of organic solvent [439]. Although this was applied to the extraction of hydrocarbons, chlorinated pesticides, and phthalate esters, at trace levels, with average recoveries of 90% after three consecutive extractions, the use of such an apparatus is not often described in the environmental literature.

3.3.4.2
Advantages and Drawbacks

The main advantages of LLE are its simplicity and the use of simple and inexpensive equipment. However, it is not free from practical problems such as the formation of emulsions, which are sometimes difficult to break up [377]. The evaporation of large solvent volumes, and the disposal of toxic and often flammable solvents, are also inherent to the method. LLE requires several sample-handling steps and contamination and loss must be avoided at every step. The glassware must be carefully washed or annealed and stored under rigorous conditions. The organic solvents used must be pure pesticide-grade when extracting traces of pesticides from water.

Carrying out LLE in the field is not easy and large water samples are usually transported and stored in laboratories. Alternatively, large volumes of water can be pumped through adsorptive cartridges and the analytes desorbed in the laboratory by extraction (LLE) or direct vaporization. Automation of the whole procedure of extraction and concentration requires the use of robotics, so it is typically an off-line procedure. Loss during the transfer and evaporation steps always occurs, although to a minor extent. Internal standards are therefore often added before LLE and then the recoveries are calculated from standard peaks by assuming that the losses are similar for solutes and standards. Solubilization of the standards in the samples should be assessed carefully. Losses due to adsorption on vessels are frequently encountered, especially for polar solutes. All these factors explain why LLE is often described as tedious, time-consuming, and costly.

3.3.5
Solid-Phase Extraction

Solid-phase extraction (SPE) or liquid-solid extraction is a sample preparation method, which is especially well adapted to the handling of water samples. SPE has been widely used and reported in several research articles [30, 440–444]. Trace organics are trapped by a suitable sorbent packed in a so-called extraction column through which the water passes and are later recovered by elution with a small volume of organic solvent. Extraction and concentration are therefore performed at the same time. This technique appears less straightforward than LLE, because there is a large choice of sorbents and because the recoveries depend on the sample volume. In fact, SPE is simple when one considers that it is based on the well-established separation principles of liquid chromatography. SPE can be used (1) off-line, the sample preparation being completely separated from the subsequent chromatographic analysis, or (2) on-line by direct connection to the chromatographic system (typically GC).

3.3.5.1
Off-Line Methods

In off-line methodologies [366], the samples are percolated through a sorbent, packed in disposable columns or cartridges, or enmeshed in an inert matrix of a membrane-based extraction disk. Disposable prepacked columns or cartridges are available from many manufacturers and the containers and reservoirs are generally made of polypropylene. The sorbent bed varies from 100 mg to 1000 mg and is retained between two porous frits. The volumes above the packing vary from 1 ml to 20 ml in columns designed with large capacity reservoirs. For larger volume samples, the reservoirs can be attached to the columns via an adapter, or directly to the cartridges. Single samples can be processed by attaching a syringe to the SPE column or reservoir for application and elution. The sample may also be aspirated through the column by vacuum. Another method of application is to use centrifugation by inserting the SPE cartridges into an appropriate centrifugal system. Various vacuum manifolds allow batches of up to 24 samples to be prepared simultaneously. The application of samples and solvents in a SPE process can thus be performed semi-automatically, with no risk of sample contamination.

Compared with LLE-based sample preparation, off-line SPE offers reduced processing times and produced substantial solvent savings. Percolation of samples can be performed in the field and good storage of the adsorbed analytes is generally observed [445]. The problem of transport and storage of voluminous samples is avoided, which is particularly useful when samples have to be taken from remote sites. Automation is possible, using robotic or special sample preparation units that sequentially extract the samples and clean them up for automatic injections. Nevertheless, a certain amount of tedious labor remains and off-line procedures have the inherent disadvantages of loss in sensitivity due to injection of an aliquot, of losses in the evaporation step and some risks of contamination, so that internal standards are required.

3.3.5.2
On-Line Methods

On-line coupling of the SPE sample preparation to GC or liquid chromatographic (LC) separation avoids many of the problems mentioned above. On-line approaches coupling SPE to LC are performed particularly easily in any laboratory and are known as column switching, precolumn technology, or on-line multidimensional chromatography. This was developed extensively and reported by several workers [445–450]. The extraction precolumn is placed in the sample-loop position of a six-port liquid switching valve. After conditioning, sample application, and eventual cleaning via a low-cost pump, the precolumn is coupled to an analytical column by switching the valve to the inject position. The adsorbed compounds are then eluted directly from the precolumn onto the analytical column by a suitable mobile phase which also enables the chromatographic separation of trapped compounds. One can expect more accurate quantitative results, as there is no sample manipulation between the preconcentration and analysis. Automation is easy and several devices are now commercialized. In contrast with off-line SPE, the entire sample is transferred and analyzed, which allows the handling of smaller sample volumes.

In general, the chemistry and principles are essentially identical for both off-line and on-line SPE. SPE can be considered as a simple chromatographic process, the sorbent being the stationary phase. The mobile phase is the water of the aqueous sample during the extraction step, or the organic solvent during the desorption step. Retention of organic compounds occurs to the extent that they are not eluted by water during the extraction step. Reversed-phase materials are widely used because, in reversed phase chromatography, water is the less mobile phase for neutral organic compounds. The highest enrichment factors are obtained when there is a high retention of analyte by water and a low retention by the desorbing organic solvent. With pure organic solvents, desorption occurs for a volume close to the void volume of the column. From a practical point of view, to obtain high enrichment factors one should select the sorbent that gives the highest retention of analyte in water.

Puig et al. [450] determined ng/l levels of priority methyl-, nitro-, and chlorophenols in river water samples by an automated on-line SPE technique, followed by liquid chromatography-mass spectrometry (LC-MS) using atmospheric pressure chemical ionization (APCI) and ion spray interfaces.

3.3.6
Column Extraction

Before using state-of-the-art analysis and characterization techniques, most environmental sample extracts are separated or fractionated prior to analysis [1, 53–55, 366, 451]. There are many fractionation schemes reported in the literature, and isolation of lipid fractions generally incorporates thin layer chromatography or column chromatography using alumina, silica gel, or a combination of both. A simple scheme that can be used to obtain various lipid fractions was developed by eluting them from the chromatographic column with solvent mixtures of increasing polarity such as *n*-hexane, dichloromethane, and methanol [e.g., 1, 53–55]. Once the fractions containing the compounds of interest have been separated, further fractionation can be made by urea adduction or molecular sieving to separate linear compounds from branched and cyclic compounds.

3.3.7
Comparative Extraction Studies

Ideally, the pollutants to be determined should be removed from the matrix as completely as possible with a minimum amount of the other non-target components. This type of selectivity was certainly anticipated from supercritical fluid extraction. However, trace organic pollutants cover a wide range of polarity, volatility, and molecular size, making selective extraction very difficult to achieve. Currently the most popular extraction methods are Soxhlet [191, 400, 402–404], blending [189, 408, 409, 411–455], liquid column extraction and ultrasonic extraction [456], and more recently supercritical fluid extraction [386, 456–463].

The main comparisons between extraction methods have been made between the Soxhlet, ultrasonication, and supercritical fluid extraction [377, 398, 456, 461, 462]. This has primarily been prompted by the need to evaluate critically the relative merits of SFE as an alternative to the more established methods. Richards and Campbell [456] made a comparison between SFE, Soxhlet, and sonication methods for the determination of some priority pollutants in soil. The SFE apparatus was the same, relatively standard system as described by Campbell et al. [457] with the addition of a CO_2 cryogenic trap to improve the trapping of the more volatile extractants in dichloromethane. The priority pollutants selected were chlorobenzenes, chlorophenols, *sym*-dichloroethyl ether, and naphthalene. With a 2% methanol modifier in CO_2 at 390 bar and 80 °C, the recoveries ranged from 70% for phenol to 83% for 2,4,6-trichlorophenol. The Soxhlet extraction used a 1:1 mixture of acetone and hexane for 16 h with recoveries ranging from 54% for 1,3-dichlorobenzene to 81% for 2,4-dichlorophenol. The sonication method used 1:1 acetone and dichloromethane and had recoveries in the range of 46% for 1,3-dichlorobenzene to 75% for hexachlorobenzene.

Onuska and Terry [461] examined the extraction of tetrachlorodibenzodioxin (TCDD) from sediments. They found that either CO_2 or N_2O with 2% methanol as modifiers gave the highest recovery at 310 bar and 40 °C. They also studied the effect of extracting wet sediment, as opposed to the dry material, and found that when the sediment was moist, the recovery diminished by 20% for the same extraction time. However, the same efficiency could be achieved with the wet sediment if the 40 min extraction time was doubled. Soxhlet extraction of the same dried sediment with *n*-hexane/acetone (1:1) (150 ml) and 2,2,4-trimethylpentane (25 ml) for 18 h was only around 65% of the SFE recovery. The Soxhlet extraction was considerably more variable (22–90%, n = 3), but since the Soxhlet actually recovered 90% of the TCDD, this means that the method can be efficient but erratic. This variability was almost certainly a function of the heterogeneity of the matrix surface and/or the wettability of the sediment.

A comparison was made between the in situ analysis using supersonic jet-laser induced fluorescence spectroscopy (SSF/LIF) and hot Soxhlet extraction for the determination of PAHs in sediment [369]. The study highlighted two aspects. First, there was good agreement between the measurements made for both benzo[*a*]pyrene and pyrene in a marine sediment by both methods. The sediment used was a reference material prepared by the National Research Council of Canada. However, the second soil from the coal gasification plant did not show the same agreement. Considerably less PAHs were detected in the soil using the hot Soxhlet extraction (dichloromethane for 24 h at 90 °C). Both Renkes et al. [464] and Junk and Richards [465] found inconsistencies in the recovery of PAHs when prolonged extraction times were used at elevated temperatures. This comparison clearly indicates the need to optimize fully the extraction conditions. Extended extraction time and extreme temperatures do not necessarily improve recovery and losses can occur through degradation.

Schuphan et al. [466] used the "Bleidner" vapor phase extraction technique [467] for the determination of organochlorine pesticides (OCPs) and PCBs in lake sediment and compared the results with traditional Soxhlet extraction. The advantage of the "Bleidner" distillation is that it avoids the time-consuming steps of drying, conventional extraction, and clean up. The thawed sediment was mixed with distilled water and an antifoaming agent, then the aqueous phase was distilled into a flask containing iso-octane, which was subsequently used to extract the distillate. Direct measurement of the OCPs and the PCBs were made by capillary ECD-GC. The Soxhlet method required the sediment to be dried. The pore water was separated from the solid particles and extracted with *n*-hexane/toluene (9:1). The moist sediment was dried with phosphorus pentoxide prior to Soxhlet extraction with *n*-hexane/toluene (9:1) for 20 h. The recoveries for the PCBs by the "Bleidner" technique declined with increasing chlorination (PCB28 98% to PCB180 43%) and was likely a function of the decrease in volatility of the congeners. The low recovery of γ-HCH (43%) was a result of the higher water solubility of this compound and of 4,4′-DDT (10%) and 4,4′-DDD (42%) was interpreted as rapid metabolism to DDD and to dichlorobenzophenone, respectively. The low recoveries of the DDTs by the "Bleidner" extraction are probably due to the stronger binding of these compounds to the sediment which is not reversed by simple steam distillation. The method, therefore, although rapid for some volatile, non-bound hydrophobic organic compounds, is not suitable for wide application as an extraction technique.

3.3.8
Micro-Extraction Methods

The mass of sample taken for analysis is primarily dependent on four factors: (1) the amount of material available, (2) the concentration of the analyte, (3) the heterogeneity of the sample, and (4) the method of analysis. Most conventional solvent extraction techniques currently start with more sample than is required, use more extraction solvent than is necessary, and ultimately only analyze 0.1% of the material prepared, e.g., 1 µl from 1 ml. Micro-extraction techniques [468] can be used in conjunction with "on-line" LC-GC or LC-MS to utilize the whole extract in the final determinations. This approach can significantly reduce the size of sample required and the volume of solvent used. Many workers have reported the use of solid phase microextraction (SPME) in different environmental matrices for various pollutants [288, 342, 345, 469–477].

3.4
Clean-Up Techniques

Normally an extraction technique is selected to give the highest recovery for a wide range of pollutants. Therefore, the extract will most likely contain a high proportion of co-extracted material. Many of the clean-up techniques have been tailored into a series of multi-residue schemes in order to maximize the use of each sample [189, 402, 453, 454, 478–481]. This is of particular value when the maximum amount of chemical information is required for each sample.

The main requirement for any clean-up and group separation scheme is that it effectively removes not only the bulk of the co-extractants, such as lipids, sulfur, carotenoids, and other pigments, but also those compounds that may potentially interfere in the final determination. There are three main ways in which co-extracted material may interfere in the final determination if not removed:

1. Gross contamination can overload the HPLC or GC columns with obvious and usually rapid deterioration of chromatographic performance. This can occur with so called "rapid" techniques where the detector is used as a filter, e.g., selected ion monitoring (SIM) MS, or where the clean-up method has been overloaded (e.g., excess of lipid). This problem can be overcome by using and monitoring more selective clean-up techniques.

2. Interferences caused by inadequate chromatographic separation during the final determination (e.g., no prior group separation of PCBs or OCPs). This can be improved by multi-dimensional GC or multi-dimensional preparative LC.
3. Interference occurs when compounds co-elute with the analytes and are not detected directly by a specific detector. The effect is to create negative peaks or an erratic response for the analyte. This problem can be identified by using a non-specific detector such as an ion trap MS detector, an MS in the electron impact ionization mode, or a flame ionization GC detector.

These problems are overcome by applying a tailored LC separation prior to the final determination and having a built-in feedback to monitor the success of the separation or to give a warning of any failure. The following are the most commonly used clean-up techniques in organic analysis of environmental pollutants.

3.4.1
Measurement of Extractable Lipids/Bitumen

Solid samples are normally examined to determine the extractable organic residue (also called lipids or bitumen). For sediments and soils, it is possible to compare the levels obtained in the organic extract with the total organic carbon determined by combustion techniques to verify the efficiency of the extraction. The lipids from other extraction methods are also measured gravimetrically. Some workers [191, 481] take an aliquot of the extract to determine the extractable lipids while others [189] evaporate the whole extract and re-dissolve the oil after weighing for the remaining analysis. The advantage of the latter method is that: (1) it is likely to be more precise for low levels of lipids; (2) none of the sample is lost; (3) the solvent can be changed; and (4) the lipid determination is made on the actual fraction that is analyzed. The disadvantage of the method is that volatiles can be lost during evaporation to dryness.

The value of the extractable lipid measurement is two-fold. First, it indicates how much lipid has to be removed in the subsequent clean-up process and second, it allows the levels of organic pollutants in the matrix to be expressed on a lipid basis. This normalization reduces the differences among samples purely as a result of the lipid in the sample and the effect of external factors that affect lipid levels.

3.4.2
Removal of Lipids/Bitumen

There are various methods for removal of co-extracted lipids and a brief description follows.

3.4.2.1
Saponification

The bulk of the triglycerides and wax esters can be removed by saponification of an extract with 5% potassium hydroxide in methanol [53–55, 482]. However, this relatively harsh treatment is suitable only for the most chemically resistant pollutants. Most organophosphorus pesticides are hydrolyzed by this method. Both 2,4′- and 4,4′-DDT are dehydrochlorinated to the corresponding 2,4′- and 4,4′-DDE and hexachlorocyclohexanes are also degraded. Saponification has been used successfully for the determination of some PCBs [405, 483] but the more chlorinated PCBs are prone to loss of chlorine especially if the reaction is undertaken at too high a temperature, e.g., >70 °C for extended time (>1 h) [191]. Therefore, it is essential to check the recovery of each analyte under the specific conditions of the reaction if this technique is used. Although this may be a disadvantage for the higher chlorinated PCBs, it is possible to make use of these hydrolysis reactions as confirmation of the presence of more reactive compounds.

3.4.2.2
Sulfuric Acid

The other chemical method used to remove the bulk of the co-extractants is the dehydration and oxidation reactions with concentrated sulfuric acid. This method is only suitable for the most robust chemical groups such as organohalogens without an oxygen bridge, i.e., not suitable for Dieldrin, Endrin, and Aldrin [189, 191, 408, 484]. The initial methods involved shaking the analyte extract in an alkane solvent with the concentrated acid. However, this reaction is more manageable if the acid is adsorbed onto silica gel. Up to 40% of sulfuric acid (w/w) can be loaded onto silica. The value of this method is that up to 20 g of lipid can effectively be denatured by passing an extract through a column containing 50 g of the 40% H_2SO_4 on SiO_2 and eluting with dichloromethane [191, 484]. It is possible to automate this clean-up in a batch process using a gravity column or a low pressure flow-through system.

3.4.2.3
Solid Phase Clean-Up

Liquid chromatographic clean up [441, 443, 450] has been used either in normal phase flow using alumina, silica, or florisil [22, 189, 403, 481, 484] or with reverse-phase (RP) columns [409, 452, 480]. In most cases these techniques are well established and are used in an "off-line" mode, primarily to remove the bulk of co-extracted materials prior to a more refined clean-up prior to the final determination. These columns may be prepared in the laboratory [22, 403–405] or commercial solid phase extraction (SPE) cartridges can be used [409, 452, 463, 470, 485, 486]. In both cases, the normal phase cartridges and column materials are disposable since many of the polar co-extractants bind firmly to the substrate surface and are difficult to remove. This has been overcome to some extent using RP materials where the polar compounds are eluted prior to the non-polar materials. These columns and cartridges can be regenerated in some circumstances by flushing with methanol, but quite often the gross contamination from the co-extractants precludes their re-use. Similar LC clean-up and separations are used "on-line" for less contaminated samples [366, 440].

3.4.2.4
Gel Permeation Chromatography

Gel permeation chromatography (GPC) with SX3 Biobeads (200–400 mesh) in a range of column sizes and solvents is used by most workers [189, 366, 400, 402, 453, 454, 487–489]. Separation has been made primarily between lipid material >500 Å which is the first to elute from such columns followed by the smaller molecules which include most of the organic pollutants that accumulate in sample matrices.

GPC or size exclusion chromatography (SEC) has several key advantages over all other methods currently available. The method is non-destructive and, unlike saponification or concentrated sulfuric acid clean up, can be used to isolate less robust pollutants (e.g., organophosphorus pesticides) [402, 453]. It is also more applicable to the isolation of unknown pollutants or alteration products where there is little information on the polarity or chemical functionality of the molecule. Adsorption chromatography is not able to isolate groups of compounds with different polarities or structures in a single small fraction. GPC is also considerably more tolerant of handling a large mass of lipid in each sample. Columns (50 cm × 25 mm i.d.) can cope with up to 500 mg of lipid, whereas the adsorption columns are limited to 50 mg/g of lipid. It is possible to increase the size of the adsorption column to remove 250 mg of lipid, but larger volumes of solvent are required to elute the more polar organics.

One main disadvantage of the GPC system is the difficulty to remove completely all traces of lipids [438]. Since triglycerides elute prior to the smaller pollutants, the "tail" of the lipid peak carries over into the second fraction. The amount of lipid in the "tail" becomes significant when a relatively large mass of triglyceride has to be removed relative to the concentration of the pollutants. Grob and Kalin [438] found that much of the tailing was caused by lipids trapped in the injection port and the connecting tubing of the HPLC. Although this contamination was reduced by appropriate switching, the lipids were not completely eliminated. Even a 0.01% carryover from 1 g of lipid will leave an unacceptably high level of co-extractant in an extract. Until this inherent problem can be solved, the low molecular weight fraction usually requires further clean-up to remove the trace lipids (e.g., SiO_2) prior to analysis.

Jansson et al. [189] were able to use the SX3 Biobeads and a mobile phase of dichloromethane in hexane (1:1) to make a further separation of the chloroparaffins from lipids and other organochlorine pollutants. Using diethylhexyl phthalate (DEHP) as a marker, they collected the appropriate fractions to isolate the chloroparaffins and other pollutants.

3.4.2.5
Supercritical Fluid Clean-Up

Most supercritical fluid extraction (SFE) studies have focused on obtaining a complete separation between the bulk matrix and the small organic pollutants (<500 Da) in situ. With a few exceptions [376, 460] the SFE removes some or all of the soluble lipophilic material along with the trace organic pollutants. The difficulties in selecting the optimum SFE parameters to obtain a lipid free sample has been a limitation of this method and of the hyphenated SFE-SFC. Lohleit and Bachmann [459] and Ali and Cole [380] used adsorbents such as Tenax, Carbopack C, Spherosil XOA200, florisil, and reverse phase C_{18} sorbent to trap organics and subsequently desorb them using SFE with CO_2. These adsorbents can be used to trap the pollutants from air, but also from SFE extraction. Carbopack was unstable when used with supercritical fluids and high molecular weight artifacts were extracted from Tenax.

3.4.2.6
Sulfur Removal

Elemental sulfur is present in most soils and sediments (especially anaerobic), and is sufficiently soluble in most common organic solvents that the extract should be treated to remove it prior to analysis by ECD-GC or GC-MS. The most effective methods available are: (1) reaction with mercury or a mercury amalgam [466] to form mercury sulfide; (2) reaction with copper to form copper sulfide; or (3) reaction with sodium sulfite in tetrabutyl ammonium hydroxide (Jensen's reagent) [490]. Removal of sulfur with mercury or copper requires the metal surface to be clean and reactive. For small amounts of sulfur, it is possible to include the metal in a clean-up column. However, if the metal surface becomes covered with sulfide, the reaction will cease and it needs to be cleaned with dilute nitric acid. For larger amounts of sulfur, it is more effective to shake the extract with Jensen's reagent [478].

3.5
Automation

Automation does not always remove the problems of time and effort associated with manual methods. A critical evaluation of both the manual methods to be replaced and the automated alternative should be made before embarking on a new scheme. New, improved, and rapid methods described in the literature may not always be appropriate [366]. The following is a summary of the most common automation techniques.

3.5.1
Robotics

Regardless of the pretreatment method, simple manipulations in sample preparation remain one of the most labor-intensive areas of analytical work [491].

There are many applications of auto-injection, multi-dimensional chromatographic separations and data analysis, but sample preparation has not had the same level of automation in most laboratories. The key advantages of automation are unattended repetitive tasks (time saving), greater accuracy, consistency, reliability, less analyst fatigue than manual methods, continuous operation possible with toxic solvents (dichloromethane) and corrosive materials (SiO_2/sulfuric acid and fine powder adsorbents) (safety). Automated systems may require isolation but not fume exhaust hoods (saving space and cost).

Robotic systems in a small analytical laboratory have the greatest application in the intermediate sample manipulation steps. The removal of excess solvent with the Zymark evaporator [492], for example, can be closely controlled, fully automated, and operate in parallel (up to six samples per instrument). This technique has considerable advantages over rotary evaporation, which is prone to loose volatile organic compounds (e.g., chlorobenzenes) under vacuum and rapid vaporization. Automated repetitive manipulations are well served by a robotic system [492].

3.5.2
On-Line Automation

Although on-line automation systems offer considerable attractions, such techniques need to be fully investigated before applying them to an analytical manipulation. The transition from a manual to an automatic method is more easily made if each step in the existing method is readily amenable to such a change. For example, column extraction or SFE are good candidates for automation, but combining them would only be suitable with extensive robotics. Most LC methods, whether gravity columns, SPE, or HPLC, can be automated and connected on-line to the final GC or GC-MS stage. However, there are three main unresolved problems with the on-line LC-GC approach for multi-residue analysis, which can be summarized as follows:

- Although the separation between some unwanted co-extractants and the analytes is well suited to an on-line system, high lipid or elemental sulfur loading is more effectively removed off-line. Most on-line systems at present work most effectively with low lipid contents [493, 494], although some applications have overcome the problem of lipid removal.
- The LC-GC is used to isolate analytes in a separate fraction from other interferences, usually by heart cutting, and then to chromatograph that fraction by GC. However, difficulties arise when multiple fractions must be isolated from each sample by the LC, for example different groups of compoundssuch as PCBs and OCPs. The sample should also be separated into fractions when similar compounds are present at considerably different concentrations or where chromatographic overlap is to be avoided. Under such conditions, the multiple fractions produced by the on-line LC cannot be analyzed directly by linking a single GC. This difficulty may, however, be overcome by two configurations. The first is by using an on-line heart cut into the GC autosampler,

so that each fraction can be taken sequentially into the GC. There are two disadvantages to this approach. First, the GC column phases may have to be different for each fraction to obtain the appropriate separation and second the inherent sensitivity of the "on-line" system is lost. The second is by an "on-line" LC heart cut into separate parallel GCs. This can be a practical option for laboratories which use multiple GCs that are optimized for the analysis of each fraction in a multi-residue scheme.
- Some form of stop-flow LC and sequential GC analysis.

Despite the difficulties of "on-line" automation, the need to develop such systems is considerable. The increase in the number of different compounds that must be determined and the number of samples required for a meaningful survey or laboratory study make it essential to improve the quality and throughput of samples. There are a number of stages in fully automating trace organic analysis. Autosampler LC or GC-data systems as GC-MS or GC-ion trap detector (ITD) are well established and require no further elaboration here [191, 203, 495].

The early developments of on-line LC-GC have been reviewed by Davies et al. [496] and Koenigbauer and Major [497]. The selectivity characteristics of the mobile and stationary phases can be optimized to give both a cleaned-up sample and group separation by heart-cutting the desired fraction prior to GC analysis. The LC is usually interfaced to the GC by an uncoated, deactivated GC capillary precolumn to transfer the heart-cut from the LC. This heart-cut from the LC is vaporized to focus the solute at the head of the GC column [498]. The volume of the GC precolumn, the volume of the heart-cut, the GC oven temperature, and carrier gas flow for the concurrent solvent evaporation are carefully matched [499, 500].

The following examples highlight the progress and pitfalls of on-line LC-GC applications in environmental pollutant analysis:

1. Maris et al. [493] determined PCBs in sediment by on-line narrow bore LC-GC. One key advantage of using the narrow bore columns is the inherent low LC flow rates of 5–50 μl/min which are more comparable to coupling to the GC. The column (150 mm × 1.1 mm i.d.) was packed with 5-μm Li Chrosorb Si60 and coupled to a 20 mm × 0.7 mm i.d. Li Chrosorb AloxT guard column. The sediment was extracted in a Soxhlet apparatus and the sulfur removed with Jensens' reagent [490]. Hexane was used as the LC solvent and the PCBs were heart-cut into the GC between 5 min and 10 min after injection onto the LC. A comparison was made between the LC-GC and the off-line alumina-silica clean-up, and the data obtained for the two methods for seven monitored PCBs 28, 52, 101, 118, 138, 153, and 180 were quite similar; however, the following observations highlight some difficulties that may occur with this method:
   a) The sediment was extracted with hexane. A more polar solvent, e.g., dichloromethane, may extract more PCBs, and almost certainly more co-extractants which would have to be removed by the LC.
   b) The LC alumina guard column deteriorated quickly with multiple samples, and even with back flushing had to be replaced regularly.

c) The GC column was wide bore (0.32 mm) with N_2 as carrier gas. Such a system has a considerably lower resolution, as evidenced by the chromatograms, than would normally be required of the high-resolution separation with H_2 or He and a 0.22 mm i.d. column [405, 495]. The lower performance of this widebore column may mask any band broadening at the LC-GC interface. The separation of the PCBs from pesticides and other organic residues was not possible with this "on-line" system and seriously interfered with the determination of the OPCs themselves and other PCBs.

The LC-GC technique clearly has the advantages of speed and improved sensitivity since the whole sample extract is used. After some development to improve the resolution of the final determination, it may be appropriate, for example, for the analysis of a small number specific PCBs in a routine monitoring program.

2. Rene et al. [494] used SPE cartridges in an automatic sample preparation with extraction columns system coupled with capillary GC-ECD to determine OCP and pyrethroid insecticides. Hexane (2 ml) extracts of drinking water or surface water were passed through the Bakerbond SPE cartridges containing 100 mg of silica. The OCPs were eluted with *n*-hexane/iso-propanol (99.9:0.1) and injected into the GC using a 6-mm fused silica retention gap, the sample introduction time being matched to give concurrent solvent evaporation. The samples were effectively cleaned-up on the SPE and 17 OCPs were isolated from co-extracted material and the pyrethroids determined. Since the GC analysis time was 70 min the sequential sample was prepared by the automatic system in parallel to the GC determinations. The recovery of 23 pesticides tested ranged from 95% to 107% with a CV% between 7.5% and 11.8% and a limit of detection between 3 ng/l and 30 ng/l.
3. Kapila et al. [501] used an on-line SFE-LC to determine chlorinated phenols in wood chips over the concentration range 1–500 mg/kg. Following the extraction, the sample was loaded into a sample loop of the HPLC and chromatographed using a conventional packed LC column and UV detector.
4. Neilen et al. [502] coupled an SFE system with a GC-ECD for "on-line" determination of PCBs which had been trapped onto solid adsorbents such as Tenax. Their application was primarily to determine organic compounds in the atmosphere, but such a system could be adapted to trap a cleaned-up extract from biological tissue prior to analysis by GC-ECD or MS.

3.6 Multi-Residue Schemes

Multi-residue schemes are used by a number of workers for the determination of very different compounds [189, 402, 405, 453, 454, 474, 478, 480, 481] and each of the methods of extraction and clean-up discussed in the earlier part of this chapter have been incorporated into an overall analytical scheme. At present the on-line approach is difficult to incorporate fully into the multi-residue scheme [189, 479] in which a large number of compounds are separated into groups and determined in parallel. The value of the multi-residue method permits: extensive analyses of costly and sometimes irreplaceable samples, especially those taken from remote sites (e.g., open ocean) or from specific experiments; correlation of data of different analytes within a single analysis to reduce variability; and the reduction of analytical effort at the sample preparation stages.

Jansson et al. [189] used the conventional approach of blending the solid particles with solvent after which an aliquot was taken to determine the volatile compounds (e.g., phenols and chlorobenzenes). A second fraction was taken after the lipid removal for determination of compounds sensitive to concentrated sulfuric acid. The bulk lipids were removed by oxidative dehydration with SiO_2/H_2SO_4 and further cleaned-up with GPC. The chloroparaffins were isolated at this stage. Separation on silica isolated the OCPs, and the organochlorines and organobromines were finally fractionated on active charcoal.

Krahn et al. [479] developed a similar multi-residue scheme for the determination of organochlorines and PAHs in sediments. In this scheme, the preparation is semi-automated with GPC to separate the biogenic material and the sulfur from both the PAHs and organochlorines in the samples. The sterols were separated and purified with an amino-cyano HPLC column prior to derivatization with bis(trimethylsilyl)trifluoroacetamide (BSTFA).

4 Identification and Characterization of Organic Pollutants

In the past few years, the number of organic pollutants which have been identified from various sources has increased dramatically due to the extensive analytical research by numerous scientists [e.g., 53–56, 60, 61, 63, 66, 68–73]. The major reason for this marked increase is a result of analytical development in the detection and identification of organic markers, in particular gas chromatography-mass spectrometry (GC-MS) and associated data systems. The development of high-resolution capillary columns has led to their routine usage in most GC-MS systems. The improved separation of complex organic mixtures (COMs) through the use of high-resolution capillary columns has led to the identification of additional molecular markers. Along with the increase in gas chromatographic resolution, fast-scanning mass spectrometers, both quadrupole and magnetic sector instruments, are able to obtain spectra on relatively small and narrow chromatographic peaks. The present situation is completely different compared to before, when it was necessary to isolate compounds in pure crystalline form in order to enable structural determinations to be made by classical chemical techniques.

A dramatic change in instrumental development for environmental chemical analysis and specifically for molecular pollutants has occurred over the last 25 years [48, 503–518]. The following sections will review the different techniques and instrumental development currently used for the characterization of organic pollutants.

4.1 Gas Chromatography

In 1975, gas chromatography (GC) with glass capillary columns provided the best means for resolving complex organic mixtures of pollutants [519–521]. Currently the available capillary columns are made of flexible fused silica with low activity, which eliminates many of the problems previously associated with glass capillary columns [366, 522, 523]. The latest development is columns with the liquid phase actually bonded to the fused silica. These columns have a much longer life and can be washed with solvents if peak shapes degenerate as a result of the accumulation of polar compounds on the column [524, 525]. Furthermore, the columns can be taken to a much higher final temperature with low levels of column bleed. Thus, the number of organic compounds currently resolvable on capillary columns is much greater than those resolved in the seventies [512, 518, 520–523, 525–529].

High-temperature high-resolution gas chromatography (HTGC) is an established technique for the separation of complex mixtures of high molecular weight compounds (HMW) which do not elute when analyzed on conventional GC columns [530]. The combination of this technique with mass spectrometry (i.e., HTGC-MS) is not so common, however, Elias et al. [530] used this novel application to evaluate and identify the occurrence of HMW tracers ($>C_{40}$) from smoke aerosols.

4.2 Gas Chromatography-Mass Spectrometry

Mass spectrometers use the difference in mass-to-charge ratio (m/z) of ionized atoms, molecular fragments, or whole molecules to differentiate between them. Mass spectrometry is therefore useful for quantitation of atoms or molecules and also for determining chemical and structural information about them [329, 531–533]. Molecules have distinctive fragmentation patterns which provide information to identify structural components. The general operation of a mass spectrometer is to: (1) create gas-phase ions, (2) separate the ions in space or time based on their mass-to-charge ratio, and (3) measure the quantity of ions of each mass-to-charge ratio. The ion separation power of a mass spectrometer is described by the resolution, which is defined as:

$$R = \left(\frac{m}{\Delta m}\right)$$

where:

m = the ion mass, and

Δm = the difference in mass between two resolvable peaks in a mass spectrum (e.g., a mass spectrometer with a resolution of 1000 can resolve an ion with an m/z of 100.0 from an ion with an m/z of 100.1).

In general, a mass spectrometer consists of an ion source, a mass-selective analyzer, and an ion detector. Since mass spectrometers create and manipulate gas-phase ions, they operate in a high vacuum system. The magnetic-sector, quadrupole, and time-of-flight designs also require extraction and acceleration ion optics to transfer ions from the source region into the mass analyzer. The following sections summarize a brief description of the most commonly used ionization techniques as well as the different types of available mass spectrometers.

4.2.1
Mass Spectrometry Ionization Methods

Different ionization techniques have been used for mass spectrometric identification and characterization of organic pollutants as described below.

4.2.1.1
Electron Impact

An electron impact (EI) ion source uses an electron beam, usually generated from a rhenium filament, to ionize gas-phase atoms or molecules. Electrons from the beam (usually 70 eV) knock an electron from a bond of the atoms or molecules creating fragments and molecular ions [366, 534, 535]. Several factors contribute to the popularity of EI ionization in environmental analyses such as stability, ease of operation, simple construction, precise beam intensity control, relatively high efficiency of ionization, and narrow kinetic energy spread of the ions formed.

4.2.1.2
Chemical Ionization

Chemical ionization (CI) has proven to be a useful technique for the MS analysis of many pollutants [533–537]. CI uses a reagent ion to react with the analyte molecules to form ions by either a proton or hydride transfer:

$$MH + C_2H_5^+ \rightarrow H_2^+ + C_2H_4$$
$$MH + C_2H_5^+ \rightarrow M^+ + C_2H_6$$

The reagent ions are produced by introducing a large excess of reagent gas (e.g., methane) relative to the analyte into an electron impact (EI) ion source. Electron collisions produce CH_4^+ and CH_3^+ which further react with methane to form CH_5^+ and $C_2H_5^+$:

$$CH_4^+ + CH_4 \rightarrow CH_5^+ + CH_3$$
$$CH_3^+ + CH_4 \rightarrow C_2H_5^+ + H_2$$

4.2.1.3
Electrospray Ionization

The electroscopy ionization (ESI) technique is widely used in environmental analysis [75, 83, 90, 538–541, 543]. In most ESI techniques, the source consists of a fine needle and a series of skimmers. A sample solution is sprayed into the source chamber to form droplets. The droplets carry a charge when they exit the capillary, and as the solvent evaporates the droplets disappear, leaving highly charged analyte molecules.

4.2.1.4
Fast-Atom Bombardment

In fast-atom bombardment (FAB) a high-energy beam of neutral atoms, typically Xe or Ar, strikes a solid sample causing desorption and ionization [366, 534, 535]. FAB is used for large organic molecules that are difficult to mobilize into the gas phase. FAB causes little fragmentation and usually gives a large molecular ion peak, making it useful for molecular weight determination. The atomic beam is produced by accelerating ions from an ion source through a charge-exchange cell. The ions pick up an electron in collisions with neutral atoms to form a beam of high-energy atoms.

4.2.1.5
Plasma and Glow Discharge

A plasma is a hot, partially-ionized gas that effectively excites and ionizes atoms [366, 534, 535]. A glow discharge is low-pressure plasma maintained between two electrodes. It is particularly effective at sputtering and ionizing material from solid surfaces.

4.2.1.6
Field Ionization

Molecules can lose an electron when subjected to a high electric potential resulting in field ionization (FI) [366, 534, 535]. High fields can be created in an ion source by applying a high voltage between a cathode and an anode called a field emitter. A field emitter consists of a wire covered with microscopic carbon dendrites, which greatly amplify the effective field at the carbon points.

4.2.1.7
Laser Ionization Mass Spectrometry

A laser pulse can ablate material from the surface of a sample, and create a microplasma which ionizes some of the sample components. The laser pulse accomplishes both vaporization and ionization of the sample [366, 534, 535]. This method is called laser ionization mass spectrometry (LIMS).

4.2.1.8
Matrix-Assisted Laser Desorption Ionization

Matrix-assisted laser desorption ionization (MALDI) is a LIMS method for vaporizing and ionizing large organic molecules such as proteins or DNA fragments [78, 287, 536–550]. The biological molecules are dispersed in a solid matrix such as nicotinic acid. A UV laser pulse ablates the matrix which carries some of the large molecules into the gas phase in an ionized form so they can be detected in the mass spectrometer.

4.2.2
Types of Mass Spectrometers

The different kinds of mass spectrometers of utility in environmental research and monitoring are described in the following:

4.2.2.1
Quadrupole Mass Spectrometry

A quadrupole mass spectrometer consists of a mass filter with four parallel metal polarity rods [366, 534, 535, 551]. Opposing rods have an applied potential of $(U + V\cos(wt))$ and the other two rods have a potential of $-(U + V\cos(wt))$, where U is a direct current voltage and $V\cos(wt)$ is an alternating current voltage. The applied voltages affect the trajectories of the ions traveling down the flight path centered between the four rods. For given direct and alternating current voltages, only ions of a certain mass-to-charge ratio pass through the quadrupole filter and all others are deflected from their original path. A mass spectrum is obtained by monitoring the ions passing through the quadrupole filter as the voltages on the rods are varied.

4.2.2.2
Magnetic-Sector Mass Spectrometry

In the case of magnetic sector mass spectrometry, the ion optics in the ion-source chamber of a mass spectrometer extract and accelerate ions to a kinetic energy of 70 eV [534, 535]. In the flight tube they are separated between the poles of the magnetic field according to mass. Only ions of mass-to-charge ratio that have equal centrifugal and centripetal forces pass through the flight tube. The accuracy is adequate to utilize this method mainly for high-resolution mass spectrometry [e.g., 552]

4.2.2.3
Ion-Trap Mass Spectrometry

The ion-trap mass spectrometer uses three electrodes to trap ions in a small volume. The mass analyzer consists of a ring electrode separating two hemispherical electrodes. A mass spectrum is obtained by changing the electrode voltages to eject the ions from the trap. The advantages of the ion-trap mass spectrometer include compact size and the ability to trap and accumulate ions thus increasing the signal-to-noise ratio of a measurement [534, 535, 551, 553].

4.2.2.4
Time-of-Flight Mass Spectrometry

A time-of-flight mass spectrometer (TOF-MS) uses the differences in transit time through a drift region to separate ions of different masses. It operates in a pulsed mode so ions must be produced or extracted in pulses. An electric field accelerates all ions into a field-free drift region with a kinetic energy of qV, where q is the ion charge and V is the applied voltage. Since the ion kinetic energy is $0.5\ mv^2$, lighter ions have a higher velocity than heavier ions and reach the detector at the end of the drift region sooner. TOF-MS has been used widely for different environmental applications [79, 80, 360, 534, 535; 539–541, 543, 544, 547–549, 554–562].

4.2.2.5
Fourier-Transform Mass Spectrometry

Fourier-transform mass spectrometry takes advantage of ion-cyclotron resonance to select and detect ions [366, 534, 535, 563–565].

4.2.3
Fragmentation Pattern and Environmental Applications

The combination of high-resolution capillary columns with fast scanning quadrupole or magnetic sector mass spectrometers provides an excellent method for the identification of a large proportion of the compounds in complex organic materials (COMs). It is worth mentioning that GC-MS analysis of environmental samples has the added advantage over GC of providing structural information on many unknown components responsible for the chromatographic peaks, as well as components that appear to be hidden in the baseline of the chromatogram [515, 566–568]. The basis for GC-MS detection of molecular markers is the fact that in the ion source of the mass spectrometer many molecular markers fragment in a systematic manner to produce one or more characteristic (key) ions which can be used to detect the particular organic marker in question.

The best example to explain the fragmentation pattern and data interpretation in GC-MS is provided by the ubiquitous molecular markers with the hopane-type structure (Fig. 1; Structures VII and VIII). The molecular weight varies according to the substituent R at C-21 which has been shown to range from H to $C_{13}H_{27}$. Generally, hopanes are a very important class of biomarkers in petroleum and environmental pollution studies [68–73]. Hopanes fragment in the mass spectrometer producing two major ions as shown in Fig. 26. The first is at m/z 191 from the *A/B* ring fragment and the second at m/z 148 + R from the *D/E* ring fragment where the mass will vary depending on the substituent R. The relative intensities of the ions at m/z 191 and m/z 148 + R vary depending on the stereochemistry at the C-17 and C-21 positions. However, by monitoring only variations in intensity of these characteristic ions (i.e., SIM) rather than acquiring a complete mass spectrum at each scan, the sensitivity of the mass spectrometer for detecting hopanes is increased by several orders of magnitude.

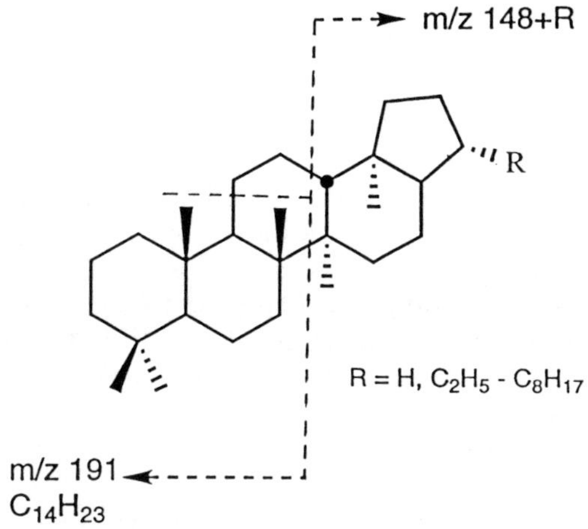

Fig. 26. Hopane fragmentation pattern

The most popular ionization technique used in GC-MS for environmental geochemical studies is electron impact, normally at an electron energy of 70 eV. A small number of papers have demonstrated that chemical ionization is of some use in molecular marker studies [569]. Negative ion electron impact MS has been utilized for synthetic organic mixtures (e.g., PCB) and generally yields useful fragmentation patterns for electronegative compounds (e.g., nitro, oxo, halogen substituted aromatics) [517, 570, 571]. Other ionization techniques such as field ionization, field desorption, and fast atom bombardment (FAB) have not as yet found widespread applications in environmental studies.

GC-MS using high-resolution capillary columns and low-resolution mass spectrometers has been a popular analytical technique in environmental organic geochemistry [507, 572, 573]. However, additional analytical techniques have been used very recently to extend the capabilities available for the determination of molecular markers. Such advances are discussed in the next few paragraphs, which show the alternative approaches to increase GC-MS sensitivity and specificity:

- Mackenzie et al. [574] described the determination of molecular markers in unfractionated crude oils using a combination of GC with high resolution mass spectrometry which has a number of advantages over GC-MS using low resolution mass spectrometry in the MID mode. In the latter, nominal masses are used to monitor for various classes of organic compounds, but this can be confusing since a number of nominal masses can be characteristic of more than one class of compound. This was explained by the following example: the ion at *m*/*z* 253 is found in mass spectra of both alkanes and monoaromatic steranes and therefore a sample with a high concentration of *n*-alkanes when analyzed for monoaromatic steranes would be dominated by the *n*-alkane distribution of *m*/*z* 253. However, using high-resolution mass spectrometry and accurate mass assignments, the fragment ion for the *n*-alkane is *m*/*z* 253.29 and that for monoaromatic steranes is *m*/*z* 253.20. Thus, the distribution of the monoaromatic steranes can be determined with the aid of the higher resolution mass spectrometer and in turn this permits the analysis of the total extract [503, 560, 575, 576], eliminating the time-consuming fractionation steps.

- Warburton and Zumberge [577] proposed that an increase in sterane specificity, or other biomarkers, could be achieved by monitoring the spontaneous (unimolecular) fragmentation of sterane parent ions in the first field free region of a double focussing mass spectrometer. Analysis of the sterane distributions, for instance, by conventional GC-MS and MID methods showed the distribution of steranes to be a complex mixture of C_{27}, C_{28}, and C_{29} stereoisomers, some of which could not be resolved. However it is possible to observe separately the sterane metastable parent ion transitions corresponding to $M^+ \rightarrow 217^+$ during a single GC-MS run, where M^+ is the molecular ion for the steranes and *m*/*z* 217 is the major fragment ion. This observation is made by using a programmable power supply to vary the accelerating voltage while holding the magnetic and electrostatic fields at appropriate constant values. The technique is valuable in that it yields a simplified sterane fingerprint.
- Tandem mass spectrometry (i.e., MS-MS) is another technique that has recently become popular for the direct analysis of individual molecular markers in complex organic mixtures [87, 505, 509, 578–583]. This technique provides a rapid method for the direct analysis of specific classes of molecular markers in whole sample extracts. In this approach the system is set up to monitor the parent ions responsible for a specific daughter ion as described above and the distribution of parent ions obtained under these conditions should provide the same information as previously obtained by GC-MS [505, 582]. Even greater specificity can be achieved by a combination of GC-MS-MS [516, 584]. In view of the complexity of COM samples and the need to detect the presence of individual organic compounds or classes of compounds, it would seem that MS-MS, especially coupled with GC, would be extremely valuable in future environmental organic geochemistry studies.

4.3 Liquid Chromatography-MS

Other combinations of chromatography techniques with MS which may be useful in environmental studies are the coupling of high performance liquid chromatography (LC) with MS [84, 384, 504, 506, 530, 585–593], LC with MS-MS [181, 594–599], LC with atmospheric pressure chemical ionization MS (LC-APCI-MS) [600], and Fourier transform infrared spectroscopy-fast atom bombardment coupled to LC-MS (FTIR-FAB-LC-MS) [514].

LC-MS has been used to study various aromatic fractions from coal derived liquids, and there are also a number of reports on its use in the analysis of porphyrin mixtures [601, 602]. The early work by Dark et al. [601] using LC-MS for coal-derived liquids was mainly concerned with the separation and identification of polycyclic aromatic components. However, it is interesting to note that developments in the field of fused silica capillary columns for GC has been so rapid that most of the aromatic compounds with six or seven aromatic rings can now be passed through a GC eliminating the need for LC [603]. Nevertheless, the role for LC in the future of petroleum and environmental geochemistry may again be directed at examining higher molecular weight and more polar molecules.

4.4
Isotope Ratio Mass Spectrometry

Isotope ratio mass spectrometry (IRMS) is a branch of analytical mass spectrometry which for many years was a highly specialized subdiscipline with a somewhat low profile among the general mass spectrometry community [604]. High-precision IRMS, meaning measurement of deviations of isotope abundance ratios from an agreed standard by only a few parts per thousand for C, H, N, O, S, and Cl, is now possible and in some cases coupled on-line to chromatographic separations. The following is a brief discussion about the theory and applications of the IRMS technique in different environmental fields.

4.4.1
Environmental Reviews

Stable isotope analysis has long been realized to be a valuable technique to investigate the sources and behavior of organic contaminants since, by definition, all organic pollutants contain carbon [605–611]. Moreover, virtually all organic pollutants of environmental concern also contain hydrogen [610, 612, 613], while many may also contain elements such as chlorine (e.g., chlorinated solvents [614]), oxygen (e.g., the gasoline additive methyl *tert*-butyl ether, MTBE [611, 613]), and nitrogen and/or sulfur (e.g., Atrazine and various pesticides [615, 616]). Hence, the potential exists to use multiple stable isotope analyses of a single individual pollutant that can provide additional discriminants to investigate its sources and the behavior in both surface and subsurface environments. Since the publication of the review by Brenna [604], other excellent overviews of the subject have been published [617–619]. Newman [620] has presented an interesting perspective on the relationship between the IRMS of greatest interest to organic analytical chemists with the techniques used for other elements [604, 617–619].

4.4.2
Theory

Here, the main theory of so-called gas phase IRMS, which is directed toward determination of variations in stable isotope compositions of elements (e.g., C, H, N, O, S, Cl) by analyzing the gases (i.e., CO_2, H_2, N_2, O_2, SO_2, etc.) are reviewed. Lichtfouse and Budzinski [621] have presented a brief description of the techniques and an account of their applications in organic geochemistry. Although the most demanding applications of IRMS involve determinations of variations in natural isotopic abundances, the same techniques are also applicable to metabolic studies using stable isotope-enriched substrates. Such studies normally employ conventional GC-MS with GC-IRMS techniques [622–624].

The principal advantage of conventional GC-IRMS in this context is its ease of use and lower sample size requirements, while the IRMS approach provides superior accuracy and precision, particularly at lower enrichments. IRMS data are usually expressed using the $\delta^y X$ notation, given as

$$\delta^y X_{\mathrm{spl}}(‰) = \left[\frac{(R_{\mathrm{SPL}} - R_{\mathrm{STD}})}{R_{\mathrm{STD}}}\right] \cdot 1000, \quad \text{where} \quad R_X = \frac{n\,(^yX)}{n\,(^zX)}$$

where:

^yX = the minor isotope (e.g., ^{13}C),
^zX = the major isotope (e.g. ^{12}C),
SPL and STD to sample and standard, respectively,
R_X = the measured ratio of numbers of atoms of the two isotopes, and
δ = referred to as "per mil".

For $X = C$, the accepted standard is a sample of carbonate rock from the Pee Dee formation in South Carolina (Pee Dee Belemnite, or PDB) with relatively high ^{13}C content, $R_{PDB} = 0.0112372 \pm 0.0000009$. A sample with $^{13}C_{PDB} = -1$ corresponds to $R_{SPL} = 0.0112260$, and terrestrial plants have values in the range $-40 < {}^{13}C_{PDB} < -10$, with non-overlapping ranges for plants based on C_3-C_4 photosynthetic pathways. The practical advantage of using the $^{13}C_{PDB}$ notation is that precision and accuracy are usually <0.4, so the use of this notation eliminates unchanging leading digits in R_C values.

A discussion of reference standards for IRMS of other elements is given by Ehleringer and Rundel [625]. For oxygen and hydrogen, the accepted standard is Standard Mean Ocean Water (i.e., SMOW). For nitrogen, air is acceptable because the isotopic composition of atmospheric N_2 is sufficiently uniform in space and time, while Canyon Diablo Troilite (CDT) is the standard for sulfur. In their role as standards, all of these are assigned a value of zero on their respective $\delta^y X$ scales, though their absolute R_X values are a subject of ongoing research [625]. For chlorine, the accepted standard is Standard Mean Ocean Chloride (i.e., SMOC) [626].

4.4.3
Sample Preparation and Handling

As for all trace-level analyses, sample preparation and handling are of crucial importance. In addition to all the usual problems of GC-MS, measurements of isotope ratios must ensure that none of these steps introduce any isotope discrimination. Any chemical reactions, including conversion of the organic sample molecules to the simple gases which are those actually analyzed, must be quantitative (100% conversion) to avoid kinetic isotope effects [627]. Until relatively recently, all gas IRMS experiments employed a dual-inlet system to permit switching between sample and standard CO_2 contained in two bellows containers. The pressures in the two bellows are adjusted to be equal and, equally important, sufficiently high that the flows into the electron impact (EI) ion source are viscous flows, not molecular flows for which flow rates vary as $M^{-1/2}$. Similarly, for samples introduced by GC, the isotope effect on GC retention times must be taken into account [627–630].

Since IRMS is a single-parameter chromatographic detector incapable of speciation, GC-IRMS places severe demands on the chromatographic resolution to guarantee peak purity [631, 632]. This is sometimes checked by splitting the effluent between the IRMS and a conventional EI mass spectrometer [633].

Compound-specific isotope analysis (CSIA) by GC-IRMS became possible in 1978 due to work of Mathews and Hayes [634], based on earlier low-precision work of Sano et al. [635]. The key innovation was the development of a catalytic combustion furnace based on Pt with CuO as oxygen source, placed between the GC exit and the mass spectrometer. The high pressure of helium (99.999% purity or better) ensures that all gas flows are viscous. After being dried in special traps avoiding formation of HCO_2^+ (i.e., interferes with $^{13}CO_2^+$) by ion-molecule reactions in the ion source, the CO_2 is transmitted to a device that regulates pressure and flow and then into the ion source [604].

4.4.4
On-Line Coupling of IRMS

The following section provides information about on-line coupling of IRMS to both GC and MS for environmental applications. For work of the highest accuracy and precision using GC coupled to IRMS through a combustion furnace (GC-C-IRMS), on-line isotopic calibration is essential. The standard practice is to introduce pulses of an isotopically standardized gas (e.g., CO_2) via an independent inlet directly into the ion source. The disadvantage of this calibration method is that it fails to compensate for any discrimination effects experienced by the analyte in its passage through the analytical train (e.g., the chromatographic isotope effect, [618, 636]. Methods of data analysis that compensate for this effect are also available [637].

Goodman and Brenna [636] introduced the idea of adding an internal standard compound of known isotopic composition to the analyte mixture, prior to injection on the GC column. The advantage thus gained can be lessened or even lost if the standard and/or analytes elute late in the chromatogram, with inevitable peak tailing and other distortions, which reduce attainable accuracy and precision in measuring the isotope ratios. A gas inlet device designed to introduce reference gases of known isotopic composition (inert or combustible) into a GC-C-IRMS instrument between the column end and combustion furnace has been described [619, 638]. This introduces reference gas pulses at any chosen position in the chromatogram, which combines the user-friendly external standard approach with the advantage of data acquisition under identical conditions for analyte and standard passing through the GC-IRMS combination. This approach does not address the chromatographic isotope effect [619].

Both Merritt and Hayes [639, 640] and Merritt et al. [641] have investigated the statistical limits to attainable precision for GC-C-IRMS techniques. For carbon isotope ratio measurements with precision not limited by counting statistics, 0.1–1 nmol/compound must be injected on-column, and this increases to ~ 5 nmol for nitrogen, mainly caused by the lower abundance of nitrogen atoms in most organic molecules.

4.4.5
Applications

Various analytical applications of different element isotopes are discussed in the following.

4.4.5.1
Carbon Isotope Analysis

Carbon isotope analysis of CO_2 and/or dissolved inorganic carbon is used successfully to investigate contaminant behavior [542, 642–646] and confirm its original source [638, 647, 648]. However, modern biological production of CO_2 from organic matter can interfere with the interpretation of these results, although analysis of ^{14}C concentrations helps to correct for this interference [644]. The ability to measure the carbon isotopic composition of the contaminant itself has been facilitated by GC, which permits the measurement of the carbon isotopic composition of individual compounds within a complex mixture IRMS [649, 650]. Samples can be prepared for GC-IRMS analysis by a number of techniques including direct injection of gas [545], pentane extraction from aqueous solution [545, 651], rapid extraction from either gas or aqueous solution by solid-phase microextraction [373] or extracts and fractionated extracts of compounds [638]. This permits isotope analysis of contaminants with sub-ppm concentrations with a typical analytical uncertainty of ±0.5‰ [652]. These techniques can be used to measure the isotopic composition of many types of contaminants, and GC-C-IRMS has been and can potentially be applied to study the sources and behavior of organic compounds and various contaminants, such as paraffin dirt [648], biomarker hydrocarbons [638], individual long-chain alkanes and alkanoic acids [647], monoaromatic hydrocarbons [652], polycyclic aromatic hydrocarbons [650, 651], chlorinated solvents [614, 655], polychlorinated biphenyls [656], crude oils and other refined hydrocarbon products [97], and aerosol organic compounds [647].

The ability to measure the isotopic composition of individual compounds can be used to determine whether or not some contaminants are being affected by surface/subsurface processes. In comparison, monitoring the isotopic composition of total carbon can only provide information concerning the processes affecting the contaminant mixture as a whole rather than individual compounds.

4.4.5.2
Nitrogen Isotope Analysis

An early study investigated the organic geochemistry of a Chilean paraffin dirt and determine its main source(s) [648]. These authors reported the bulk $\delta^{15}N$ values for the lipid, humate and kerogen fractions of this organic matter to be

–13.8‰, –14.2‰, and –10.8‰, respectively, indicating a carbon source from natural gas seepage (i.e., +8 to –15‰) rather than from higher plant (terrigenous, i.e., ≅+2‰) or marine/lacustrine (i.e., ≅+12‰) organic matter sources.

Now, GC-IRMS can be used to measure the nitrogen isotopic composition of individual compounds [657]. Measurement of nitrogen isotope ratios was described by Merritt and Hayes [639], who modified a GC-C-IRMS system by including a reduction reactor (Cu wire) between the combustion furnace and the IRMS, for reduction of nitrogen oxides and removal of oxygen. Preston and Slater [658] have described a less complex approach which provides useful data at lower precision. Similar approaches have been described by Brand et al. [657] and Metges et al. [659]. More recently Macko et al. [660] have described a procedure, which permits GC-IRMS determination of $^{15}N/^{14}N$ ratios in nanomole quantities of amino acid enantiomers with precision of ±0.3–0.4‰. A key step was optimization of the acylation step with minimal nitrogen isotope fractionation [660].

4.4.5.3
Hydrogen Isotope Analysis

GC-IRMS systems that allow for the measurement of the hydrogen isotopic composition of individual contaminant compounds have recently become commercially available. The difficult problem of measuring $^2H/^1H$ ratios in an excess of helium carrier gas has been tackled by Prosser and Scrimgeour [661], who designed a flight tube with two spurs to provide a large mass separation between $^1H^1H^+$ and $^2H^1H^+$ and at the same time to prevent any tailing from $^4He^+$ into the Faraday cup set for *m/z* 3. The original application was for water-derived samples [661]. Recently this system was adapted to permit simultaneous measurements of δ^2H and $\delta^{18}O$ from on-line reduction of water samples, with a precision of 4‰ and 0.5‰, respectively [662, 663]. Rennie et al. [664] have used the same approach to measure natural-abundance ^2H values from on-line converted organic compounds with a precision of ±2–4‰.

Tobias et al. [665] have described a method in which the GC effluent is passed into a combustion furnace to convert the organic hydrogen content into water, which is then selectively reduced to hydrogen in a reduction furnace containing Ni metal. The final stream is transmitted to the IRMS via a heated Pd filter, which passes only hydrogen isotopes to the ion source. For a benzene sample a precision of <5‰ was obtained for δ^2H_{SMOW}, which approaches the performance of off-line techniques and the requirements for studies of natural variability. This already meets requirements for analysis of D-labeled compounds used in tracer studies [666, 667].

4.4.5.4
Oxygen Isotope Analysis

The isotopic analysis of oxygen in organic materials was first based on catalytic pyrolysis, but in 1987 Santrock and Hayes [668] adapted the Unterzaucher procedure (pyrolysis followed by equilibration with carbon to form CO, which is then oxidized to CO_2 by I_2O_5), already well developed for elemental analysis of oxygen, for the determination of $^{18}O/^{16}O$ ratios. The classical determination of oxygen isotope ratios in water uses equilibration with CO_2 and determination of the isotopic composition of CO_2 in the conventional way [669]. An alternative method has been described whereby conversion of water to CO_2 with guanidine hydrochloride in sealed tubes permits reduction of sample sizes by a factor of 5 [670].

Brand et al. [657] investigated on-line conversion of water samples to CO by carbon fibers and by diamond. Begley and Scrimgeour [662, 663] developed the use of nickelized carbon in a tubular microfurnace for on-line reduction of water samples to CO and H_2, for simultaneous determinations of δ^2H and δ^{18}O with a precision of 4‰ and 0.4‰, respectively. The technique was also extended to some volatile organic compounds. Werner et al. [671] and Koziet [672] have described bulk analyses of solid materials using similar methods. Very recently, Bréas et al. [673] adapted an elemental analyzer to pyrolyze bulk organic samples, followed by catalytic conversion of the pyrolysis products to CO prior to IRMS analysis, and demonstrated the utility of this procedure for the determination of the geographical origins of vegetable oils based on their δ^{18}O contents.

None of the foregoing methods considered the complications arising from the presence of nitrogen in the organic samples. Farquhar et al. [674] developed a method for oxygen isotope analysis incorporating pyrolysis over nickelized carbon and conversion to CO. The CO in the resulting gases is then separated from N_2 by GC before analysis by IRMS, permitting a precision in δ^{18}O values of ±0.2‰. At present this approach has been employed only for solid organic samples or for water samples (i.e., not to GC effluents). However, Farquhar et al. [674] pointed out that the approach has several advantages in addition to its applicability to nitrogenous samples, including avoidance of possible isotopic discrimination in oxidation of $CO \rightarrow CO_2$, and faster pump-out and equilibration times for CO in the ion source. On the other hand, Barkan and Luz [675] have improved the procedure for the determination of isotope ratios in gaseous O_2 by conversion to CO_2 prior to IRMS analysis. Ball et al. [676] have elaborated on the method for carbon and oxygen isotope analysis of small samples (<0.1 mg) of carbonate minerals using conventional phosphoric acid digestion.

4.4.5.5
Chlorine Isotope Analysis

Although GC-C-IRMS systems that can measure the chlorine isotopic composition of individual chlorinated hydrocarbons are currently unavailable, it is clear that chlorine isotope analysis is also a useful technique to consider for study [614, 677, 678]. Measurement of chlorine stable isotope ratios in natural samples such as rocks and waters has become routine [626, 679, 680], but few measurements of chlorine isotopes in chlorinated aliphatic hydrocarbons have been reported [614]. A chlorine isotope effect was found in *tert*-butyl chloride [681], demonstrating that ^{37}Cl is more strongly bound to carbon than is ^{35}Cl. Significant differences in the δ^{37}Cl values of some atmospheric chlorinated organic compounds were measured and reported by Tanaka and Rye [682]. More recently, δ^{37}Cl and δ^{13}C values of chlorinated aliphatic hydrocarbons [678, 683, 684] and PCBs [656] obtained from various manufacturers were also reported.

4.4.6
Modern Application Examples

The present section represents a brief introduction to modern GC-C-IRMS practice for environmental organic compound analysis, mixture characterization, and source confirmation, as follows:

- Because little has been said concerning difficulties arising from derivatization of samples to render them suitable for GC analysis, replacement of GC by HPLC for non-volatile or thermally labile compounds is a possibility. However, the demands of reproducible solvent removal for a reliable LC-C-IRMS approach are formidable. Caimi and Brenna [685, 686] have developed an instrument based on a moving wire transport system. The analytes are deposited on the wire as they elute from the HPLC column and, after solvent drying at 200 °C, are transported into an 800 °C combustion furnace loaded with CuO, where the resulting CO_2 is picked up by an He carrier stream and swept via a drying trap into the IRMS.
- Part of the success of on-line GC-C-IRMS (i.e., CSIA) methods is due to the time compression of sample introduction into the IRMS (2-s to10-s GC peaks), permitting analysis of low-nanomole and even picomole samples as a result of adequate mass flow rates into the EI source. Direct coupling of elemental analyzers to IRMS generally requires much larger sample sizes for acceptable precision in isotope ratios, because of the long durations of the peaks eluting from the analyzer. Fry et al. [447] have described an apparatus in which an elemental analyzer is coupled to an IRMS using a continuous helium flow via a set of cryogenic traps. In this way, the CO_2, SO_2, and N_2 are separately collected and then may be transmitted in turn to the IRMS under controlled-flow conditions. Samples containing >50 nmol of atomic C, N, or S could be isotopically analyzed with a precision of 0.3‰. Degassing the frozen samples very slowly into the IRMS resulted in very high precision (±0.05‰ for ^{13}C values) [447].
- Corso and Brenna [687] have described an experiment in which *intramolecular* carbon isotope distributions of chemically pure compounds are investigated by controlled pyrolysis of the analytes emerging from a GC column, followed by a second GC step to separate the pyrolysis products which are then analyzed by the combustion IRMS techniques described above.
- Holt et al. [683] and Jendrzejewski et al. [684] have described methods for simultaneous determination of isotopic distributions for carbon and chlorine to better than 0.1‰ in volatile chlorinated solvents.
- Eglinton et al. [688] described a practical approach for isolation of individual compounds from complex organic matrices for natural abundance radiocarbon measurement. This approach uses an automated preparative capillary gas chromatography (PCGC) to separate and recover sufficient quantities of

individual target compounds for ^{14}C analysis by accelerator mass spectrometry (AMS). This approach was developed and tested using a suite of samples whose ages spanned the ^{14}C time scale and which contained a variety of compound types (fatty acids, sterols, hydrocarbons). Comparison of individual compound and bulk radiocarbon signatures for the isotopically homogeneous samples revealed that ^{14}C values generally agreed well ($\pm 10\%$). Background contamination was assessed at each stage of the isolation procedure, and incomplete solvent removal prior to combustion was the only significant source of additional carbon [688].

- Mansuy et al. [97] investigated the use of GC-C-IRMS as a complimentary correlation technique to GC and GC-MS, particularly for spilled crude oils and hydrocarbon samples that have undergone extensive weathering. In their study, a variety of oils and refined hydrocarbon products, weathered both artificially and naturally, were analyzed by GC, GC-MS, and GC-C-IRMS. The authors reported that in case of samples which have lost their more volatile *n*-alkanes as a result of weathering, the isotopic compositions of the individual compounds were not found to be extensively affected. Hence, GC-C-IRMS was shown to be useful for correlation of refined products dominated by *n*-alkanes in the $C_{10}-C_{20}$ region and containing none of the biomarkers more commonly used for source correlation purposes. For extensively weathered crude oils which have lost all of their *n*-alkanes, it has been demonstrated that isolation and pyrolysis of the asphaltenes followed by GC-C-IRMS of the individual pyrolysis products can be used for correlation purposes with their unaltered counterparts [97].
- Regarding the subsurface environment, stable isotope analysis has been used to determine the organic contamination sources rather than to understand the processes affecting organic contaminant concentrations. However, the lack of studies that have used stable isotope analysis to investigate subsurface processes may be related to the lack of data available concerning the isotope fractionation factors associated with the various biotic and abiotic processes that can affect contaminant concentrations in the subsurface. While vapor pressure data of isotopically labeled compounds can provide qualitative measurements of isotope fractionation during vaporization [689], quantitative measurements are only available for benzene, toluene, ethylbenzene (carbon only [690]), trichloroethylene (carbon and chlorine [690]), dichloromethane (carbon and chlorine), tetrachloroethylene, 1,1,1-trichloroethane, carbon tetrachloride, and chloroform (carbon and chlorine [691]). Of these studies, only one has performed measurements over a range of temperatures.
- Isotope fractionation between the vapor phase and the dissolved aqueous phase has been studied only for toluene and trichloroethylene (carbon only [545, 690]). Fractionation associated with adsorption has been quantified only for toluene in regard to sample extraction using a poly(dimethylsiloxane)-coated solid-phase microextraction fiber [373] and qualified for benzene, toluene, and ethylbenzene based on high-pressure liquid chromatography analyses of isotopically labeled and unlabeled compounds (carbon and hydrogen [692]). Isotope fractionation associated with the reductive dechlorination of chlorinated ethylenes by zero-valent iron and zinc has been

studied (carbon and chlorine; [690, 693, 694]), while isotope fractionation associated with natural or microbial degradation has been studied for dichloromethane (carbon and chlorine [695]), trichloroethylene (chlorine only [614], carbon only [690]), tetrachloroethylene (chlorine only [696]), and toluene (carbon only [697]).
- Preston et al. [698, 699] have described novel approaches to ^{15}N-isotope dilution determination of ammonium and of free amino acids in natural waters, incorporating chemical derivatization and conventional GC-MS analysis.
- Sturchio et al. [614] explored the use of Cl isotope ratios for investigating the natural attenuation of trichloroethene (TCE) at a well-characterized field site in western Kentucky and ranking the site in terms of its potential for TCE anaerobic biodegradation.

4.5 Future Developments in Organic Pollutant Identification and Characterization

Most of the development work on organic pollutants has resulted from the use of GC-MS and synthesis of authentic standards or surrogate standards. However, with advances in other techniques it is clear that this field will benefit by making greater use of alternative identification and characterization methods. The following is a summary of some advances and instrument combinations:

- Fourier transform infrared spectroscopy (FTIR) can now be combined with GC to provide IR spectra on peaks eluting from a capillary column [700–702].
- A combination of GC-FTIR-MS is also being developed to provide an extremely powerful tool for identifying molecular markers [703, 704]. If sufficient quantities of individual molecular markers can be isolated, then there are various ^{1}H [705, 706] and ^{13}C nuclear magnetic resonance techniques [505, 707–710] available to assist in their structural identification.
- High-performance liquid chromatography was combined with electrospray ionization mass spectrometry (i.e., HPLC-ESI-MS) to differentiate quantitatively crude natural extracts of various environmental samples [711, 712] and with NMR (i.e., HPLC-MS-NMR) for quantitation measurement [713].
- Lewis et al. [714] combined HPLC with Tandem Electrospray Mass Spectrometry (i.e., HPLC-ESI-MS-MS) for the determination of sub-ppb levels of toxins in extracts of fish.
- Microwave-assisted extraction coupled with gas chromatography-electron capture negative chemical ionization mass spectrometry (i.e., MAE-GC-EC-NCI-MS) was described for the simplified determination of imidazolinone herbicides in soil at the ppb level [715].
- Liquid chromatography was developed to analyze carbonyl (2,4-dinitrophenyl) hydrazones with detection by diode array ultraviolet spectroscopy (DA-UV) and by atmospheric pressure negative chemical ionization (APNCI) mass spectrometry [716]. In addition, LC can be combined with electrospray ionization coupled on-line with a photolysis reactor for better detection and confirmation of photodegradation products [717].

- High-resolution gas chromatography/electron capture negative ion high-resolution mass spectrometry (HRGC-EC-NI-HRMS) has been used for quantifying chloroalkanes in environmental samples [718].
- Gas chromatography/combustion/isotope ratio mass spectrometry (CSIA) was used to determine the stable isotope composition of amino acid enantiomers by nitrogen isotope analysis [660].
- Flash pyrolysis-GC-MS has been applied recently to identify and determine various principal groups of pyrolyzed organic matter as well as other organic compounds [505, 719, 720].

5
Conclusions

Organic pollutants present in aqueous-solid phase environments and discussed in the present chapter include petroleum hydrocarbons, pesticides, phthalates, phenols, PCBs, chlorocarbons, organotin compounds, and surfactants. In order to study the chemodynamic behavior of these pollutants, it is important that: (1) suitable pre-extraction and preservation treatments are implemented for the environmental samples, and (2) specific extraction and/or cleanup techniques for each organic pollutant are carried out prior to the identification and characterization steps.

Most of the work on organic pollutant analysis and characterization has resulted from the use of GC (especially ECD-GC) and GC-MS, but with advances in other techniques it is clear that the field of environmental monitoring and analysis will benefit by making greater use of alternative identification methods, such as Fourier transform infrared spectroscopy and nuclear magnetic resonance techniques. Synthesis of authentic standards or surrogate standards is also advancing the field. Isotopic measurements can now be used to obtain information on the history and origin of a sample. It is also possible to perform stable isotopic analyses on individual organic compounds by GC-isotope ratio MS without prior isolation of components from a mixture and determine the natural isotope abundances and thus sources of the compounds. It is clear that the IRMS techniques provide the highest attainable accuracy and precision for measurement of stable isotope ratios, as required for determination of variations in natural isotopic abundances. However, for experiments in which stable isotope-enriched compounds are used, this high level of performance may not be necessary and more convenient GC-MS and LC-MS techniques may provide adequate data.

There are many areas into which environmental organic chemistry and environmental engineering can advance as a result of developments in various analytical techniques. All of this information will provide a much clearer picture on the chemodynamics of organic compounds, their biodegradation residues, and transformation products. Information such as this is important for modeling the fate and transport of organic compounds in the environment.

References

1. Aboul-Kassim TAT (1998) Ph.D. Dissertation. Department of Civil, Construction and Environmental Engineering, Oregon State University, Corvallis, Oregon, USA
2. Kankaanpää HT, Laurén MA, Saares RJ, Heitto LV, Suursaar ÜK (1997) Environ Sci Technol 31:964
3. Boehm PD, Page DS, Gilfillan ES, Bence AE, Burns WA, Mankiewicz PJ (1998) Environ Sci Technol 32:567
4. Brzuzy LP, Hites RA (1996) Environ Sci Technol 30:1797
5. Castillo M, Alonso MC, Riu J, Barceló D (1999) Environ Sci Technol 33:1300
6. Ismail Y, Lafleur AL, Giese RW (1998) Environ Sci Technol 32:2494
7. Kucklick JR, Baker JE (1998) Environ Sci Technol 32:1192
8. Macdonald RW, Ikonomou MG, Paton DW (1998) Environ Sci Technol 32:331
9. Miguel AH, Kirchstetter TW, Harley RA, Hering SV (1998) Environ Sci Technol 32:450
10. Mukherji S, Peters CA, Weber WJ Jr (1997) Environ Sci Technol 31:416
11. Nelson ED, McConnell LL, Baker JE (1998) Environ Sci Technol 32:912
12. Tanabe S, Prudente M, Mizuno T, Hasegawa J, Iwata H, Miyazaki N (1998) Environ Sci Technol 32:193
13. Yunker MB, Snowdon LR, Macdonald RW, Smith JN, Fowler MG, Skibo DN, McLaughlin FA, Danyushevskaya AI, Petrova VI, Ivanov GI (1996) Environ Sci Technol 30:1310
14. Bergen BJ, Rahn KA, Nelson WG (1998) Environ Sci Technol 32:3496
15. Durlak SK, Biswas P, Shi J, Bernhard MJ (1998) Environ Sci Technol 32:2301
16. Fernández P, Alder AC, Suter MJ-F, Giger W (1996) Anal Chem 68:921
17. Kannan K, Guruge KS, Thomas NJ, Tanabe S, Giesy JP (1998) Environ Sci Technol 32:1169
18. Næs K, Oug E (1997) Environ Sci Technol 31:1253
19. Pascoe GA, Riley MG, Floyd TA, Gould CL (1998) Environ Sci Technol 32:813
20. Roy TA, Krueger AJ, Taylor BB, Mauro DM, Goldstein LS (1998) Environ Sci Technol 32:3113
21. Colombo JC, Brochu C, Bilos C, Landoni P, Moore S (1997) Environ Sci Technol 31:3551
22. de Voogt P, Wells DE, Reutergardh L, Brinkman UAT (1990) Int J Environ Anal Chem 40:1
23. Hofelt CS, Shea D (1997) Environ Sci Technol 31:154
24. Jeremiason JD, Eisenreich SJ, Baker JE, Eadie BJ (1998) Environ Sci Technol 32:3249
25. Karamanev DJ, Samson R (1998) Environ Sci Technol 32:994
26. Madenjian CP, Hesselberg RJ, DeSorcle TJ, Schmidt LJ, Stedman RM, Quintal RT, Begnoche LJ, Passino-Reader DR (1998) Environ Sci Technol 32:886
27. McFarland VA, Clarke JU (1989) Environ Health Perspect 81:225
28. Meadows JC, Echols KR, Huckins JN, Borsuk FA, Carline RF, Tillitt DE (1998) Environ Sci Technol 32:1847
29. Dachs J, Bayona JM, Raoux C, Albaigés J (1997) Environ Sci Technol 31:682
30. Di Corcia A, Crescenzi C, Guerriero E, Samperi R (1997) Environ Sci Technol 31:1658
31. Durell GS, Lizotte RD Jr (1998) Environ Sci Technol 32:1022
32. Eljarrat E, Caixach J, Rivera J (1997) Environ Sci Technol 31:2765
33. Field JA, Reed RL (1996) Environ Sci Technol 30:3544
34. Krueger CJ, Radakovich KM, Sawyer TE, Barber LB, Smith RL, Field JA (1998) Environ Sci Technol 32:3954
35. Krueger CJ, Barber LB, Metge DW, Field JF (1998) Environ Sci Technol 32:1134
36. Simcik MF, Zhang H, Eisenreich SJ, Franz TP (1997) Environ Sci Technol 31:2141
37. Agrell C, Okla L, Larsson P, Backe C, Wania F (1999) Environ Sci Technol 33:1149
38. Alcock RE, Jones KC (1996) Environ Sci Technol 30:3133
39. Fattore E, Benfenati E, Mariani G, Fanelli R, Evers EHG (1997) Environ Sci Technol 31:1777
40. Kimbrough RA, Litke DW (1996) Environ Sci Technol 30:908
41. Kolpin DW, Barbash JE, Gilliom RJ (1998) Environ Sci Technol 32:558
42. Senseman SA, Lavy TL, Mattice JD, Gbur EE, Skulman BW (1997) Environ Sci Technol 31:395

43. Thurman EM, Goolsby DA, Aga DS, Pomes ML, Meyer T (1996) Environ Sci Technol 30:569
44. Van Metre PC, Callender E, Fuller CC (1997) Environ Sci Technol 31:2339
45. González-Mazo E (1997) Environ Sci Technol 31:504
46. Järnberg UG, Asplund LT, Egebäck A-L, Jansson B, Unger M, Wideqvist U (1999) Environ Sci Technol 33:1
47. Müller MD, Buser H-R (1997) Environ Sci Technol 31:1953
48. Müller SR, Berg M, Ulrich MM, Schwarzenbach RP (1997) Environ Sci Technol 31:2104
49. Potter CL, Glaser JA, Chang LW, Meier JR, Dosani MA, Herrmann RF (1999) Environ Sci Technol 33:1717
50. Rockne KJ, Strand SE (1998) Environ Sci Technol 32:3962
51. Shang DY, Macdonald RW, Ikonomou MG (1999) Environ Sci Technol 33:366
52. Sugai SF, Lindstrom JE, Braddock JF (1997) Environ Sci Technol 31:1564
53. Aboul-Kassim TAT, Simoneit BRT (1995) Environ Sci Technol 29:2473
54. Aboul-Kassim TAT, Simoneit BRT (1995) Mar Poll Bull 30:63
55. Aboul-Kassim TAT, Simoneit BRT (1996) Mar Chem 54:135
56. Simoneit BRT (1978) In: Riley JP, Chester R (eds) Chemical oceanography, 2nd edn. Academic Press, New York, p 233
57. Aga DS, Thurman EM, Yockel ME, Zimmerman LR, Williams TD (1996) Environ Sci Technol 30:592
58. Allen JO, Durant JL, Dookeran NM, Taghizadeh K, Plummer EF, Lafleur AL, Sarofim AF, Smith KA (1998) Environ Sci Technol 32:1928
59. Arnold CG, Berg M, Müller SR, Dommann U, Schwarzenbach RP (1998) Anal Chem 70:3094
60. Bailey NJL, Jobson AM, Rogers MA (1973) Chem Geol 11:203
61. Bailey NJL, Krouse HR, Evans CR, Rogers MA (1973) Am Assoc Petrol Geol Bull 57:1276
62. Bedard DL, May RJ (1996) Environ Sci Technol 30:237
63. Brassell SC, Eglinton G, Maxwell JR, Philp RP (1978) In: Hutzinger O, Van Lelyveld IH, Zoeteman BCJ (eds) Aquatic pollutants. Pergamon Press, Oxford, p 69
64. Castillo M, Oubiña A, Barceló D (1999) Environ Sci Technol 32:2180
65. Jackson AW (1996) Environ Sci Technol 30:1139
66. Kvenvolden KA, Rapp JB, Bourell JH (1985) In: Magoon LB, Claypool GE (eds) Alaska North Slope oil/rock correlation study. American Association of Petroleum Geologists, Tulsa, OK, Study in Geology, No 20, p 593
67. Mössner SG, Wise SA (1999) Anal Chem 71:58
68. Peters KE, Moldowan JM (eds) (1993) The biomarker guide interpreting molecular fossils in petroleum and ancient sediments. Prentice Hall, Englewood Cliffs, NJ, p 363
69. Philp RP (ed) (1985) Fossil fuel biomarkers methods in geochemistry and geophysics, vol 23. Elsevier, New York, p 292
70. Simoneit BRT, Kaplan IR (1980) Mar Environ Res 3:113
71. Simoneit BRT (1982) Int J Environ Anal Chem 12:124
72. Simoneit BRT (1985) Int J Environ Anal Chem 22:203
73. Simoneit BRT (1986) In: Johns RB (ed) Biological markers in the sedimentary record. Elsevier, New York, p 43
74. Abad JM, Pariente F, Hernández L, Abruña HD, Lorenzo E (1998) Anal Chem 70:2848
75. Apffel A, Chakel JA, Fischer S, Lichtenwalter K, Hancock WS (1997) Anal Chem 69:1320
76. Baltruschat H, Kamphausen I, Oelgeklaus R, Rose J, Wahlkamp M (1997) Anal Chem 69:743
77. Bringmann G, Günther C, Schlauer J, Rückert M (1998) Anal Chem 70:2805
78. Dale MG, Knochenmuss R, Zenobi R (1996) Anal Chem 68:3321
79. Hankin SM, John P (1999) Anal Chem 71:1100
80. Hankin SM, John P, Smith GP (1997) Anal Chem 69:2927
81. Ingram JC, Groenewold GS, Appelhans AD, Delmore JE, Olson JE, Miller DL (1997) Environ Sci Technol 31:402
82. Kala SV, Lykissa ED, Lebovitz RM (1997) Anal Chem 69:1267
83. Kamel AM, Brown PR, Munson B (1999) Anal Chem 71:968

84. Koeber R, Bayona JM, Niessner R (1999) Environ Sci Technol 33:1552
85. Kostiainen R, Kotiaho T, Mattila I, Mansikka T, Ojala M, Ketola RA (1998) Anal Chem 70:3028
86. Leblanc WG, Gilbert R, Hubert J (1999) Anal Chem 71:78
87. Plomley JB, March RE, Mercer RS (1996) Anal Chem 68:2345
88. Wiberg K, Letcher R, Sandau C, Duffe J, Norstrom R, Haglund P, Bidleman T (1998) Anal Chem 70:3845
89. Hernández F, Hidalgo C, Sancho JV, López FJ (1998) Anal Chem 70:3322
90. Lee C-Y, Shiea J (1998) Anal Chem 70:2757
91. Murayama M, Dasgupta PK (1996) Anal Chem 68:1226
92. Pérez S, Ferrer I, Hennion M-C, Barceló D (1998) Anal Chem 70:4996
93. Bergqvist P-A, Strandberg B, Ekelund R, Rappe C, Granmo Å (1998) Environ Sci Technol 32:3887
94. Easterling ML, Mize TH, Amster IJ (1999) Anal Chem 71:624
95. Gillette JS, Luthy RJ, Clemett SJ, Zare RN (1999) Environ Sci Technol 33:1185
96. Koivusalmi E, Haatainen E, Root A (1999) Anal Chem 71:86
97. Mansuy L, Philp R, Allen J (1997) Environ Sci Technol 31:3417
98. Eglinton G, Calvin M (1967) Sci Am 216:32
99. Simoneit BRT, Cardoso JN, Robinson N (1990) Chemosphere 21:1285
100. Simoneit BRT, Sheng G, Chen X, Fu J, Zhang J (1991) Atmosph Environ 25A:2111
101. Albaigés J, Cuberes MR (1980) Chemosphere 9:539
102. Albaigés J, Grimalt J, Bayona JM, Risebrough R, de Lappe B, Walker W (1984) Org Geochem 6:237
103. Albaigés J (1980) In: Albaigés J (ed) Analytical techniques in environmental chemistry. Pergamon Press, Oxford, p 69
104. Grimalt J, Albaigés J (1990) Mar Geol 95:207
105. Grimalt J, Marfil C, Albaigés J (1984) Anal Chem 18:183
106. Grimalt J, Bayona JM, Albaigés J (1986) Journ Etud Pollut CIESM 7:533
107. Grimalt J, Olivé J, Gómez-Belinchon JI (1990) Anal Chem 28:305
108. Saliot A, Marty JC, Scribe P, Sicre MA, Viets TC, de Leeuw JW, Schenck PA (1990) Org Geochem 17:239
109. Hedges JI, Prahl FG (1993) In: Engel MH, Macko SA (eds) Organic geochemistry – principles and applications. Plenum Press, NY, p 237
110. Saliot A, Laureillard J, Scribe P, Sicre MA (1991) Mar Chem 36:233
111. Philp RP (1993) In: Engel MH, Macko SA (eds) Organic geochemistry: principles and applications. Plenum Press, NY, p 445
112. Simoneit BRT, Mazurek MA (1982) Atmosph Environ 16:2139
113. Simoneit BRT (1984) Atmosph Environ 18:51
114. Simoneit BRT (1986) Int J Environ Anal Chem 23:207
115. Simoneit BRT (1986) In: Johns RB (ed) Biological markers in the sedimentary record. Elsevier, New York, p 43
116. Kennicutt MC II, Wade TL, Presley BJ, Requejo AG, Brooks JM, Denoux G (1994) Environ Sci Technol 28:1
117. Mazurek M, Simoneit BRT (1983) In: Keith LH (ed) Identification, analysis of organic pollutants in air. Ann Arbor Science/Butterworth Publishers, Woburn, MA, p 353
118. Seifert WK, Moldowan JM (1979) Geochim Cosmochim Acta 43:111
119. Simoneit BRT, Mazurek MA, Cahill TA (1980) J Am Pollut Contr Assoc 30:387
120. Albaigés J, Albrecht P (1979) In: Frei RW (ed) Recent advances in environmental analysis. Gordon and Breach, London, p 261
121. Farran A, Grimalt J, Albaigés J, Botello AV, Macko SA (1987) Mar Poll Bull 18:284
122. Simoneit BRT (1982) In: Thompson JAJ, Jamieson WD (eds) Marine chemistry into the eighties. National Research Council of Canada, Ottawa, p 82
123. Simoneit BRT (1982) Intern J Environ Anal Chem 12:71
124. Aquino Neto FR, Restle A, Connan J, Albrecht P, Ourisson G (1982) Tetrahedron Letters 23:2027

125. Aquino Neto FR, Trendel JM, Restle A, Connan J, Albrecht P (1983) In: Bjorøy M (ed) Advances in organic ceochemistry. Wiley, Chichester, p 695
126. Moldowan JM, Seifert WK, Gallegos EJ (1983) Geochim Cosmochim Acta 47:1531
127. Trendel JM, Restle A, Connan J, Albrecht P (1982) J Chem Soc Chemical Communications 30:4
128. Simoneit BRT (1985) Can J Earth Sci 22:1919
129. Simoneit BRT (ed) (1990) Organic matter in hydrothermal systems – petroleum generation, migration and biogeochemistry. Appl Geochem 5:1
130. Simoneit BRT (1994) In: Mottl M, Davis E, Fisher A, Slack J (eds) Proceedings of the ocean drilling program, scientific results. Ocean Drilling Program, College Station 139:447
131. Coleman PJ, Lee RJM, Alcock RE, Jones KC (1997) Environ Sci Technol 31:2120
132. Cornelissen G, Rigterink H, Ferdinandy MMA, van Noort PCM (1998) Environ Sci Technol 32:966
133. Escartin E, Porte C (1999) Environ Sci Technol 33:406
134. Jenkins BM, Jones AD, Turn SQ, Williams RB (1996) Environ Sci Technol 30:2462
135. Koganti A, Spina DA, Rozett K, Ma B-L, Weyand EH, Taylor BB, Mauro DM (1998) Environ Sci Technol 32:3104
136. Ramaswami A, Luthy RG (1997) Environ Sci Technol 31:2260
137. Ramaswami A, Ghoshal S, Luthy RG (1997) Environ Sci Technol 31:2268
138. Sanders G, Hamilton-Taylor J, Jones KC (1996) Environ Sci Technol 30:2958
139. Simoneit BRT (1998) In: Neilson AH (ed), The handbook of environmental chemistry. Springer, Berlin Heidelberg New York, chap 5, p 176
140. IARC (1989) Diesel and gasoline engine exhausts and some nitroarenes. International Agency for Research on Cancer (IARC), vol 46
141. IARC (1990). Some flame-retardants and textile chemicals, and exposures in the textile manufacturing industry. International Agency for Research on Cancer (IARC), vol 48
142. IARC (1997). Chemicals and industrial processes associated with cancer in humans. International Agency for Research on Cancer (IARC), Supplement (1), vols 1–20
143. Rushdi AI, Simoneit BRT (2000) Appl Geochem (submitted)
144. Alexander R, Kagi RI, Sheppard PN (1983) J Chrom 267:367
145. Rowland SJ, Alexander R, Kagi RI (1984) J Chrom 294:407
146. Radke M, Garrigues P, Wilsch H (1990) Org Geochem 15:17
147. Garrigues P, de Sury R, Angelin NL, Bellocq J, Oudin JL, Ewald M (1988) Geochim Cosmochim Acta 52:375
148. US-EPA (1987) US Environmental Protection Agency, Agricultural chemicals in ground water: proposed pesticide strategy. US EPA, Washington, DC, p 1
149. US-EPA (1990) National pesticide survey phase I report, PB91–125765, US Environmental Protection Agency, National Technical Information Service, Springfield, VA
150. US-EPA (1992) Another look: national survey of pesticides in drinking water wells. Phase II Report, EPA 570/9–91–020, US Environmental Protection Agency, National Technical Information Service, Springfield, VA
151. Ware GW (1989) The pesticide book, 3rd edn. Thomson Publications, Fresno CA
152. Buser H-R (1990) Environ Sci Technol 24:1049
153. Premazzi G (1983) Evaluation of the impact of Malathion on the aquatic environment. Commission of the European Communities, Joint Research Centre, Ispra, p 147
154. Zen J-M, Jeng S-H, Chen H-J (1996) Anal Chem 68:498
155. Milne GWA (1995) CRC handbook of pesticides. CRC Press, Boca Raton, Florida, p 402
156. Howard PH, Neal M (1992) Dictionary of chemical names and synonyms. Lewis Publishers, Chelsea, Michigan, p 1125
157. Schottler SP, Eisenreich SJ (1997) Environ Sci Technol 31:2616
158. Xing B, Pignatello JJ, Gigliotti B (1996) Environ Sci Technol 30:2432
159. Becker RL, Herzfeld D, Stamm-Kalovich EJ, Ostlie KR (1991) Am Nurseryman 173:108
160. Brach J (1989) Agriculture, water quality: best management practice for Minnesota. Minnesota Pollution Control Agency, St Paul, Minnesota, p 64

161. Buchanan GA, Hiltbold AE (1973) Weed Sci 21 : 413
162. California Department of Food, Agricuelture (Cal-DFA) (1986) Pesticide use American report of food and agriculture, agriculture chemicals and feed, Sacramento, California, p 110
163. Cartwright N, Clark L, Bird P (1991) The impact of agriculture on water quality. Outlook Agric 20 : 1157
164. Wehtje GR, Leavitt JRC, Spalding RF, Mielke LN, Schepers JS (1981) Sci Total Environ 21 : 47
165. Fielding MD, Barcelo A, Helweg S, Galassi L, Torstensson P, van Zoonen R, Wolter G, Angeletti K (1992) Pesticides in ground, drinking water. Water Pollution Research Rep 27, Commission of the European Communities, Brussels, p 1
166. Durand G, Bouvot V, Barcelo D (1992) J Chromatogr 607 : 319
167. Tronczynski J, Munschy C, Durand G, Barcelo D (1993) Sci Total Environ 132 : 327
168. Vincent G (ed) (1991) In: Organic micropollutants in the aquatic environment, Lisbon Symposium. Kluwer, Dordreeht, p 285
169. Mallat E, Barzen C, Klotz A, Brecht A, Gauglitz G, Barceló D (1999) Environ Sci Technol 33 : 965
170. Buser H-R, Müller MD (1997) Environ Sci Technol 31 : 1960
171. Buser H-R, Müller MD (1998) Environ Sci Technol 32 : 626
172. Reighard TS, Olesik SV (1997) Anal Chem 69 : 566
173. Sayles GD, You G, Wang M, Kupferle MJ (1997) Environ Sci Technol 31 : 3448
174. Maruya KA, Lee RF (1998) Environ Sci Technol 32 : 1069
175. Harper FD, Weisskopf CP, Cobb GP (1998) Anal Chem 70 : 3329
176. Rodriguez M, Orescan DB (1998) Anal Chem 70 : 2710
177. Eykholt GB, Davenport DT (1998) Environ Sci Technol 32 : 1864
178. Kalkhoff SJ, Kolpin DW, Thurman EM, Ferrer I, Barcelo D (1998) Environ Sci Technol 32 : 1738
179. Schultz MM, Parris RM, Wise SA, Won HT, Turle R (1992) Chemosphere 24 : 1687
180. Mangiapan S, Benfenati E, Grasso P, Terreni M, Pregnolato M, Pagani G, Barceló D (1997) Environ Sci Technol 31 : 3637
181. Vargo JD (1998) Anal Chem 70 : 2699
182. Agg Ba AR, Zabel TF (1990) J Inst Water Environ Manage 4 : 44
183. IAEA-FAO-UNEP (1991) MEDPOL workshop on the assessment of pollution by herbicides and fungicides. Monaco, p 362
184. Munch DJ, Graves RL, Maxey RA, Engel TM (1990) Environ Sci Technol 24 : 1446
185. Munch DJ, Frebis CP (1992) Environ Sci Technol 26 : 921
186. Jensen S, Jansson B (1976) Ambio 5 : 257
187. Jensen S, Sundstrom G (1974) Ambio 3 : 70
188. Bandh C, Ishaq R, Broman D, Näf C, Rönquist-Nii Y, Zebühr Y (1996) Environ Sci Technol 30 : 214
189. Jansson B, Andersson R, Asplund L, Bergman A, Litzen K, Wyland K, Reutergardh L, Sellstrom IJ, Uvemo UB, Wahlberg C, Wideqvist U (1991) Fresenius Z Anal Chem 340 : 439
190. Lawruk TS, Lachman CE, Jourdan SW, Fleeker JR, Hayes MC, Herzog DP, Rubio FM (1996) Environ Sci Technol 30 : 695
191. Schulz DE, Petrick G, Duinker JC (1989) Environ Sci Technol 23 : 852
192. Wells DE, Echarri I (1992) Int J Environ Anal Chem 47 : 75
193. Stow CA, Qian SS (1998) Environ Sci Technol 32 : 2325
194. de Voogt P, Brinkman UAT (eds) (1989) In: Halogenated biphenyls, terphenyls, naphthalenes, dibenzodioxins and related products. Elsevier, Amsterdam, p 34
195. Bremle G, Larsson P (1997) Environ Sci Technol 31 : 3232
196. Burgess RM (1996) Environ Sci Technol 30 : 2556
197. Burgess RM (1996) Environ Sci Technol 30 : 1923
198. Grundy SL, Bright DA, Dushenko WT, Reimer KJ (1996) Environ Sci Technol 30 : 2661
199. Kannan K, Maruya KA, Tanabe S (1997) Environ Sci Technol 31 : 1483
200. Kannan N, Yamashita N, Petrick G, Duinker JC (1998) Environ Sci Technol 32 : 1747

201. Pearson RF, Hornbuckle KC, Eisenreich SJ, Swackhamer DL (1996) Environ Sci Technol 30:1429
202. Campfens WJ (1997) Environ Sci Technol 31:577
203. de Boer J (1988) Chemosphere 17:1811
204. Gustafsson Ö, Gschwend PM, Buesseler KO (1997) Environ Sci Technol 31:3544
205. Erickson MD (1986) Analytical chemistry of PCBs. Ann Arbor Science Pub, p 287
206. Steinwandter H (1984) Fresenius Z Anal Chem 317:867
207. Steinwandter H, Bruce H (1983) Fresenius Z Anal Chem 314:160
208. Morrison HA, Gobas FAPC, Lazar R, Whittle DM, Haffner GD (1998) Environ Sci Technol 32:3862
209. Safe S, Hutzinger O (1987) Environmentai toxin series. 1. PCBs: mammalian and environmental toxicity. Springer, Berlin Heidelberg New York
210. Safe S, Bandeira S, Sarsyer T, Robertson L, Safe L, Parkinson A, Thomas PE, Ryan DE, Reik LM, Levin W, Denomme MA, Fujita T (1985) Environ Health Perspect 60:47
211. Asplund L, Graf'strom A-K, Haglund P, Jarnberg B, Jansson U, Mace D, Strandell M, de Wit C (1990) Chemosphere 20:1481
212. Huckins JN, Schwartz TR, Petty JD, Smith LM (1988) Chemosphere 17:1995
213. Schulz-Bull DE, Petrick G, Duinker JC (1991) Mar Chem 36:365
214. Shelton DR, Boyd SA, Tiedje JM (1984) Environ Sci Technol 18:93
215. Giam CS, Atlas A, Powers JMA, Leonard JE (1984) In: Hutzinger O (ed) The handbook of environmental chemistry. Springer, Berlin Heidelberg New York, p 67
216. Peakall DB (1975) Residue Rev 54:1
217. Ejlertsson J, Alnervik M, Jonsson S, Svensson BH (1997) Environ Sci Technol 31:2761
218. Nielsen E, Larsen PB (1996) Toxicological evaluation and limit values for DEHP and phthalates other than DEHP. Danish Environmental Protection Agency, p 452
219. Schmitzer JL, Scheunert I, Korte F (1988) J Agric Food Chem 36:210
220. Staples CA, Peterson DR, Parkerton TF, Adams WJ (1997) Chemosphere 35:667
221. Beliles R, Salinas JA, Klune WM (1989) Drug Metab Rev 21:3
222. Schulz CO (1989) Drug Metab Rev 21:111
223. Siddiqui A, Srivastava SP (1992) Bull Environ Contam Toxicol 48:115
224. Zurmühl T, Durner W, Herrmann R (1991) J Contam Hydrol 8:111
225. Kurane R, Suzuki T, Takahara Y (1979) Agric Biol Chem 43:421
226. Kurane R, Suzuki T, Fukuoka S (1984) Appl Microbiol Biotechnol 20:378
227. Saeger VW, Tucker ES (1976) Appl Environ Microbiol 31:29
228. Efroymson RA, Alexander M (1994) Environ Toxicol Chem 13:405
229. Irvine RL, Earley JP, Kehrberger GJ, Delaney BT (1993) Environ Prog 12:39
230. Shanker R, Ramakrishna C, Seth PK (1985) Environ Pollut 39:1
231. Subba-Rao RV, Rubin HE, Alexander M (1982) Appl Environ Microbiol 43:1139
232. Michael PR, Adams WJ, Werner AE, Hicks O (1984) Environ Toxicol Chem 3:377
233. Douglas GR, Hungenholtz AP, Blakey DH (1986) Environ Health Perspect 65:255
234. Kageyama K, Onoyama Y, Nakajima T, Otani S, Yano I, Hotta H, Yuasa I, Kogawa H, Miwa H (1994) Anticancer Res 14:2769
235. Iturbe R, Moreno G, Elefsiniotis P (1991) Environ Sci Technol 12:783
236. Thebault P, Cases JM, Fiessinger F (1981) Water Res 15:183
237. Gardner D, Dostal K, Osantowski R, Dampsey C (1987) Proc Ind Wastes Symp 31:4
238. Wang X, Leslie Grady CP Jr (1994) Water Res 28:1247
239. Saeger VW, Tucker ES (1973) Plast Eng 29:46
240. Gibbons JA, Alexander M (1989) Environ Toxicol Chem 8:283
241. Inman JC, Strachan SD, Sommers LE, Nelson DW (1984) J Environ Sci Health B19: 245
242. Johnson BT, Heitkamp MA, Jones JR (1984) Environ Pollut 8:101
243. Kurane R (1986) Microbiol Sci 3:92
244. O'Grady DP, Howard PH, Werner AF (1985) Appl Environ Microbiol 49:443
245. Sakagami Y, Tanaka K, Watanabe K, Odagiri M, Mori K, Murooka H (1982) Bull Jpn Soc Sci Fish 48:633

246. Sugatt RH, O'Grady DP, Banerjee S, Howard PH, Gledhill WE (1984) Appl Environ Microbiol 47 : 601
247. Walker WW, Cripe CR, Pritchard PH, Bourquin AW (1984) Chemosphere 13 : 1283
248. Horowitz A, Shelton DR, Cornell CP, Tiedje JM (1982) Dev Indust Microbiol 23 : 435
249. O'Conner OA, Rivera MD, Young LY (1989) Environ Toxicol Chem 8 : 569
250. Ejlertsson J, Meyerson U, Svensson BH (1996) Biodegradation 7 : 345
251. Farran A, de Pablo J, Barcelo D (1988) J Chromatogr 445 : 163
252. Fayyad MK, Alawi MA, El-Ahmed RL (1989) Chromatographia 28 : 465
253. Durand G, de Sertrand N, Barcelo D (1991) J Chromatogr 554 : 233
254. Wennrich L, Popp P, Möder M (2000) Anal Chem 72 : 546
255. Bueno M, Astruc A, Astruc M, Behra P (1998) Environ Sci Technol 32 : 3919
256. Blunden SJ, Evans CJ (1990) In: Hutzinger O (ed) The handbook of environmental chemistry. Springer, Berlin Heidelberg New York, vol 3, part E, p 1
257. Fent K (1996) Crit Rev Toxicol 26 : 1
258. Abel R (1996) In: Champ MA, Seligman PF (eds) Organotin, environmental fate and effects. Chapman & Hall, London, p 27
259. Champ MY, Wade L (1996) In: Champ MA, Seligman PF (eds) Organotin: environmental fate and effects. Chapman & Hall, London, p 55
260. Green GA, Cardwell R, Brancato MS (1997) Environ Sci Technol 31 : 3032
261. Hall LW, Bushong SJ (1996) In: Champ MA, Seligman PF (eds) Organotin: environmental fate and effects. Chapman & Hall, London, p 157
262. Laughlin RB, Thain J, Davidson B, Valkirs AO, Newton FC (1996) In: Champ MA, Seligman PF (eds) Organotin: environmental fate and effects. Chapman & Hall, London, p 191
263. Le LTH, Takahashi S, Saeki K, Nakatani N, Tanabe S, Miyazaki N, Fujise Y (1999) Environ Sci Technol 33 : 1781
264. Batley G (1996) In: de Mora J (ed) Tributyltin: case study of an environmental contaminant. Cambridge University Press, Cambridge, p 139
265. Davies IM, Bailey SK, Harding MJC (1998) J Mar Sci 55 : 34
266. Alzieu C, Sanjuan J, Michel P, Birel M, Dreno JP (1989) Mar Pollut Bull 20 : 22
267. Alzieu C, Michel P, Sanjuan J, Averty B (1990) Appl Organomet Chem 4 : 55
268. Alzieu C, Michel P, Tolosa I, Bacci E, Mee LD, Readman JW (1991) Mar Environ Res 32 : 261
269. Bryan GW, Burt GR, Gibbs PE, Pascoe PL (1993) J Mar Biol Assoc 73 : 913
270. Dirkx W, Lobinski R, Ceulemans M, Adams F (1993) Sci Total Environ 136 : 279
271. Maguire RJ (1996) In: de Mora SJ (ed) Tributyltin: case study of an environmental contaminant. Cambridge University Press, Cambridge, p 94
272. Michel P, Averty B (1997) Annu Inst Oceanogr Paris 73 : 25
273. Ritsema R, Laane RWP, Donnard OFX (1991) Mar Environ Res 32 : 243
274. Ritsema R (1994) Appl Organomet Chem 8 : 5
275. Seligman PF, Grovhoug JG, Valkirs AO, Stang PM, Fransham R, Stallard MO, Davidson B, Lee RF (1989) Appl Organomet Chem 3 : 31
276. Waite ME, Waldock MJ, Thain JE, Smith DJ, Milton SM (1991) Mar Environ Res 32 : 89
277. Kubilai L, Yemenicioglu S, Tugrul S, Salihoglu I (1996) Mar Pollut Bull 32 : 238
278. Hashimoto S, Watanabe M, Noda Y, Hayashi T, Kurita Y, Takasu Y, Otsuki A (1998) Mar Environ Res 45 : 169
279. Yamada H, Tagayanagi K, Tateishi M, Tagata H, Ikeda K (1997) Environ Pollut 96 : 217
280. Iwata H, Tanabe S, Mizuno T, Tatsukawa R (1995) Environ Sci Technol 29 : 2959
281. Kim GB, Tanabe S, Iwakiri R, Tatsukawa R, Amano M, Miyazaki N, Tanaka H (1996) Environ Sci Technol 30 : 2620
282. Kim GB, Lee JS, Tanabe S, Iwata H, Tatsukawa R, Shimazaki K (1996) Mar Pollut Bull 32 : 558
283. Ten Hallers-Tjabbes CC, Kemp JF, Boon JP (1994) Mar Pollut Bull 28 : 311
284. Takahashi S, Tanabe S, Kubodera T (1997) Environ Sci Technol 31 : 3103
285. Kannan K, Senthilkumar K, Giesy JP (1999) Environ Sci Technol 33 : 1776

286. Laniewski L, Borén H, Grimvall A (1998) Environ Sci Technol 32:3935
287. Li H, Gong B, Matsumoto K (1996) Anal Chem 68:2277
288. Poerschmann J, Kopinke F-D, Pawliszyn J (1997) Environ Sci Technol 31:3629
289. Traas TP, Stäb JA, Roel P, Kramer G, Cofino WP, Aldenberg T (1996) Environ Sci Technol 30:1227
290. Beaumont AR, Newman PB (1986) Mar Pollut Bull 17:457
291. Alzieu C (1991) Mar Environ Res 32:7
292. Alzieu C, Sanjuan J, Deltreil JP, Bovel M (1986) Mar Pollut Bull 17:494
293. Gibbs PE, Bryan GW, Pascoe PL, Burt GR (1990) J Mar Biol Assoc 70:639
294. Bryan GW, Gibbs PE, Hummerstone LG, Burt GR (1986) J Mar Biol Assoc 66:611
295. Bryan GW, Gibbs PE, Burt GR, Hummerstone LG (1987) J Mar Biol Assoc 67:525
296. Kannan K, Tanabe S, Iwata H, Tatsukawa R (1995) Environ Pollut 90:279
297. Fent K, Muller MD (1991) Environ Sci Technol 25:489
298. Maguire RJ, Chan YK, Bengert GA, Hale EJ, Wong PTS, Kramar O (1982) Environ Sci Technol 16:698
299. Quevauviller P, Donald OF, Etcheber H (1994) Environ Pollut 84:89
300. Kan-atireklap S, Tanabe S, Tabucanon S, Hungspreugs M (1997) Environ Pollut 97:79
301. Maguire RJ (1984) Environ Sci Technol 18:291
302. Fent K, Stegeman J (1993) J Aquatic Toxicol 24:219
303. Guruge KS, Tanabe S, Iwata H, Tatsukawa R, Yamagishi S (1996) Arch Environ Contam Toxicol 31:210
304. Guruge KS, Tanabe S, Fukuda M, Yamagishi S, Tatsukawa R (1997) Toxicol Environ Chem 58:197
305. Iwata H, Tanabe S, Mizuno T, Tatsukawa R (1997) Appl Organometal Chem 11:257
306. Kannan K, Corsolini S, Focardi S, Tanabe S, Tatsukawa R (1996) Arch Environ Contam Toxicol 31:19
307. Chau YK, Maguire RJ, Brown M, Yang F, Batchelor SP (1997) Water Qual Res J Can 32: 453
308. Carlier-Pinasseau C, Astruc A, Lespes G, Astruc MJ (1996) Chromatogr A 750:317
309. Donard OFX, Lalère B, Martin F, Lobinski R (1995) Anal Chem 67:4250
310. Schmitt VO, Szpunar J, Donard OFX, Lobinski R (1997) Can J Anal Sci Spectrosc 42:41
311. Hermosin MC, Piedad M, Cornejo J (1993) Environ Sci Technol 27:2606
312. Arnold CG, Weidenhaupt A, David MM, Müller SR, Haderlein SB, Schwarzenbach RP (1997) Environ Sci Technol 31:2596
313. Weidenhaupt A, Arnold CG, Müller SR, Haderlein SB, Schwarzenbach RP (1997) Environ Sci Technol 31:2603
314. Aboul-Kassim TAT, Simoneit BRT (1993) Part I, CRC – Critical Reviews Environ Sci Technol 23:325
315. Aboul-Kassim TAT (1992) In: Garber WF, Neves RJJ, Roberts PJW (eds) Marine disposal systems. Water Sci Technol 25:93
316. Brownawell BJ, Chen H, Collier JM, Westall JC (1990) Env Sci Technol 24:1284
317. Zutic V, Cosovic B, Marcenko E, Bihari N, Krsinic F (1981) Mar Chem 10:505
318. Jafvert CT (1991) Environ Sci Technol 25:1039
319. Jafvert CT, Heath JK (1991) Environ Sci Technol 25:1031
320. Black JG, Howes D (1980) In: Gloxhuber C (ed) Anionic surfactants: biochemistry, toxicology, dermatology. Dekker, New York, p 51
321. Karsa DR (1987) Industrial applications of surfactants: an overview. In: Proceedings of a Symposium Organized by the North West Region of the Industrial Division of the Royal Society of Chemistry, University of Salford, 15–17th April, special publication no 59
322. Di Corcia A, Capuani L, Casassa F, Marcomini A, Samperi R (1999) Environ Sci Technol 33:4119
323. Di Corcia A, Casassa F, Crescenzi C, Marcomini A, Samperi R (1999) Environ Sci Technol 33:4112
324. Takada H, Ogura N (1992) Mar Chem 37:257
325. González-Mazo E, Forja JM, Gómez-Parra A (1998) Environ Sci Technol 32:1636

326. Nielsen AM, Britton LN, Beall CE, McCormick TP, Russell GL (1997) Environ Sci Technol 31:3397
327. Tabor CF, Barber LB II (1996) Environ Sci Technol 30:161
328. Tolls J, Haller M, de Graaf I, Thijssen MATC, Sijm DTHM (1997) Environ Sci Technol 31:3426
329. Zeng EZ, Yu CC (1996) Environ Sci Technol 30:322
330. Cross JT (1970) In: Jungermann E (ed) Cationic surfactants. Surfactant science series, vol 4. M Dekker, New York, p 419
331. Lucy CA, Underhill RS (1996) Anal Chem 68:300
332. Richmond JM (ed) (1990) Cationic surfactants, organic chemistry. Surfactant science series, 34, M Dekker, New York, p 382
333. Visek KE (1990) In: Richmond JM (ed) Cationic surfactants, organic chemistry. Surfactant science series, chap 1. M Dekker, New York, p 1
334. Friedli FE (1990) In: Richmond JM (ed) Cationic surfactants, organic chemistry. Surfactant science series, chap 2. M Dekker, New York, p 51
335. Earl GW (1990) In: Richmond JM (ed) Cationic surfactants, organic chemistry. Surfactant science series, chap 3. M Dekker, New York, p 102
336. Smith KR (1990) In: Richmond JM (ed) Cationic surfactants, organic chemistry. Surfactant science series, chap 4. M Dekker, New York, p 145
337. Reck RA (1990) In: Richmond JM (ed) Cationic surfactants, organic chemistry. Surfactant science series, chap 5. M Dekker, New York, p 163
338. Bailey BR III (1990) In: Richmond JM (ed) Cationic surfactants, organic chemistry. Surfactant science series, chap 6. M. Dekker, New York, p 187
339. Gadberry JF (1990) In: Richmond JM (ed) Cationic surfactants, organic chemistry. Surfactant science series, chap 7. M Dekker, New York, p 221
340. Berger DR (1990) In: Richmond JM (ed) Cationic surfactants, organic chemistry. Surfactant science series, chap 8. M Dekker, New York, p 243
341. Saver JD (1990) In: Richmond JM (ed) Cationic surfactants, organic chemistry. Surfactant science series, chap 9. M Dekker, New York, p 275
342. Boyd-Boland AA, Pawliszyn JB (1996) Anal Chem 68:1521
343. Ding W, Fritz JS (1997) Anal Chem 69:1593
344. Fujita Y, Reinhard M (1997) Environ Sci Technol 31:1518
345. Heinig K, Vogt C, Werner G (1998) Anal Chem 70:1885
346. Jones FW, Westmoreland DJ (1998) Environ Sci Technol 32:2623
347. Li J, Carr PW (1997) Anal Chem 69:2550
348. Lye CM, Frid CLJ, Gill ME, Cooper DW, Jones DM (1999) Environ Sci Technol 33:1009
349. Rudel RA, Melly SJ, Geno PW, Sun G, Brody JG (1998) Environ Sci Technol 32:861
350. Turmine M, Macé C, Millot F, Letellier P (1999) Anal Chem 71:196
351. Zimmerman JB, Kibbey TCG, Cowell MA, Hayes KF (1999) Environ Sci Technol 33:169
352. Bock KJ, Stache H (1982) In: Hutzinger O (ed) The handbook of environmental chemistry, anthropogenic compounds, 3 (part B). Springer, Berlin Heidelberg New York, p 36
353. Takano S, Tsuji K (1983) Analysis of cationic and amphoteric surfactants. III. Structural analysis of imidazolinium cationic surfactants. J Am Oil Chem Soc 60:870
354. Bluestein BR, Hilton CL (eds) (1982) Amphoteric surfactants. Surfactant science series. M Dekker, New York, p 343
355. Rosen MJ, Zhao F, Murphy DS (1987) J Am Oil Chem Soc 64:439
356. Aboul-Kassim TAT, Simoneit BRT (2001) Part II, CRC – Crit Rev Environ Sci Technol (in preparation)
357. Bokern SM, Harms HH (1997) Environ Sci Technol 31:1849
358. Benarde MA, Koft BW, Horvath R, Shaulis L (1965) Appl Microbiol 13:103
359. Bird JA (1972) PhD Thesis, University of Newcastle upon Tyne, p 163
360. Bird JA, Cain RB (1972) Proc Biochem Soc Biochem J 127:46
361. Bird JA, Cain RB (1974) Biochem J 140:121
362. Chester TL, Pinkston JD, Raynie DE (1996) Anal Chem 68:487
363. Chester TL, Pinkston JD, Raynie DE (1998) Anal Chem 70:301

364. Onuska FI (1989) J High Resolut Chromatogr 12:4
365. Subra P, Hennion M-C, Rosset R (1989) Analusis 17:163
366. Clement RE, Yang PW, Koester CJ (1997) Anal Chem 69:251
367. Berthod A, Carda-Broch S, Alvarez-Coque MCG (1999) Anal Chem 71:879
368. Remberger M, Hynning RA, Neilson AH (1988) Environ Toxicol Chem 7:795
369. Lai JK, Filseth SV, Sadowski CM, Morgan FJ (1990) Int J Environ Anal Chem 40:99
370. Kuehl DW, Butterworth BC, Libal J, Marquis P (1991) Chemosphere 22:849
371. Patterson DG Jr, Lapez CR Jr, Barnhart ER, Groce DF, Burse VW (1989) Chemosphere 19:127
372. Uthe JF, Chou CL (1988) Sci Total Environ 71:67
373. Dias RF, Freeman KH (1997) Anal Chem 69:944
374. Price DJ, Birge WJ (1999) Environ Sci Technol 33:1137
375. Liikala TL, Olsen KP, Teel SS, Lanigan DC (1996) Environ Sci Technol 30:3441
376. Anon K (1990) Oslo and Paris Commission, Principles and Methodology of the Joint Monitoring Program, London
377. David MD, Seiber JN (1996) Anal Chem 68:3038
378. Schantz MM, Nichols JJ, Wise SA (1997) Anal Chem 69:4210
379. Zosel K (1978) Angew Chem 90:148
380. Ali MY, Cole RB (1998) Anal Chem 70:3242
381. Benner BA Jr (1998) Anal Chem 70:4594
382. Dressman SF, Simeone AM, Michael AC (1996) Anal Chem 68:3121
383. Maio G, von Holst C, Wenclawiak BW, Darskus R (1997) Anal Chem 69:601
384. Papilloud S, Haerdi W, Chiron S, Barcelo D (1996) Environ Sci Technol 30:1822
385. Song S, Ashley DL (1999) Anal Chem 71:1303
386. Sterzenbach D, Wenclawiak BW, Weigelt V (1997) Anal Chem 69:831
387. Sterzenbach D, Wenclawiak BW, Weigelt V (1997) Anal Chem 69:965
388. van Bavel B, Järemo M, Karlsson L, Lindström G (1996) Anal Chem 68:1279
389. Vejrosta J, Karásek P, Planeta J (1999) Anal Chem 71:905
390. Lancas FM, de Martinis BS, da Matta MHR (1990) J High Chromatogr 13:838
391. Jacobson GB, Moulder R, Lu L, Bergström M, Markides KE, Långström B (1997) Anal Chem 69:275
392. Levy JM, Storoznsky E, Ravey RM (1991) J High Resoult Chromatogr 14:661
393. McNally MEP, Wheeler JR (1988) J Chromatogr 435:63
394. McNally MEP, Wheeler JR (1988) J Chromatogr 447:53
395. Engelhardt H, Gross H (1988) J High Resolut Chromatogr Commun 11:726
396. Lopez-Avila V, Dodhiwala NS, Beckert WF (1990) J Chromatogr Sci 28:468
397. Young TM, Weber WJ Jr (1997) Anal Chem 69:1612
398. Yu X, Wang Z, Bartha R, Rosen JD (1990) Environ Sci Technol 24:1732
399. Tena MT, Luque de Castro MD, Valcárcel M (1996) Anal Chem 68:2386
400. Hale RC, Bush E, Gallagher K, Gundersen JL, Mothershead RF (1991) J Chromatogr 539:149
401. Heemken OP, Theobald N, Wenclawiak BW (1997) Anal Chem 69:2171
402. Porte C, Barcelo D, Albaiges J (1992) Chemosphere 24:735
403. van der Valk F, Wester PG (1991) Chemosphere 22:57
404. Wester PG, van der Valk F (1990) Environ Contam Toxicol 45:6
405. Wells DE, de Boer J, Tuinstra LGMT, Reutergardh L, Griepink B (1988) Fresenius Z Anal Chem 322:591
406. García-Ayuso LE, Sánchez M, Fernández de Alba A, Luque de Castro MD (1998) Anal Chem 70:2426
407. Bouwman H, Cooppan RM, Reinecke AJ (1989) Chemosphere 19:1563
408. Jensen S, Reutergardh L, Jansson B (1983) FAO Fish Tech Paper 212:21
409. Redondo MJ, Pico Y, Server-Carrio J, Manes J, Font G (1991) J High Resolut Chromatogr 14:597
410. Schwab P, Su J, Wetzel S, Pekarek S, Banks MK (1999) Environ Sci Technol 33:1940
411. Brezny R, Joyce TW (1992) Chemosphere 24:1031

412. Kung K-H, McBride MB (1991) Environ Sci Technol 25:702
413. Paasivirta J, Tenhola H, Palm H, Lammi R (1992) Chemosphere 24:1253
414. Martinsen K, Kringstad A, Carlberg GE (1988) Water Sci Technol 20:13
415. Barcelo D (1991) Analyst 116:681
416. Grob RL, Kaiser MA (1982) Environmental problem solving using gas and liquid chromatography. Elsevier, Amsterdam, p 87
417. Poole SK, Dean TA, Oudsema JW, Poole CF (1990) Anal Chim Acta 236:3
418. Shen Y, Jönsson J, Mathiasson L (1998) Anal Chem 70:946
419. Webb RG (1987) Int J Environ Anal Chem 5:239
420. Yoshida M, Akane A (1999) Anal Chem 71:1918
421. Murray AP, Gibbs GF, Kavanagh PE (1983) Int J Environ Anal Chem 16:167
422. HMSO (1980) Organophosphorus pesticides in river and drinking water tentative method. HMSO, London
423. HMSO (1986) Chlorophenoxy acid herbicides, trichlorobenzoic acid, chlorophenols, triazines and glyphosate in water. HMSO, London
424. Saliot A, Andrie C, Ho R, Marty JC (1985) Int J Environ Anal Chem 22:25
425. Tronczyzski J, Marty J-C, Scribe P, Saliot A Int (1986) J Environ Anal Chem 23:169
426. Bruchet AL, Cognet J, Mailevialle L (1983) Rev Fr Sci Eau 2:297
427. Czuczwa J, Leuenberger C, Tremp J, Giger W, Ahel M (1987) J Chromatogr 403:233
428. Foster GD, Rogerson P (1990) Int J Environ Anal Chem 41:105
429. Peters TL (1982) Anal Chem 54:1913
430. Colgrove SG, Svec HJ (1981) Anal Chem 53:1731
431. Cruz I, Wells DE (1992) Int J Environ Anal Chem 48:101
432. Mohnke H, Rhode KH, Brugmann L, Franz P (1986) J Chromatogr 364:323
433. Olivier BG, Nicol KD (1986) Int J Environ Anal Chem 25:275
434. Ming WR, Yman J, Chun G, Shu Ming B, Shi Jun Q (1985) Int J Environ Anal Chem 11:5
435. Ogan K, Katz E, Salvin W (1979) Anal Chem 51:1315
436. Abrahamsson K, Xie TM (1983) J Chromatogr 279:199
437. Durand G, Barcelo D (1989) Toxicol Environ Chem 25:1
438. Grob K, Kalin I (1991) J High Resolut Chromatogr 14:451
439. Murray DAD (1979) J Chromatogr 177:135
440. Crescenzi C, Di Corcia A, Guerriero E, Samperi R (1997) Environ Sci Technol 31:479
441. Ferrer I, Barceló D, Thurman EM (1999) Anal Chem 71:1009
442. Martin P, Morgan ED, Wilson ID (1997) Anal Chem 69:2972
443. Warren ME, Brockman AH, Orlando R (1998) Anal Chem 70:3757
444. Wübert J, Reder E, Kaser A, Weber PC, Lorenz RL (1997) Anal Chem 69:2143
445. Neilen MWF, Frei RW, Brinkman UAT (1989) In: Frei RW, Zeich K (eds) Selective sample handling and detection in high-performance liquid chromatography. Elsevier, Amsterdam, 39 A, p 5
446. Bjarnason B, Chimuka L, Ramström O (1999) Anal Chem 71:2152
447. Fry B, Garritt R, Tholke K, Neill C, Michener RH, Mersch FJ, Brand W (1996) Rapid Commun Mass Spectrom 10:953
448. Hagen DF, Markell CG, Schmitt GA, Blevins DD (1990) Anal Chim Acta 236:157
449. Lacorte S, Barceló D (1996) Anal Chem 68:2464
450. Puig D, Silgoner I, Grasserbauer M, Barceló D (1997) Anal Chem 69:2756
451. Tan LK, Liem AJ (1998) Anal Chem 70:191
452. Bolygo E, Hadfield ST (1990) J High Resolut Chromatogr 13:457
453. Holstege DM, Scharberg DL, Richardson ER, Moller G (1991) J Anal Chem 74:394
454. Larsson BK, Pyysalo M, Sauri Z (1988) Lebensm Unters Forsch 187:546
455. Wahlberg C, Renberg L, Wideqvist U (1990) Chemosphere 20:179
456. Richards M, Campbell RM (1991) LC-GC 9:358
457. Campbell RM, Meunier DM, Cortes HJ (1989) J Microcol Sep 11:302
458. Dooley KM, Ghonasgi D, Knopf FC (1990) Environ Prog 4:197
459. Lohleit M, Bachmann K (1990) J Chromatogr 505:227
460. Nam KS, Kapila S, Yanders AF, Puri RK (1990) Chemosphere 20:873

461. Onuska FI, Terry KA (1989) J High Resolut Chromatogr 12:357
462. Pinkston JD, Delaney TE, Bowling DJ, Chester TL (1991) J High Resolut Chromatogr 14:401
463. Wild SR, Waterhouse KS, McGrath SP, Jones KC (1990) Environ Sci Technol 24:1706
464. Renkes GD, Walters SN, Woo CS, Iles MK, D'Silva AP, Fassel VA (1981) Anal Chem 55:2229
465. Junk GA, Richards JJ (1980) Anal Chem 58:962
466. Schuphan I, Ebing W, Holthofer J, Krempler R, Lanka E, Ricking M, Pachur H-J (1990) Fresenius J Anal Chem 336:564
467. Bleidner WE, Backer HM, Levitesky M, Lowen WK (1954) J Agric Food Chem 2:476
468. Steinwandter H (1992) In: Cairns T, Sherma J (eds) Emerging strategies for pesticide analysis. CRC Press, Boca Raton, FL, p 338
469. Ali J (1997) Anal Chem 69:3260
470. Ali J (1997) Anal Chem 69:1230
471. Chen W, Poon K-F, Lam MHW (1998) Environ Sci Technol 32:3816
472. DeBruin LS, Josephy PD, Pawliszyn JB (1998) Anal Chem 70:1986
473. Dean JR, Tomlinson WR, Makovskaya V, Cumming R, Hetheridge M, Comber M (1996) Anal Chem 68:130
474. Hageman KJ, Mazeas L, Grabanski CB, Miller DJ, Hawthorne SB (1997) Anal Chem 69:801
475. Liu Y, Lee ML, Hageman KJ, Yang Y, Hawthorne SB (1997) Anal Chem 69:5001
476. Llompart M, Li K, Fingas M (1998) Anal Chem 70:2510
477. Nguyen A-L, Luong JHT (1997) Anal Chem 69:1726
478. Cuiz C, Halman R, Li K, Thomas RS, Lao RC (1986) Chemosphere 15:1091
479. Krahn MH, Brown DW, Wigen CA, Burrows DC, MacLeod WD Jr, Chan SL (1989) Oceans '89 2:397
480. Torreti L, Simonella A, Dossena A, Torreti E (1992) J High Resolut Chromatogr 15:99
481. Wells DE, Cowan AE, Christie AEG (1985) J Chromatogr 328:372
482. Pan L, Pawliszyn J (1997) Anal Chem 69:196
483. van der Valk F, Dao QT (1988) Chemosphere 17:1735
484. Storr-Sansen E, Cleemann M, Cederberg L, Jansson B (1992) Chemosphere 24:323
485. Garrigues P, Bellocq J (1989) J High Resolut Chromatogr 46:400
486. Hopper ML (1992) In: Cairns L, Sherma J (eds) Emerging strategies for pesticide analysis. CRC Press, Boca Raton, FL, p 39
487. Haglund P, Asplund L, Larnberg U, Jansson B (1990) Chemosphere 20:887
488. Haglund P (1991) Isolation and determination methods for halogenated polycyclic aromatic compounds. Swedish EPA Rep 3905, p 145
489. Tuinstra LG, van Rhijn JA, Roos AH, Traag WA, van Mazijk RJ, Kolkman PJW (1990) J High Resolut Chromatogr 13:797
490. Jensen S, Renberg L, Reutergardh L (1977) Anal Chem 49:316
491. Felder RA (1991) Drugs Pharmacol Sci 47:185
492. Law I, Jones RN (1987) In: Strimaitis IR, Hawk GL (eds) Advances in laboratory automation robotics, vol 4. Zymark Corp Hopkinton, p 13
493. Maris FA, Noroozian E, Otten RR, van Dijck RCJM, de Jong GJ, Brinkman JA (1988) J High Resolut Chromatogr 11:197
494. Rene G, van der Itott K, Gort SM, Baumann RA, van Zoonen P (1991) J High Resolut Chromatogr 14:465
495. de Boer J, Dao QT (1991) J High Resolut Chromatogr 14:593
496. Davies IL, Markides KE, Miller L, Raynor MW, Bartle KD (1989) J High Resolut Chromatogr 12:193
497. Koenigbauer MJ, Major RE (1990) LC-CC 5:510
498. Grob K (1987) On-column injection in capillary gas chromatography: basic techniques, retention times and solvent effects. Springer, Berlin Heidelberg New York, p 67
499. Grob K, Laubli T (1987) J High Resolut Chromatogr 8:499
500. Grob K, Walder C, Schilling B (1986) J High Resolut Chromatogr 9:95

501. Kapila S, Nam KS, Liu MH, Puri RK, Yanders AF (1992) Chemosphere 25:11
502. Neilen MWF, Sanderson JT, Frei RW, Brinkman UAT (1989) J Chromatogr 474:388
503. Bourne DJ (1998) 9th International Symposium on Marine Natural Products, Townsville, Australia, 5–10 July
504. Cappiello A, Famiglini G, Palma P, Berloni A, Bruner F (1995) Environ Sci Technol 29:2295
505. Cardoso AM, Barros CMF, Ferrer Correia AJ, Cardoso JM, Cortez A, Carvalho F, Baldaia L (1997) J Am Soc Mass Spectrom 8:365
506. Edler B, Zwiener C, Frimmel FH (1997) Fresenius J Anal Chem 359:288
507. Eriksson M, Swartling A, Dalhammar G (1998) Applied Microbiol Biotech 50:129
508. Espadaler I, Caixach J, Om J, Ventura F, Cortina M, Paune F, Rivera J (1997) Water Res 31:1996
509. Geerdink RB, Berg PJ, Kienhuis PGM, Niessen WMA, Brinkman UAT (1996) Int J Environ Anal Chem 64:265
510. Guthrie EA, Bortiatynski JM, van Heemst JDH, Richman JE, Hardy KS, Kovach EM, Hatcher PG (1999) Environ Sci Technol 33:119
511. Hachimi A, Krier G, Poitevin E, Schweigert MC, Peter S, Muller JF (1996) Int J Environ Anal Chem 62:219
512. Kanno S, Sugimoto M (1995) Ind Health 33:207
513. Lau B, Weber D, Andrews P (1996) Chemosphere 32:1021
514. Ohkura T, Takechi T, Deguchi S, Ishimaru T, Maki T, Inouye H (1994) Jap J Toxicol Environ Health 40:266
515. Puig IA, Cebrian GN, Sarasa AJ, Martinez NMC, Ormad MMP, Mutuberria CMS, Ovelleiro NJL (1996) Water Res 27:1167
516. Pyle SM, Marcus AB, Robertson GL (1998) Environ Sci Technol 32:3213
517. Rothweiler B, Berset J-D (1999) Chemosphere 38:1517
518. Sacks R, Akard M (1994) Environ Sci Technol 28:428 A
519. Blumer M (1975) Angew Chemie 14:507
520. Eiceman GA, Hill HH Jr, Gardea-Torresdey J (1998) Anal Chem 70:321
521. Eiceman GA, Hill HH Jr, Davani B, Gardea-Torresdey J (1996) Anal Chem 68:291
522. Gohda H, Hatano H, Hanai T, Miyaji K, Takahashi N (1993) Chemosphere 27:9
523. Skrbic BD, Vojinovic-Miloradov MB (1994) Water Sci Technol 30:91
524. Akapo SO, Dimandja J-MD, Matyska MT, Pesek JJ (1996) Anal Chem 68:1954
525. Natangelo M, Romagnano S, Guzzella L, Benfenati E (1995) Int J Environ Anal Chem 58:55
526. Bruckner CA, Prazen BJ, Synovec RE (1998) Anal Chem 70:2796
527. Leonard C, Grall A, Sacks R (1999) Anal Chem 71:2123
528. Woolfenden E (1995) J Air Waste Manage Assoc 47:20
529. Wylie PL, Knipe CR, Gere DR (1994) Am Environ Lab 6:18
530. Elias VO, Simoneit BRT, Pereira AS, Cabral JA, Cardoso JN (1999) Environ Sci Technol 33:2369
531. Carpenter RA, Hollowell RH, Hill KM (1997) Anal Chem 69:3314
532. Troost JR, Olavesen EY (1996) Anal Chem 68:708
533. Wester PG, de Boer J (1996) Environ Sci Technol 30:473
534. Burlingame AL, Boyd RK, Gaskell SJ (1996) Anal Chem 68:599
535. Burlingame AL, Boyd RK, Gaskell SJ (1998) Anal Chem 70:647
536. Solouki T, Reinhold BR, Costello CE, O'Malley M, Guan S, Marshall AG (1998) Anal Chem 70:857
537. Xu N, Lin Y, Hofstadler SA, Matson D, Call CJ, Smith RD (1998) Anal Chem 70:3553
538. Bullock J, Chowdhury S, Johnston D (1996) Anal Chem 68:3258
539. Caprioli RM, Farmer TB, Gile J (1997) Anal Chem 69:4751
540. Hung KC, Ding H, Guo B (1999) Anal Chem 71:518
541. Hurst JB, Weaver K, Doktycz MJ, Buchanan MV, Costello AM, Lidstrom ME (1998) Anal Chem 70:2693
542. Jackson AW, Pardue JH, Araujo R (1996) Environ Sci Technol 30:1139

543. Juhasz P, Roskey MT, Smirnov IP, Haff LA, Vestal ML, Martin SA (1996) Anal Chem 68:941
544. Katta V, Chow DT, Rohde MF (1998) Anal Chem 70:4410
545. Slater GF, Dempster HS, Lollar BS, Ahad J (1999) Environ Sci Technol 33:190
546. McCarley TD, McCarley RL, Limbach PA (1998) Anal Chem 70:4376
547. O'Connor PB, Duursma MC, van Rooij GJ, Heeren RMA, Boon JJ (1997) Anal Chem 69:2751
548. Schriemer DC, Li L (1997) Anal Chem 69:4176
549. Whittal RM, Schriemer DC, Li L (1997) Anal Chem 69:2734
550. Wingerath T, Kirsch D, Spengler B, Kaufmann R, Stahl W (1997) Anal Chem 69:3855
551. Jonscher KR, Yates JR III (1996) Anal Chem 68:659
552. Simoneit BRT, Smith DH, Eglinton G, Burlingame AL (1973) Arch Environ Contam Toxicol 1:193
553. Kornienko O, Ada ET, Hanley L (1997) Anal Chem 69:1536
554. Andrawes F, Valcarcel T, Haacke G, Brinen J (1998) Anal Chem 70:3762
555. Brock A, Rodriguez N, Zare RN (1998) Anal Chem 70:3735
556. Guevremont R, Siu KWM, Wang J, Ding L (1997) Anal Chem 69:3959
557. Orea JM, Bescós B, Montero C, Ureña AG (1998) Anal Chem 70:491
558. Potyrailo RA, Hieftje GM (1998) Anal Chem 70:1453
559. Potyrailo RA, Hieftje GM (1998) Anal Chem 70:3407
560. Prazen BJ, Bruckner CA, Synovec RE, Kowalski BR (1999) Anal Chem 71:1093
561. Preisler J, Foret F, Karger BL (1998) Anal Chem 70:5278
562. Rashidzadeh H, Guo B (1998) Anal Chem 70:131
563. Badman ER, Wells EM, Bui HA, Cooks RG (1998) Anal Chem 70:3545
564. Behm JM, Hemminger JC, Lykke KR (1996) Anal Chem 68:713
565. Crabtree HJ, Kopp MU, Manz A (1999) Anal Chem 71:2130
566. Infante AP, Guajardo NC, Alonso JS, Navascues MCM, Melero MPO (1993) Water Research 27:1167
567. Nilsson M, Ingemarsson A, Pedersen JR, Olsson JO (1999) Chemosphere, 38:1469
568. Thompson TS, Miller BD (1998) Chemosphere 36:2867
569. Mackenzie AS, Hoffmann CF, Maxwell JR (1981) Geochim Cosmochim Acta 45:1345
570. Bowie JH (1979) In: Johnstone RAW (ed) Mass spectrometry, vol 5. The Chemical Society, London, p 279
571. Hites RH (1988) Biomed Environ Mass Spectrom 17:311
572. Marchand M, Termonia M, Caprais JC, Wybauw M (1994) Analusis 22:326
573. Pichler H, Gans O, Krska R, Grasserbauer M (1997) Fresenius J Anal Chem 359:293
574. Mackenzie AS, Disko V, Rullkotter J (1983) Org Geochem 5:57
575. Mueller S, Efer J, Engewald W (1997) Fresenius J Anal Chem 357:558
576. Young D, Becerra M, Kopec D, Echols S (1998) Chemosphere 37:711
577. Warburton GA, Zumberge JE (1983) Anal Chem 55:123
578. Asam R, Ray KA, Glish GA (1998) Anal Chem 70:1831
579. Hauer CR, Leimbacher W, Hunziker P, Neuheiser F, Blau N, Heizmann CW (1992) Biochem Biophys Res Commun 182:953
580. Price WD, Schnier PD, Williams ER (1996) Anal Chem 68:859
581. Reilly PTA, Gieray RA, Yang M, Whitten WBJ, Ramsey M (1997) Anal Chem 69:36
582. Richardson SD, Thruston AD, McGuire JM, Weber EJ (1993) Structural characterization of reactive dyes using liquid secondary ion mass spectrometry/tandem mass spectrometry. Environmental Protection Agency (EPA) Rep EPA600J93423, p 99
583. Schroeder HF (1996) Water Sci Technol 34:21
584. Colleen K, Van Pelt P, Haggarty J, Brenna T (1998) Anal Chem 70:4369
585. Dorsey JG, Cooper WT, Siles BA, Foley JP, Barth HG (1996) Anal Chem 68:515
586. Dorsey JG, Cooper WT, Siles BA, Foley JP, Barth HG (1998) Anal Chem 70:591
587. Durand AY, Brown RG (1997) Chemosphere 31:3595
588. Katagi M, Tatsuno M, Tsuchihashi H (1994) Jap J Toxicol Environ Health 40:357
589. LaCourse WR, Dasenbrock CO (1998) Anal Chem 70:37

590. Lankmayer EP, Hayes MJ, Karger BL, Vouros P, McGuire JM (1983) Int J Mass Spec and Ion Physics 46:177
591. Linscheid M (1992) Int J Environ Anal Chem 49:1
592. Suter MJ-F, Riediker S, Giger W (1999) Anal Chem 71:897
593. Vestal ML (1983) Int J Mass Spectrom Ion Phys 46:193
594. Cai J, Henion J (1996) Anal Chem 68:72
595. Driskell WJ, Hill RH Jr, Shealy DB, Hull RD, Hines CJ (1996) Bull Environ Contam Toxicol 56:853
596. Hirayama K, Yuji R, Yamada N, Kato K, Arata Y, Shimada I (1998) Anal Chem 70:2718
597. Lim CK, Yuan Z, Jones RM, White INH, Smith LL (1997) J Pharm Biomed Anal 15:1335
598. Matuszewski BK, Constanzer ML, Chavez-Eng CM (1998) Anal Chem 70:882
599. Schilling JB, Cepa SP, Menacherry SD, Bavda LT, Heard BM, Stockwell BL (1996) Anal Chem 68:1905
600. Nishikawa M, Nakajima K, Tatsuno M, Kasuya F, Igarashi K, Fukui M, Tsuchihashi H (1994) Forensic Sci Int 66:149
601. Dark WA, McFaddden WH, Bradford DL (1977) J Chromatog Sci 15:454
602. McFadden WH, Bradford DC, Eglinton G, HajIbrahim S, Nicolatides N (1979) J Chromatog Sci 17:518
603. Nondek L, Kuzilek M, Krupicka S (1993) Chromatographia 37:381
604. Brenna JT (1994) Acc Chem Res 27:340
605. Aravena R, Robertson WD (1998) Ground Water 36:975
606. Aravena R, Evans ML, Cherry JA (1993) Ground Water 31:180
607. Goodman KJ (1998) Anal Chem 70:833
608. Hayes JM (1998) Anal Chem 70:2737
609. Mariotti A, Landreau A, Simon B (1988) Geochim Cosmochim Acta 52:1869
610. Pirard J-P (1997) Anal Chem 69:2030
611. Yang W (1998) Anal Chem 70:5159
612. Brenna JT (1997) Anal Chem 69:3148
613. Scrimgeour CM (1997) Anal Chem 69:1530
614. Sturchio NC, Clausen JL, Heraty LJ, Huang L, Holt BD, Abrajano TA Jr (1998) Environ Sci Technol 32:3037
615. Andrusevich V (1997) Anal Chem 69:926
616. Borgerding MF (1999) Anal Chem 71:1083
617. Barrie A, Debney S, Workman CT, Pullan C (1994) Proc Int Symp Nuclear Relat Technol, Vienna, Austria, p 29
618. Brand WA (1996) J Mass Spectrom 31:225
619. Meier-Augenstein W (1997) LC-GC 15:244
620. Newman A (1996) Anal Chem 68:373A
621. Lichtfouse E, Budzinski H (1995) Analysis 23:364
622. Brunengraber H, Kelleher JK, Des Rosiers C (1997) Annu Rev Nutr 17:559
623. Patterson BW (1997) Metabolism 46:322
624. Pont F, Duvillard L, Maugeais C, Athias A, Perségol L, Gambert P, Vergès B (1997) Anal Biochem 248:277
625. Ehleringer JR, Rundel RW (1988) In: Rundel RW, Ehleringer JR, Nagy KA (eds) Stable isotopes: history units and instrumentation. Springer, Berlin Heidelberg New York, pp 1–15
626. Long A, Eastoe CJ, Kaufmann RS, Martin JG, Wirt L, Finley JB (1993) Geochim Cosmochim Acta 57:2907
627. Matucha M, Jokisch W, Verner P, Anders G (1991) J Chromatogr 588:251
628. Cherrah Y, Falconnet JB, Desage M, Brazler JL, Zimi R, Tillement JP (1987) Biomed Environ Mass Spectrom 14:653
629. Goromaru T, Maeda H (1994) Biol Pharm Bull 17:1635
630. Meier-Augenstein W, Hoffman GF, Holmes B, Jones JL, Nyhan WL, Sweerman L (1993) J Chromatogr B 615:127
631. Meier-Augenstein W (1995) J High Resolut Chromatogr 18:28

632. Nitz S, Weinreich B, Drawert J (1992) J High Resolut Chromatogr 15 : 387
633. Meier-Augenstein W, Brand W, Rating D (1994) Biol Mass Spectrom 23 : 376
634. Mathews DE, Hayes JM (1978) Anal Chem 50 : 1465
635. Sano M, Yotsui Y, Abe H, Sasaki S (1976) Biomed Mass Spectrom 3 : 1
636. Goodman KJ, Brenna JT (1994) Anal Chem 66 : 1294
637. Ricci MP, Merritt DA, Freeman KH, Hayes JM (1994) Org Geochem 21 : 561
638. Simoneit BRT, Schoell M, Dias RF, Aquino Neto FR (1993) Geochim Cosmochim Acta 57 : 4205
639. Merritt DA, Hayes JM (1994) Anal Chem 66 : 2336
640. Merritt DA, Hayes JM (1994) J Am Soc Mass Spectrom 5 : 387
641. Merritt DA, Freeman KH, Ricci MP, Studley SA, Hayes JM (1995) Anal Chem 67 : 2461
642. Aggarwal PK, Hinchee RE (1991) Environ Sci Technol 25 : 1178
643. Aggarwal PK, Fuller ME, Gurgas MM, Manning JF, Dillon MA (1997) Environ Sci Technol 31 : 590
644. Conrad ME, Daley PF, Fischer ML, Buchanan BB, Leighton T, Kashgarian M (1997) Environ Sci Technol 31 : 1463
645. Landmeyer JE, Vroblesky DA, Chapelle FH (1996) Environ Sci Technol 30 : 1120
646. Suchomel KH, Kreamer DK, Long A (1990) Environ Sci Technol 24 : 1824
647. Simoneit BRT (1997) Atmos Environ 31 : 2225
648. Simoneit BRT, Didyk, BM (1978) Chem Geol 23 : 21
649. Freeman KH, Hayes JM, Trendel J-M, Albrecht P (1990) Nature 343 : 254
650. Hayes JM, Freeman KH, Popp BN, Hoham CH (1990) Org Geochem 16 : 1115
651. Dempster HS, Lollar BS, Feenstra S (1997) Environ Sci Technol 31 : 3193
652. Kelley CA, Hammer BT, Coffin RB (1997) Environ Sci Technol 31 : 2469
653. Hammer BT, Kelley CA, Coffin RB, Cifuentes LA, Mueller JG (1998) Chem Geol 152 : 43
654. O'Malley VP, Abrajano TA Jr, Hellou J (1996) Environ Sci Technol 30 : 634
655. Conrad ME, DePaolo DJ, Song DL, Neher E (1999) Proceedings of the 9th VM Goldschmidt Conference, Cambridge, MA
656. Jarman WM, Hilkert A, Bacon CE, Collister JW, Ballschmiter K, Risebrough RW (1998) Environ Sci Technol 32 : 833
657. Brand WA, Tegtmeyer AR, Hilkert A (1994) Org Geochem 21 : 585
658. Preston T, Slater C (1994) Proc Nutr Soc 53 : 363
659. Metges CC, Petzke KJ, Hennig U (1996) J Mass Spectrom 31 : 367
660. Macko SA, Uhle ME, Engel MH, Andrusevich V (1997) Anal Chem 69 : 926
661. Prosser SJ, Scrimgeour CM (1995) Anal Chem 67 : 1992
662. Begley IS, Scrimgeour CM (1996) Rapid Commun Mass Spectrom 10 : 969
663. Begley IS, Scrimgeour CM (1997) Anal Chem 69 : 1530
664. Rennie MJ, Meier-Augenstein W, Wyatt PW, Patel A, Begley IS, Scrimgeour CM (1996) Biochem Soc Trans 24 : 928
665. Tobias HJ, Goodman KJ, Blacken CE, Brenna JT (1995) Anal Chem 67 : 2486
666. Tobias HJ, Brenna JT (1996) Anal Chem 68 : 3002
667. Tobias HJ, Brenna JT (1997) Anal Chem 69 : 3148
668. Santrock J, Hayes JM (1987) Anal Chem 59 : 119
669. Epstein S, Mayeda TK (1953) Geochim Cosmochim Acta 4 : 213
670. Wong WW, Lee LS, Klein PD (1987) Anal Chem 59 : 690
671. Werner RA, Kornexl BE, Robmann BE, Schmidt H-L (1996) Anal Chim Acta 319 : 159
672. Koziet J (1997) J Mass Spectrom 32 : 103
673. Bréas O, Guillou C, Reniero F, Sada E, Angerosa F (1998) Rapid Commun Mass Spectrom 12 : 188
674. Farquhar GD, Henry BK, Styles JM (1997) Rapid Commun Mass Spectrom 11 : 1554
675. Barkan E, Luz B (1996) Anal Chem 68 : 3507
676. Ball JD, Crowley SF, Steele DF (1996) Rapid Commun Mass Spectrom 10 : 987
677. Beneteau K, Aravena R, Frape SK, Abrajano TA (1997) Proceedings of the Geological Society of America, Salt Lake City, UT

678. van Warmerdam EM, Frape SK, Aravena RJ, Drimmie RJ, Flatt H, Cherry J (1995) Appl Geochem 10:547
679. Eggenkamp HGM (1994) PhD thesis, Universitet Utrecht
680. Magenheim A, Spivack A, Volpe C, Ransom B (1994) Geochim Cosmochim Acta 57:3117
681. Bartholomew R, Brown F, Lousburry M (1954) Can J Chem 32:979
682. Tanaka R, Rye D (1991) Nature 357:707
683. Holt BD, Sturchio NC, Abrajano TA, Heraty LJ (1997) Anal Chem 69:2727
684. Jendrzejewski N, Eggenkamp HGM, Coleman ML (1997) Anal Chem 69:4259
685. Caimi RJ, Brenna JT (1993) Anal Chem 65:3497
686. Caimi RJ, Brenna JT (1995) J Mass Spectrom 30:466
687. Corso TN, Brenna JT (1997) Proc Natl Acad Sci USA 94:1049
688. Eglinton TI, Aluwihare LI, Bauer JE, Druffel ERM, McNichol AP (1996) Anal Chem 68:904
689. Jancso G, Van Hook WA (1974) Chem Rev 74:689
690. Slater GF, Dempster HD, Lollar SB, Spivack J, Brennan M, Mackenzie P (1998) Proceedings of the 1st International Battelle Conference on Remediation of Chlorinated and Recalcitrant Compounds, Monterey, CA
691. Sturchio NC, Heraty LJ, Holt BD, Abrajano T (1999) Proceedings of the 9th Annual VM Goldschmidt Conference, Cambridge, MA
692 Poulson SR, Drever JI, Colberg PJS (1997) Chemosphere 35:2215
693. Dayan H, Abrajano T, Heraty L, Huang L, Sturchio NC (1997) Proceedings of the Geological Society of America, Salt Lake City, UT
694. Vaillancourt J, Frape S, Aravena R (1998) Proceedings of the Geological Society of America, Toronto, Ontario
695. Heraty LJ, Fuller ME, Huang L, Abrajano T Jr, Sturchio NC (1999) Org Geochem
696. Coleman ML, McGenity TJ, Isaacs MCP (1999) Proceedings of the 9th Annual VM Goldschmidt Conference, Cambridge, MA
697. Morin JP, Kelley CA, Coffin RB, Cifuentes LA (1996) In: Spargo BJ (ed) In situ bioremediation and efficacy monitoring. NRL/PU/6115-96-317 Naval Research Laboratory, p 271
698. Preston T, Bury S, Présing M, Moncoiffe G, Slater C (1996) Rapid Commun Mass Spectrom 10:959
699. Preston T, Bury S, McMeekin B, Slater C (1996) Rapid Commun Mass Spectrom 10:965
700. Lijour Y, Gentric E, Deslandes E, Guezennec J (1994) Anal Biochem 220:244
701. Mossoba MM, Adams S, Roach JA, Trucksess MW (1996) J Am Oil Assoc 79:1116
702. Ojanperae I, Pihlainen K, Vuori E (1998) J Anal Toxicol 22:290
703. Gueven KC, Okus E, Dogan E, Uenlue S, Gezgin T, Burak S (1997) Turk J Mar Sci 3:123
704. Malins DC, Gunselman SJ (1994) Proceedings of the National Academy of Sciences, USA, 91:13,038
705. Fan TW-M, Colmer TD, Lane AN, Higashi RM (1993) Anal Biochem 214:260
706. Holmes E, Nicholls AW, Lindon JC, Ramos S, Spraul M, Neidig P, Connor SC, Connelly J, Damment SJ, Haselden J, Nicholson JK (1998) NMR Biomed 11:235
707. Ahmad VU, Jassbi AR, Pannahi MSC (1999) J Essential Oil Research 11:107
708. Gilmour I, Swart PK, Pillinger CT (1984) Org Geochem 6:29
709. Hama T (1988) Deep-Sea Res 35:91
710. Knulst JC, Boerschke RC, Loemo S (1998) Environ Sci Technol 32:8
711. Gaus HJ, Owens SR, Winniman M, Cooper S, Cummins L (1997) Anal Chem 69:313
712. Julian RK Jr, Higgs RE, Gygi JD, Hilton MD (1998) Anal Chem 70:3249
713. Godejohann M, Preiss A, Mügge C (1998) Anal Chem 70:590
714. Lewis RJ, Jones A, Vernoux J-P (1999) Anal Chem 71:247
715. Stout SJ, daCunha AR, Allardice DG (1996) Anal Chem 68:653
716. Grosjean E, Green PG, Grosjean D (1999) Anal Chem 71:1851
717. Volmer DA, Lay JO Jr, Billedeau SM, Vollmer DL (1996) Anal Chem 68:546
718. Tomy GT, Stern GA, Muir DCG, Fisk AT, Cymbalisty CD, Westmore JB (1997) Anal Chem 69:2762
719. Biber MV, Gülaçar FO, Buffle J (1996) Environ Sci Technol 30:3501
720. Cheng TMH, Malawer EG (1999) Anal Chem 71:468

CHAPTER 2

Interaction Mechanisms Between Organic Pollutants and Solid Phase Systems

Tarek A.T. Aboul-Kassim[1], Bernd R.T. Simoneit[2]

[1] Department of Civil, Construction and Environmental Engineering, College of Engineering, Oregon State University, 202 Apperson Hall, Corvallis, OR 97331, USA
e-mail:Tarek.Kassim@orst.edu

[2] Environmental and Petroleum Geochemistry Group, College of Oceanic and Atmospheric Sciences, Oregon State University, Corvallis, OR 97331, USA
e-mail: simoneit@oce.orst.edu

The chemical and structural constitutions of solid phase surfaces in the environment make them active sorbing sites for various organic pollutants. The mineral/humic/organic matter coatings of these solid phases (e.g., soils, sediments, suspended solids, colloids, and biocolloids/biosolids) interact with organic pollutants in different ways. The most important are adsorption and partitioning. Different factors can affect the interaction mechanisms at the pollutant-solid phase interface. These include interfacial tension, cosolvency, precipitation, pH, colloidal stability, functional groups, and cation exchange capacity. In addition, dissolved humic substances present in the aqueous environment can play a major role during these interactions. They can help reduce the tendency for such interaction mechanisms to occur with regard to pollutant solubilization, hydrolysis, and photosensitization processes.

Keywords. Sorption, Interaction mechanisms, Organic pollutants, Solid phases, Adsorption, Partitioning, Humic substances, Humus, Organic matter

The Handbook of Environmental Chemistry Vol. 5 Part E
Pollutant-Solid Phase Interactions: Mechanism, Chemistry and Modeling
(by T. A. T. Aboul-Kassim, B. R. T. Simoneit)

List of Abbreviations

CMC Critical micelle concentration
COMs Complex organic mixtures
DHS Dissolved humic substances

DOM Dissolved organic matter
DP_{HS} Dissolved phase humic substances
DTA Differential thermal analysis
ESR Electron spin resonance
FA Fulvic acids
HA Humic acids
HS Humic substances
K_{OC} Organic carbon partition coefficient
K_{OW} Octanol-water partition coefficient
Kp Partition coefficient
OC Organic carbon
PAHs Polycyclic aromatic hydrocarbons
POM Particulate organic matter
SOM Solid organic matter
SP_{HA} Solid phase humic acids
SP_{HS} Solid phase humic substances
SP_{OM} Solid phase organic matter
SWMs Solid waste materials
TOC Total organic carbon

1 Introduction

The mechanism of sorption and/or desorption for various toxic organic pollutants (see Chap. 1) by various solid phases has long been a subject of profound interest because of its direct impact on the mobility and activity of the organic pollutants in both soils and aquatic sediments [1–11]. A sorption-desorption transformation mechanism, a part of the environmental chemodynamic changes to environmental organic pollutants, occurs when a thermodynamically favorable reaction occurs to these pollutants [1]. A transformation process denotes a change in the target organic pollutant, whether by adding and/or removing a substituent group, or rearranging, breaking, or forming bonds. A sorption-desorption transformation process is not necessarily the same as degradation, but is often a step toward degradation because it represents a modification of the target pollutant in a step toward its ultimate mineralization.

A sorption/desorption mechanism is an important transformation process which occurs mainly at phase interfaces [9]. Most organic pollutant transformations and reactions that occur in aqueous media take place at phase discontinuities, such as the air-water or solid-water interfaces [12–14]. Sorption onto a solid surface (i.e., sediments, soils, colloids, suspended particles, and biocolloids/biosolids) can alter the configuration or energy status of an organic pollutant molecule in such a way to enable a reaction to occur. The physical process of organic pollutant adsorption onto a solid surface causes changes in the conformation or arrangement of the bonds in the adsorbed species [15, 16]. Such changes may increase the rate of a reaction and thus be considered a catalytic effect. The catalysis of organic chemical reactions by certain surfaces is an important process. Clay minerals are reported to catalyze some reactions involving organic chemicals [17, 18]. In addition to catalysis, concentration of material at a surface can increase the effective concentration of reactants and thus enable reactions that might not be possible in dilute systems.

Adsorption of an organic pollutant is its concentration on the external surface of any solid phase material at an interface, while absorption usually describes the movement of something into the interior of a matrix (Fig. 1). Because of the difficulties in discerning the boundaries of aqueous-solid phase interfaces, the more general term "*sorption*" has frequently been adopted to describe both adsorption and absorption. Sorption is a more generally applicable term, which encompasses both processes and simply relates interfacial flux. In practice, sorption of an organic pollutant(s) to an aqueous-solid phase interface, or pollutants leached from solid waste materials (SWMs) of complex organic mixtures (COMs) to other solid phases [1], usually indicate the movement from the free or mobile phase (gas or liquid) into or onto the fixed phase. Desorption is used to denote the movement of a certain pollutant(s) from the fixed phase back into the mobile phase (e.g., leachates of COMs from various SWM landfills). In actuality, sorption involves two main processes: (1) the movement from one phase to another involves changes in both phases, and (2) the overall systems of both phases will reflect the event. In the case of adsorption from an aqueous medium, it is usually considered that the process is competitive and that something must be removed (desorbed) or rearranged to accommodate the newly sorbed species. Absorption, on the other hand, can involve the movement from one liquid into another without the necessity of removing a sorbed species or competing for a site on the surface [19, 20].

The main objectives of the present chapter are to: (1) discuss in detail the compositions of the different solid phase systems covered in this volume which include soils, sediments, suspended matter, colloids, and biocolloids/biosolids, (2) review the various interaction mechanisms between organic pollutants and

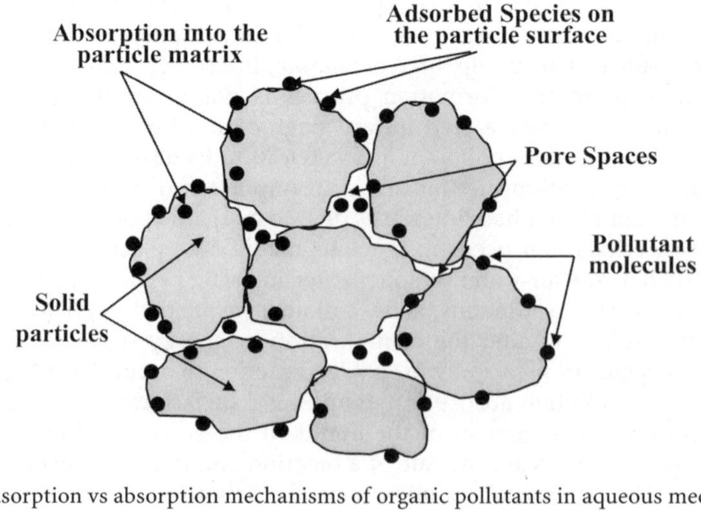

Fig. 1. Adsorption vs absorption mechanisms of organic pollutants in aqueous media

solid phases, (3) describe the various factors controlling sorption mechanisms at the aqueous-solid interface, (4) provide some evidence for pollutant-solid phase interactions in different environmental multimedia, and (5) illustrate the roles of humic substances and colloids in the interaction mechanisms.

2 Solid Phase Compositions

Before discussing the various interaction mechanisms between organic pollutants and solid phase systems, it is important to describe briefly the compositions of such solids mentioned in this chapter and throughout the volume. This can provide insight about the possible interaction mechanisms and their mode of chemical interactions. These phases include soils, sediments, suspended solids, colloids, and biocolloids (i.e., biosolids).

2.1 Soils, Sediments, and Suspended Solids

Solid surfaces such as soils, sediments, and suspended solids are composed mainly of mineral and organic matter (OM) associations. Their compositions will be described in detail in the next few sections.

2.1.1 *Clay Fraction*

Much of the early work in characterizing the environmental behavior of chemicals was accomplished in the area of agricultural chemistry. Work surrounding the behavior of plant nutrients in the soil has provided a large base of information about the processes of environmental chemistry. Workers investigating the effectiveness of soil-applied herbicides determined that the herbicidal activity of organic chemicals varied with soil properties. It was determined that clay and OM contents of the soil were related to the ability of a soil to diminish the effectiveness of an organic herbicide applied [17, 21].

The clay fraction, which has long been considered as a very important and chemically active component of most solid surfaces (i.e., soil, sediment, and suspended matter) has both textural and mineral definitions [22]. In its textural definition, clay generally is the mineral fraction of the solids which is smaller than about 0.002 mm in diameter. The small size of clay particles imparts a large surface area for a given mass of material. This large surface area of the clay textural fraction in the solids defines its importance in processes involving interfacial phenomena such as sorption/desorption or surface catalysis [17, 23]. In its mineral definition, clay is composed of secondary minerals such as layered silicates with various oxides. Layer silicates are perhaps the most important component of the clay mineral fraction. Figure 2 shows structural examples of the common clay solid phase minerals.

Because of isomorphic substitution of ions in the crystalline lattice of layer silicates, many clay surfaces have a net negative charge which results in the abi-

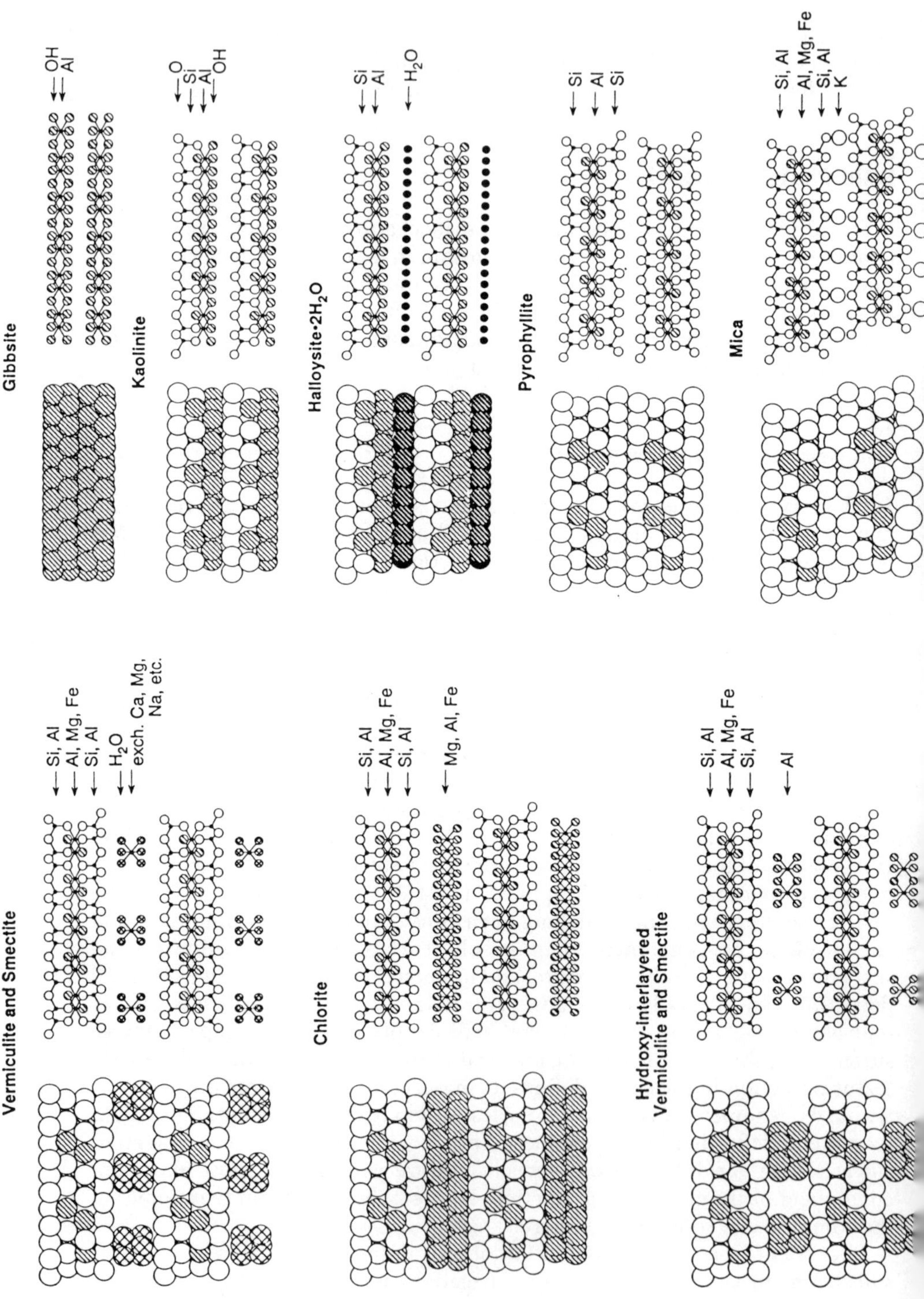
Gibbsite
OH
Al
Kaolinite
O
Si
Al
OH
Halloysite·2H2O
Si
Al
H2O
Pyrophyllite
Si
Al
Si
Mica
Si, Al
Al, Mg, Fe
Si, Al
K
Vermiculite and Smectite
Si, Al
Al, Mg, Fe
Si, Al
H2O
exch. Ca, Mg, Na, etc.
Chlorite
Si, Al
Al, Mg, Fe
Si, Al
Mg, Al, Fe
Hydroxy-Interlayered Vermiculite and Smectite
Si, Al
Al, Mg, Fe
Si, Al
Al

lity of such minerals to exchange cations from the solid solution. The cation exchange capacity varies from about 3 meq/100 g to 200 meq/100 g [17]. In addition to the existence of a static charge on the clay surface resulting from intracrystalline charge imbalances (isomorphic substitution), solid minerals may acquire charge from the pH-influenced dissociation of surface hydroxyl groups. The magnitude of this type of cation exchange capacity will tend to increase with pH.

The sphere of influence or extent of impact of the charged clay surface on the structure of ions of the solution will be, to some extent, determined by the ionic strength of the solution according to the double-layer effects. In addition to the presence of phyllosilicate minerals which exist in crystalline layers and frequently possess a net surface charge from isomorphic substitution, other products of mineral weathering and dissolution may be present in the clay fraction which do not exist in layers and do not possess an intrinsic charge. These are sometimes called accessory minerals and some of these minerals may have a pH-dependent charge. Accessory minerals consist of uncharged oxides, hydroxides, and hydroxyoxides of aluminum, iron, and titanium. Finely divided grains of these accessory minerals coat the surfaces of other mineral grains in the solid phase.

The weathering of minerals forms particles with a size continuum from ions to grains. Mineral dissolution and precipitation occur more or less continuously as a function of ambient conditions. Particles of the clay textural fraction may be suspended in solution as colloids as well as occurring as part of the stationary solids.

It is reasonable to assume that clay colloids exhibit a similar surface chemistry as clay which is sorbed, bonded, or precipitated in the stationary solid phase. Mineral colloids may be formed when precipitation or dissolution generate particles which are resistant to settling. These particles may be formed by any number of conditions whereby the solubility of a particular solute is exceeded or a stable solid is disrupted mechanically [21, 24].

The composition of the mineral fraction of the solid phase, being extensively composed of oxygen and silicon bonded with various metals, will lend a relatively polar nature to the surface of most of the inorganic solid components. This polar/ionic nature will create a natural affinity between solid and ionic or polar solutes. In addition to sorption of ionic and polar solutes onto clay in the solution and solid phases, the clay fraction has been shown to be important in the sorptive behavior of various neutral hydrophobic organic compounds from water [17, 19]. The large surface area of the clay fraction offers a large sorptive interface upon which hydrophobic bonding may occur [6]. For such nonpolar compounds, polar/ionic attraction is generally secondary to hydrophobic effects in sorption on most sediments and soils.

2.1.2 *Organic Matter and Humus*

Similar to inorganic components, solid organic matter (SOM) plays a significant role in affecting the chemistry of solid phase surfaces. Humus and SOM can be considered as synonyms, and include the total organic compounds in solid phases excluding undecayed plant and animal tissues, their partial decomposition products, and the solid biomass [25]. Humus includes humic substances (HS) plus resynthesis products of microorganisms, which are stable, and a part of the solid phase itself. HS include a series of relatively higher molecular weight, brown-to-black-colored substances formed by secondary synthesis reactions. This term is used as a generic name to describe the colored material or its fractions obtained on the basis of solubility characteristics. The physical nature of humus is that of an amorphous, brownish material with a density somewhat lower than that of mineral solids. In its natural state, humic material is somewhat variable in composition and form depending on its source [22, 26–32].

2.1.2.1
Formation and Complex Composition

Several mechanisms have been proposed to explain the formation and complex composition of solid phase humic substances (i.e., SP_{HS}, Fig. 3). Selman Waksman's classical theory [17], the so called "*Lignin Theory*", was that HS are modified lignins which remain after microbial attack (pathway 4, Fig. 3), and undergo further modifications yielding first humic acids (HA) and then fulvic acids (FA). Pathway 1 (Fig. 3), which is not considered significant, assumes that HS form from sugars [25]. The contemporary view of HS genesis is the "*Polyphenol Theory*" (pathways 2 and 3, Fig. 3) which involves quinones. In pathway 3 (Fig. 3) lignin is an important component of HS creation, but phenolic aldehydes and acids which are released from lignin during microbial attack

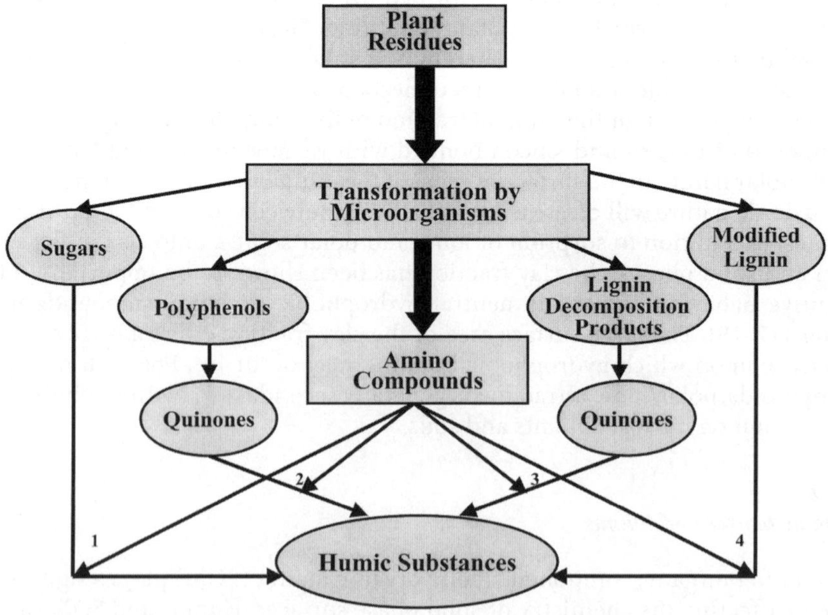

Fig. 3. Mechanisms for the formation of solid humic substances

are altered enzymatically to quinones, which polymerize to form *humic-like* macromolecules. Pathway 2 is analogous to pathway 3 (Fig. 3) except that the polyphenols are synthesized microbially from non-lignin carbon sources (i.e., cellulose), and oxidized by enzymes to quinones and then polymerize to HS (25).

The processes of formation of SOM are more or less unique to a particular geographic environment on the large scale [17, 33]. Since the material is derived from plant remains (Fig. 3), which have been more or less degraded by detritovores, it is reasonable to assume that there would be some differences between humic material from different biomes. The vegetation, solid minerals, climate, and microbial population are some of the variables which might act to create differences in the OM of a different area. In spite of reported differences, the variations are not so great as to preclude comparisons between humus from one biome to another. The process of plant growth is essentially through photosynthesis (i.e., the sunlight-driven reduction of oxidized carbon). The reduced carbon takes many forms and combines with many elements in a complex array of chemical compounds. When this living material dies, the chemical energy it contains is exploited by heterotrophic organisms for their own life processes. Humus is considered to be produced from that portion of the reduced carbon which was resistant to degradation either as a function of its intrinsic nature (e.g., polyphenols such as lignins and tannins) or of ambient conditions which restrict the oxidative processes of degradation (e.g., cold, anaerobic environments).

In addition to the reported differences between and within bioregions (Fig. 4), it is reported that humic material of terrestrial origin is different from

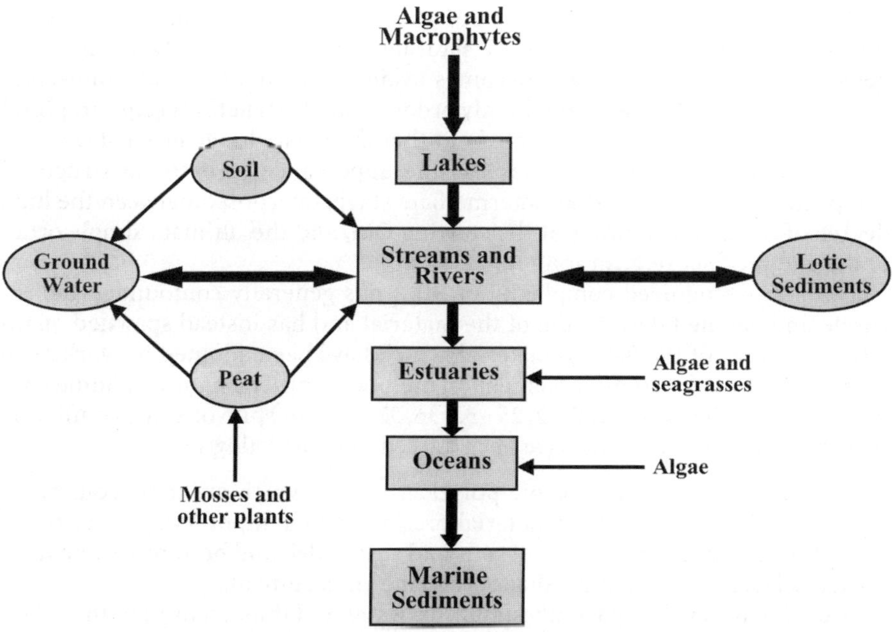

Fig. 4. Diagram of the numerous possible environmental flowpaths of humic substances

the humic material in freshwater streams. The humic material in aqueous sediments is reported to be less polar (as judged by a lower oxygen percentage) than the solid phase organic matter (SP_{OM}) [17, 33]. There are numerous paths that HS can take in the environment (Fig. 4). Water is obviously the most important medium which affects the transport of HS. A host of environmental conditions affect HS, ranging from oxic to anoxic environments and from particulate to dissolved humic substances (DHS). Additionally, the time range that HS remain in the environment is wide (i.e., from weeks to months for HS in surface waters of lakes, streams, and estuaries to hundreds of years in soils and deep aquifers) [17].

Humus, in general, undergoes changes as it ages. Humus, which exists as solid particles, is the result of extensive alteration of the original component materials and is subject to degradation. Under different conditions, humus undergoes diagenesis and transformation in response to the ambient conditions. Humus buried deep in the subsurface is subject to different processes and will accordingly become kerogen after the passage of time. Peat, coal, and shales are examples of OM that has undergone extensive diagenesis [34, 35].

Diagenesis increases with depth and time of burial [34]. Maturation (also termed catagenesis) is the result of elevated heat and pressure acting on OM, and interactions with mineral surfaces and complexed metals may also be involved [17, 34, 35]. Thermodynamic stabilization of OM occurs during both diagenesis and catagenesis. The least stable and most reactive components or their substituents are gradually eliminated. This process leads, with increasing age and depth of burial, to a gradual stabilization, not necessarily of each individual compound but of the sedimentary OM as a whole. In terms of structures the transformation of open chains to saturated rings and finally to aromatic networks is favored; hydrogen becomes available for inter- or intramolecular reduction processes. Eventually, highly ordered, stable structures (e.g., graphite) may be formed. It should be pointed out that the most characteristic feature of organic diagenesis and catagenesis is the appearance of extreme structural complexity and disorder at an intermediate stage, interposed between the high degree of biochemical order of the starting OM, and the ultimate simple order of the end product (e.g., graphite) [34, 35].

The long-recognized complexity of SP_{OM} has generally confounded the accurate and detailed description of the material and has instead spawned qualitative divisions of the natural material, which have been adopted by workers in the field to allow for some agreement on methodology. The nature of humus has been studied extensively [17, 22, 25–33, 36, 37] and, in spite of some conflicting reports, a number of points have been agreed upon as follows:

- Humic acids (HA) are organic polyelectrolytes, which are most commonly identified with the organic material present in contemporary solid particles. HS are present in practically all soils and suspended and bottom sediments of rivers, lakes, estuaries, and shallow marine environments.
- Humic materials are partially soluble in water and thus occur in both surface and groundwaters.

Humus/SOM enter into a wide variety of physical and chemical interactions, including sorption, ion exchange, free radical reactions, and solubilization. The water holding capacity and buffering capacity of solid surfaces and the availability of nutrients to plants are controlled to a large extent by the amount of humus in the solids. Humus also interacts with solid minerals to aid in the weathering and decomposition of silicate and aluminosilicate minerals. It is also adsorbed by some minerals.

2.1.2.2
Chemical Nature

The chemical nature of humus is the subject of variable and sometimes conflicting reports in the literature [17, 22, 25–29, 31, 32]. For the purpose of this volume, it is important to point out that humus is chemically reactive and has variable chemistry, manifesting both polar and nonpolar tendencies. In general, humus contains a number of chemical functional groups associated with a polycyclic aromatic matrix of varying sizes. In terms of structural moieties, humus contains numerous polymerized substances, aromatics, polysaccharides, amino acids, polymers of uronic acids, and phosphorus- and sulfur-containing components. Chemical degradation has shown that the basic building blocks of humic acids are benzene carboxylic acid groups, substituted phenolic groups, and quinone groups [30]. Figure 5 shows the % composition of the different humic substance fractions in terms of their carbon, hydrogen, nitrogen, sulfur, and oxygen contents (data were taken from [17]).

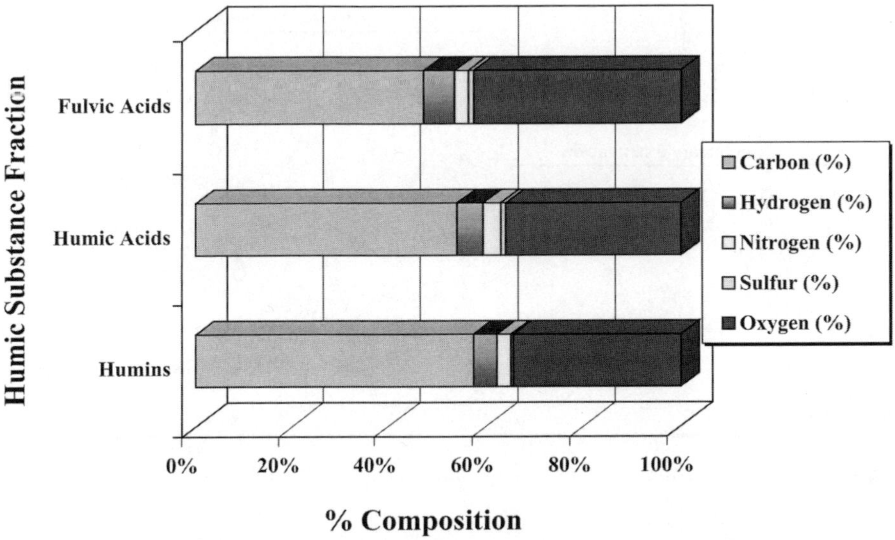

Fig. 5. Elemental composition of humic substances representing various solid phase organic matter (SP_{OM}, number of samples = 52)

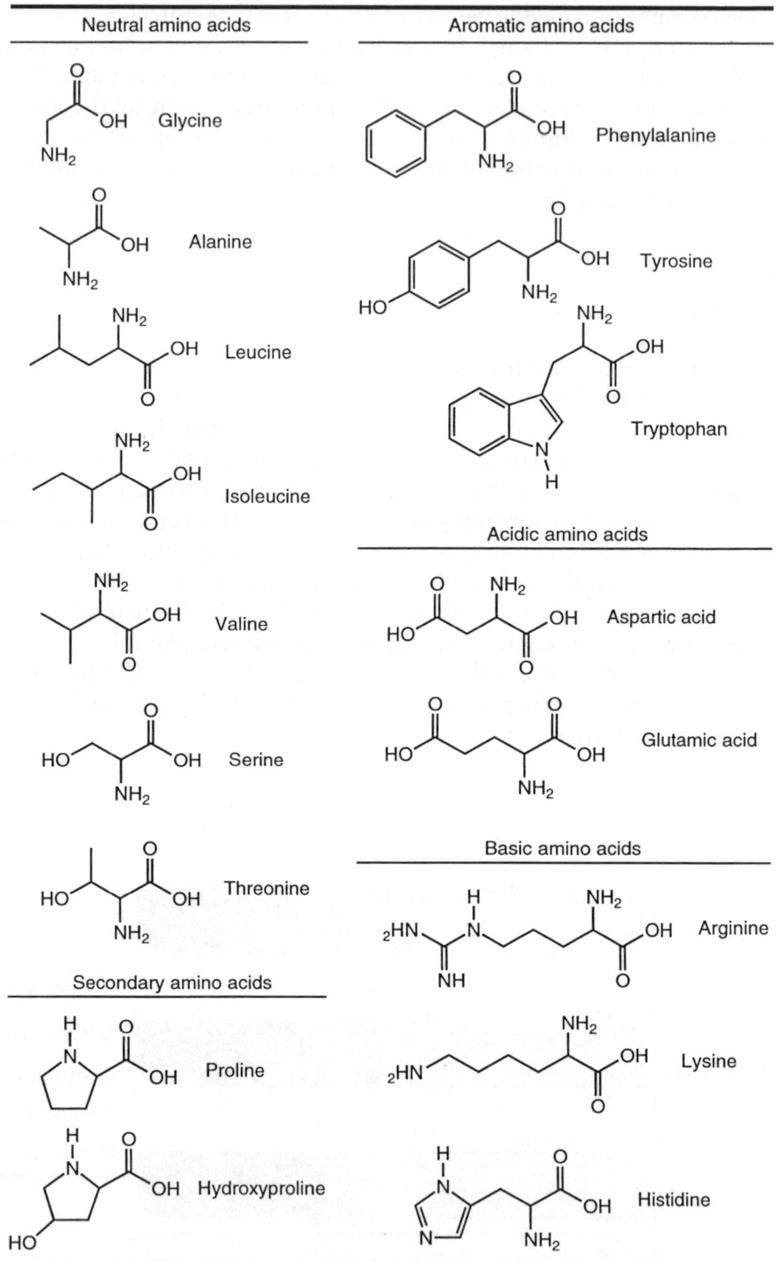

Fig. 6. Chemical structures of some protein amino acids found in soils

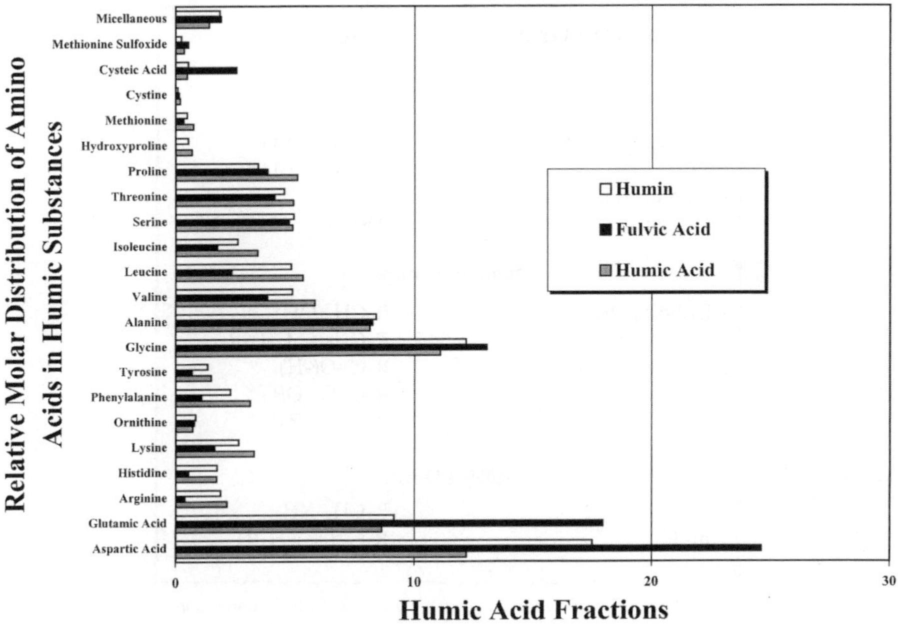

Fig. 7. Relative molar distribution of amino acids in humic substances (data were collected from Ghosh and Schnitzer [37] and Schnitzer [38])

The chemical structures of the amino acids found in soil-solids are shown in Fig. 6 [25] while the quantities of amino acids found in HS extracted from various solid phases are represented in Fig. 7 (data were collected from Ghosh and Schnitzer [37] and Schnitzer et al. [38]). High levels of amino acid nitrogen were found in HA, FA, and humin fractions, indicating incorporation of common acidic and some neutral amino acids, particularly glycine, alanine, and valine.

2.1.2.3
Bonding

The mechanism(s) bringing about the aggregation or disaggregation of humic substances (HS) are a consequence of the charge and functional group distributions on the exposed surfaces. A number of workers have shown that humic materials contain abundant polar functional groups [17, 22, 25–33, 37]. The highly polar nature of some of the functional groups of SP_{OM} (Fig. 8) makes dipole and hydrogen bonding probable active mechanisms of structural change. Reports that humic materials contain both electron rich and electron deficient sites provide evidence that polar bonding may likely to occur [30, 39].

In addition to hydrogen bonding, coulombic attraction of charged particles will also create bonds in humus. The charged sites on a polyelectrolyte molecule may arise in several different ways. Ionic compounds will dissociate in solution, producing molecules with charged sites. These charged sites might also result

Functional Groups	Structure*
Acidic groups	
Carbonyl	R-C=O-(OH)
Enol	R-CH=CH-OH
Phenolic compounds	AR-OH
Quinone	AR=O
Neutral groups	
Alcoholic OH	R-CH_2-OH
Ether	R-CH_2-O-CH_2-R
Aldehyde	R-C=O(-H)
Ester	R-C=O(-OR)
Ketone	R-C=O(-R)
Basic groups	
Amine	R-CH_2-NH_2
Amide	R-C=O(-NH-R)

* R = aliphatic hydrocarbon group, AR = aromatic hydrocarbon group

Fig. 8. Some important functional groups of solid organic matter

from charge-transfer reactions such as the transfer of an electron from a carbanion or a radical anion to another molecule.

Humus is also capable of forming covalent bonds with aqueous solutes [40]. Humus is the site of considerable microbial activity where living and dead organisms and extracellular enzymes are typically associated with it as part of the material [25, 30]. The presence of enzymes can catalyze reactions. It is also reported that humus contains stable free radicals, which make it very reactive and able to form covalent bonds or create ions. It is reasonable to assume that charge-transfer reactions between free radicals are important in the aggregation of humic materials in light of the high concentrations of free radicals which have been detected in both soils and aqueous humic acid preparations. The free radicals detected in soils and humic acids may arise from the reduction of a diamagnetic molecule by a solvated electron, enzymatic reactions, or photolysis [26, 27, 29, 41].

The diverse nature of chemical bonding arrangements exhibited by humus enables the formation of associations both with non-humic materials and with other humic materials to create a dynamic structure. Such a structure is capable of undergoing inter- and intra-molecular bonding to add or lose constituents or change configuration in response to ambient conditions. The chemically diverse and highly reactive nature of the humic matrix imparts the ability of humus both to lose and to acquire molecular moieties in a dynamic manner.

2.1.2.4
Fractionation

A fractionation procedure has been established and widely applied to studies of humic materials [42–44]. The procedure begins with natural OM (i.e., humus) and uses an aqueous basic solution (e.g., 0.1–0.5 mol/l NaOH and Na_2CO_3) to solubilize a fraction of the OM. The basic extract is then acidified which causes a precipitate to form, i.e., humic acids (HA). The fraction, which remains in solution, is called fulvic acids (FA). Humin is the name given to the insoluble organic fraction that remains after extraction of humic and fulvic acids. At near-neutral pH (pH 5–8), which is characteristic of most natural water, the FA are the most water soluble of these three fractions. HA are somewhat less soluble, with their solubility increasing as the pH increases. Humin is insoluble at all pH values.

Because alkali extractions can dissolve silica, contaminating the humic fractions, and dissolve protoplasmic and structural components from organic tissues, milder extractants (e.g., $Na_4P_2O_7$ and EDTA, dilute acid mixtures with HF, and organic solvents) can also be employed; however, they will also reduce the amount of soil organic matter extracted [25]. In addition, gel permeation chromatography, ultrafiltration membranes, adsorption on hydrophobic resins (XAD, non-ionic methylmethacrylate polymer), adsorption on ion exchange resins, charcoal and Al_2O_3, and centrifugation are also used for SP_{HS} fractionation [45].

FA are soluble in water and so are the majority of the salts of these acids [17, 43, 44, 46, 47]. The aquatic FA fraction contains substances with molecular weights ranging from 500–2000 and is monodispersed. Aquatic FA are dissolved rather than colloidal and contain traces of branched, cyclic, and linear alkanes, as well as fatty acids [43, 46, 47].

HA are pictured as being made up of a hierarchy of structural elements (Fig. 9) [48]. At the lowest level in this hierarchy are simple phenolic, quinoid, and benzene carboxylic acid groups. These groups are bonded covalently into small particles. The molecules of HA are reported to be nonspherical, or more probably, nonspherical and hydrated or rigid spherocolloids in solution [49–55]. Ghosh and Schnitzer [37] concluded that the configurations of HA and FA molecules are not unique – they vary with changes in the environment. These authors report that both HA and FA molecules are flexible linear colloids at low concentrations, provided hydrogen ion and neutral salt concentrations are not too high. As these factors increase, the macromolecules assume coiled configurations similar to those of uncharged polymers or rigid spherocolloids. HA can have an amorphous structure and furthermore their size and molecular weight can vary as a function of ambient solution conditions. It has been postulated that the molecular weight of the HA species varies from 1000 to 50,000 and consists of particles capable of aggregation or dissociation.

HA are larger than FA and form polydisperse systems [43]. Precipitation is used to isolate HA from solids. The HA must aggregate to precipitate, and therefore the resulting polydisperse systems confirm that HA exist as aggregates of various sizes. HA are mixtures of a limited number of more or less chemically

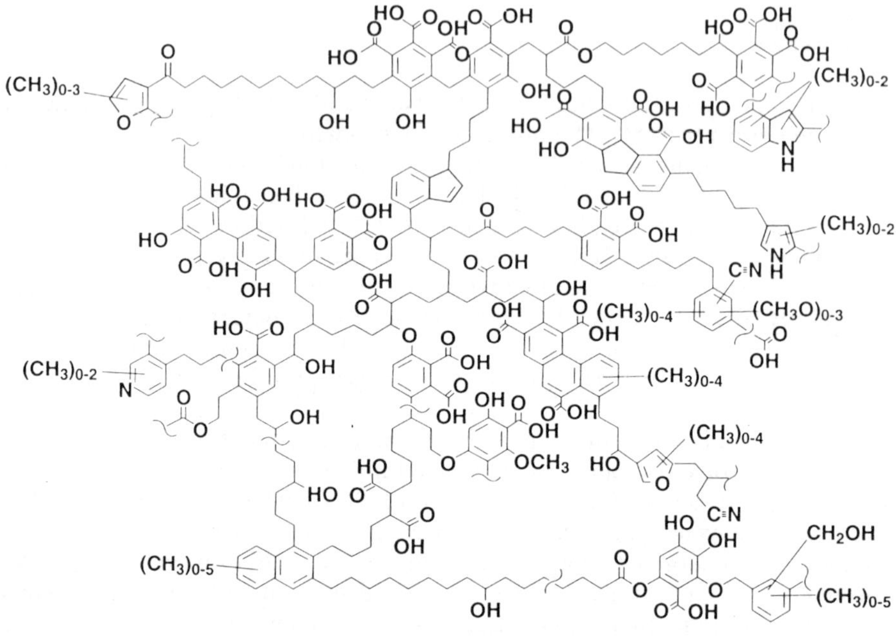

Fig. 9. Schematic of humic acid structure. From Schulten and Schnitzer [48], with permission

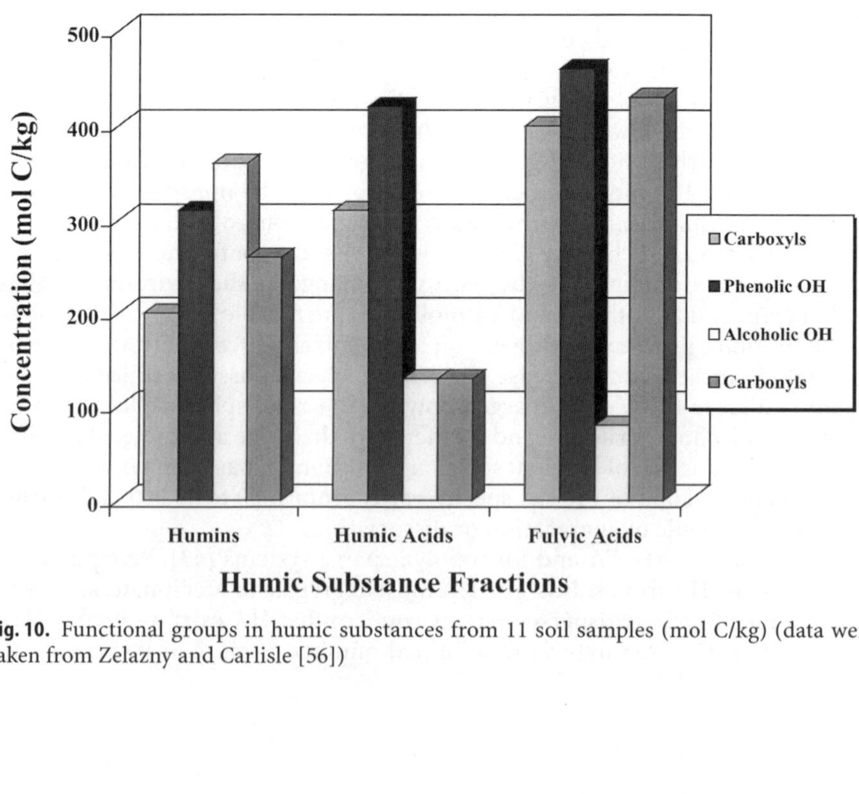

Fig. 10. Functional groups in humic substances from 11 soil samples (mol C/kg) (data were taken from Zelazny and Carlisle [56])

distinct fractions of relatively low molecular weight OM which form molecular aggregates in solution. Particles of similar chemical structures are thought to be linked together by weak bonds to form homogeneous aggregates. Two or more different types of aggregates may be linked together to form mixed aggregates [43].

The concentrations of HS fractions, i.e., fulvic acids, humic acids, and humin expressed as mol *C*/kg of SP_{HS} in terms of various functional groups such as carboxyl, phenolic OH, alcoholic OH, and carbonyl are shown in Fig. 10.

2.1.2.5
Existence

Humus can exist in solution as well as in the solid phase. The behavior of water-soluble humic materials is of great relevance to the discussion of solubility enhancement of aqueous pollutants, both organic and inorganic. Freshwater aquatic HS originate from soil humic material and decomposing terrestrial and aquatic plants. In surface waters these compounds generally account for 30–50% of the dissolved organic matter (DOM) [43]. Systems that contain naturally high levels of DOM include bogs, swamps, and interstitial waters of sediments. Interstitial water (porewater) is formed by the entrapment of water during sedimentation, which isolates it from the overlying water. Porewater is considered to be in equilibrium with the sedimentary solid phase and separate from the overlying water column, or bulk water. In sediments with high organic carbon, dissolved organic carbon (DOC) in porewater can exceed 100 mg/l, whereas overlying waters typically contain less than 5 mg/l of DOC [44, 57]. The molecular weight of most HS in water is less than 10,000.

2.1.2.6
Humic/Mineral Associations

In addition to the ability of HS to form associations with hydrophobic organic species, humic material also reacts readily to form associations with inorganic minerals as well as polar and ionic organic materials. These types of associations are involved in colloid formation with a wide variety of materials [58–61].

FA can interact with clay minerals and are known to form stable complexes with metal ions and hydrous oxides [59, 61]. The operational technique for isolation of HA involves a pH-induced precipitation and it is likely that accessory minerals may be associated with the precipitation process. Complexes of HA and clay minerals are also formed, the increased ash content of HA suggesting that amorphous silica, iron hydroxides, and clay may aggregate with the HA fraction [58, 60, 61].

The amphiphilic nature of dissolved humic substances (DHS) lends them the ability to associate with both hydrophobic organics and polar or ionic species [62–64]. Inorganic ions or mineral colloids in solution will interact with the electrically active surface of humic material in solution or in the solid phase according to the same bonding forces which lead to the association between SP_{OM} and the solid mineral matrix. Humic matter in water is associated with various metal ions, clays, and amorphous oxides of iron and aluminum [19, 65]. In aqueous environments, oxide mineral surfaces are generally covered with hydroxyl groups. Organic macromolecules can sorb onto these surfaces both by ligand exchange and by van der Waals forces to create a strong association.

Humus can form stable complexes such as chelates with polyvalent cations. SOM is capable of strong polydentate binding to transition metals in a chelate [17, 19, 45, 65–67]. The complexation of metal ions by SOM is extremely important in affecting the retention and mobility of metal contaminants in solid phases and waters [45]. Several different types of SOM/humus-metal reactions can occur (Fig. 11), and include reactions between DOC-metal ions, complexation reactions between SOM-metal ions, and bottom sediments-metal ions. The functional groups of SOM (Fig. 10) have different affinities for metal ions as shown below:

$$[-\bar{O}] > [-NH_2] > [-N{=}N] > [\text{'N'}] > [COO^-] > [-O-] > [-C{=}O]$$

(enolate) (amine) (azo compound) (ring N) (carboxylate) (ether) (carbonyl)

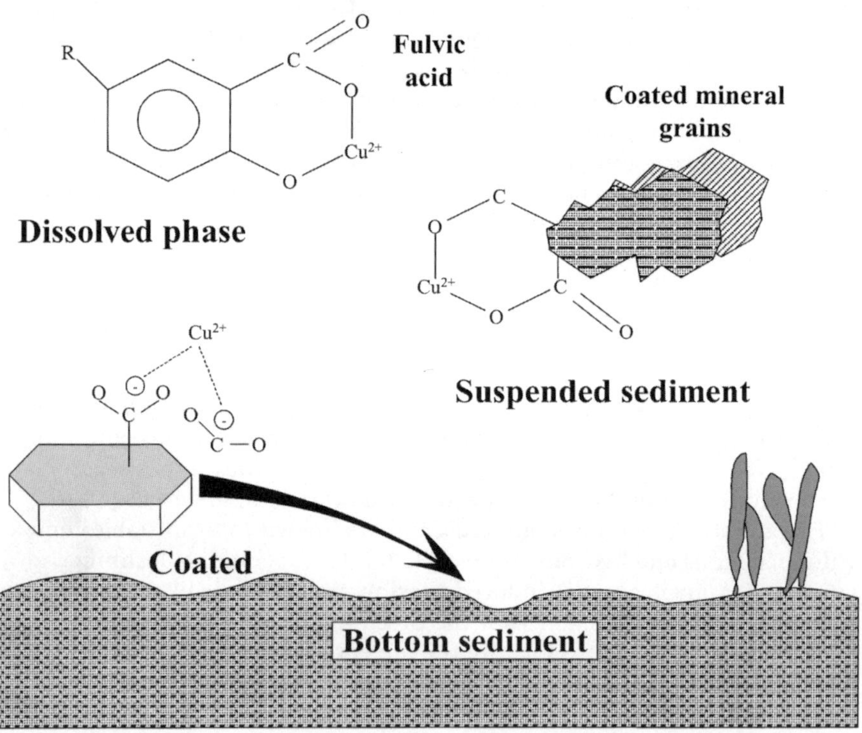

Fig. 11. Complexation of metal ions by organic matter in suspended sediments, bottom sediments, colloidal and dissolved phases

2.1.2.7 Properties

It appears that for highly hydrophobic molecules, the organic carbon (i.e., OC) content of the sorbent is of greater importance than the mineral surface itself [68–71]. However, in the low carbon environments characteristic of the subsurface, mineral sorption may play an important role in affecting hydrophobic pollutant mobility [41, 72–75]. It has been reported that if the solid contains more than about 0.2% organic carbon, all of the sorption of hydrophobic organic compounds appears to be due to the organic carbon [17, 71]. If the solid phase contains <0.2% organic carbon, the sorption of hydrophobic organic compounds from the aqueous phase may be attributed to the clay fraction [41, 72, 73].

In early work with sorption of aromatic hydrocarbons by sediments, it was reported by Karickhoff et al. [76] that the ratio of individual partition coefficients (K_d) for the sorption of the organic compounds to the organic carbon contents of the sediments (% OC) yields a unique constant (K_{OC}) (Eq. 1), which was independent of sediment properties and dependent only upon the nature of the organic analytes:

$$K_{OC} = \left(\frac{K_d}{\% \mathrm{OC}} \right) \tag{1}$$

These authors reported a significant correlation between the K_{OC} values obtained from the sorption of organic compounds on three sediments and the partition constants (K_{OW}) for the partitioning of the compounds between octanol and water (Eq. 2):

$$\log K_{OC} = 1.00 \log K_{OW} - 0.21 \tag{2}$$

A number of similar empirical expressions have been developed for relating the partitioning behavior of an organic compound between water-organic carbon to the octanol-water partition coefficient for the organic chemical itself [19, 77–80]. It has been noted that the tendency of an organic compound to partition into the organic phase of the soil/sediment is inversely related to the water solubility of the compound [81–83]. Hence, the tendency of an organic compound to be sorbed by soil or sediment organic matter (i.e., SOM) will be a function of its hydrophobicity. The octanol-water partitioning (K_{OW}) behavior of a compound has also been related to its intrinsic hydrophobicity [84–87]. The SOM fraction has been determined to be of considerable importance in the environmental behavior of organic pollutants.

2.2 Colloids

The presence of colloids in natural aqueous systems acts to influence the distribution and behavior of organic pollutants [24, 88–94]. Colloids are formed by some physical and chemical processes. These physical and chemical processes govern the distribution of pollutants in natural and perturbed systems to and from the colloidal phase. The composition and behavior of colloids are complex and difficult to define rigorously. In the absence of effective models for colloidal systems in the natural world, we must rely on descriptions of the processes in a more conceptual sense. Several workers have investigated the effects of colloids on the interaction mechanisms between pollutant and various solid phases in the environment [95–102].

2.2.1
Definition

Colloids are suspended particles in a solution medium and will not settle out over time. They are common in natural waters and can enhance the apparent solubility of a wide range of water pollutants, both organic and inorganic. Colloids may be considered as an extension of the solid and aqueous phases and are formed by conditions that can be quite variable in time and space; hence colloids can be dynamic. The composition of colloids can vary with the composition of the solid and aqueous phases. Colloids can be made up of organic, inorganic, or a mixture of materials.

A colloidal solution is defined as a solution intermediate in character between a suspension and a true solution. Particles with diameters $<10\ \mu m$ are usually called colloids [19, 65], although the distinction based on size is arbitrary. The size of particles is a continuum and the point at which large macromolecules end and small colloids begin is subject to judgment, as is the upper end of the size continuum, where colloids and suspended particles merge. The tendency of suspended particles to settle out of solution is not really a function of size alone, rather the relative density of the particles and the motion of the water will determine what is suspended and what settles.

2.2.2
Presence

Colloids are present in natural waters (i.e., surface and groundwaters). Surface systems receive terrestrial input as runoff, which carries solid-derived materials into streams, rivers, lakes, or estuaries. Groundwater receives leachates from land fills and percolation water and is frequently well connected with surface water bodies. Colloids may also be formed in situ by native processes of precipitation and dissolution, suspension, or biological activity [103, 104].

Colloids in solution represent a highly dispersed suspended particle phase. Because of the sorptive behavior of interfaces, the higher surface area of dispersed colloids tends to make colloids a more effective adsorbent on a mass basis than an equivalent mass of precipitated or solid material. Colloids act to enhance the solubility of slightly soluble pollutants, whether they are organic or inorganic. Hydrophobic organics, and slightly soluble inorganics, have been associated with colloids in apparent solution.

2.2.3
Solubility Enhancement

The importance of the colloidal phase in the distribution of water pollutants is a relatively recent issue in the environmental literature [4, 105, 106]. The phenomenon of colloidal solubility enhancement was detected by workers in several fields and was largely unexplained. The concept was apparently developed and forwarded by working with partitioning behavior of water pollutants in water/sediment systems.

O'Connor and Connolly [107] found that equilibrium sorption partition coefficients of several pollutants into Texas River sediments declined as sediment concentration increased in isothermal studies. This has been interpreted as an indication that colloids in solution were competing with the sediment for sorbate and that the concentration of colloids increased as the concentration of sediment increased.

The importance of colloids was first recognized by Voice et al. [108] when they discussed what they called the *particle concentration effect*, a term coined to describe the observation that the partition coefficient for strongly sorbed or slightly soluble solutes varied with the concentration of the soil/sediment used in the experiment. They proposed that the observed change in partitioning behavior due to solid concentrations could be attributed to a transfer of sorbing, or solute binding, material from the solid phase to the liquid phase during the course of the partitioning experiment. This material, whether dissolved, macromolecular, or microparticulate in nature, was not removed from the liquid phase during the separation procedure and was capable of stabilizing the pollutant of interest in solution. The amount of material contributed to the liquid phase was thought to be most likely proportional to the amount of solid phase present, and thus the capacity of the liquid phase to accommodate solute would depend upon the concentration of solids in the system. The overall effect can be viewed either as a two-phase system, where the properties of one phase (liquid) vary with the mass of the other (solids), or as a three-phase system consisting of water, solids, and a third phase which is not separated from the water but possesses a higher capacity for the solute than the water itself. This is the colloidal phase.

Measurements of "dissolved" sorbing phase (e.g., weight of dissolved solids, turbidity, and DOC) demonstrate the increased loading of nonsettling microparticles or macromolecules in the supernatants of batch equilibrium experiments as the solids-to-water ratio increases. It is clear that nonsettling microparticles or macromolecules vary regularly with suspended solid concentration.

The observation that dissolved colloidal material was increasing the apparent solubility of pollutants in laboratory studies led to the attempt to wash the sediment in order to try to remove these materials (i.e., colloids). Successive washings reduced the amount of material in solution, but failed to remove it. After five successive washes, the nonsettling microparticles or macromolecule content dropped about an order of magnitude, yet remained at an amazingly high level of 100 mg/l [109]. Walters et al. [110] confirmed the report that aqueous colloids could not be satisfactorily removed by washing or centrifugation.

Particle size distributions of natural sediments and soils are undoubtedly continuous and do not drop to zero abundance in the region of typical centrifugation or filtration capabilities. Additionally, there is some evidence to indicate that dissolved and particulate organic carbon in natural waters are in dynamic equilibrium, causing new particles or newly dissolved molecules to be formed when others are removed. Experiments with soil columns have shown that natural soils can release large quantities of DOC into percolating fluids [109].

Colloids have repeatedly been shown to be important in enhancing the apparent solubility of hydrophobic organic chemicals [4, 19, 62, 96, 105, 106, 111–113]. The solid phase is the source of dissolved or suspended colloidal material which acts as the third phase. It is observed that the solution phase is in dynamic equilibrium with the solid phase.

2.3
Biocolloids or Biosolids

The term "biocolloids or biosolids" is frequently applied to microbes in solution. Bacteria, algae, protozoans, and many other biological agents present in the aqueous phase can be considered to exhibit colloidal behavior [114–124]. Insofar as these species are able to sorb pollutants like other colloids, the distinction between living and nonliving colloids is relatively unimportant. It is also known that biological exudates or subcellular fragments may exist in colloidal solution [124].

The sorptive nature of bacterial or algal exterior membranes is well-documented [118–122]. Biological particles can influence the distribution of heavy metals in natural waters because the functional groups on the cell surfaces are able to bind certain metal ions [124].

Microbes are ubiquitous in the subsurface environment and as such may play an important role in groundwater solute behavior. Microbes in the subsurface can influence pollutants by solubility enhancement, precipitation, or transformation (biodegradation) of the pollutant species. Microbes in the groundwater can act as colloids or participate in the processes of colloid formation. Bacterial attachment to granular media can be reversible or irreversible and it has been suggested that extracellular enzymes are present in the system. Extracellular exudates (slimes) can be sloughed-off and act to transport sorbed materials [122]. The stimulation of bacterial growth in the subsurface may be considered as in situ formation of colloids.

In the same way as described for subsurface water, inputs of DOM, which constitute reduced carbon to the surface, tend to stimulate microbial activity because DOM can be utilized as a substrate. Microbial activity associated with inputs of organic substrate will consume oxygen and create reducing conditions if oxygen demand exceeds supply [125, 126].

3 Interaction Mechanisms in the Pollutant-Solid Phase Interface

The significance of interactions between solid phase-humic substances (SP_{HS})/ solid phase organic matter (SP_{OM}) and organic pollutants which are present in aqueous systems will be discussed from a mechanistic point of view. Emphasis will be given mainly to sorption mechanisms, with some background information about solubilization effects, hydrolysis, catalysis, and photosensitization. It should be noted at this point that the chemical properties and behavior of the dissolved and solid-phase fractions of HS may be sufficiently different that these two fractions will interact differently with a given pollutant and that various chemical properties of organic pollutants will result in several interaction mechanisms that may frequently operate in combination. The following are the different modes of interactions.

3.1 Adsorption

Adsorption mechanisms represent probably the most important interaction phenomena exerted by solid surfaces on the environmental fate of organic pollutants [65, 127–130]. Adsorption controls the quantity of free organic components in solution and thus determines their persistence, mobility, and bioavailability. The extent of adsorption depends on the amount and properties of both solid phase-humic substances (SP_{HS}) and organic pollutants. Once adsorbed on an SP_{HS}, an organic pollutant may be easily desorbed, desorbed with difficulty, or not at all. Thus sorption phenomena may vary from complete reversibility to total irreversibility.

The most important properties of an organic pollutant which determine its mode of interaction with SP_{HS}/SP_{OM} are the chemical character of the molecule, shape and configuration, acidity (pK_a) or basicity (pK_b), water solubility, polarity, molecular size, polarizability, and charge distribution.

3.1.1 *Isotherms*

Construction and use of adsorption isotherms from equilibrium sorption data has been employed by numerous researchers to describe adsorption of organic pollutants on a solid matrix [131–137]. An isotherm represents a relation between the amount of solute (i.e., the pollutant of interest) adsorbed per unit weight of solid adsorbent (i.e., soil, sediment) and the solute concentration in solution at equilibrium.

Adsorption can be described by four general types of isotherms (S, L, H, and C) shown in Fig. 12 and as follows:

- With an S-type isotherm, the slope initially increases with adsorptive concentrations, but eventually decreases and becomes zero as vacant adsorbent sites are filled. This type of isotherm indicates that at low concentrations the surface has a low affinity for the adsorbate which increases at high concentrations.

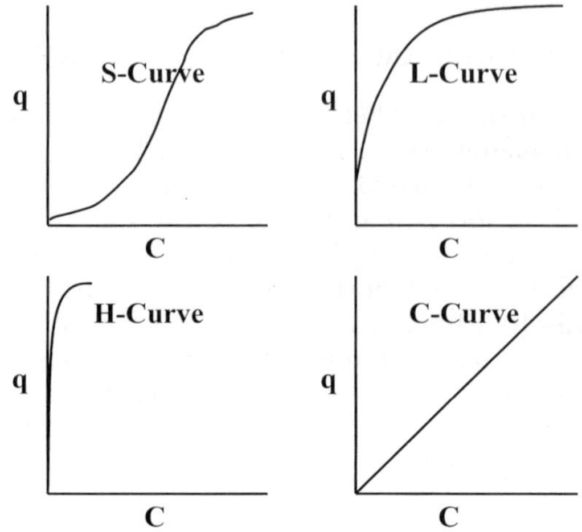

Fig. 12. The four general categories of adsorption isotherms

- The L-type (Langmuir) isotherm is characterized by a decreasing slope as concentration increases since vacant adsorption sites decrease as the adsorbent becomes covered. Such adsorption behavior could be explained by the high affinity of the adsorbent for the adsorbate at low concentrations, which then decreases as concentration increases.
- The H-type (high affinity) isotherm is indicative of strong adsorbent-adsorbate interactions such as inner sphere complexes.
- The C-Type isotherm indicates partitioning mechanisms whereby adsorptive ions or molecules are distributed or partitioned between the interfacial phase and the bulk solution phase without any specific bonding between the adsorbent and the adsorbate.

Lemke et al. [21] reported that adsorption of zearalenone by organophilic montmorillonite clay gave an S-shaped curve with at least two plateaus, suggesting additional mechanisms of adsorption. On the other hand, Grant and Philip [135] and Valverde-Garcia et al. [16] reported that binding of aflatoxins on phyllosilicate clay and pesticides (such as thiram and dimethoate) on soils explained an L-shape isotherm.

One should realize that adsorption isotherms are purely descriptions of macroscopic data and do not definitively prove a reaction mechanism. Mechanisms must be gleaned from molecular investigations (e.g., the use of spectroscopic techniques). Thus the conformity of experimental adsorption data to a particular isotherm does not indicate that this is a unique description of the experimental data, and that only adsorption is in operation.

In general, there is an array of equilibrium-based mathematical models which have been used to describe adsorption on solid surfaces. These include the widely used Freundlich equation, a purely empirical model, and the Langmuir equation as discussed in the following sections. More detailed modeling approaches of sorption mechanisms are discussed in more detail in Chap. 3 of this volume.

3.1.1.1
Freundlich Equation

The Freundlich model, which was first used to describe gas phase adsorption and solute adsorption, is an empirical adsorption model that has been widely used in environmental chemistry [138]. It can be expressed as

$$q = K_\mathrm{d} \cdot C^\mathrm{n} \tag{3}$$

where q is the amount of adsorption (adsorption/unit mass of adsorbent), C is the equilibrium concentration of the material in solution, K_d is an equilibrium constant indicative of sorption strength, and n is the degree of non-linearity (when $n > 1$, there is no limit to the amount sorbed other than its solubility, which is not expected with a true adsorption process).

A linear form of this relation is

$$\log q = \log K_\mathrm{d} + n \cdot \log C \tag{4}$$

It is widely used in the analysis of environmental data. If log q is plotted as a function of $\log C$, a straight line should be obtained with an intercept on the ordinate of $\log K_\mathrm{d}$ and slope n.

Normally, within a reasonable range of adsorbate concentrations, the logarithmic form of Eq. (3) is linear with n being constant. The K_d value may be considered as an index of the degree of adsorption of various adsorbates by different organic surfaces, assuming the determinations are made at the same concentration range. In general, adsorption of various pesticides and polycyclic aromatic hydrocarbons (PAHs) on different soil and bottom sediment surfaces fit the Freundlich equation reasonably well with an exponent $n = 1$ reduction to a linear equation [1, 2, 18, 19, 139].

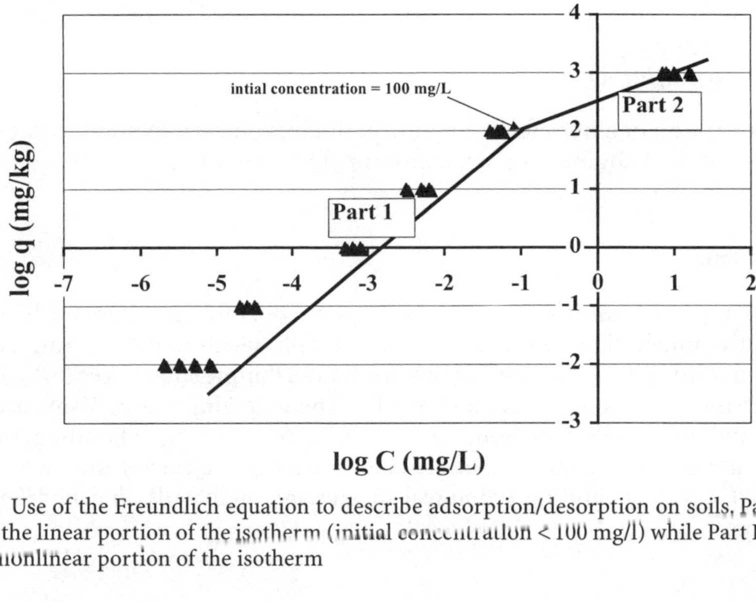

Fig. 13. Use of the Freundlich equation to describe adsorption/desorption on soils, Part I refers to the linear portion of the isotherm (initial concentration < 100 mg/l) while Part II refers to the nonlinear portion of the isotherm

One of the major disadvantages of the Freundlich equation is that it does not predict an adsorption maximum. The single K_d term in the Freundlich equation implies that the energy of adsorption on a homogeneous surface is independent of surface coverage. While researchers have often used the K_d and n parameters to make conclusions concerning mechanisms of adsorption, and have interpreted multiple slopes from the Freundlich isotherms (Fig. 13) as evidence of different binding sites, such interpretations are speculative.

3.1.1.2
The Langmuir Equation

Another widely used sorption model is the Langmuir equation. It was developed by Irving Langmuir [140] to describe the adsorption of gas molecules on a planar surface. It was first applied to soils by Fried and Shapiro [141] and Olsen and Watanabe [142] to describe phosphate sorption on soils. Since that time, it has been heavily employed in many environmental fields to describe sorption on various solid surfaces [19, 65]. The general Langmuir model is

$$q = \frac{QbC}{(1 + bC)} \tag{5}$$

where q and C are as defined previously, Q is a constant related to binding strength, and b is the maximum amount of adsorbate that can be adsorbed (monolayer coverage). In many of the environmental literature x/m, i.e., the weight of the adsorbate/unit weight of adsorbent, is plotted in lieu of q. An advantage of the Langmuir model is that it can approach Henry's law at low concentrations. The constants in the Langmuir equation can be determined by plotting $\left(\frac{C}{q}\right)$ vs C and making use of Eq. (5) rewritten as

$$\frac{C}{q} = \frac{1}{Qb} + \frac{C}{Q} \tag{6}$$

The adsorption of a number of organic pollutants on various solid surfaces was found to fit the Langmuir-model isotherm [139, 143–145].

3.1.2
Mechanisms

Several types of sorption mechanisms often operate simultaneously in the adsorptive interaction between aqueous-solid phase interfaces, i.e., interactions between solid systems containing organic matter/humic substances (SP_{OM}/SP_{HS}) and organic pollutants in aqueous media. The following mechanisms are proposed and discussed: ionic bonding (ion exchange), hydrogen bonding, van der Waals attractions, ligand exchange, charge-transfer (electron donor-acceptor process), covalent binding (chemical or enzyme-mediated), and hydrophobic bonding.

3.1.2.1
Ionic Bonding (Ion Exchange)

Adsorption by this mechanism applies only to a relatively small number of organic pollutants, which are cations in solution or can accept a proton (i.e., protonate) to become cationic. Adsorption via cation exchange or ionic bonding operates through ionized carboxylic and phenolic hydroxyl functional groups of SP_{HS} [17]. For example, Diquat and Paraquat, being divalent cationic pesticides, can react with more than one negatively charged site on an SP_{HS} (e.g., two COO- groups or a COO- plus a phenolate ion).

On the basis of an IR study of some *s*-triazines and HA systems, several authors reported that ionic bonding took place between a protonated secondary amino group of the *s*-triazine and a carboxylate anion on the HA [17, 146, 147]. Successive studies, mainly conducted by IR spectroscopy, confirmed previous results and also provided evidence for the possible involvement of the acidic phenol-OH of HA in the proton exchange of the *s*-triazine molecule [17, 146–150]. Differential thermal analysis (DTA) curves measured by Senesi and Testini [146, 147] showed an increased thermal stability of the HA-*s*-triazine complexes, thus confirming that ionic binding took place between the interacting products.

Amitrole (i.e., a weakly basic pesticide) and the insecticide Dimefox (tetramethyl phosphorodiamidic fluoride) have been shown to be adsorbed by HA through ionic bonding [17, 151–153]. The interaction between the cationic pesticide Chlorodimeform and SP_{HA} was studied and, based on IR data, Maqueda et al. [154] indicated that an ion exchange bonding mechanism occurred.

3.1.2.2
Hydrogen Bonding

Hydrogen bonding is an important polar interaction in aqueous media. This importance often leads to the consideration that hydrogen bonding is a special or unique bond type. It can be considered as an extreme manifestation of a dipole-dipole interaction, which typically arises when hydrogen is attached to very electronegative atoms. Hydrogen bonding also occurs in some other polar liquids such as alcohols.

The presence of oxygen and nitrogen containing functional groups, as well as hydroxyl and amino groups, on SP_{HS} strongly suggests that hydrogen bonding represents an important adsorption mechanism for organic pollutants containing similar complementary groups [2, 19, 25, 65, 155]. The organic pollutant, however, will be in competition with water molecules for such bonding sites.

A large body of evidence for hydrogen bonding was obtained from IR and DTA studies [17, 19, 146, 147, 149, 150, 153, 156]. Hydrogen bonding plays an important role as an adsorption mechanism for substituted ureas, phenyl carbamates, and other nonionic polar organic pollutants (i.e., Alachlor and Cycloate, Metolachlor, Malathion, Bromacil, and dialkylphthalates) which possess functional groups that can form hydrogen bonds with SP_{HS}-sites [17, 25,

147, 148, 156]. Acidic or anionic pesticides (e.g., chlorophenoxyalkanoic acids and esters, Asulam, and Dicamba) were reported to be adsorbed by hydrogen bonding onto SP_{HS} at pH values below their pK_a in nonionized forms through their -COOH and -COOR groups [151–153, 157].

3.1.2.3
Van der Waals Attractions

Ionic species can induce a dipole in a nonpolar molecule over a short range. London forces exist between instantaneous and induced dipoles, and are operative between all bodies when they are close together. For molecular systems they are also commonly called van der Waals attractive forces after the Dutch physicist (J.D. van der Waals) who described these forces as being active in crystals [65]. The London/van der Waals force is also frequently referred to as the dispersion force and is important in the solution phase.

The nature of the London force is that it is proportional to the molecular volume and the number of polarizable electrons of the species experiencing the force. Even nonpolar neutral species undergo momentary imbalances in electron distribution. The forces, which exist between instantaneous dipoles, are responsible for much of the interactive cohesion in solutions of nonpolar liquids. The impact of the London force on sorption mechanisms from solution tends to become pronounced when large molecules are involved; larger molecules have a larger molecular volume and more electrons. It is thought that the essence of the van der Waals force is the attraction of electrons of one molecule for the atomic nuclei of another [19, 158]. The ability of species to engage in van der Waals bonding is related to the number of electrons and to the ability of those electrons to accommodate the close approach of the bonding partner's electrons. This latter ability is called polarizability and may be thought of as the ease of inducing a dipole moment in a species.

As a result of the nature of the intermolecular interaction giving rise to the van der Waals force, it is active only at very close range. The molecules must approach one another closely before the attraction, which results in sorption, can exert itself. It is generally believed that the force of the van der Waals attraction between two molecules is proportional to the square of the polarizability and varies inversely with the sixth power of the distance between the molecules [65]:

$$Q \propto \left(\frac{n^2}{r^6}\right) \tag{7}$$

where Q is the van der Waals attraction force between molecules, n is the polarizability, and r is the distance between the molecules.

The variation of the energy of attraction attributed to the van der Waals force as a function of distance between sorbate and sorbent may be described graphically with a hypothetical plot of potential energy vs. distance.

At distances greater than a few molecular diameters, the energy of attraction is negligible. As the molecules approach, the force of attraction increases (the potential energy decreases) as natural or induced dipoles begin to interact. As the molecules approach even closer, steric factors come into play and the potential increases dramatically. The point of minimum potential energy, then, is the point of maximum attraction and relates to the point of closest approach.

Van der Waals forces, although very weak, operate in all adsorbent-adsorbate interactions, and result from short-range dipole-dipole, dipole-induced dipole, or induced dipole-induced dipole attractions. Although van der Waals interactions are forces acting universally, they assume particular importance in the adsorption of nonionic and non-polar molecules or portions of molecules on similar sites of the adsorbent molecule [17, 159]. These forces are additive, and thus their contribution increases with the size of the molecule and with its capacity to adapt to the adsorbent surface. Van der Waals attractions have often been invoked in case of difficulties in explaining adsorption of an organic pollutant onto SP_{HS}, but the experimental evidence has not always been convincing.

Van der Waals forces were considered to be involved in the physical adsorption of Carbaryl, Parathion, Alachlor, Picloram, and 2,4-D by SP_{HS} [17, 25, 152, 160–162].

These bonding interactions just described are frequently considered to be representative of the major types of forces which exist between species, although there is some disagreement about their nature and magnitude involved. It is probable that combined or hybrid forces come into play in real material interactions. It is also probable that multiple types of attractive and repulsive adsorptive forces are operative [158, 160–162].

3.1.2.4
Ligand Exchange

A ligand is an atom, functional group, or molecule that is attached to the central atom of a coordination compound. The types of interactions between metal ions and complexing agents such as inorganic (anions) and organic (e.g., carboxyl and phenolic groups of SOM) ligands is called a ligand exchange adsorption [65]. Adsorption by this mechanism involves the replacement of water of hydration or other weak ligands partially holding transition metal ions bound to SP_{HS} functional groups by suitable adsorbent molecules [17, 25, 160–163]. This type of mechanism was reported to be involved in the binding of *s*-triazines on incompletely coordinated transition metals of SP_{HA} [160].

3.1.2.5
Electron Donor-Acceptor Interaction (Charge Transfer)

The presence of groups possessing an electron-deficient acceptor (i.e., quinones) and an electron-rich donor (e.g., nitrogen or activated aromatic rings) in SP_{HS} and the existence of organic pollutants possessing the same characteristics provides the possibility of an interaction based on the formation of electron "*donor-acceptor*" or "*charge-transfer*" systems between a suitable organic pollutant and HS structural moieties [39, 164–167].

The feasible formation of charge-transfer complexes between various organic compounds (e.g., pesticides and herbicides), PAHs and polychlorinated biphenyls (PCBs), and SP_{HS} phases were postulated by several authors [17, 25, 147, 150, 153, 154, 156, 160]. Electron spin resonance (ESR) studies [17, 25, 150, 160–162] confirmed the approach of the electron donor-acceptor interaction mechanism, showing that free radical concentrations increased in the interaction products between a number of *s*-triazines and HA of different origin and nature [147, 156]. This effect was explained assuming that electron-deficient quinone-like structures in the HA molecule are able to remove electrons from the electron-rich donating amine or heterocyclic nitrogen atoms of the triazine molecules. Such an electron transfer is assumed to occur by single-electron donor-acceptor processes through the formation of semiquinone free radical intermediates.

Senesi and Testini [147, 156] and Senesi et al. [150, 153] showed by ESR the interaction of HA from different sources with a number of substituted urea herbicides by electron donor-acceptor processes involves organic free radicals which lead to the formation of charge-transfer complexes. The chemical structures and properties of the substituted urea herbicides influence the extent of formation of electron donor-acceptor systems with HA. Substituted ureas are, in fact, expected to act as electron donors from the nitrogen (or oxygen) atoms to electron acceptor sites on quinone or similar units in HA molecules.

The importance of charge-transfer interactions in HA chemistry under the conditions prevailing in aqueous-solid systems has been emphasized and confirmed by UV spectroscopy [168–170]. Specific charge-transfer interactions between the condensed ring structures of PAHs and PCBs with HA molecules have also been suggested and confirmed by the fluorescence quenching studies [39, 158, 168, 171].

3.1.2.6
Covalent and Enzyme-Mediated Binding

Formation of covalent bonds which lead to stable, mostly irreversible incorporation of organic pollutants into SP_{HS}, or more likely of their degradation reaction intermediates or products have been shown to occur. Covalent bond formation is often mediated by chemical, photochemical, or enzymatic catalysts. Enzyme-mediated oxidative coupling reactions, which are universally recognized to be important in the synthesis of HS, may also be responsible for the reactive degradation intermediates into SP_{OM} and play a role in their hydrogen bonding sorption mechanism, thus determining the fate of many organic pollutants in various solid phase systems. The following is a summary of some evidence for such an interaction mechanism [17, 19, 25, 40, 65, 160, 172–178]:

- Phenylamide, phenylurea, and analogous herbicides are known to be biodegraded in soil with the release of free chloroaniline residues. These residues were shown to be prevalently immobilized by chemical binding to the SOM without the intervention of microbial activity. The chemical attachment of chloroanilines to SP_{HS} was proposed to occur by at least two mechanisms involving carbonyl, quinone, and carboxyl groups of SP_{HS} and leading to hydrolyzable (e.g., anilinoquinone) and to non-hydrolyzable (e.g., heterocyclic rings or ether) bound forms.

- The study on the reaction of several ring-substituted anilines with HS in aqueous solutions showed that the free radical intermediates were converted to stable products by self-coupling or cross-coupling with other radical species (i.e., indigenous humic free radicals becoming incorporated into the HS macromolecule).
- Clear evidence that covalent binding of substituted phenols and aromatic amines occurs to SP_{HS} was provided by the presence of various phenoloxidases.
- An extracellular lactase enzyme, isolated from the fungus *Rhizoctonia praticola*, was shown to mediate cross-coupling between phenolic constituents of HS and 2,4-dichlorophenol formed during the decomposition of 2,4-D, thus leading to the incorporation of this xenobiotic into SP_{OM}.
- The fungal enzyme from *R. praticola* was able to catalyze the oxidative coupling of pentachlorophenol (PCP) and syringic acid, a representative of phenol carboxylic acids from lignin occurring in HS structures.
- A lactase from the fungus *Trameles aversicolor* was shown to catalyze the co-polymerization of syringic acid and 2,6-xylenol, a major pollutant in streams and other water resources, which is known to be toxic to fish and other organisms.
- Various chloro- and alkyl-substituted anilines, which represent the aromatic base of a large number of organic pollutants, were shown to react with phenolic humic constituents in the presence of a phenoloxidase isolated from *R. praticola*, while no reaction occurred when only the anilines were incubated with the fungal lactase.
- Chlorocatechols, known intermediates in the decomposition of 2,4-D,2,4,5-T, and other pesticides, were shown to be incorporated by enzymatic polymerization into HA when reacted with purified horseradish peroxidase.

Because practically all aromatic organic pollutants that release phenols or anilines in the course of their degradation could bind HS through enzymatic catalysis, methods employing enzyme-catalyzed polymerization reactions minimizing their presence by partial removal in aquatic and terrestrial environments might be utilized in pollution control. This can have a remarkable effect in environmental engineering practice.

3.2 Partitioning

Reversible physical adsorption of hydrophobic pollutants with dissolved-phase and solid-phase HS (i.e., DP_{HS} and SP_{HS}, respectively) is a well established and fundamental interaction affecting the equilibrium distribution and rate of an organic pollutant between soil/sediment, water, and air [82, 181–184]. There has been – and still is – continuing literature discussion regarding the physical association of hydrophobic organic pollutants with sediment and soil involving a process of adsorption or partitioning [77, 103, 108, 113, 130, 185–188].

It should be noted that Chiou et al. [77, 81, 189, 190] suggested early that the controlling sorptive mechanism of nonionic organic compounds from water consists primarily of solute partition, rather than adsorption into the solid humus. This concept was based principally on their results from sorption studies performed for a number of chlorinated hydrocarbons, benzene derivatives, and PCBs on various solid phases. The results showed, in fact, that sorption isotherms were linear over a wide range of aqueous concentrations relative to solute solubility, and that solid uptake of organic solutes exhibited a small heat effect with a lack of apparent solute competition. Nevertheless, Chiou et al. [191] considered that the solute partition coefficient (K_{OM}) might be affected by the network and polarity of SP_{HS} and by variations of HS properties with different soil conditions.

Thus, we now focus on the partitioning mechanism in terms of both thermodynamic and modeling approaches as follows.

3.2.1
Thermodynamic (Free Energy) Approach

Partitioning is governed mainly by free energy change. The net free energy describes the overall tendency of the system to make a specific change. The concept is in accord with the laws of thermodynamics and assumes that it is the natural tendency of a system to seek spontaneously a condition of minimum energy and maximum disorder [65, 192–194]. The most common form of the equation is

$$\Delta G = \Delta H - (T \cdot \Delta S) \qquad (8)$$

where ΔG is the change in free energy associated with the event, ΔH is the change in enthalpy, T is the absolute temperature, and ΔS is the change in entropy which accompanies the event.

The consideration of net free energy is associated with a specified change and demands clear definitions of the system under consideration, both before and after the change. The value of the free energy relation is that spontaneous reactions must always be associated with a negative change in free energy (i.e., $\Delta G < 0$). If $\Delta G > 0$, the reverse reaction is thermodynamically favored. The free energy of a sorption process can, in principle, be determined from K (the slope of the linear isothermal plot) according to

$$\Delta G = RT \cdot \ln K_d \qquad (9)$$

where R is the gas constant and T is the absolute temperature.

The quantitative application of free energy data requires rigorous definitions of the system. Since the equilibrium constant for the distribution between the bulk and surface phases (i.e., K_d) is not well defined due to the uncertainty in the thickness (i.e., volume) of the adsorption layer, the values of ΔG are only approximate [186].

The tendency of the system to minimize its energy is accounted for by considering the energy (enthalpy, H) contained in the bonds or forces of association between the system components before and after the specified change. If the net energy of bonds is lower in the system after the change, the change is considered to be favorable from the aspect of net enthalpy.

The free energy concept accounts for the tendency of the system to maximize disorder through the entropy term (S). The entropy of the system is directly related to the numbers of system components and the freedom of random motion of the system before and after the specified change.

It is incorrect to assume that adsorption always represents a decrease in system entropy. Adsorption at the solid surface by a solute component may require the removal of another species which is adsorbed to the surface, hence the increased order or disorder of the system accompanying competitive adsorption from solution is not so clear cut as might be the case of adsorption of a gas molecule from a near-vacuum.

The transfer of a hydrophobic solute from an aqueous solution across a phase boundary into an immiscible liquid phase is reported to represent an increase in entropy. Two major sources of entropy increase have been suggested. One is that hydrophobic solutes lead to increased structuring of water. Decreased structuring when the solute leaves the aqueous phase would increase randomness in water and therefore increase entropy. Another cause of increase in entropy is greater conformational freedom of hydrophobic molecules in non-aqueous media than in water. The increase in structural conformation leads to an increase in randomness and an increase in entropy [65, 112]. Entropy changes in complex systems may be difficult to enumerate. In fact, spontaneous events (i.e., those with $\Delta G < 0$) are observed to display variations in both magnitude and sign for enthalpy (H) and entropy (S) changes [186, 195]. It is the combination of these two parameters, along with the consideration of the temperature (T), which describes the net free energy, and hence the opportunity for a spontaneous event.

In any case it is useful to remember that the existence of a favorable free energy gradient ($\Delta G < 0$) does not guarantee that an event will occur within any time frame. Kinetics is not considered in the free energy determination, nor is the existence of activation energy. An event may have a favorable free energy gradient and yet be limited by the kinetics or activation energy requirements.

3.2.2
Modeling Approach

Partitioning interaction can be modeled as an equilibrium reaction, similarly to the partitioning of a solute between two immiscible solvents [19, 65, 78, 80, 81, 158, 196, 197]. In other words, HS both in the solid- and dissolved-phase (i.e., SP_{HS} and DP_{HS}, respectively), are treated as a nonaqueous solvent into which the organic pollutant can partition from water. The distribution of organic pollutant between aqueous solution and organic carbon component of solid phases (i.e., soils, sediments, and suspended matter) may be described, therefore, by the use of partitioning equilibrium constants (K_{OC} or K_{OM}). The partitioning of a pollutant in the organic phase, X_{OC}, can be given by

$$X_{OC} = \left(\frac{P_{OC} \cdot K_{OC}}{1 + P_{OC} \cdot K_{OC}} \right) \tag{10}$$

where P_{OC} is the humic organic carbon-to-water weight ratio [198], and K_{OC} is the partition coefficient.

An important advantage of this approach is the fact that K_{OC} values can be closely correlated with K_{OW} and water solubilities, thus facilitating the estimation of K_{OC} values that have not been experimentally determined [81].

Several studies have shown that sorption of various organic compounds on solid phases could be depicted as an accumulation at hydrophobic sites at the OM/water interface in a way similar to surface active agents. In addition Hansch's constants [19, 199–201], derived from the partition distribution between l-octanol and water, expressed this behavior better than other parameters. Excellent linear correlations between K_{OC} and K_{OW} were found for a variety of nonpolar organic compounds, including various pesticides, phenols, PCBs, PAHs, and halogenated alkenes and benzenes, and various soils and sediments that were investigated for sorption [19, 76, 80, 199–201].

Various methods by which the K_{OW} of PAHs could be calculated are based on their molecular structures, i.e., a quantitative structure-property relationship (QSPR) approach [1, 199, 200]. One of the most famous techniques involves summation of hydrophobic fragmental constants (or f-values) for all groups in a molecule of a specific compound. On the other hand, Aboul-Kassim [1] and Aboul-Kassim et al. [202, 203] introduced a modeling technique based on the molecular connectivity indices of various PAHs, ranging from low- to high-molecular weight compounds. More details are given in Chap. 4 of this volume.

3.2.3
Critical Evaluation

The evidence presented in the literature on the dominance of a partition mechanism in the process of adsorption of a nonionic organic pollutant onto SOM does not mean, for instance, that the physical adsorption model based on weak chemical forces of interaction can be ignored or excluded [82, 99, 107, 109, 114, 115, 183, 192, 204–218]. The following summary is a critical evaluation for reconsidering the universal applicability of the partitioning model to various nonionic compounds onto SP_{OM} [82, 84, 92, 103, 113, 130, 182, 184, 185, 187, 193, 219, 220, 222–226]:

- Many thermodynamic arguments, reported in the literature, were oversimplified and cannot be used to distinguish between adsorption and partitioning. This is mainly due to the fact that enthalpy (ΔH) and entropy (ΔS) values may vary in magnitude and sign. The use of *in-depth* thermodynamic models is more appropriate.
- The observed linearity of adsorption isotherms in various data sets in the literature and the absence of competitive effects are not evidence for partitioning alone, because such behavior can also be consistent with a physical adsorption model.
- Since SOM is not uniform in all solid phases, it cannot be *universally* treated as a well-defined organophilic phase. The appreciable variation of reported K_{OM} values for many nonionic compounds between soils/sediments with change in SOM composition is a strong argument limiting the universality of the partitioning model.

- Deviations of an order of magnitude or more from the calculated fit between K_{OM} and solubility or K_{OW} exist in the literature, and thus the use of these parameters to predict K_{OM} should be treated cautiously.
- The molecular structure and conformation of an organic pollutant is a property which may affect adsorption onto a solid surface and/or partition into its organic lipid phase *differently*, thus hindering the expected correlation between K_{OM} and K_{OW}.
- Correlations between solubility, liquid-liquid partition, and solid sorption have been shown to be insufficient proof of a partition process and do not allow predictions to be applied to *all diverse groups* of organic pollutants and solid phases.
- The complexity of many uptake processes on solid surfaces cannot be simply defined as adsorption or partitioning and based just on isotherm equations and modeling approaches, but rather should be viewed as a summation of the many possible interaction mechanisms, which can be determined by the structural and chemical parameters of the adsorbates and adsorbents.
- Estimations of partitioning models based on K_{OM} correlations with K_{OW} or solubility are acceptable as long as the limitations of these correlations are taken into consideration.

4 Factors Affecting Sorption Interaction Mechanisms

The following sections describe various factors that affect interaction mechanisms between various organic pollutants and solid phase systems.

4.1 Interfacial Tension

Water molecules at the air-water interface experience unbalanced attraction for both water and the air phases [227–229]. This is a manifestation of the polar nature of water in contact with a nonpolar phase (i.e., the air). The water molecules are drawn together, resulting in a phenomenon called "surface tension". The contact area between the water and the nonpolar phase is a region of relatively high interfacial tension and the system will naturally tend to minimize such contact. This polar structure of water will also make the aqueous medium relatively inhospitable to nonpolar, neutral (i.e., uncharged) molecules [230–234].

A nonpolar neutral species in a polar medium such as water experiences interfacial tension. Solvophobic theory is a general statement of hydrophobic theory, which has been developed to explain the tendency of neutral organic species to flee the water phase. It has been reported that the solution of nonelectrolytes in water is attended by a net decrease in entropy [65, 158]. This has been attributed to an increased structuring of water molecules in the vicinity of the solute. The process may be conceptually rationalized by considering that a solute must occupy space in a cohesive medium. The solute must create a "cavity" in the water milieu and then occupy that cavity [19, 65, 158]. The very high cohesive density of water creates considerable interfacial tension in the region of contact with a nonpolar solute and is responsible for the magnitude of the hydrophobic effect. This interfacial tension has also been called the internal pressure and it creates a driving force for the nonelectrolyte to flee the solution as the system tries to minimize the area of contact between the water and the nonpolar solute. The hydrophobic concept has been of great utility in explaining the behavior of organic chemicals in water. Hydrophobic forces can drive nonpolar neutral solutes across an interfacial boundary into an adjacent immiscible nonpolar liquid [235–237]. A substantial part of the driving force for this reaction may be a positive entropy change that was described above.

Hydrophobic bonding is largely the extension of solvophobic behavior to create a partitioning event such as adsorption onto a solid material. The hydrophobic bond is not so much a special type of bond as a way for the system to minimize the area of the polar and nonpolar interface [238, 239]. If the site of sorption is itself hydrophobic, sorption of a nonelectrolyte onto such a site will be attended by a proportionally greater reduction in the overall system interfacial tension and the driving force will be that much greater. Upon sorption, London forces are certainly involved and so bonding per se is occurring, but the solvophobic tendency is providing a considerable gradient for the sorption event. A direct consequence of hydrophobic theory is manifested in Traube's rule [19, 239, 240] which states that the water solubilities of a homologous organic series decrease as the length of the carbon chain increases. As the length of the nonpolar carbon chain increases, so does the nonpolar surface area of the molecule. While a functional group may be relatively polar, the nonpolar surface area creates the interfacial tension in aqueous solution and thus the water solubility will decrease as the chain length increases.

Traube's rule accommodates the balance between hydrophobicity and hydrophilicity. It has been extended somewhat and formalized with the development of quantitative methods to estimate the surface area of molecules based on their structures [19, 237]. The molecular surface area approach suggests that the number of water molecules that can be packed around the solute molecule plays an important role in the theoretical calculation of the thermodynamic properties of the solution. Hence, the molecular surface area of the solute is an important parameter in the theory. In compounds other than simple normal alkanes, the functional groups will tend to be more or less polar and thus relatively compatible with the polar water matrix [227, 240]. Hence, the total surface area of the molecule can be subdivided into "functional group surface area" and "hydrocarbonaceous surface area". These quantities may be determined for simple compounds as an additive function of constituent groups with subtractions made for the areas where intramolecular contact is made and thus no external surface is presented.

4.2 Cosolvency

Polar neutral organics can be very miscible in water due to their compatibility with the polar water molecules. For example, dipole-dipole interactions such as those interactions between short-chain alcohols and water give rise to essentially complete miscibility. In contrast to the increase in surface tension accompanying solutions of salts, miscible organics in solution tend to decrease the surface tension of the aqueous medium [143, 241, 158, 242–245].

Miscible organic solutes modify the solvent properties of the solution to decrease the interfacial tension and give rise to an enhanced solubility of organic chemicals in a phenomenon often called "*cosolvency*". According to theory, a miscible organic chemical such as a short chain alcohol will have the effect of modifying the structure of the water in which it is dissolved. On the macroscopic scale, this will manifest itself as a decrease in the surface tension of the solution [238, 246].

It has been generally considered that there is an exponential increase in the solubility of a solute as the fraction of the cosolvent increases linearly. The only requirement for the log linear relationship seems to be that the solute must be less polar than the mixed solvent [19]. The validity of the log-linear nature of the cosolvent process has been well validated in the literature [110, 188, 247–249, 262, 263]. The effect of a cosolvent on solubility can be calculated according to

$$\ln S_m = f_c \cdot \ln S_C + (1 - f_c) \cdot \ln S_W \tag{11}$$

where S_m is the molar solubility of a nonpolar solute, f_c is the nominal cosolvent volume fraction, S_C is the molar solubility in pure cosolvent, and S_W is the molar solubility in pure water.

This model assumes the absence of specific solute-solvent interactions and is based upon a linear relationship between the free energy of solution and solute surface area. It assumes that the overall solubility is simply the sum of the solubilities in the individual solvent components. This model treats the cosolvent and the water as distinct entities and neglects any interaction between them [19, 145, 226, 253, 261].

More recent work with cosolvency in dilute systems seems to indicate that the magnitude of the solubility enhancement is linear up to some 10–20% cosolvent fraction [55, 172, 184, 250–262]. At very low concentrations of cosolvent, the assumption of non-interaction between the cosolvent and water cannot hold. In dilute solutions the individual cosolvent molecules will be fully hydrated and, as a result, will disrupt the water network structure. If the total volume disrupted is regarded as the extended hydration shell, and if S_C^* is the average solubility within this shell, then the overall solubility S_m in the water-cosolvent mixture will be approximated by

$$S_m = f_c \cdot V_H \cdot S_C^* + (1 - f_c \cdot V_H)\,; \quad f_c \cdot V_H < 1 \tag{12}$$

where V_H is the ratio of the hydration shell volume to the volume of the cosolvent.

In dilute solutions, the solute will, on average, contact only one hydrated cosolvent molecule at a time, and the degree of solubilization should be a linear rather than a logarithmic function of cosolvent content. Thus, it is expected that the log-linear relationship between S_m and f_c that applies at high cosolvent concentrations will become linear at low cosolvent levels due to a change in the mechanism of solubilization. If S^+ is defined as solubility enhancement

$(S_m - S_W)$, then the relative solubility enhancement at low cosolvent concentration will be given by

$$\left(\frac{S^+}{S_W}\right) = f_c \cdot V_H \cdot \left(\frac{S_C^*}{S_{W-1}}\right) \tag{13}$$

In a natural water system where cosolvent was present at sufficient levels to influence pollutant solubility, the cosolvent itself would probably constitute a pollutant. In a contaminated groundwater, however, such a cosolvent concentration may be realistic to create, thereby significantly enhancing the degradation of the target pollutant. If the cosolvent were itself biodegradable, the resulting effect would be the removal of the pollutant without adverse long-term effects on the resource.

The log-linear solubility enhancement by cosolutes may be important in characterizing concentrated leachate plumes or chemical spills, but will be of little importance in characterizations of the dilute aqueous systems that predominate in nature [19, 55, 143, 145, 158, 184, 226, 241–247, 249–263].

4.3 Micelles

Organic pollutants can be quite variable in structure (i.e., polar vs nonpolar moieties) and properties (i.e., hydrophobic vs hydrophilic tendencies). If a molecule containing a hydrophilic region also has a significant hydrophobic region, such as a long carbon chain, the water solubility will be diminished. This diminished solubility can be manifested in several ways. The chemical can sorb onto a surface and thereby diminish the interfacial tension with the water or it can form a separate, immiscible bulk phase. A third possibility exists, whereby the nonpolar moiety can undergo association with the nonpolar regions of other molecules to form smaller subunits within the water matrix. Such an organizational arrangement minimizes the contact between the hydrophobic moieties and the water while allowing the hydrophilic (polar/ionic) moieties to contact the water. Such an aggregate arrangement is frequently referred to as a "*micelle*" [65, 206–208, 225, 246, 268].

Typically, organic pollutants having both polar and nonpolar moieties can form micelles. Such pollutants are often referred to as "*amphiphiles*" or described as being "*amphipathic*", which refers to the dual affinity of such species for both polar and nonpolar media [269, 270]. Surfactants and soaps are amphiphiles. They are often characterized by having a polar or ionic end (or "*head*") and a nonpolar "hydrocarbonaceous" end (or "*tail*"). These molecules in solution are subjected to the forces of interfacial tension or polar affinity [271, 272]. The polar or ionic end will be readily solvated by water, which will repel the nonpolar end. Micelles arise when these molecules undergo intermolecular association of the hydrophobic moieties and form a droplet of material that has a hydrophobic interior and a hydrophilic exterior [273, 274]. The interfacial tension between the water and the hydrophobic end is thus minimized and the droplet may be solvated by its outer shell of polar or charged ends in association with the polar water phase. This arrangement has been called a "*pseudophase*" denoting the existence of a hydrophobic interior of the droplets suspended by the interaction of the hydrophilic moieties with the polar water [19, 158, 273, 274].

It has been observed that the association of homogeneous surfactant monomers to form micelles is characterized by some critical concentration of dissolved monomers before true micelle formation (micellization) occurs. A commonly described parameter associated with micelle formation is the *Critical Micelle Concentration* (i.e., CMC). The onset of micellization, which occurs at CMC, is typically accompanied by some well-defined or observable change at that point [273–276]. It is commonly reported that the addition of surfactant monomer to water can cause the surface tension of the solution to decline steadily until CMC is attained, after which continued addition of monomer produces no more drop in the measured surface tension. The transition is typically a sharp one. Experimentally, it is often found that micelles are undetectable in dilute solutions of the monomers, and become detectable over a narrow range of concentrations as the total concentration of solute is increased, above which nearly all additional solute material forms micelles [277–281]. The concentration at which the micelles first become detectable depends on the sensitivity of the experimental apparatus used to observe the change in surface tension. The concentration range over which the fraction of additional solute which forms micelles changes from nearly zero to nearly unity depends on such factors as the number of monomers in the micelle, the chain length of the monomer, the properties of counter ions, and other details affecting the monomer-micelle equilibrium. An approximate rule is that the higher the CMC value, the broader is the concentration range over which this transition takes place, in absolute value as well as in relative value in comparison with the CMC [280]. Since different experimental methods may reflect this transition to different extents, some systematic variations in operationally defined CMCs are expected [277–279, 281].

The impact of salt concentration on the formation of micelles has been reported and is in apparent accord with the interfacial tension model discussed in Sect. 4.1, where the CMC is lowered by the addition of simple electrolytes [19, 65, 280, 282]. The existence of a micellar phase in solution is important not only insofar as it describes the behavior of amphipathic organic chemicals in solution, but the existence of a nonpolar pseudophase can enhance the solubility of other hydrophobic chemicals in solution as they partition into the hydrophobic interior of the micelle. A general expression for the solubility enhancement of a solute by surfactants has been given by Kile and Chiou [253] in terms of the concentrations of monomers and micelles and the corresponding solute partition coefficients, giving

$$\left(\frac{S_W^*}{S_W}\right) = (1 + X_{mn} \cdot K_{mn} + X_{mc} \cdot K_{mc}) \tag{14}$$

where S_W^* is the apparent solute solubility, X is the total stoichiometric surfactant concentration, S_W is the intrinsic solubility in pure water, X_{mn} is the concentration of the surfactant as monomers, X_{mc} is the concentration of the surfactant in micellar form, K_{mn} is the partition constant between monomers and water, and K_{mc} is the partition constant between micelles and water.

The separation of the concentration terms (X_{mn} and X_{mc}) accounts for differences in the partition efficiency of the solute with monomers and micelles.

While CMC is assumed to be an observable and definite value in the case of surfactant monomers, there are frequent reports in the literature of the formation of *aggregates* or *micelle-like* associations in solutions of organic solutes so dilute as to preclude apparently the formation of micelles [208, 267–269, 272, 275, 278]. Work with different types of commercial surfactants has indicated that molecularly non-homogeneous surfactants do not display the sharp inflection in surface tension associated with CMC in molecularly homogeneous monomers, but rather the onset of aggregation is broad and indistinct [253, 267, 268]. The lack of well-defined CMCs for non-homogeneous surfactants is speculated to result from the successive micellization of the heterogeneous monomers at different stoichiometric concentrations of the surfactant, which results in a breadth of the monomeric-micelle transition zone.

It has been reported that molecularly non-homogeneous surfactants are able to enhance the solubility of very hydrophobic chemicals, e.g., DDT, at surfactant concentrations well below the CMC. This is attributed to the successive micellization of the heterogeneous monomer species [271, 273, 274, 276, 278]. Examination of the solubility enhancement with different types of commercial surfactants reveals that molecularly homogeneous surfactants show relatively insignificant (but linear) solubility enhancement below CMC. Molecularly non-homogeneous surfactants, on the other hand, show a much greater solubility enhancement at concentrations below the CMC.

The presence of water-soluble macromolecules in solution at submicellar concentrations has been reported to enhance the water solubility of hydrophobic organic chemicals in several instances [19, 106, 113]. The presence of macromolecules in solution can enhance the apparent solubility of solutes by sorptive interactions in the solution phase. The processes by which macromolecules enhance the solubility of pollutants are probably variable as a function of the particular physical and chemical properties of the system. A macromolecule possessing a substantial nonpolar region can sorb a hydrophobic molecule, thereby minimizing the interfacial tension between the solute and the water.

4.4 pH

The pH is a fundamental property which can have an impact on the solubility of organic and inorganic solutes. The pH can have an effect on reaction equilibrium if the reaction, or a related reaction, consumes or produces H^+ or OH^-. The dependence of sorption mechanisms on pH have been reported by several authors [283–286].

Hydroxide and carbonate typically form insoluble precipitates with polyvalent cations in natural waters. The activity of both of these species increases with pH. The presence of surface functional groups that are capable of exchanging a proton creates pH dependent-charge, whereby the ionic character of the surface increases with pH [158, 284, 285].

The molecular configuration of polyelectrolytes may be influenced by pH as the molecules coil and uncoil as the pH decreases or increases. In such a situation, charged sites such as acidic hydroxyl groups or amines can lose or acquire charge as a result of changes in solution pH. In a large, flexible, polyfunctional molecule, intramolecular self-association is thought to occur in the absence of electrostatic repulsion. The tendency to form such intramolecular bonds will vary as charged sites are created or satisfied by pH changes. In such a situation, decreases in pH will satisfy the charge on the surface of the molecule, thereby lowering the hydrophilicity of the surface and also decreasing the coulombic repulsion of the molecular chain for itself and permitting intramolecular bonding [65, 283, 285, 286].

4.5 Colloid Stability

It is commonly reported that dissolved humic substances (i.e., DHS) tend to coat mineral particles and thereby affect the surface chemistry of those materials. DHS coat the surfaces of solid particles even when they are present at very low concentrations. They furthermore impart a negative charge to the surfaces which they coat. The organic coating is expected to have a great significance on subsequent adsorption of various pollutants [88, 91–93, 287, 288].

The ability of OM to coat mineral particles enhances the cation exchange capacity of the solid minerals. A thin organic coating may tend to increase the disperse nature of small mineral particles by imparting a net negative charge and creating a repulsion between the particles [25]. The pH-dependent nature of the charge on such coated particles can create a pH-dependent dispersion tendency; as the pH drops and the surface functional groups of the OM become electrically neutral, the particles coated with this OM would become less mutually repulsive and intraparticle collisions might result in the formation of van der Waals bonds. Such an event might result in flocculation of the particles. The intraparticle repulsion of such coated minerals will also diminish as the ionic strength of the solution increases [60, 61]. This is in accord with the model of double layer compression at higher ionic strength, which allows closer approach between particles. Organic coated particles coagulate much slower than particles with OM coatings removed. They will also resist sorption onto the stationary phase if the stationary phase is also coated with negatively charged OM [24, 59]. Such behavior is important in environmental chemodynamics and management practices.

The dispersal and sedimentation of clay minerals and other mineral colloids may be influenced appreciably by sorbed humic matter. While humic matter may keep clay particles in a dispersed state under conditions otherwise conducive to flocculation, humic matter could conceivably "*cement*" clay particles together, as a polyelectrolyte bridge. This forms stable aggregates, as in soil, thereby promoting deposition of clay in a hydraulic regime in which individual colloids would be kept in suspension [17, 48]. The sorptive nature of the colloidal surface creates the possibility for aggregation between colloids. Aggregation or other processes as coagulation or flocculation can cause settling of the colloids as the particle densities increase. The tendency of colloids to coagulate is a function of conditions such as pH, ionic strength, solution composition, and repulsion between colloids. In natural and polluted waters, these conditions causing flocculation can change and the aggregated particles can disperse back into the solution. The stability of colloids in natural waters cannot be explained by electrostatic theory alone, but must be considered as a combination of electrical, kinetic, and purely chemical forces [58, 65].

DOM in solution will influence the sorption chemistry and aggregation behavior of mineral particles in aqueous systems. The presence and nature of suspended and dissolved minerals, in turn, will influence the behavior of the DOM. The aqueous phase will thus contain suspended and dissolved mineral/organic colloids at greater or lesser concentration as a function of ambient chemistry and physical conditions [45, 65–67]. Organic material can form colloids when aggregates or micelles form. Mineral/organic colloids can exist when mixed aggregates coprecipitate or agglomerate in solution, or when conditions bring mixed material into apparent solution [58–61].

4.6 Functional Groups of Pollutants

Functional groups (Fig. 8) are chemically reactive atoms or groups of atoms bound to the structure of an organic compound that are either acidic or basic. It was reported that adsorption of dissolved organics in the liquid phase by solid phase particles is dependent mainly on the nature of the functional groups present in the organic molecule, and is also a function of shape, size, configuration, polarity, polarizability, as well as water solubility [17, 19, 42, 160]. The chemical properties of the functional group types influence the surface acidity of the solid-soil and/or solid-sediment particles, which is vital because surface acidity is critical in determining the adsorption of ionizable organic molecules by solid-system particles. It is the major factor in the adsorption by solid-system particles of amines, triazines, amides, and substituted ureas where protonation takes place on the carbonyl group [17, 160].

In the case of an organic pollutant or mixtures of organic pollutants leached from SWMs, the nature of the functional groups of such pollutants will influence their characteristics and their abilities to interact with solid phase constituents. For instance, depending on how these functional groups are situated, they will determine the mechanisms of interaction, persistence, and ultimate fate of such compounds in both surface and subsurface environments. The following is a summary of some important functional groups and their effects on the chemical interactions between pollutant-solid phase constituents.

4.6.1 *The Hydroxyl Group*

The hydroxyl (OH) group is the dominant reactive functional group on the surface of many solid phase particles, amorphous silicate minerals, metal oxides, oxyhydroxides, and hydroxides [17, 25, 160]. In the case of various organic pollutants in the aqueous environment and different leachates of COMs, the (OH) group is represented by two broad compound classes, as discussed below.

4.6.1.1 *Alcohols*

Alcohols are hydroxylated alkyl-compounds (R-OH) which are neutral in reaction due to their unionizable (OH) group (e.g., methanol, ethanol, isopropanol, and *n*-butanol). The hydroxyl of alcohols can displace water molecules in the primary hydration shell of cations adsorbed onto soil-solid and sediment-solid clay particles. The water molecule displacement depends mainly on the polarizing power of the cation. The other adsorption mechanisms of alcohol hydroxyl groups are through hydrogen bonding and cation-dipole interactions [19, 65].

4.6.1.2 *Phenols*

The phenolic functional group consists of a hydroxyl attached directly to a carbon atom of an aromatic ring. The OH group can also be the consequence of further oxidation or combination with other pollutants such as pesticides, aldehydes, and alcohols (i.e., 2,4-D, cyclic alcohols, cresols, naphthols, quinones, nitrophenols, and pentachlorophenol compounds) forming new more toxic compounds [17, 42, 160].

4.6.2 *The Carbonyl Group*

Compounds possessing a carbonyl group, called carbonyl compounds, include both aldehydes (-CHO), and ketones (=C=O). Since the carbonyl group consists of a carbon atom bonded to an oxygen atom by two pairs of electrons, most carbonyl compounds have dipole moments because the electrons in the double bond are shared unsymmetrically. Whilst they can accept protons, the stability of complexes between carbonyl groups and protons is considered to be weak. One of the remaining two valences of carbon bears a hydrogen atom in the aldehydes, and the carbonyl group is attached to two carbon atoms in the ketones. The adsorption mechanism for ketones is hydrogen bonding between an (OH^-) group of the adsorbent and the carbonyl group of the ketone, or via a water bridge [17, 42]. The nature of the exchangeable cation and hydration status of clay particles affects adsorption of ketones.

4.6.3 *The Carboxyl Group*

The carboxyl group (-COOH) of organic acids interacts either directly with the interlayer cation or by forming a hydrogen bond with the water molecules coordinated to the exchangeable cation on the soil-solid and sediment-solid clay particles [17, 160]. Adsorption of organic acids depends on the cation polarizing power. Some organic acids can be physically adsorbed onto the clay particles of any solid phase and water bridging is an important mechanism in the adsorption process. In addition to coordination and hydrogen bonding, organic acids can be adsorbed through the formation of salts with the exchangeable cations. It has been noted that anions can be adsorbed by weak bonding of the carboxyl group to the positive sites of the oxide surfaces of goethite [17, 42].

4.6.4
The Amino and Sulfoxide Groups

The amino group (-NH_2) is found in primary amines, which are organic bases that form stable salts with strong acids. They may be aliphatic, aromatic, or mixed. Depending on the nature of the functional groups, they are classified as (1) primary: methylamine (primary aliphatic), aniline (primary aromatic), (2) secondary: dimethylamine (secondary aliphatic), diphenylamine (secondary aromatic), and (3) tertiary: trimethylamine (tertiary aliphatic), triphenylamine (tertiary aromatic).

The sulfoxide group (-SO_2) is one of the more polar organic functional groups that form complexes through either sulfur or oxygen atoms. Sulfoxide groups readily form complexes with transition metals or with an exchangeable cation, and/or form a water bridge hydrogen bond between the sulfoxide oxygen and an exchangeable cation [17, 38, 42, 289].

4.7
Cation Exchange Capacity

Many organic pollutants are positively charged by protonation (adding hydrogen) and are adsorbed on solid phase clay particles depending on the particle cation exchange capacity. Generally, exchange reactions produce no net change in energy and are independent of temperature. Solid clay minerals differ in their cation exchange capacity and in their adsorption capacity for organic cations [25]. Organic cation adsorption on clays, which is a cation-dependent process, is related to the molecular weight of the organic cations. Higher molecular weight organic cations are adsorbed more strongly by clays than inorganic cations because of their sizes and high weights [17]. However, acid-base type reactions predominate in sorption interaction mechanisms involving short-range forces between solid particle surfaces and organic ions [23, 25].

4.8
Carrying Capacity of Subsurface Soil

Migration and transport of organic pollutant(s) and/or mixtures of complex organic pollutants (such as SWMs leachates of COMs) through the subsurface soil environment can lead to eventual groundwater contamination [1, 17, 23, 65]. Pollutants in organic leachate will interact with the soil constituents through different processes, resulting in a pollutant accumulation in and/or by the subsurface soil. For the subsurface soil to act as a proper buffer for "COM leachate" transport to the groundwater, it is important to understand how such pollutants may be held by the soil particles, which in turn tells us how strongly and permanently they may be fixed to the soil. Intuitively, one would expect the soil system to have a limited capacity for retaining pollutants within its system. Thus, if continued COM pollutants are released, there will be the danger that the usefulness of the subsurface soil system as a buffer diminishes or even ceases. Thus, the determination of the carrying capacity of a subsurface soil site is important in order to understand the short and long term chemical and physical compatibility between leachate and soil clay liners, and to predict the chemodynamics of the various target leachate pollutants in the subsurface soil.

It is the chemical buffering system which contributes significantly to the carrying capacity of a soil [17]. In general, any soil cannot completely adsorb all the pollutants from the liquid solution. There is an equilibrium between solvent and solution phases. The amount left in solution gradually increases as the buffer capacity of the soil is approached.

5 Role of Dissolved Humic Substances in Pollutant-Solid Phase Interactions

Dissolved humic substances (DHS) are the main constituents of the dissolved organic carbon (DOC) pool in surface waters (freshwaters and marine waters), groundwaters, and soil porewaters and commonly impart a yellowish-brown color to the water system. Despite the different origins responsible for the main structural characteristics of DHS, they all constitute refractory products of chemical and biological degradation and condensation reactions from plant or animal residues and play a crucial role in many biogeochemical processes.

DHS can significantly affect the environmental behavior of hydrophobic organic compounds and lower the possibility of direct contact of such organic compounds with various solid phases. The rate of chemical degradation, photolysis, volatilization, transfer to sediments/soils, and biological uptake may be different for the fraction of organic pollutant that is bound to DHS. If this is the case, the distribution and total mass of a pollutant in an ecosystem depends, in part, on the extent of humic matter-hydrophobic binding.

Organic pollutants may be bound to DHS through abiotic or biological processes whereby the formation of bound residues usually results in detoxification of these pollutants. Therefore, enhancing the binding of xenobiotic chemicals to humic matter can serve as a means to reduce toxicity as well as migration of the toxic compounds [51, 52, 64, 67, 166, 290]. Binding of a pollutant to DHS, clays, or other materials would be expected to decrease its toxic effects. Binding can reduce the amount of a compound available to the biota and, as the quantity of an available xenobiotic is reduced, toxicity also declines. Below are some possible ways of binding with DHS:

- Complex formation between DHS and aquatic organic pollutants can occur by an oxidative coupling reaction leading to oligomeric and polymeric products. Bollag and Bollag [164] reported the effect of phenoloxidazes (i.e., peroxidases, tyrosinases, and lactases) on the binding of substituted phenols and aromatic amines to humic matter. Copolymerization largely depends on the chemical reactivity of the substrates involved. Certain phenolic humic matter constituents (such as synergic acid, guaiacol, ferulic acid, etc.) are highly reactive in the presence of phenoloxidases. When one of these compounds was incubated together with a phenoloxidase with less or even non-reactive phenols, anilines, or other compounds, a synergistic reaction took place, resulting in an increased formation of bound residues of these compounds.

- A wide variety of xenobiotics can become cross-linked to naturally occurring humic matter by the action of phenyloxidases. These xenobiotics include phenols (e.g., mono-, di-, and tri-substituted chlorophenols, 2,6-xylenol), and anilines (e.g., 4-chloroaniline, 3,4-dichloroaniline, and 2,6-diethylaniline) [17, 160].
- The addition of a highly reactive humic matter component (e.g., syringic acid) to a phenoloxidase-containing system can initiate the effective polymerization and/or binding of a molecule which by itself is only poorly transformed [160]. The enzyme-induced oxidation of naturally-occurring phenols yields free radical quinonoid structures, a common pathway in the phenoloxidase-catalyzed polymerization and binding of both naturally-occurring and man-made compounds. Another pathway is the decarboxylation of a highly reactive compound such as syringic acid and the formation of a covalent bond at that site to generate phenolic oligomers [17, 25].

DHS have been shown to effect the interaction mechanisms between various organic pollutants and solid phases. The following paragraphs will discuss how DHS can significantly decrease the chance of interactions in the pollutant-solid phase interface. This includes solubilization, hydrolysis, catalysis, and photosensitization effects.

5.1 Solubilization

The DHS fraction plays an important role in the solubility enhancement of various organic pollutants [169, 226, 253]. If the mechanism of solubility enhancement of hydrophobic organics is one of surface sorption, it might be expected that partition coefficients of aquatic HS may be less than those of the OM on particles, since macromolecules in solution must be relatively hydrophilic [19, 109]. This view is supported by the reports describing heteroatom compositional differences between FA [46, 47] and HA [49, 50, 52–55, 291, 292] recovered from natural waters. The smaller, more water-soluble FA have higher oxygen-to-carbon ratios compared to the larger humic acids [44, 64]. Thus, smaller, more water-soluble macromolecules can be more polar sorbents (i.e., exhibit relatively lower K_{OC}s) than related larger macromolecules and particulate matter.

Chiou et al. [189] were the first to consider the mechanism for water solubility enhancement of nonionic organic solutes by DOM of soil and bottom sediments. Such enhancement effects were effectively explained in terms of *a par-*

tition-like interaction of solutes with dissolved high molecular weight humic materials on the basis of the properties of the solutes and humic materials. The observed solubility enhancement of the solute by DOM can be expressed by

$$S_W^* = S_W(1 + X \cdot K_{DOM}) \tag{15}$$

where S_W^* is the apparent water solubility in the solution, S_W is the apparent water solubility in pure water, X is the concentration of DOM, and K_{DOM} is the partition coefficient between DOM and water.

The difference in values of K_{DOM} for a solute with different types of fractionated humic materials has been explained in terms of the polarity, molecular size, and molecular configuration of the humic materials. It gives a reasonable estimate of the relative enhancing effects among humic extracts.

The compositions and structures of DHS in aquatic systems can be significantly different because of environmental factors such as sources, water pH, biological processes, and the presence of other chemical species that affect the concentration of humic materials [46, 47, 49, 50, 53, 54, 291]. In more acidic streams or rivers, there appears to be a tendency for the humic material to contain a larger percentage of oxygen compared to samples from neutral or basic waters. A decrease in oxygen content of the humic materials from acidic to neutral water can also be accompanied by an increase in carbon content. The solubility enhancement effects of individual humic samples appear to be closely correlated with the polarity of the materials, suggesting that differences in molecular sizes of humic materials are not as much a critical factor as their polarity in affecting the partition interaction with organic solutes [52, 55, 189, 292].

Solubility enhancement cannot be explained by the cosolvency theory because the magnitude of the solubility enhancement is greater than that which would be predicted from cosolvent effects alone [19, 110, 129, 247, 249]. This was early investigated by Chiou et al. [105] who used phenylacetic acid, synthetic organic polymers [poly(acrylic acid)], and dissolved HA and FA (i.e., D_{HA}, D_{FA}, respectively) to assess the solubility enhancement effects on different organic compounds. They found significant solubility enhancements of relatively water-insoluble solutes by DHS of soil and aquatic origins. The concentrations of the DHS varied from 0 mg/l to 94 mg/l. They observed that the apparent solute solubilities increased linearly with DHS concentration and showed no competitive effect between solutes. With a given DHS sample, the solute partition coefficient increased with a decrease of the solute's water solubility or with an increase of the solute's octanol-water partition coefficient (K_{OW}). The K_{OW} values of solutes with soil-derived HA were approximately four times greater than with soil FA, and five to seven times greater than with D_{HA} and D_{FA}. The effectiveness of DHS in enhancing solute solubility appeared to be largely controlled by the molecular size and polarity of the material.

On the other hand, the organic acid and polymer (molecular weight varied from 2000 to 90,000) created no observable solubility enhancement. The investigation of phenyl acetic acid as a cosolute, with concentration > 600 mg/l, shows a slight enhancement for the most hydrophobic DDT. The magnitude of the DDT "*solubility enhancement/unit mass*" [19, 249] for phenylacetic acid was much smaller than with the D_{HA} or D_{FA}. They found that the solubility enhancement exhibited by the DHS may be described in terms of a *partition-like* interaction of the solutes with a "*microscopic nonpolar organic environment*" associated with the high-molecular-weight humic species. The relative inability of high-molecular-weight poly(acrylic acids) to enhance solute solubility was attributed to their high polarities and extended chain structures that do not permit the formation of a sizable intramolecular nonpolar environment.

This observed "*partition-like*" interaction between hydrophobic organic solutes and DHS has led to the proposition that humic micelles may exist in solution [19]. Such humic micelle materials are pictured as existing as *membrane-like aggregates* which are made up of partially decomposed plant-derived compounds, held together in the aggregates by weak bonding mechanisms (e.g., *pi* bonding, hydrogen bonding, and hydrophobic interactions). The humic membrane-like structure consists of polar hydrophilic exterior surfaces with hydrophobic interiors. Polar compounds will interact with the exterior polar groups of the humic structures, while hydrophobic compounds will partition into the hydrophobic interiors of the structures.

This *humic-micelle model* is consistent with much of the reported information in the literature, especially with regard to the *membrane-like* behavior. Such a structure might explain the following findings reported by several authors [19, 44, 254, 293–297]:

- Enhancement of cholesterol solubility by high molecular weight DHS in river water, where solvent extraction of the radiolabeled cholesterol was ineffective as a means of recovery unless the OM content was altered by UV radiation.
- Enhancement of the solvent recovery of various sorbed hydrophobic organics by a digestion technique, which degraded the DHS. Simple adsorption onto the nonpolar region of humic molecules by van der Waals forces and hydrophobic interfacial tension would probably hinder solvent recovery of adsorbed hydrophobic organics.

It has also been shown [254] that a commercial petroleum sulfonate surfactant which consists of a diverse admixture of monomers does not exhibit behavior typically associated with micelle formation (i.e., a sharp inflection of solvent properties as the concentration of surfactant reaches CMC). These surfactants exhibit gradual change in solvent behavior with added surfactant. This gradual solubility enhancement indicates that micelle formation is a gradual process instead of a single event (i.e., CMC does not exist as a unique point, rather it is a continuous function of molecular properties). This type of surfactant can represent humic material in water, and may indicate that DHS form molecular aggregates in solution, which comprise an important *third phase* in the aqueous environment. This phase can affect an increase in the apparent solubility of very hydrophobic chemicals.

Application of pollutant chemodynamic models, which neglect the DHS phase, may result in inaccurate estimations of apparent solubility and transport parameters. The impact of a DHS solubility enhancement is most pronounced for the least water-soluble solutes. The affinity of a solute for a DHS is a function of the same properties, which drive a complex organic mixture(s) to sorb onto the stationary solid phase, namely bonding interactions and hydrophobicity.

Hence, DHS will manifest the greatest solubility enhancement for those pollutants which are the least soluble in water or the most attracted to the solid phase. Organic pollutants, which are soluble in water, are less likely to be sorbed onto the solid or colloidal phase in the absence of specific bonding interactions.

5.2 Hydrolysis

The catalytic effects of DHS on interaction mechanisms at the aqueous-solid phase interface have been extensively studied although their mechanisms are not completely understood. Evidence for such catalysis effects by DHS on the rate of hydrolysis of other organic pollutants has been reported by several authors [19, 217, 298–302] and will be summarized in the following paragraphs:

- Aquatic HS inhibited the base-catalyzed hydrolysis of the *n*-octyl ester of 2,4-D (i.e., 2,4-DOE). The hydrolysis rate of 2,4-DOE at pH 9–10 decreased by a factor equal to the fraction of the ester associated with the DHS. These observations are consistent with an unreactive humic-bound 2,4-DOE in equilibrium with reactive aqueous-phase 2,4-DOE. Thus, association between D_{HA} and 2,4-DOE inhibited the base-catalyzed hydrolysis reaction.
- A general mechanism for the effects of DHS on the hydrolysis kinetics of hydrophobic organic pollutants was proposed and derived by a combination of equations that separately describe partitioning equilibria, general acid-base catalysis and micellar catalysis. The resulting model predicted that the overall effect of DHS in modifying hydrolysis reaction rates of an organic pollutant can be attributed to partitioning equilibria and micellar catalysis, with only a minor effect due to general acid-base catalysis. General acid-base catalysis by DHS is predictable by the model to be relatively unimportant, and to remain insignificant even in the presence of rather high concentrations of DHS (e.g., >200 mg/l) when other processes such as partitioning or association equilibria may become significant for hydrophobic pollutants.
- HS may alter the reactivities of bound substrates in a way similar to that of anionic surfactants (inhibiting base-catalyzed and accelerating acid-catalyzed reactions). These effects were attributed to electrostatic stabilization of the transition state for the acid catalysis in which the substrate becomes more positively charged, and to destabilization of the transition state for base-catalyzed hydrolysis in which the substrate becomes more negatively charged.
- In natural waters, the base-catalyzed hydrolysis rate of a weakly HS-associated pollutant (e.g., Parathion) was not significantly affected by HS, while for more strongly associated pollutants (e.g., DDT) the effect of HS was clearly potentially significant in this reaction.
- In conditions where much higher concentrations of DHS are possible (i.e., in sewage sludge or in sediment/soil interstitial water), the impact of DHS on organic pollutant hydrolysis kinetics was predicted to be larger.
- The inhibition effect exerted by D_{HA} on hydrolytic enzymes in soils was regarded as an additional mechanism by which D_{HA} may indirectly influence hydrolysis reactions.

5.3 Photosensitization

DHS are known to be among the most important natural components of solid phase surfaces and aquatic environments which absorb sunlight, and constitute about half of the organic and nearly all of the colored matter in all of the different natural environments [303–305]. Soil humic substances generally differ from freshwater humic substances in their elemental and functional group composition; they typically have higher molecular weights, lower carboxylic and higher phenolic contents, and the ratio of extractable humic to fulvic acid is frequently higher [303]. Freshwater humic substances contain stronger acidic functions due to the presence of keto acid and aromatic carboxyl-group structures [306–308], and marine humic substances lack lignin constituents and have an aliphatic and peptide origin derived from non-lignin-containing biota [309]. Despite these structural differences, all humic substances contain a variety of active chromophores at wavelengths found in the solar spectrum; most prominent are aromatic systems as well as conjugated carbonyl derivatives.

Natural DHS present in ecosystems undergo a complex array of primary and secondary photoprocesses when exposed to sunlight. Numerous studies have been performed to assess the environmental relevance of photochemical degradation pathways for xenobiotics and natural organic matter (e.g., [310, 311]). DHS are known to affect the photodegradation of pollutants, either acting as a photosensitizer or as absorbing (and light attenuating) chromophore [38, 312–314] depending on their chemical structure [315, 316]. In general, a significant portion of the solar radiation adsorbed by aquatic DHS results in the formation of electronically excited molecules (HS*) which are capable of greatly accelerating or even determining a number of light induced transformations that organic pollutants can undergo in natural aqueous environments [53, 146, 147, 156, 317, 318]. In surface waters DHS can act as sensitizers or precursors for the production of singlet oxygen ($^1O_2$), humic-derived peroxy radicals ($ROO^{\bullet}$), hydrogen peroxide, and solvated electrons (e^-_{aq}), and as the scavenger which controls their lifetimes [53, 319–321].

A proposed mechanism taking place when an excited sensitizer (HS*) interacts with an energy acceptor can be described by the key energy-transfer steps depicted in the following scheme:

$$HS^* \xrightarrow{h\nu} {}^1HS^* \rightarrow {}^3HS^* \tag{16}$$

$${}^3HS^* \rightarrow HS + \text{heat} \tag{17}$$

$${}^3HS^* + TOC \rightarrow TOC^* + HS \tag{18}$$

$$TOC^* \rightarrow \text{photoproducts} \tag{19}$$

$${}^3HS^* + O_2 \rightarrow HS + {}^1O_2 \tag{20}$$

$${}^1O_2 + TOC \rightarrow (TOC - O_2) \tag{21}$$

Light absorption promotes the photosensitizer molecules (HS) to their first excited states $^1HS^*$, which are short-lived and transform in part to excited triplet states $^3HS^*$ (Eq. 16), which are in turn considerably longer-lived. Such triplets may in part decay to the ground state (Eq. 17), or transfer energy to the substrate (TOC) forming its triplet state (TOC*, Eq. 18), which then produces its photoproducts (Eq. 19), or transfer energy to ground state triplet oxygen producing excited singlet molecular oxygen $^1O_2$ (Eq. 20), which is a powerful oxidant and may in turn decay back to its groundstate or react rapidly with an acceptor (TOC) thus producing its photooxidation products (Eq. 21).

Extensive research has been carried out to investigate the photosensitization effect of natural DHS on the fate and transport of various toxic pollutants. The following is a summary of the findings reported by various authors [53, 146, 147, 156, 315–332]:

- DHS with higher specific light absorption exhibit somewhat lower quantum efficiencies. However, no significant relationship with a DHS-molecular weight fraction was found.
- The occurrence of singlet oxygen is important for the elimination of dissociated forms of some pollutants such as phenolic, cyclic diene, and sulfur compounds.
- Hydroxyl radicals, which are important for the elimination of refractive micropollutants, are consumed predominantly by fast scavenging reactions of the DHS present in natural waters.
- Different types of aquatic DHS were shown to exhibit comparable rate constants for trapping hydroxyl radicals. Peroxy radical photooxidants (i. e., a mixture of different HS-derived species) were shown to be important for the elimination of alkylphenols, which are typical compounds classified as antioxidants.
- Direct photo-ionization or photo-induced electron transfer from marine and terrestrial DHS to a variety of polyaromatic electron acceptors have been documented by time-resolved and steady-state laser flash kinetic spectroscopy studies under conditions which facilitate extrapolation to the environment.
- Because the formation rate of solvated electrons from DHS photolysis is extremely low, it was considered to be relevant only for the elimination of highly refractive compounds.
- DHS can photosensitize reactions involving hydrogen atom transfer, which likely involve triplet state intermediates. For example, hydrogen transfer from the nitrogen of aniline to the sensitizer occurs at much higher rates than observed in the aniline photoreaction in distilled water.
- Quantitative kinetic data showed that photosensitized oxygenations of various pollutants (e. g., 2,5-dimethylfuran and the insecticide Disulfoton) in air-saturated natural water samples containing aquatic HS and in distilled water containing soil-extracted or commercial HA/FA were at least one order of magnitude faster than those in distilled water.
- DHS could act as a photosensitizer of some previously bound substances, which can undergo detoxification stimulated by light and oxygen:

$$\text{HS} + \text{TOC} \xrightarrow{h\nu} (\text{HS-TOC}) \xrightarrow{h\nu} \text{photoproducts} \tag{22}$$

This occurs by a mechanism called *static photosensitization*, analogous to that followed by biologically acting photosensitizers like riboflavin.
- ESR studies have suggested that visible and UV light irradiation of DHS may enhance the indigenous free radical contents of DHS, which are highly susceptible to free-radical mediated interaction of HS with organic pollutants. ESR monitored free radical increase in many donor-acceptor systems, such as HA-*s*-triazine and HA-urea herbicides. This has also been suggested to be important to the unpairing of electrons originating from the formation of charge-transfer complexes under the effect of light.
- D_{HA} were significantly less active than aquatic D_{HS} in the photosensitization reaction of various pollutants.

6 Conclusions

The chemical and structural nature of humic substances coating solid phase surfaces makes them active in the environmental fate and transport of organic pollutants. The presence of bound enzymes and free radicals in the material allows it to form covalent bonds with a variety of molecules. The existence of nonpolar regions of the humic matter introduces the possibility of intramolecular sorptive partitioning of nonpolar organic compounds into the humic matrix. The extent and polarizability of the humic matter surface enable it to bind to materials by van der Waals forces. The existence of electrostatic charges on the surface of the substance makes it reactive with respect to water, ions, and mineral surfaces. The nature of the surface chemistry grants humic matter a surface charge which is pH-dependent. Hence, the tendency to flocculate or disperse is more or less a function of pH and ionic character of the solution.

The humic/organic matter coatings of different solid phases (i.e., SP_{HS}/SP_{OM}), such as soils, sediments, suspended solids, colloids, and biocolloids/biosolids, interact with organic pollutants in aqueous systems in various ways. Adsorption is an important interaction mode. The reversibility and/or irreversibility of the adsorption processes is of major importance. The question whether the bound residues of pollutants are to be considered definitely inactivated has been the focus of extensive research. This question was posed as follows. Have the adsorbed pollutants become common components incorporated into the humic polymer coating of solid phases (i.e., being absorbed), or are they only momentarily inactivated in reversibly bound forms thus representing a possible source of pollution by a time-delayed release of toxic units?

Several factors can dramatically affect the rate at which organic pollutants can interact with various solid phase surfaces. These include interfacial tension of aqueous systems, cosolvency effect, micelle formation, pH of the surrounding medium, colloidal concentration and stability, variations in organic pollutant functional groups, cation exchange capacity at the aqueous-solid phase interface, and the carrying capacity of the subsurface soil solids. Such factors can increase and/or decrease the rates of sorption/desorption interaction mechanisms. Thus, detailed study of these processes and factors, with what controls them, is extremely important for environmental engineering and management purposes.

Dissolved humic substances (DHS) are the main constituents of the dissolved organic carbon (DOC) pool in surface, ground, and soil pore waters. DHS can significantly affect the environmental behavior of hydrophobic organic compounds and lower the possibility of the direct contact of such organic compounds with various solid phases. The rate of chemical degradation, photolysis, solubilization, transfer to sediments/soils, and biological uptake may be different for the fraction of organic pollutant that is bound to DHS. If this is the case, the distribution and total mass of a pollutant in an ecosystem depends, in part, on the extent of humic matter-hydrophobic binding.

The sources of SP_{HS} and their diverse macromolecular sizes and chemical properties are extremely important in determining the mode and extent of interaction with organic pollutants. The importance of improving our understanding of the interacting HS/OM and the nature of their interaction with organic pollutants is recognized but needs further research by advanced techniques, including: (1) nuclear magnetic resonance (NMR), for the identification of structural features of TOC-bound residues; (2) electron spin resonance (ESR), for the investigation of chemical, enzymatic, and photochemical HS-organic pollutant interactions involving free radical species as starting reagents and/or intermediates, or products of reactions; and (3) fluorescence spectrometry, for the study of a number of chemical and functional modifications which occur upon interaction between SP_{HS} and organic pollutants in situ, without separation of the interacted organic pollutant molecules from the free. These methods provide important yet scarcely exploited means for the investigation of organic pollutant and SP_{HS} interactions.

References

1. Aboul-Kassim TAT (1998) Ph. D. Dissertation, Department of Civil, Construction and Environmental Engineering, College of Engineering, Oregon State University, Corvallis, Oregon
2. Al-Bashir B, Hawari J, Leduc R, Samson R (1994) Water Res 28:1817
3. Alvarez J, Carton A, Isla T, Herguedas A (1995) 3rd International Conference on Water Pollution: Modeling, Measuring and Prediction. Computational Mechanics, Billerica, MA
4. Aminabhavi TM, Naik HG (1998) J Hazardous Mater 60:175
5. Aminabhavi TM, Naik HG (1999) J Hazardous Mater 64:251
6. Aochi YO, Farmer WJ, Sawhney BL (1992) Environ Sci Technol 26:329
7. Arocha MA, Jackman AP, McCoy BJ (1996) Environ Sci Technol 30:1500
8. Kibbey TCG, Hayes KF (1997) Environ Sci Technol 31:1171
9. Luthy RG, Aiken GR, Brusseau ML, Cunningham SD, Gschwend PM, Pignatello JJ, Reinhard M, Traina SJ, Weber WJ Jr, Westall JW (1997) Environ Sci Technol 31:3341
10. Moreau C, Mouvet C (1997) J Environ Quality 26:416
11. Noegrohati S, Hammers WE (1992) Toxicol Environ Chem 34:187
12. O'Connor BI, Voss RH (1992) Environ Sci Technol 26:556
13. René FY, Harris JM (1996) Anal Chem 68:1651
14. Scheidegger AM, Sparks DL (1996) Soil Sci 161:813
15. Seybold CA, Mersie W (1996) J Environ Qual 25:1179
16. Valverde-Garcia A, Gonzalez-Pradas E, Villafranca-Sanchez M, Rey-Bueno F, del Garcia-Rodriguez A (1988) Soil Sci Soc Am J 52:1571
17. Sparks DL (1995) Environmental soil chemistry. Academic Press, 267 pp
18. Zytner RG (1994) J Hazard Mater 38:113

19. Schwarzenbach RP, Gschwend PM, Imboden M (1993) Environmental organic chemistry. Wiley, p 681
20. Zhang ZZ, Sparks DL, Scrivner NC (1993) Environ Sci Technol 27:1625
21. Lemke SL, Grant PG, Phillips TD (1998) J Agric Food Chem 46:3789
22. Fitch A, Du J (1996) Environ Sci Technol 30:12
23. Schultze DG (1989) In: Dixon JB Weed SB (eds) Minerals in soil environment. SSSA Book Ser. No.1. Soil Sci Soc Am, Madison, Wisconsin, pp 1–34
24. Gschwend PM, Reynolds MD (1987) J Contaminant Hydrology 1:309
25. Stevenson FJ (1982) Humus chemistry: genesis, composition, reactions. Wiley, NY, 443 pp
26. Benedetti MF, Van Riemsdijk WH, Koopal LK (1996) Environ Sci Technol 30:1805
27. Engebretson RR, Amos T, von Wandruszka R (1996) Environ Sci Technol 30:990
28. Govi M, Sarti A, Di Martino E, Ciavatta C, Rossi N (1995) Soil Sci 161:265
29. Jones KD, Tiller CL (1999) Environ Sci Technol 33:580
30. Kim J-E, Fernandes E, Bollag J-M (1997) Environ Sci Technol 31:2392
31. Govi M, Sarti A, Di Martino E (1996) Soil Science 161:265
32. Mobed JJ, Hemmingsen SL, Autry JL, McGown LB (1996) Environ Sci Technol 30:3061
33. Chang S, Berner RA (1998) Environ Sci Technol 32:2883
34. Hunt JM (1997) Petroleum geochemistry and geology, 2nd edn. WH Freeman, 743 pp
35. Tissot BP, Welte DH (1984) Petroleum formation and occurrence. Springer, Berlin Heidelberg New York, 699 pp
36. Simoneit BRT (1978) In: Riley JP, Chester R (eds) Chemical oceanography, vol 7. Academic Press, p 233
37. Ghosh K, Schnitzer M (1980) Soil Science 129:266
38. Schnitzer M (1985) In: Aiken GR, McKnight DM, Wershaw RL (eds) Humic substances in soil, sediments and water. Wiley
39. Melcer ME, Zalewski MS, Hassett JP (1989) In: Suffet IH, MacCarthy P (eds) Aquatic humic substances: Influence on fate and treatment of pollutants. Advances in Chemistry Series, ACS, Washington DC, vol 219, p 173
40. Parris GE (1980) Environ Sci Technol 14:1099
41. Benoit P, Barriuso E, Houot S (1996) Eur J Soil Sci 47:567
42. Hayes MHB, Swift RS (1978) In: Greenland DJ, Hayes MHB (eds) Chemistry of soil constituents. Wiley, p 179
43. Thurman EM, Wershaw RL, Malcolm RL, Pinkney DJ (1982) Org Geochem 4:27
44. Thurman EM, Malcolm RL (1981) Environ Sci Technol 15:463
45. Thurman EM (1985) In: Aiken GR, McKnight DM, Wershaw RL MacCarthy P (eds) Humic substances in soil, sediment and water. Wiley, New York, p 87
46. Leenheer JA, Brown JK, MacCarthy P, Cabaniss SE (1998) Environ Sci Technol 32:2410
47. Nanny MA, Bortiatynski JM, Hatcher PG (1997) Environ Sci Technol 31:530
48. Schulten HR, Schnitzer M (1993) Naturwissenschaften 80:29
49. Laor Y, Rebhun M (1997) Environ Sci Technol 31:3558
50. Nielsen T, Siigur K, Helweg C, Jrgensen O, Hansen PE, Kirso U (1997) Environ Sci Technol 31:1102
51. Rebhun M, De Smedt F, Rwetabula J (1996) Water Res 30:2027
52. Rebhun M, Meir S, Laor Y (1998) Environ Sci Technol 32:981
53. Schmitt-Kopplin P, Hertkorn N, Schulten H-R, Kettrup A (1998) Environ Sci Technol 32:2531
54. Sein LT Jr, Varnum JM, Jansen SA (1999) Environ Sci Technol 33:546
55. Takimoto K, Ito K, Mukai T, Okada M (1998) Environ Sci Technol 32:3907
56. Zelazny LW, Carlisle VW (1974) In: Aandall AR, Buol SW, Hill OE, Barely HH (eds) Histosols – their characteristics, classification and use. SSSA Spec. Publ. 6, Am Soc Agron, Madison, WI, p 63
57. Caron GI, Suffet IH (1989) In: Suffet IH, MacCarthy P (eds) Aquatic humic substances: influence on fate and treatment of pollutants. Advances in Chemistry Series, ACS, Washington DC, vol 219, p 117
58. Al-Kanani T, MacKenzie AF (1991) Canadian J Soil Science 71:327

59. Kinniburgh DG, Milne CJ, Benedetti MF, Pinheiro JP, Filius J, Koopal LK, Van Riemsdijk WH (1996) Environ Sci Technol 30:1687
60. Mortland MM (1985) In: Ward CH, Giger W, McCarty PL (eds) Groundwater quality. Wiley, p 370
61. Onken BM, Traina SJ (1997) J Environ Qual 26:132
62. Hassett JP, Milicic E (1985) Environ Sci Technol 19:638
63. McCarthy JF, Jimenez BD, Barbee T (1985) Aquatic Toxicol 7:15
64. Traina SJ, McAvoy DC, Versteeg DJ (1996) Environ Sci Technol 30:1300
65. Stumm W, Morgan JJ (1981) Aquatic chemistry, 2nd edn. Wiley
66. Hering JG, Morel FMM (1990) Environ Sci Technol 24:242
67. Manahan SE (1989) In: Suffet IH, MacCarthy P (eds) Aquatic humic substances: influence on fate and treatment of pollutants. Advances in Chemistry Series, ACS, Washington DC, vol 219, p 83
68. Rochette EA, Harsh JB, Hill HH Jr (1996) Environ Sci Technol 30:1220
69. Schlebaum W, Badora A, Schraa G, van Riemsdijk WH (1998) Environ Sci Technol 32:2273
70. Weber WJ Jr, Huang W, Yu Hong (1998) J Contaminant Hydrology 31:149
71. Xing B, Pignatello JJ (1998) Environ Sci Technol 32:614
72. Banerjee S, Piwoni MD, Ebeid K (1985) Chemosphere 14:1057
73. Ganaye VA, Keiding K, Vogel TM, Viriot M-L, Block J-C (1997) Environ Sci Technol 31:2701
74. Graber ER, Borisover MD (1998) Environ Sci Technol 32:258
75. Huang E, Weber WJ Jr (1997) Environ Sci Technol 31:2562
76. Karickhoff SW, Brown DS, Scott TA (1979) Water Res 13:241
77. Chiou CT, Porter PE, Schmedding DW (1983) Environ Sci Technol 17:227
78. Karickoff SW (1981) Chemosphere 10:833–846
79. Means JC, Woods SG, Hassett JJ, Banwartw L (1980) Environ Sci Technol 14:1524
80. Schwarzenbach RP, Westall J (1981) Environ Sci Technol 15:1360
81. Chiou CT, Peter LJ, Freed VH (1979) Science 206:831
82. Chiou CT, Schmedding DW, Manes M (1982) Environ Sci Technol 16:4
83. Mackay D, Bobra AM, Shiu WY, Yalkowsky SH (1980) Chemosphere 9:701
84. Banerjee S, Yalkowsky SH, Valvani SC (1980) Environ Sci Technol 14:1227
85. Kleineidam S, Rügner H, Ligouis B, Grathwohl P (1999) Environ Sci Technol 33:1637
86. Nam K, Chung N, Alexander M (1998) Environ Sci Technol 32:3785
87. Piatt JJ, Brusseau ML (1998) Environ Sci Technol 32:1604
88. Buffle J, Wilkinson KJ, Stoll S, Filella M, Zhang J (1998) Environ Sci Technol 32:2887
89. Burgess RM (1996) Environ Sci Technol 30:1923
90. Burgess RM (1996) Environ Sci Technol 30:2556
91. Celis R, Cornejo J, Hermosin MC (1997) Soil Sci Soc Am J 61:436
92. Celis R, Hermosín MC, Cox L, Cornejo J (1999) Environ Sci Technol 33:1200
93. Celis R, Koskinen WC, Hermosen MC (1999) J Agri Food Chem 47:776
94. Grout H, Wiesner MR, Bottero J-Y (1999) Environ Sci Technol 33:831
95. Johnson PR, Sun N, Elimelech M (1996) Environ Sci Technol 30:3284
96. Means JC, R Wijayaratne (1982) Science 215:968
97. Seaman JC, Bertsch PM, Strom RN (1997) Environ Sci Technol 31:2782
98. Stordal MC, Santschi PH, Gill GA (1996) Environ Sci Technol 30:3335
99. Villholth KG (1999) Environ Sci Technol 33:691
100. Wan J, Tokunaga TK (1997) Environ Sci Technol 31:2413
101. Wijayaratne RD, JC Means (1984) Environ Sci Technol 18:121
102. Williams PS, Xu Y, Reschiglian P, Giddings JC (1997) Anal Chem 69:349
103. Lee CM, Meyers SL, Wright CL Jr, Coates JT, Haskell PA, Falta RW Jr (1998) Environ Sci Technol 32:3574
104. Sigleo AC, Hoering TC, GR Helz GR (1982) Geochim Cosmochim Acta 46:1619
105. Chiou CT, Malcolm RL, Brinton TI, Kile DE (1986) Environ Sci Technol 20:502
106. Pignatello JJ, Xing B (1996) Environ Sci Technol 30:1

107. O'Connor DJ, Connolly JP (1980) Water Res 14:1517
108. Voice TC, Rice CP, Weber WJ (1983) Environ Sci Technol 17:513
109. Gschwend PM, Wu S (1985) Environ Sci Technol 19:90
110. Walters RW, Ostazeski SA, Guiseppi-Elie A (1989) Environ Sci Technol 23:480
111. Hassett JP, Anderson MA (1979) Environ Sci Technol 13:1526
112. Hassett JP, Anderson MA (1982) Water Res 16:681
113. Whitehouse B (1985) Estuarine Coastal Shelf Sci 20:393
114. Axelman J, Broman D, Näf C (1997) Environ Sci Technol 31:665
115. Bowman BT, Sans WW (1983) J Environ Sci Health B18:667
116. Eisele M, Schorer M (1997) Vom Wasser. Weinheim Vom Wasser 89:139
117. Falandysz J (1996) Environ Sci Technol 30:3362
118. Herman DC, Lenhard RJ, Miller RM (1997) Environ Sci Technol 31:1290
119. Kefford B, Skjelleberg SK, Marshall KC (1982) Arch Microbiol 133:257
120. Marchesi JR, Russell NJ, White GF, House WA (1991) Appl Environ Microbiology 57:2507
121. Mueller HR, Oberbremer A, Meier R, Wagner F (1990) In: Arendt F, Hinsenveld M, Van Den Brink WJ (eds) Contaminated soil '90. Karlsruhe, Federal Republic of Germany, vol II, 3rd International KFK/TNO Conference on Contaminated Soil, p 491
122. Skoglund RS, Swackhamer DL (1999) Environ Sci Technol 33:1516
123. Skoglund RS, Stange K, Swackhamer DL (1996) Environ Sci Technol 30:2113
124. Xue H-B, Stumm W, Sigg L (1988) Water Res 22:917
125. Parkin TB (1987) Soil Sci Soc Am J 51:1194
126. Douglas GS, Quinn JG (1989) In: Suffet IH, MacCarthy P (eds) Aquatic humic substances: influence on fate and treatment of pollutants. Advances in Chemistry, p 387
127. Schlebaum W, Schraa G, van Riemsdijk WH (1999) Environ Sci Technol 33:1413
128. Sheng G, Xu S, Boyd SA (1996) Environ Sci Technol 30:1553
129. Woodburn KB, Lee LS, Rao PSC, Delfino JJ (1989) Environ Sci Technol 23:407
130. Xia G, Ball WB (1999) Environ Sci Technol 33:262
131. Chiou CT, Kile DE (1998) Environ Sci Technol 32:338
132. di Toro DM, Horzempa LM (1982) Environ Sci Technol 16:594
133. Fesch C, Simon W, Haderlein SB, Reichert P, Schwarzenbach RP (1998) J Contaminant Hydrology 31:373
134. Gaston LA, Selim HM (1994) Water Res Research 30:3013
135. Grant PG, Phillips TD (1998) J Agric Food Chem 46:599
136. Huang W, Young TM, Schlautman MA, Yu H, Weber WJ Jr (1997) Environ Sci Technol 31:1703
137. Thibaud-Erkey C, Guo Y, Erkey C, Akgerman A (1996) Environ Sci Technol 30:2127
138. Freundlich H (1926) Colloid and capillary chemistry. Methuen, London
139. Sannino F, Violante A, Gianfreda L (1997) Pest Sci 51:429
140. Langmuir I (1918) J Am Chem Soc 40:1361
141. Fried M, Shapiro G (1956) Soil Sci Soc Am Proc 20:471
142. Olsen SR, Watanabe FS (1957) Soil Sci Soc Am Proc 21:144
143. Brown MJ, Burris DR (1996) Ground Water 34:734
144. Taylor RW, Bleam WF, Tu SI (1996) Commun Soil Sci Plant Anal 27:2713
145. Werkheiser WO, Anderson SJ (1996) J Environ Qual 25:X809
146. Senesi N, Testini C (1982) Geoderma 28:129
147. Senesi N, Testini C (1983) Ecol Bull Stockolm 35:477
148. Senesi N, Testini C (1980) Soil Sci 10:314
149. Kalouskova N (1986) J Environ Sci Health B21:251
150. Senesi N, Testini C, Miano TM (1987) Org Geochem 11:25
151. Senesi N, Miano TM, Testini C (1986) In: Pawlowski L, Alaerts G, Lacy WJ (eds) Chemistry for protection of the environment 1985. Studies in Environmental Science 29, Elsevier, Amsterdam, p 183
152. Senesi N, Padovano G, Loffredo E, Testini C (1986) Proc 2nd Int Conf Environmental Contamination, Amsterdam, p 169

153. Senesi N, Miano TM, Testini C (1987) In: Giovannozzi-Sermanni G, Nannipieri P (eds) Current perspectives in environmental biogeochemistry. CNR-IPRA, Rome, p 295
154. Maqueda C, Perez Rodriguez RJL, Martin F, Hermosin MC (1983) Soil Sci 136:75
155. Xiao KP, Bühlmann P, Umezawa U (1999) Anal Chem 71:1183
156. Senesi N, Testini C (1983) Pestic Sci 14:79
157. Senesi N, Testini C, Metta D (1984) Proc Int Conf Environmental Contamination, London, CEP Cons Ltd, Edinburgh, p 96
158. Buffle J, Stumm W (1994) In: Buffle J, Devitre RR (eds) Chemical and biological regulation of aquatic systems. CRC Press, Boca Raton, Florida, p 1
159. Burchill S, Hayes MHB, Greenland DJ (1981) In: Greenland DJ, Hayes MHB (eds) The chemistry of soil processes. Wiley, New York, chap 6, p 18
160. Cheng HH (ed) (1990) Pesticides in the soil environment: processes, impacts and modeling. SSSA Book Ser, No 2. Soil Sci Soc Am, Madison, Wisconsin
161. Huang PM, Schnitzer M (eds) (1986) Interaction of soil minerals with natural organics and microbes. SSSA Spec Publication No 17. Soil Sci Soc Am, Madison, Wisconsin, 312 pp
162. Schnitzer M (1986) In: Interaction of soil minerals with natural organics and microbes. SSSA Spec Publication No 17. Soil Sci Soc Am, Madison, Wisconsin
163. Senesi N, Steelink C (1989) In: Hayes MHB, MacCarthy P, Malcolm RL, Swift RS (eds) Humic substances: In search of structure. Wiley, NewYork, p 46
164. Bollag J-M, Bollag WB (1990) Int J Environ Anal Chem 39:147
165. Odencrantz JE, Bae W, Rittmann BE, Valocchi AJ (1990) J Contaminant Hydrology 6:37
166. Park J-W, Dec J, Kim J-E, Bollag J-M (1999) Environ Sci Technol 33:2028
167. Scott DT, McKnight DM, Blunt-Harris EL, Kolesar SE, Lovley DR (1998) Environ Sci Technol 32:2984
168. Gauthier TD, Booth KA, Grant CL, Seitz WR (1987) Am Chem Soc, Div Environ Chem 27:246–248
169. Gauthier TD, Seitz WR, Grant CL (1987) Environ Sci Technol 21:243
170. Lindqvist I (1983) Swed J Agric Res 13:201
171. Lindqvist I (1982) Swed J Agric Res 12:105
172. Banerjee S, Yalkowsy SH (1988) Anal Chem 60:2153
173. Berry DF, Boyd SA (1984) Soil Sci Soc Am J 48:565
174. Bollag JM (1983) In: Christman RF, Gjessing ET (eds) Aquatic and terrestrial humic materials. Ann Arbor Sci Publ, Ann Arbor, Michigan, p 33
175. Bollag JM (1987) Am Chem Soc-Div Environ Chem 27:289
176. Bollag JM, Liu S-Y, Minard RD (1980) Soil Sci Soc Am J 44:52
177. Bollag JM, Minard RD, Liu S-Y (1983) Environ Sci Technol 17:72
178. Liu S-Y, Bollag JM (1985) Water Air Soil Pollut 25:97
179. Stott DE, Martin JP, Focht DD, Haider K (1983) Soil Sci Soc Am J 47:66
180. Thorn KA, Pettigrew PJ, Goldenberg WS (1996) Environ Sci Technol 30:2764
181. Chiou CT, McGroddy SE, Kile DE (1998) Environ Sci Technol 32:264
182. Escher BI, Schwarzenbach RP (1996) Environ Sci Technol 30:260
183. Hunchak-Kariouk K, Schweitzer L, Suffet IH (1997) Environ Sci Technol 31:639
184. Ko S-O, Schlautman MA (1998) Environ Sci Technol 32:2776
185. Chiou CT, Porter PE, Shoup TD (1984) Environ Sci Technol 18:295
186. Mingelgrin U, Gerstl Z (1983) J Environ Quality 12:1
187. Pedersen JA, Gabelich CJ, Lin C-H, Suffet IH (1999) Environ Sci Technol 33:1388
188. Woodburn KB, Rao PSC, Fukui M, Nkedi-Kizza P (1986) In: Macalady DL (ed) Transport and transformations of organic contaminants. J Contaminant Hydrol 1:277
189. Chiou CT, Kile DE, Brinton TI, Malcolm RL, Leenheer JA, MacCarthy P (1987) Environ Sci Technol 21:1231
190. Chiou CT, Shoup TD, Porter PE (1985) Org Geochem 8:9
191. Chiou CT, Porter PE, Schmedding DW (1983) Environ Sci Technol 17:227
192. Jang M, Kamens RM, Leach KB, Strommen MR (1997) Environ Sci Technol 31:2805
193. Luehrs DC, Hickey JP, Nilsen PE, Godbole KA, Rogers TN (1996) Environ Sci Technol 30:143

194. Pospeev VE, Logvinenko LM, Ovchinnikova ZG, Mikhailiants RS (1985) Agrokhimiia 22:97
195. Opperhuizen A, Serne P, Van der Steen JMD (1988) Environ Sci Technol 22:286
196. Karickoff SW (1980) In: Baker RA (ed) Contaminants and sediments. Ann Arbor Sci Publ, Ann Arbor, Michigan, p 193
197. Yalkowski SH, Valvani SC (1979) J Chem Eng Data 24:127
198. Perdue EM (1987) Am Chem Soc-Div Environ Chem 27:448
199. Mackay D, Peterson S (1990) In: Karcher W, Devillers J (eds) Practical applications of quantitative structure–activity relationships (QSAR) in environmental chemistry and toxicology, Kluwer Academic Publishers, Dordrecht, Holland, p 433
200. Mackay D, Stiver WH (1991) In: Grover R, Lessna AJ (eds) Environmental chemistry of herbicides. CRC Press, Boca Raton, Florida, vol II, p 281
201. Mackay D (1991) Multimedia environmental models. The fugacity approach. Lewis Publishers, Chelsea, Michigan, p 553
202. Aboul-Kassim TAT, Williamson KJ, Simoneit BRT (1999) V – Modeling the joint toxic effect of multicomponent PAH mixtures. Environ Sci Technol (submitted)
203. Aboul-Kassim TAT, Williamson KJ, Simoneit BRT (1999) Part VI – $\sum$PAH model for fresh water algal toxicity estimation. Environ Sci Technol (submitted)
204. Chiou CT (1981) In: Hazard assessment of chemicals current developments. Academic Press, New York, p 117
205. Doucette WJ, Andren AW (1987) Environ Sci Technol 21:521
206. Guha S, Jaffé PR (1996) Environ Sci Technol 30:1382
207. Guha S, Jaffé PR (1996) Environ Sci Technol 30:605
208. Guha S, Jaffé PR, Peters CA (1998) Environ Sci Technol 32(15):2317–2324
209. Hawker DW (1989) Chemosphere 19:1586
210. Hawker DW, Connell DW (1988) Environ Sci Technol 22:382
211. Isnard P, Lambert S (1989) Chemosphere 18:1837
212. Mailhot H, Peters RH (1988) Environ Sci Technol 22:1479
213. Means JC (1995) Mar Chem 51:3
214. Miller CT, Weber W (1984) Ground Water 22:584
215. Miller MM, Ghodbane S, Wasik SP, Tewari YB, Martire DE (1984) J Chem Eng Data 29:184
216. Miller MM, Wasik SP, Huang GL, Shiu WY, Mackay D (1985) Environ Sci Technol 19:522
217. Perdue EM (1983) In: Christman RF, Gjessing ET (eds) Aquatic and terrestrial humic materials. Ann Arbor Sci Publ, Ann Arbor-Michigan, p 441
218. Tomlinson E, Hafkenscheid TL (1986) In: Dunn WJ III, Block JH, Pearlman RS (eds) Partition coefficient, determination and estimation. Pergamon Press, New York, p 101
219. Berthod A, Carda-Broch S, Alvarez-Coque MCG (1999) Anal Chem 71:879
220. Chiou CT (1985) Environ Sci Technol 19:57
221. Chiou CT (1989) In: MacCarthy P, Malcolm RL, Clapp E, Bloom P (eds) Humic substances in soil and crop sciences. American Society of Agronomy, Madison, Wisconsin, p 214
222. De Bruijn J, Hermens J (1990) Quant Struct Act Relat 9:11
223. De Bruijn J, Busser G, Seinen W, Hermens J (1989) Environ Toxicol Chem 8:499
224. Doucette WJ, Andren AW (1988) Chemosphere 17:345
225. Woodrow BN, Dorsey JG (1997) Environ Sci Technol 31:2812
226. Zimmerman JB, Kibbey TCG, Cowell MA, Hayes KF (1999) Environ Sci Technol 33:169
227. Aratono M, Toyomasu T, Shinoda T, Ikeda N, Takiue T (1997) Langmuir 13:2158
228. Yarranton HW, Masliyah JH (1996) J Phys Chem 100:1786
229. Semmler A, Ferstl R, Kohler H-H (1996) Langmuir 12:4165
230. Kwok DY, Lee Y, Neumann AW (1998) Langmuir 14:2548
231. de Hoog EHA, Lekker HNW (1999) J Phys Chem 103:5274
232. Moura Ramos JJ (1997) Langmuir 13:6607
233. Lord DL, Hayes KF, Demond AH, Salehzadeh A (1997) Environ Sci Technol 31:2045
234. Ermoshkin AV, Semenov AN (1996) Macromolecules 29:6294
235. Freitas AA, Quina FH, Carroll FA (1997) J Phys Chem 101:7488
236. Cai B-Y, Yang J-T, Guo T-M (1996) J Chem Eng Data 41:493

237. Goebel A, Lunkenheimer K (1997) Langmuir 13:369
238. Rosen MJ, Mathias JH, Davenport L (1999) Langmuir 16:4332
239. Chatterjee J, Nikolov A, Wasan DT (1998) Ind Eng Chem Res 37:2301
240. Svitova T, Hoffmann H, Hill RM (1996) Langmuir 12:1712
241. Abdul AS, Gibson TL, Ang CC, Smith JC, Sobcynski RE (1992) Ground Water 30:219
242. DiVincenzo JP, Dentel SK (1996) J Environ Qual 25:1193
243. Guha S, Jaffé PR, Peters CA (1998) Environ Sci Technol 32:930
244. Hayworth JS, Burris DR (1997) Environ Sci Technol 31:1277
245. Hayworth JS, Burris DR (1997) Environ Sci Technol 31:1284
246. Celis R, Barriuso E, Houot S (1998) Chemosphere 37:1091
247. Nkedi-Kizza P, Rao PSC, Hornsby AG (1985) Environ Sci Technol 19:975–979
248. Nkedi-Kizza P, Rao PSC, Hornsby AG (1987) Environ Sci Technol 21:1107
249. Rao P, Suresh C, Lee LS, Pinal R (1990) Environ Sci Technol 24:647
250. Ji W, Brusseau ML (1998) Water Resources Research 34:1635
251. Johnson JC, Sun S, Jaffé PR (1999) Environ Sci Technol 33:1286
252. Kilduff JE, Wigton A (1999) Environ Sci Technol 33:250
253. Kile DE, Chiou CT (1989) In: Suffet IH, MacCarthy P (eds) Aquatic humic substances: influence on fate and treatment of pollutants. Advances in Chemistry Series, ACS, Washington DC, vol 219, p 131
254. Kile DE, Chiou CT, Helburn RS (1990) Environ Sci Technol 24:205
255. Li Z, Bowman RS (1998) Environ Sci Technol 32:2278
256. Nzengung VA (1996) Environ Sci Technol 30:89
257. Nzengung VA, Nkedi-Kizza P, Jessup RE, Voudrias EA (1997) Environ Sci Technol 31:1470
258. Sahoo D, Smith JA, Imbrigiotta TE, McLellan HM (1998) Environ Sci Technol 32:1686
259. Smith JA, Sahoo D, McLellan HM, Imbrigiotta TE (1997) Environ Sci Technol 31:3565
260. Tiehm A, Stieber M, Werner P, Frimmel FH (1997) Environ Sci Technol 31:2570
261. Zachara JM, Ainsworth CC, Schmidt RL, Resch CT (1988) J Contaminant Hydrol 2:343
262. Banerjee S, Castrogivanni MA (1987) J Chromatog 396:169
263. Walters RW, Guiseppi-Elie A (1988) Environ Sci Technol 22:819
264. Ashby KD, Das K, Petrich JW (1997) Anal Chem 69:1925
265. Bury SJ, Miller CA (1993) Environ Sci Technol 27:104
266. Chien YY, Kim E-G, Bleam WF (1997) Environ Sci Technol 31:3204
267. Katsuta S, Saitoh K (1998) Anal Chem 70:1389
268. Merica SG, Banceu CE, Jdral W, Lipkowski J, Bunce NJ (1998) Environ Sci Technol 32:1509
269. Geetha B, Mandal AB (1997) Langmuir 13:2410
270. Khougaz K, Zhong XF, Eisenberg A (1996) Macromolecules 29:3937
271. Lin S-Y, Lin Y-Y, Chen E-M, Hsu C-T, Kwan C-C (1999) Langmuir 15:4370
272. Chang H-C, Lin Y-Y, Chern C-S, Lin S-Y (1998) Langmuir 14:6632
273. Shiloach A, Blankschtein D (1997) Langmuir 13:3968
274. Shiloach A, Blankschtein D (1998) Langmuir 14:4105
275. Florenzano FH, Dias LG (1997) Langmuir 13:5756
276. Huibers PDT, Lobanov VS, Katritzky AR, Shah DO, Karelson M (1996) Langmuir 12:1462
277. Wong JE, Duchscherer TM, Pietraru G, Cramb DT (1999) Langmuir 15:6181
278. Ranganathan R, Peric M, Bales BL (1998) J Phys Chem 102:8436
279. Perez-Rodriguez M, Prieto G, Rega C, Varela LM, Sarmiento F, Mosquera V (1998) Langmuir 14:4422
280. Cifuentes A, Bernal JL, Diez-Masa JC (1997) Anal Chem 69:4271
281. Lesemann M, Thirumoorthy K, Kim YJ, Jonas J, Paulaitis ME (1998) Langmuir 14:5339
282. Maeda H, Muroi S, Kakehashi R (1997) J Phys Chem 101:7378
283. Arnold CG, Weidenhaupt A, David MM, Müller SR, Haderlein SB, Schwarzenbach RP (1997) Environ Sci Technol 31:2596
284. Barranco FT Jr, Dawson HE (1999) Environ Sci Technol 33:1598
285. Che M, Loux MM, Traina SJ (1992) J Environ Qual 21:698
286. Gundersen JL, MacIntyre WJ, Hale RC (1997) Environ Sci Technol 31:188

287. Dalang F, Buffle J, Haerdl W (1984) Environ Sci Technol 18:135
288. Roy SB, Dzombak DA (1997) Environ Sci Technol 31:656
289. Sowden FJ, Griffith SM, Schnitzer M (1976) Soil Biol Biochem 8:55
290. Fukushima M, Oba K, Tanaka S, Nakayasu K, Nakamura H, Hasebe K (1997) Environ Sci Technol 31:2218
291. Keoleian GA, Curl RL (1989) In: Suffet IH, MacCarthy P (eds) Aquatic humic substances: influence on fate and treatment of pollutants. Advances in Chemistry Series, ACS, Washington DC, vol 219, p 231
292. Kilduff JE, Karanfil T, Weber WJ Jr (1996) Environ Sci Technol 30:1344
293. Bouchard DC, Enfield CG, Piwoni MD (1989) In: Sawhney BL, Brown K (eds) Reactions and movement of organic chemicals in soils. Soil Sci Soc Am: American Society of Agronomy, Madison, Wisconsin, Series: SSSA special publication, p 349
294. Carlson DJ, Mayer LM, Brann ML, Mague TH (1985) Mar Chem 16:141
295. Caron G, Suffet IH, Belton T (1985) Chemosphere 14:993
296. Fish CL, Driscoll MS, Hassett JP (1989) In: Suffet IH, MacCarthy P (eds) Aquatic humic substances: influence on fate and treatment of pollutants. Advances in Chemistry Series, ACS, Washington DC, pp 219–223
297. Wershaw RL (1986) J Contaminant Hydrology 1:29
298. Macalady DL, Wolfe NL (1984) In: Krueger RF, Seiber JN (eds) Treatment and disposal of pesticides wastes. ACS Symp Series N 259, p 221
299. Macalady DL, Wolfe NL (1985) J Agric Food Chem 33:167
300. Macalady DL, Wolfe NL (1987) Am Chem Soc-Div Environ Chem 27:12
301. Malini de AR, Pospisil F, Vockova K, Kutacek M (1980) Biol Plant 22:167
302. Perdue EM, Wolfe NL (1982) Environ Sci Technol 16:847
303. Steinberg C, Muenster U (1985) In: Aiken GR, McKnight DM, Wershaw RL, MacCarthy P (eds) Humic substances in soil, sediment and water. Wiley, NY
304. Thurman EM (1986) In: Organic geochemistry of natural waters. Martinus Nijhoff/Junk Publishers, Dordrecht
305. Malcolm RL (1990) Anal Chim Acta 232:19
306. Leenheer JA, Werschaw RL, Reddy MM (1995) Environ Sci Technol 29:393
307. Leenheer JA, Werschaw RL, Reddy MM (1995) Environ Sci Technol 29:399
308. Averett RC, Leenheer JA, McKnight DM, Thorn KA (1987) Humic substances in the Suwannee river, Georgia: interactions, properties, and proposed structures, Open-File Report 87–557, US Geological Survey, Denver, CO
309. Shevchenko SM, Bailey GW (1996) Crit Rev Environ Sci Technol 26:95
310. Choudhry GG (1984) In: Humic substances-photophysical and free radical aspects and interactions with environmental chemicals. Gordon and Breach Science Publishers, New York, 215 pp
311. Frimmel FH (1994) Environ Int 20:373
312. Hermann R, Ziechmann WZ (1988) Pflanzenernähr Bodenk 151:219
313. Minero C, Pramauro E, Pelizzeti E, Dolci M, Marchesini A (1992) Chemosphere 24:1597
314. Klöpffer W (1992) Sci Total Environ 123/124:145
315. Schmitt P, Freitag D, Sanlaville Y, Lintelmann J, Kettrup AJ (1995) Chromatogr A 709:215
316. Aguer JP, Richard C, Andreux F (1996) J Photochem Photobiol 103:163
317. Shin HS, Moon H (1996) Soil Sci 161:250
318. Valentine RL, Zepp RG (1993) Environ Sci Technol 27:409
319. Canonica S, Jans U, Stemmler K, Hoigné J (1995) Environ Sci Technol 29:1822
320. Cooper WJ, Zika RG, Pestane RG, Fischer AM (1987) In: Suffet IH, MacCarthy P (eds) Aquatic humic substances, influence on fate and treatment of pollutants. Advances in Chemistry Series 219, ACS, Washington, DC, p 333
321. Wetzel RG, Hatcher PG, Bianchi TS (1995) Limnol Oceanogr 40:1369
322. Amalay M, Bussieres D (1996) In: Clapp CE, Hayes MHB, Senesi N, Griffith SM (eds) Humic substances and organic matter in soil and water environments. International Humic Substances Society, University of Minnesota, Madison, p 251
323. Bothwell ML, Sherbot DMJ, Pollock CM (1994) Science 265:97

324. Canonica S, Hoigné J (1995) Chemosphere 30:2365
325. Croasmun WR, Carlson RMK (eds) (1994) In: Two-dimensional NMR spectroscopy. VCH Publishers, Weinheim
326. Dahlen J, Bertilsson S, Petterson C (1996) Environ Int 22:501
327. Garrison AW, Schmitt PH, Kettrup A (1995) Water Res 29:2149
328. Kieber JK, Zhou X, Mopper K (1990) Limnol Oceanogr 35:1503
329. Mopper K, Zhou X, Kieber RJ, Kieber DJ, Sikorski RJ, Jones RD (1991) Nature 353:60
330. Power JF, Sharma DK, Langford CH, Bonneau R, Joussot-Dubien J (1986) J Agric Food Chem 45:1012
331. Schindler DW, Curtis PJ, Parker BR, Stainton MP (1996) Nature 379:705
332. Schmitt P, Garrison AW, Freitag D, Kettrup A (1997) Water Res 31:2037

CHAPTER 3

Sorption/Desorption of Organic Pollutants from Complex Mixtures: Modeling, Kinetics, Experimental Techniques and Transport Parameters

Tarek A. T. Aboul-Kassim[1], Bernd R. T. Simoneit[2]

[1] Department of Civil, Construction and Environmental Engineering, College of Engineering, Oregon State University, 202 Apperson Hall, Corvallis, OR 97331, USA
e-mail: Tarek.Kassim@orst.edu

[2] Environmental and Petroleum Geochemistry Group, College of Oceanic and Atmospheric Sciences, Oregon State University, Corvallis, OR 97331, USA,
e-mail: simoneit@oce.orst.edu

Sorption/desorption is one of the most important processes influencing movemement of organic pollutants in natural systems. Sorption with reference to a pollutant is its transfer from the aqueous phase to the solid phase; on the other hand, desorption is its transfer from the solid phase to the aqueous phase. Similar to all interphase mass-transfers, the sorption/desorption process can be defined by the final-phase equilibrium of the pollutant at the aqueous-solid phase interface and the time required to approach final equilibrium.

The main goal of this chapter is to review the most widely used modeling techniques to analyze sorption/desorption data generated for environmental systems. Since the definition of sorption/desorption (i.e., a mass-transfer mechanism) process requires the determination of the rate at which equilibrium is approached, some important aspects of chemical kinetics and modeling of sorption/desorption mechanisms for solid phase systems are discussed. In addition, the background theory and experimental techniques for the different sorption/desorption processes are considered. Estimations of transport parameters for organic pollutants from laboratory studies are also presented and evaluated.

An important and recently reported issue, namely slow sorption/desorption rates, their causes at the intra-particle level of various solid phases, and how these phenomena relate to contaminant transport, bioavailability, and remediation, is also discussed and evaluated. A case study showing the environmental impact of solid waste materials which are mainly complex organic mixtures and/or their reuse/recycling as highway construction and repair materials is presented and evaluated from the point of view of sorption/desorption behavior and data modeling.

Keywords. Organic pollutants, Aqueous-solid phase systems, Sorption, Desorption, Kinetics, Modeling, Transport parameters, Solid waste materials, Slow sorption/desorption, Highway materials, Remediation

The Handbook of Environmental Chemistry Vol. 5 Part E
Pollutant-Solid Phase Interactions: Mechanism, Chemistry and Modeling
(by T. A. T. Aboul-Kassim, B. R. T. Simoneit)

List of Abbreviations

BET Brunauer-Emmett-Teller
COMs Complex organic mixtures
IAS Ideal adsorbed solution
K_d Partition coefficient
K_d^{app} Apparent sorption distribution coefficient
K_{OC} Organic carbon partition coefficient
K_{OW} Octanol-water partition coefficient
QSAR Quantitative structure-activity relationship
SCAM Simplified competitive equilibrium adsorption model

SCS Single component system
SP_{OM} Solid phase organic matter
SWMs Solid waste materials
TOC Total organic carbon

1 Introduction

Chemodynamic studies of organic pollutant(s) and/or solid waste material (SWM) leachates of complex organic mixtures (COMs) examine the fate and transport of these pollutants in various environmental compartments. Many of these pollutants have been shown to be toxic, genotoxic, and/or carcinogenic, in both surface/subsurface and aquatic environments, by external and internal interactions, resulting in reactions occurring between these pollutants and/or SWM leachates with solid phase components [1–5]. These reactions include various chemical, physical, and biological processes. During transport of pollutants and/or SWM leachates, it is difficult to identify and/or categorize fully the contribution made by each process to all the reactions established between pollutant- and/or leachate-solid phase constituents. For instance, the thermodynamic reactions occurring within the subsurface environment are generally considered to be instantaneous, i.e., equilibrium is attained almost instantly in chemical reactions. This is known to be highly unlikely in field situations because of lack of contact with all surfaces.

During pollutants and/or SWM leachate transport through the surface/subsurface environments, physical and chemical processes can result in the accumulation of pollutants on the solid phase constituents. The degree to which this accumulation renders the trapped pollutants immobile is of vital interest in considerations for modeling the proposed pollutant fate and transport.

The processes controlling transfer and/or removal of pollutants at the aqueous-solid phase interface occur as a result of interactions between chemically reactive groups present in the principal pollutant constituents and other chemical, physical and biological interaction sites on solid surfaces [1]. Studies of these processes have been investigated by various groups (e.g., [6–14]). Several workers indicate that the interactions between the organic pollutants/SWM leachates at the aqueous-solid phase surfaces involve chemical, electrochemical, and physico-chemical forces, and that these can be studied in detail using both chemical reaction kinetics and electrochemical models [15–28].

The main objectives of this chapter are to: (1) review the different modeling techniques used for sorption/desorption processes of organic pollutants with various solid phases, (2) discuss the kinetics of such processes with some insight into the interpretation of kinetic data, (3) describe the different sorption/desorption experimental techniques, with estimates of the transport parameters from the data of laboratory tests, (4) discuss a recently reported issue regarding slow sorption/desorption behavior of organic pollutants, and finally (5) present a case study about the environmental impact of solid waste materials/complex organic mixtures (i.e., SWMs/COMs) and/or their recycling/reuse as highway construction and repair materials from the perspective of their sorption/desorption behavior and data modeling.

2 Modeling Techniques

A number of models have been developed to reflect the actual sorption/desorption processes that occur in the natural environment [1, 29–33]. Some models have a sound theoretical basis; however, they may have only limited experimental utility because the assumptions involved in the development of the relationship apply only to a limited number of sorption processes. Other models are more empirical in their derivation, but tend to be more generally applicable. In the latter case, the theoretical basis is uncertain.

A sorption isotherm expresses the quantity of material adsorbed per unit mass of adsorbent as a function of the equilibrium concentration of the adsorbate. The necessary data is derived from experiments where a specified mass of adsorbent is equilibrated with a known volume at a specific concentration of a chemical and the resultant equilibrium concentration is measured in solution [33].

The following sections show various sorption isotherms that can be used to model single pollutant/leachate component system adsorption. In addition some predictive models for multi-pollutants/leachate(s) component solutions are also summarized and discussed.

2.1 Single Component System Models

Single component system (SCS) adsorption models actually mean one pollutant component in aqueous system or in a SWM leachate [34]. Since water is simply assumed to be inert, and the pollutant/leachate adsorption is assumed to be unaffected by water, the system is treated as an SCS. To represent the equilibrium relation for SCS adsorption, a number of isotherm models reported in the literature are reviewed in the following.

2.1.1 *Langmuir Model*

The Langmuir adsorption model describes the equilibrium between aqueous and solid phase systems as a reversible chemical equilibrium between species [15, 27, 35]. This sorption isotherm has a sound conceptual basis and was originally developed for defining the adsorption of gases onto solid phases. In developing the isotherm the following assumptions were made: (a) the adsorption energy is constant and independent of the extent of surface coverage, (b) adsorption is on localized sites with no interaction between adsorbed molecules, and (c) the maximum adsorption possible is a complete monolayer. The adsorbent surface (i.e., solid phase) is made up of fixed individual sites where molecules of adsorbate (i.e., the organic pollutant of interest) may be chemically bound. This can be expressed mathematically by denoting an unoccupied surface site as $[-S]$ and the adsorbate in dilute leachate solution as species $[A]$, with concentration $[C]$, and considering the reaction between the two to form occupied sites $[-SA]$:

$$[-S] + [A] \leftrightarrow [-SA] \tag{1}$$

For the Langmuir adsorption isotherm it is assumed that this reaction (Eq. 1) has a fixed free energy of adsorption equal to ΔG_a^0, which is not dependent on the extent of adsorption and not affected by interaction among sites. In addition, each site is assumed to be capable of binding at most one molecule of adsorbate. If Q is the maximum number of moles of a pollutant adsorbed per mass adsorbent when the surface sites are saturated with an adsorbate (i.e., a full monolayer), and q is the number of moles of adsorbate per mass adsorbent at equilibrium, then according to the law of mass action Eq. (2) follows:

$$b = \frac{[-SA]}{[-S][A]} = \left[\frac{q}{(Q-q)\cdot C}\right] \tag{2}$$

where:

- $[b = e^{(-\Delta G_a^0/RT)}]$ = an equilibrium constant, and
- C = the equilibrium concentration in solution.

The rearrangement of Eq. (2) leads to:

$$q = \frac{QbC}{(1+bC)} \tag{3}$$

Correspondence of experimental data to the Langmuir model does not mean that the stated assumptions are valid for the particular system being studied, because departure from the assumptions can have a canceling effect. An advantage of this model is that it can approach Henry's law at low concentrations.

The constants in the Langmuir model can be determined by plotting $\left(\frac{C}{q}\right)$ vs C and making use of Eq. (3) rewritten as:

$$\frac{C}{q} = \frac{1}{Qb} + \frac{C}{Q} \tag{4}$$

This isotherm finds use mainly in the study of the adsorption of gases on solids; however, it can be useful in the study of adsorption of pollutants from aqueous systems, particularly onto solid phases. The heterogeneous nature of a solid surface (i.e., soils, sediments, suspended solids) would obviously invalidate the first assumption (i.e., a, above) used in developing the relationship. The third assumption (i.e., c, above) also would be invalid in a situation where one is dealing with multi-layer adsorption.

2.1.2
Double-Reciprocal Langmuir Model

The double-reciprocal Langmuir model has been extensively used in site assessment projects for elemental adsorption data. The double-reciprocal Langmuir is an adaptation of the traditional equation for elemental sorption of solid phases exhibiting two primary adsorbing surface sites. The double-reciprocal Langmuir model is as follows:

$$\frac{q}{Q} = \frac{k_1 \cdot b_1 \cdot C}{(1 + k_f \cdot C)} + \frac{k_2 \cdot b_2 \cdot C}{(1 + k_2 \cdot C)} \tag{5}$$

where:

- q and Q are as defined earlier,
- C is the concentration of solute at equilibrium,
- k_f is a constant = $[(q/Q)/C]$,
- k_1 and k_2 are constants, and
- b_1 and b_2 are constants (i.e., the maximum quantities of the compound that can be sorbed by two surfaces).

The basic assumptions for application of graphic isotherm and regression equations are that the data be derived under equilibrium conditions, constant temperature, and minimal fixation effects, and the data can be modeled as a regression function. The equations are valid only within the experimental concentration ranges used to determine the sorption.

2.1.3
Brunauer-Emmett-Teller Model

Brunauer-Emmett-Teller (BET) adsorption describes multi-layer Langmuir ad sorption. Multi-layer adsorption occurs in physical or van der Waals bonding of gases or vapors to solid phases. The BET model, originally used to describe this adsorption, has been applied to the description of adsorption from solid solutions. The adsorption of molecules to the surface of particles forms a new surface layer to which additional molecules can adsorb. If it is assumed that the energy of adsorption on all successive layers is equal, the BET adsorption model [36] is expressed as Eq. (6):

$$\frac{q}{Q} = \frac{A_m \cdot K_B \cdot C}{(C_s - C) \cdot \left[1 + (K_B - 1) \cdot \left(\frac{C}{C_s}\right)\right]} \tag{6}$$

where:

- A_m is maximum adsorption density of first layer,
- K_B is a dimensionless constant related to the free energy difference between adsorbate on the first and successive layers, and
- C_s is the saturation concentration of the adsorbate in solution.

When $K_B \gg 1$ and $(C/Cs) \ll 1$, Eq. (6) may be rearranged to a linear form as:

$$\left(\frac{C}{C_s - C}\right) = \left(\frac{1}{K_B \cdot A_m}\right) + \left(\frac{K_B - 1}{K_B \cdot A_m}\right)\left(\frac{C}{C_s}\right) \tag{7}$$

2.1.4
Freundlich Model

The Langmuir and BET models incorporate an assumption that the energy of adsorption is the same for all surface sites and not dependent on degree of coverage. Since in reality the energy of adsorption may vary because real surfaces are heterogeneous, the Freundlich adsorption model (see Chap. 2) [37] attempts to account for this:

$$q = K_f \cdot C^n \tag{8}$$

where:

- C = the equilibrium concentration of the chemical compound of interest in solution,
- K_f = an equilibrium constant indicative of sorption strength,
- n = the degree of non-linearity (when $n > 1$, there is no limit to the amount sorbed other than its solubility, which is not expected with a true adsorption process).

A linear form of Eq. (8) can be presented as shown in Eq. (9):

$$\log q = \log K_f + n \cdot \log C \tag{9}$$

If log q is plotted as a function of $\log C$, a straight line should be obtained with an intercept on the ordinate of $\log K$ and slope n.

2.1.5
Langmuir-Freundlich Model

Sips [38] modified the Langmuir adsorption model by introducing a power law expression of the Freundlich equation:

$$q = \frac{Q \cdot b \cdot C^n}{(1 + b \cdot C^n)} \tag{10}$$

This reduces to the Freundlich equation for low concentrations and exhibits saturation for high concentrations.

2.1.6
Linear Model

When the Freundlich isotherm n values approximate one, that indicates a linear relationship between the amount sorbed and the equilibrium concentration in solution. Thus, the distribution of any organic pollutant in the aqueous-solid system can be defined by a simple proportionality constant. Equation (8) can be modified as follows:

$$q = K_d \cdot C \tag{11}$$

where K_d is a simple measure of the distribution of an organic pollutant between the two phases. A variation of this relationship is used to account for the contribution of the solid phase organic matter (i.e., SP_{OM}):

$$q = K_{om} \cdot C \tag{12}$$

where the amount of the sorbed organic pollutant is expressed per unit of organic matter on the solid phase (i.e., soil, sediment, suspended matter, colloids, and biocolloids/biosolids) rather than per unit mass of solid phase. Thus, the relation between the two distribution constants (i.e., Eqs. 11 and 12) is:

$$K_{om} = \frac{(K_d) \cdot (100)}{(\% \text{ Organic Matter})} \tag{13}$$

This distribution constant may also be expressed as amount of organic pollutant sorbed per unit mass of solids organic carbon (K_{OC}), the relation between the two being defined by the following:

$$\text{Organic matter} = 1.3 \text{ (Organic carbon)} \tag{14}$$

and thus:

$$K_{OC} \cdot K_{OM} \cdot (1.3) \tag{15}$$

For the linear isotherm model, the parameter (K_d) that relates both sorbate and solute is called the *partition coefficient*. A number of studies have developed empirical relationships for partition coefficients in natural solid phases and several of these studies are summarized in Table 1.Various theoretical-based methods of partition coefficient estimations also exist (Table 1, Eqs. a–f).

Generally, it is clear how K_d can be predicted for organic hydrophobic pollutants which obey a linear isotherm relationship. First, the organic carbon partition coefficient (i.e., K_{OC}) is predicted based on either solubility or the octanol-water partition coefficient (K_{OW}). Then based on an estimate of the organic carbon fraction in the fine and coarse sediments/soils, K_d can be estimated from Eqs. (a and b) (Table 1).

For most organic pollutants, SP_{OM} is the major variable determining the extent of sorption from aqueous systems. However, when the K_d is calculated based on organic carbon (K_{OC}), a relatively constant value is obtained for each solid system, despite the fact that some variation should be expected from one solid system to another based on the characteristics of the organic matter. Thus, the K_d is dependent primarily on the SP_{OM} content, while K_{OC} and hence K_{OM} are characteristic for each organic pollutant. Sorption distribution constants based on organic matter or organic carbon will vary over a wide range for different organic pollutants [17, 32, 39–63].

The relative amount of organic pollutant sorbed on a solid phase or dissolved in an aqueous environment depends mainly on the sorbate concentration (i.e.,

Table 1. Various theoretical methods for partition coefficient estimations

Organic pollutant type	Sorbent type	Predictive models of partition coefficient, k_{oc} and k_{ow} values	Equation number
Aromatic hydrocarbons Chlorinated hydrocarbons	Natural sediments and soils	Partition coefficient based on sediment organic carbon content [43, 47, 48, 51, 53–63] $K_d = K_{OC} \cdot X_{OC}$ where K_{OC} is the partition coefficient expressed on an organic carbon basis, and X_{OC} is the mass fraction of organic carbon in sediment	a
Aromatic hydrocarbons Chlorinated hydrocarbons	Natural sediments	Partition coefficient showing the influence of particle size [43, 47, 48, 51, 53–63] $K_d = K_{OC}\,[0.2\,(1-f)\,X^{S}_{OC} + f\,X^{f}_{OC}]$ where f is the mass fraction of fine sediments ($d < 50$ mm), X^{S}_{OC} is the organic carbon content of coarse sediment fraction, and X^{f}_{OC} is organic carbon content of fine sediment fraction	b
Aromatic hydrocarbons Chlorinated hydrocarbons	Natural sediments and soils	Relationship between K_{OC} and K_{OW} [40, 42, 43, 47, 48, 51–54, 59–63] $K_{OC} = 0.63\,K_{OW}$ where K_{OW} is the octanol-water partition coefficient defined as concentration of chemical in octanol divided by concentration of chemical in water at equilibrium.	c
9-Chloro-*s*-triazine Dinitroaniline compounds		$\log K_{OC} = 0.937 \log K_{OW} - 0.006$	d
Aliphatic and aromatic hydrocarbons Aromatic acids Organochlorine and organophosphate pesticides Polychlorinated biphenyls	natural sediments and soils	Relationship between K_{OC} and aqueous solubility [43, 47, 48, 51, 53–63] $\log K_{OC} = 0.54 \log S_W + 0.44$ where S_w is the water solubility of sorbate, expressed as a mole fraction	e
		Relationship between K_{OC} and aqueous solubility [41, 42, 44–46, 49–51, 53, 54, 59–63] $\log K_{OC} = 5.00 - 0.670 \log S_W$ where S_w is the solubility (g.mol/l) *The previous equation covers more than eight orders of magnitude in solubility and six orders of magnitude in the octanol-water partition coefficient*	f

soil/sediment solids, suspended matter, colloids, and biocolloids/biosolids) and partition coefficient. At equilibrium, the relative dissolved amount of a certain organic pollutant can be given by:

$$\alpha_w = \frac{C_w}{C_T} = \frac{1}{1 + K_d \cdot S} \tag{16}$$

where:

- C_w = total dissolved pollutant phase concentration,
- $C_S = X_S$,
- $C_T = (C_w + C_S)$,
- K_d = partition coefficient.
- S = solid phase material (i.e., suspended matter, sediment or soil concentration, on a part/part basis), and
- X = mass of sorbed pollutant/mass of solid phase material).

2.1.7
Toth Model

Toth [64] has only considered adsorption of gases in his model but his idea can be extended to adsorption of solutes from dilute aqueous solution [65]. The Toth adsorption model has the form:

$$q = \frac{QC}{(b + C^M)^{1/M}} \tag{17}$$

It consists of three parameters, which are C (i.e., the equilibrium concentration of the chemical compound of interest in solution), Q (i.e., the maximum number of moles of a pollutant adsorbed per mass adsorbent), and q (i.e., the number of moles of adsorbate per mass adsorbent at equilibrium). The Toth model (Eq. 17) reduces to Henry's law at very low concentrations and exhibits saturation at high concentrations.

2.2
Multicomponent Equilibria Models

Multicomponent pollutants in an aqueous environment and/or leachate of SWMs, which are COMs, usually consist of more than one pollutant in the exposed environment [1, 66–70]. Multicomponent adsorption involves competition among pollutants to occupy the limited adsorbent surface available and the interactions between different adsorbates. A number of models have been developed to predict multicomponent adsorption equilibria using data from SCS adsorption isotherms. For simple systems considerable success has been achieved but there is still no established method with universal proven applicability, and this problem remains as one of the more challenging obstacles to the development of improved methods of process design [34, 71–76].

2.2.1
Multicomponent Langmuir Model

The Langmuir model for competitive adsorption can be used as a common model for predicting adsorption equilibria in multicomponent systems. This was first developed by Butler and Ockrent [77] and is based on the same assumptions as the Langmuir model for single adsorbates. It assumes, as in the case of the Langmuir model, that the rate of adsorption of a species at equilibrium is equal to its desorption rate. This is expressed by Eq. (18):

$$q_i = \frac{Q_i \cdot b_i \cdot C_i}{1 + \sum_{i=1}^{n} b_i \cdot C_i} \tag{18}$$

where Q_i and b_i are the Langmuir constants determined from the single solute adsorption isotherm of species i (Eqs. 3 and 4).

Because of its mathematical simplicity, the multicomponent Langmuir adsorption model is widely used [78–92]. In order to increase the performance of clean up methods at contaminated sites and improve environmental engineering/management practices, the fate and transport of various anthropogenic pollutants through the subsurface environment (i.e., soil-solids) have been investigated by several authors [83–85, 87, 89–91]. A one-dimensional solute transport model was developed by Thayumanavan [87] to predict the movement of various pollutants through a simulated subsurface environment, and to verify the model with experimentally determined breakthrough curves. Particular importance was given to the effect of low pH on desorption processes. The one-dimensional solute transport model was developed under the assumption of a one-dimension, steady-state, pollutant saturated groundwater flow through a homogeneous porous medium. In general, desorption was described by a nonlinear competitive Langmuir model, while numerical solutions of the transport equations were obtained by the forward-time, centered-space, finite difference method. Computer simulations were fitted to experimental breakthrough curves using estimates for model parameters, which could not be determined independently in experiments.

It should be mentioned that the extension of the Langmuir theory to adsorption from binary adsorbate systems is thermodynamically consistent only in the special case where $Q_1 = Q_2$. However, that thermodynamic consistency is of secondary importance if Eq. (18) provides the correct analytical description of the adsorption phenomena.

2.2.2
Modified Multicomponent Langmuir Model

Jain and Snoeyink [93] reported that if the Langmuir model for competitive adsorption satisfactorily predicts the extent of adsorption from a bisolute system when $Q_1 \neq Q_2$, it is probably due to the competition for all available sites. They have proposed a model which can be used to predict the extent of adsorption of each species from a bisolute solution if a portion of the adsorption occurs without competition. The model is based on the hypothesis that adsorption without competition occurs when $Q_1 \neq Q_2$ [88–91]. Furthermore, it was assumed that the number of sites on solid phases for which there was no competition was equal to the quantity $(Q_1 - Q_2)$, where $Q_1 > Q_2$. On this basis, the following equations were proposed:

$$q_1 = \left[\frac{(Q_1 - Q_2) \cdot b_1 \cdot C_1}{1 + b_1 \cdot C_1}\right] + \left[\frac{Q_2 \cdot b_1 \cdot C_1}{1 + b_1 \cdot C_1 + b_2 \cdot C_2}\right] \tag{19}$$

$$q_2 = \left[\frac{Q_2 \cdot b_2 \cdot C_2}{1 + b_1 \cdot C_1 + b_2 \cdot C_2}\right] \tag{20}$$

The first term on the right side of Eq. (19) is the Langmuir expression for the number of moles of species 1 which adsorb without competition on the surface area proportional to $(Q_1 - Q_2)$. The second term represents the number of moles of species 1 adsorbed on the surface area proportional to Q_2 under competition with species 2 and is based on the Langmuir model for competitive adsorption. The number of moles of species 2 adsorbed on the surface area proportional to Q_2 and under competition with species 1 can be calculated from Eq. (20).

2.2.3
Multicomponent Langmuir-Freundlich Model

The Sips [38] model (Eq. 10) can easily be extended to binary or multicomponent systems [34, 74]. The resulting expression for the multicomponent Langmuir-Freundlich adsorption model is:

$$q_i = \frac{Q_i \cdot b_i \cdot C_i^{n_i}}{1 + \sum b_i \cdot C_i^{n_i}} \tag{21}$$

The simple formula makes this method very attractive. Although not thermodynamically consistent, this expression (Eq. 21) has been shown to provide a reasonably good empirical correlation of binary equilibrium data for a number of simple gases on molecular sieve adsorbents [34, 73–75]. However, because of the lack of a proper theoretical foundation this approach should be treated with caution.

2.2.4
Ideal Adsorbed Solution Model

The most common model for describing adsorption equilibrium in multicomponent systems is the Ideal Adsorbed Solution (IAS) model, which was originally developed by Radke and Prausnitz [94]. This model relies on the assumption that the adsorbed phase forms an ideal solution and hence the name IAS model has been adopted. The following is a summary of the main equations and assumptions of this model (Eqs. 22–29).

The IAS model relates the concentration of solute i in a complex mixture (C_1) to a corresponding concentration of this solute in an single solute system (C^o) (i.e., Eq. 22):

$$C_i = (\Pi, T, Z_i) = Z_i C_i^0(\Pi, T) \tag{22}$$

where

- Z_i = the mole fraction of surface coverage by component i,
- Π = the spreading pressure on the surface, and
- T = the absolute temperature.

The spreading pressure defines the lowering of surface tension at the aqueous-solid phase (i.e., adsorbate-solution) interface:

$$\Pi = \gamma^0 - \gamma \tag{23}$$

where

- γ^0 = the surface tension of the pure solvent (water), and
- γ = the surface tension created by the mixture of solvent and solutes.

Equation (23) holds only when Π and T in the mixture are the same as those in the respective single-solute systems. Spreading pressure can be related to the characteristic adsorption equilibria of each single solute system according to the following relationship:

$$\Pi_i = \frac{RT}{A} \int_0^{C_i^0} q_i^0 \frac{dC_i^0}{C_i^0} \tag{24}$$

where

- R = the universal gas constant,
- A = the surface area per unit weight,
- C_i^0 = the liquid-phase concentration of species i in single-solute systems which gives the same spreading pressure as that of the mixture, and
- q_i^0 = the solid-phase loading corresponding to C_i^0.

Equivalence of the spreading pressures of all the solutes in the mixture gives the following equation:

$$\int_0^{C_i^0} q_i^0 \frac{dC_i^0}{C_i^0} = \int_0^{C_2^0} \frac{dC_2^0}{C_2^0} = \int_0^{C_3^0} \frac{dC_3^0}{C_3^0} = \ldots \tag{25}$$

The relationship between q_i^0 and C_i^0 is given by the single solute adsorption isotherm:

$$q_i = f_i \cdot (C_i^0) \tag{26}$$

Combining the IAS theory with the Gibbs equation for isothermal adsorption gives the relationship necessary for equilibrium calculations:

$$\frac{1}{q_T} = \sum_i^n \frac{Z_i}{q_i^0} \tag{27}$$

Other two equations required for IAS model calculations are:

$$\sum_{i}^{n} Z_{\mathrm{i}} = 1 \tag{28}$$

$$q_{\mathrm{i}} = Z_{\mathrm{i}} \cdot q_{\mathrm{T}} \tag{29}$$

Equations (22), and (25)–(29) constitute a set of simultaneous equations from which the IAS model calculation can be made.

The IAS model has received widespread use in multisolute adsorption research for a variety of reasons [15, 27, 32, 34, 65, 71, 81, 92, 95, 96]. Besides the fact that the application of the IAS model necessitates only single-solute data means that the model is flexible in that multicomponent calculations can be performed using several different single-solute isotherm relationships. In addition, this model has a solid theoretical foundation, providing a useful understanding of the thermodynamic approach to adsorption. In this regard it is similar to the Gibbs adsorption equation upon which it is based. This is in contrast to the Langmuir competitive model (Eqs. 18–20), which is founded on the same limiting assumptions as the single-solute Langmuir model (i.e., monolayer adsorption and a homogeneous adsorbent surface).

However, it should be pointed out that the IAS model for predicting multisolute adsorption is most reliable for those systems where solute adsorption loading is moderate. If solute adsorption loading is large, the deviations of the predictions from experimentally observed data may be significant. Similar to the Langmuir and other multicomponent equilibrium models, the IAS model predicts that the adsorbate more favorably adsorbed in single-solute solutions also adsorbs to a greater extent when in competition at equimolar concentration. However, this is true only when adsorption is reversible and competition for adsorption sites is ideal.

The criterion of ideal competition implies that the adsorbent is homogeneous with respect to adsorption sites and that the sites are equally accessible. However, many adsorbates (i.e., solid phases) cannot be considered homogeneous because of their extensive microporous structure and the occurrence of different organic functional groups on their surfaces. An assumption of ideal competition is therefore invalid. Some researchers have also shown that the adsorptions of some organic compounds, such as phenols, are highly irreversible [97–100]. This implies that it is difficult for components to replace each other once one of them was adsorbed prior on an adsorbent. It is evident that adsorption kinetics will affect multicomponent adsorption if the component adsorption rates are not proportional to their respective adsorptive capacities. Consequently, IAS and other existing multicomponent equilibria models fail to accurately predict solid-phase loading under system conditions which are significantly non-ideal, i.e., unequal competition and irreversible adsorption effects [76, 95, 97–101].

2.2.5 Simplified Competitive Equilibrium Model

A number of attempts have been made to modify the IAS model (Eqs. 22–29) to improve its accuracy and reduce computational efforts. Using the IAS model, DiGiano et al. [80] derived a Simplified Competitive Equilibrium Adsorption Model (SCAM). This model, which is based on the Freundlich isotherm, assumes the single-solute isotherms of all the components are equal and it utilizes average isotherm constants when this assumption is not valid. The IAS model equations have been reduced to a single expression:

$$q_i = K'^{\left(\frac{n'-1}{n'}\right)} [K_i C_i^{n_i}]^{1/n'} \left[\sum_{i=1}^{n} \left(\frac{K_i}{K} C_i^{n_i} \right)^{\frac{1}{n'}} \right]^{(n'-1)} \tag{30}$$

where

- q_i = the solid-phase equilibrium concentration of solute i,
- n_i, K_i = the empirical Freundlich constants for single solute i,
- C_i = the liquid-phase equilibrium concentration of solute i,
- n' = the average value of n_i, and
- K' = the average value of K_i.

This model significantly simplifies the computations of the IAS model, although it does not improve its accuracy [15, 27, 76, 88]. One popularized approach to modify the IAS model is to incorporate an empirical coefficient (R_i) into Eq. (29) to describe more accurately experimental equilibria [76, 95, 101, 102] as the following:

$$q_i = R_i \cdot Z_i \cdot q_T \tag{31}$$

The modification factors (R_i) are determined from multicomponent equilibrium data with a minimization procedure. This modification provides a significantly better data description. However, this improvement is the result of parameters that are determined from the multicomponent data itself.

3 Kinetics of Sorption/Desorption Processes

Most of the sorption/desorption transformation processes of various solid phases are time-dependent. To understand the dynamic interactions of organic pollutants with solid phases and to predict their fate with time, knowledge of the kinetics of these processes is important [20, 23].

There are four main processes (i.e., bulk transport; chemical reaction; film and particle diffusion) which can affect the rate of solid phase chemical reactions and can broadly be classified as transport and chemical reaction processes [10, 31, 103–107]. The slowest of these will limit the rate of a particular reaction. Bulk transport process of a certain pollutant(s), which occurs in the aqueous phase, is very rapid and is normally not rate-limiting. In the laboratory, it can be eliminated by rapid mixing. The actual chemical reaction at the surface of a solid phase (e.g., adsorption) is also rapid and usually not rate limiting. The two remaining transport or mass transfer processes (i.e., film and particle diffusion processes), either singly or in combination, are normally rate-limiting. Film diffusion involves transport of a pollutant through a boundary layer or film (water molecules) that surrounds the solid particle surface. Particle diffusion (i.e., intraparticle diffusion) involves transport of a pollutant along pore-wall solid surfaces and/or within the pores of the solid particle surface (e.g., soils, sediments).

Aboul-Kassim [1] studied the characterization, chemodynamics, and environmental impact assessment of organic leachates from complex mixtures. He reported that an important factor in controlling the rate of solid phase adsorption reactions is the type and quantity of solid phase components as well as the time period (i.e., short vs long) over which the organic contaminant has been in contact with the solid phase.

It is important to differentiate between two terms that are widely used in the literature, namely "*chemical kinetics*" and "*kinetics*". Chemical kinetics is defined as the investigation of chemical reaction rates and the molecular processes by which reactions occur where transport (e.g., in the solution phase, film diffusion, and particle diffusion) is not limiting. On the other hand, kinetics is the study of time-dependent processes. Because of the different particle sizes and porosities of soils and sediments, as well as the problem to reduce transport processes in these solid phase components, it is difficult to examine the chemical kinetics processes. Thus, when dealing with solid phase components, usually the kinetics of these reactions are studied.

3.1 Rate Laws

The main reasons for investigating the rates of solid phase sorption/desorption processes are to: (1) determine how rapidly reactions attain equilibrium, and (2) infer information on sorption/desorption reaction mechanisms. One of the important aspects of chemical kinetics is the establishment of a rate law. By definition, a rate law is a differential equation [108] as shown in Eq. (32):

$$aA + bB \rightarrow yY + zZ \tag{32}$$

The reaction rate is proportional to some power of the concentrations of reactants A and B and/or other species (C, D, etc., Eq. 32) in the system. The terms a, b, y, and z are stoichiometric coefficients, and are assumed to equal one. The power to which the concentration is raised may equal zero (i.e., the rate is independent of concentration), even for reactant A or B. Rates are expressed as a decrease in reactant concentration or an increase in product concentration per unit time. Thus, the rate of reactant A (Eq. 32), which has a concentration $[A]$ at any time (t), is $\{-d[A]/(dt)\}$ while the rate with regard to product Y having a concentration $[Y]$ at time (t) is $\{d[Y]/(dt)\}$. The rate expression for Eq. (32) is:

$$\frac{d[Y]}{dt} = -\frac{d[A]}{dt} = k[A]^{\alpha} \cdot [B]^{\beta} \tag{33}$$

where

- K = the rate constant,
- α = the partial order of the reaction with respect to reactant A, and
- β = the partial order of the reaction with respect to reactant B.

These orders are determined experimentally and are not necessarily integral numbers. The sum of all partial orders is the overall order (n) and is expressed as shown in Eq. (34):

$$n = \alpha + \beta + \dots \tag{34}$$

Once the values of α, β, etc., are determined experimentally, the rate law is defined. In reality, reaction order provides only information about the manner in which rate depends on concentration.

There are four types of rate laws that can be determined for solid phase sorption/desorption processes [109, 110]: mechanistic, apparent, transport with apparent, and transport with mechanistic rate laws, as follows:

- *Mechanistic rate laws* assume that only chemical kinetics is operational and transport phenomena are not occurring. Consequently, it is difficult to determine mechanistic rate laws for most solid phase systems due to the heterogeneity of the solid phase system caused by different particle sizes, porosities, and types of retention sites.
- *Apparent rate laws* include both chemical kinetics and transport-controlled processes. The apparent rate laws and rate coefficients indicate that diffusion and other microscopic transport processes affect the reaction rate.
- *Transport with apparent rate laws* emphasize transport phenomena and assume first-order or zero-order reactions.
- *Transport with mechanistic rate laws* describe simultaneous transport-controlled and chemical kinetics phenomena and explain accurately both the chemistry and the physics of the solid phase system.

3.2
Reaction Order and Rate Constant Determinations

The basic techniques to determine the rate laws and rate constants of a solid phase chemical reaction include initial rate, integrated equations and data plotting, and a nonlinear least square analyses [10, 23, 108, 109, 111, 112].

3.2.1
Initial Rate Equations

Assuming the following elementary reaction between species A, B, and Y (Eq. 35):

$$A + B \underset{k_2}{\overset{k_1}{\longleftrightarrow}} Y \tag{35}$$

A forward reaction rate law can be written as:

$$\frac{d[A]}{dt} = -k_1[A][B] \tag{36}$$

where k_1 is the forward rate constant, and α and β (Eq. 33) are each assumed to be 1. The reverse reaction rate law for Eq. (35) is:

$$\frac{d[A]}{dt} = +k_{-1}[Y] \tag{37}$$

Equations (36) and (37) are only applicable far from equilibrium where back or reverse reactions are insignificant. If both these reactions are occurring, Eqs. (36) and (37) must be combined such that:

$$\frac{d[A]}{dt} = -k_1[A][B] + k_{-1}[Y] \tag{38}$$

Equation (38) applies the principle that the net reaction rate is the difference between the sum of all reverse reaction rates and the sum of all forward reaction rates.

One way to ensure that back reactions are not important is to measure initial rates. The initial rate is the limit of the reaction rate as time reaches zero. With an initial rate method, one plots the concentration of a reactant or product over a short reaction time period during which the concentrations of the reactants change so little that the instantaneous rate is hardly affected. Thus, by measuring initial rates, one can assume that only the forward reaction in Eq. (35) predominates. This would simplify the rate law to that given in Eq. (36) which as written would be a second-order reaction, first-order in reactant A and first-order in reactant B. Equation (35), under these conditions, would represent a second-order irreversible elementary reaction.

3.2.2 *Integrated Rate Equations*

In general, the relationship between the rate of a chemical reaction (i.e., sorption/desorption), the concentration of a pollutant, and the reaction order, n, (i.e., 0, 1, 2), is given by:

$$r = C^n \quad \text{and} \quad \log r = n \log C \tag{39}$$

where

- r = the rate of the reaction,
- n = the order of the reaction, and
- C = concentration of pollutant.

Zero-order is defined where the rate of reaction is independent of the concentration. First-order is defined where the rate is directly proportional to the concentration. Second-order is defined where the rate is proportional to the square of the concentration. The following section presents the different reaction order equations.

3.2.2.1 *Zero-Order Reaction*

Considering the following zero-order reaction, where the single organic pollutant A is lowered in concentration, the rate of the reaction of pollutant A, according to zero-order kinetics, is:

$$-\frac{d[A]}{dt} = k_0 \tag{40}$$

where the minus sign indicates that the concentration of A is reduced with time. If C represents the concentration of A at any time t, and k_0 is the reaction rate constant then:

$$-\frac{d[C]}{dt} = k_0 \tag{41}$$

Integrating:

$$\begin{aligned} &C = -k_0 t + \text{constant} \quad \text{when} \quad C = C_0 \quad \text{at time} \quad t = 0 \\ &C - C_0 = -k_0 t \quad \text{or} \quad C = C_0 \cdot e^{(-k_0 t)} \end{aligned} \tag{42}$$

A useful measure of a pollutant of interest is its half-life time, i.e., the time it takes the pollutant to react/adsorb to 50% completion or half its initial concentration, as follows:

$$\frac{C_0}{2} - C_0 = -k_0 t \quad \text{then} \quad t_{0.5} = \frac{C_0}{2k_0} \tag{43}$$

3.2.2.2
First-Order Reaction

The rate of reaction of a pollutant A for first-order kinetics is as follows:

$$-\frac{d[C]}{dt} = k_1 \cdot C \tag{44}$$

where k_1 is the first-order rate constant and C the concentration at any time t. Integrating:

$$\ln\left(\frac{C_0}{C}\right) = k_1 t \quad \text{or} \quad \log\left(\frac{C_0}{C}\right) = \frac{k_1 t}{2.3} \tag{45}$$

The half-life constant is:

$$\ln\left(\frac{C_0}{C_0/2}\right) = k_1 t_{0.5} \quad \text{then} \quad t_{0.5} = \frac{\ln(2)}{k_1} = \frac{0.69}{k_1} \tag{46}$$

3.2.2.3
Second-Order Reaction

The rate of reaction of a pollutant A for second-order kinetics is described by:

$$-\frac{d[C]}{dt} = k_2 \cdot C^2 \tag{47}$$

where k_2 is the second-order reaction rate constant. Integrating:

$$\frac{1}{C} - \frac{1}{C_0} = k_2 t \tag{48}$$

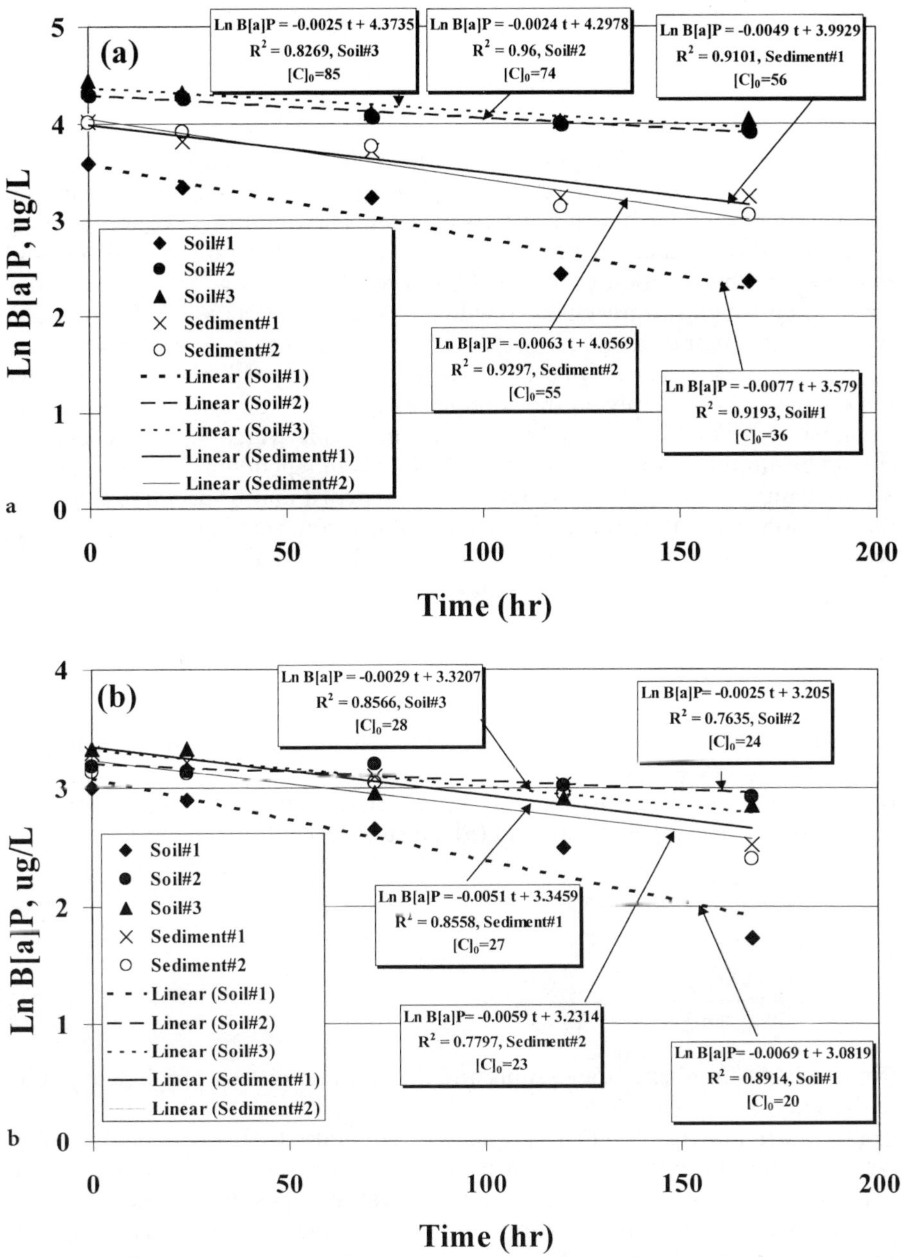

Fig. 1 a, b. Example of the first order plots of benzo[*a*]pyrene at two different concentrations: **a** high; **b** low

The half-life constant is:

$$\frac{1}{C_0/2} - \frac{1}{C_0} = k_2 t_{0.5} \quad \text{then} \quad t_{0.5} = \frac{1}{k_2 C_0} \tag{49}$$

An example of first-order plots is shown in Fig. 1 for benzo[*a*]pyrene (i.e., B[a]P) sorption on three different soils (in terms of organic matter content) and two sediment samples (marine and fresh water) at two different concentrations [1]. It can be noted that the plots are linear at both concentrations, which would indicate that the sorption process is first order. The findings that the rate constants are not significantly changed with concentration is a good indication that the reaction is first order under the experimental conditions that were imposed.

In general, it is not strictly correct to conclude that a particular reaction order fits the data based simply on the conformity of data to an integrated equation. As illustrated above, multiple initial concentrations which vary considerably should be employed to assess whether the rate is independent of concentration. Multiple integrated equations should also be tested. It may be useful to show that the reaction rate is not affected by species whose concentrations do not change considerably during an experiment; these may be substances not consumed in the reaction (i.e., catalysts) or present in large excess [23, 108].

3.2.3
Least Squares Analysis

With this method, the best straight line is fitted to a set of points that are linearly related as "$y = mx + b$", where y is the ordinate and x is the abscissa datum point, respectively. The slope (m) and the intercept (b) can be calculated by least squares analysis using Eqs. (50) and (51), respectively [23]:

$$m = \frac{n \sum xy - \sum x \sum y}{n \sum x^2 - (\sum x)^2} \tag{50}$$

$$b = \frac{\sum y \sum x^2 - \sum x \sum (xy)}{n \sum x^2 - (\sum x)^2} \tag{51}$$

where n is the number of data points and the summations are for all data points in the set.

Curvature may result when kinetic data are plotted. This may be due to an incorrect assumption of reaction order. If first-order kinetics is assumed and the reaction is really second order, downward curvature is observed. If second-order kinetics is assumed but the reaction is first-order, upward curvature is observed. Curvature can also be due to fractional, third, higher, or mixed reaction orders. Non-attainment of equilibrium often results in downward curvature. Temperature changes during the study can also cause curvature; thus, it is important for temperature to be controlled accurately during a kinetic experiment.

3.3
Temperature Effect On Reaction Rates

Temperature has a marked effect on the kinetics of reaction rates of solid phase sorption/desorption processes [113–116]. Arrhenius noted the following relationship between k and T (Eq. 52):

$$k = A_f \cdot e^{\left(-\frac{E_a}{RT}\right)} \tag{52}$$

where

- A_f = a frequency factor, and
- E_a = the energy of activation.

Converting Eq. (52) to linear form results in Eq. (53):

$$\ln k = \ln A_f - \left(\frac{E_a}{RT}\right) \tag{53}$$

A plot of ($\ln k$) vs ($1/T$) yields a linear relationship with the slope equal to ($-E_a/R$) and the intercept equal to ($\ln A_f$). Thus, by measuring (k) values at several temperatures, the (E_a) value can be determined. Low E_a values (<42 kJ mole) usually indicate diffusion-controlled transport processes, whereas higher E_a values indicate chemical reaction or surface-controlled processes [21, 25].

3.4
Kinetics Modeling Techniques

To interpret the kinetics experimental data of an organic pollutant(s) or leachate from complex organic mixtures, it is necessary to determine the adsorption/desorption process steps in a given experimental system which govern the overall adsorption/desorption rate. For instance, the adsorption process of an organic compound by a porous adsorbent can be categorized as three consecutive steps:

- The first is pollutant transport across the boundary layer or surface film to the exterior surface of the adsorbent solid phase particle (i.e., soils/sediments and their components).
- The second is pollutant transport within the pores of the adsorbent solid phase particle, from the exterior of the particle to the interior surfaces of the particle. Similarly, a pollutant may be transported along surfaces of pore walls.
- The final step is the physical or chemical binding of the organic pollutant to the interior surface of the adsorbent.

While first-order models have been used widely to describe the kinetics of solid phase sorption/desorption processes, a number of other models have been employed. These include various ordered equations such as zero-order, second-order, fractional-order, Elovich, power function or fractional power, and parabolic diffusion models. A brief discussion of these models will be provided; the final forms of the equations are given in Table 2.

Table 2. Final forms of kinetic modeling equations

Model Name	Equation
Zero Order	$C - C_0 = -k_0 t \quad \text{or} \quad C = C_0 \cdot e^{(-k_0 t)}$
First Order	$\ln\left(\frac{C_0}{C}\right) = k_1 t \quad \text{or} \quad \log\left(\frac{C_0}{C}\right) = \frac{k_1 t}{2.3}$
Second Order	$\frac{1}{C} - \frac{1}{C_0} = k_2 t$
Elovich	$q_t = \left(\frac{1}{\beta}\right) \ln\left(\frac{\alpha}{\beta}\right) + \left(\frac{1}{\beta}\right) \ln(t)$
Parabolic Diffusion	$\left(\frac{q_t}{q_\infty}\right) = \left(\frac{4}{\pi^{1/2}}\right)\left(\frac{Dt^{1/2}}{r^2}\right) - \left(\frac{Dt}{r^2}\right)$
Fractional Power or Power Function	$q = k t^{\upsilon}$
External Film Diffusion	$-\frac{dC}{dt} = K_f \cdot a \left[C - \frac{(C_0 - C)}{b \cdot [QM - (C_0 - C)]}\right]$
Internal Surface Diffusion	$\frac{\partial q(r,t)}{\partial t} = D_S \cdot \left[\frac{\partial^2 q(r,t)}{\partial r^2} + \frac{2}{r}\frac{\partial q(r,t)}{\partial t}\right]$
Linear-Driving-Force Approximation	$-\frac{dC}{dt} = K_p \cdot a \cdot \left[\frac{MQbC}{(1+bC)} + (C_0 - C)\right]$
Surface Reaction	$-\frac{dC}{dt} = K_a \left[C(QM - C_0 + C) - \frac{1}{b}(C - C_0)\right]$

3.4.1 Elovich Model

The Elovich model was originally developed to describe the kinetics of heterogeneous chemisorption of gases on solid surfaces [117]. It describes a number of reaction mechanisms including bulk and surface diffusion, as well as activation and deactivation of catalytic surfaces. In solid phase chemistry, the Elovich model has been used to describe the kinetics of sorption/desorption of various chemicals on solid phases [23]. It can be expressed as [118]:

$$q_t = \left(\frac{1}{\beta}\right) \cdot \ln\left(\frac{\alpha}{\beta}\right) + \left(\frac{1}{\beta}\right) \cdot \ln(t) \qquad (54)$$

where

- q_t = the amount of sorbate per unit mass of sorbent at time (t), and
- α and β = constants during any one experiment.

A plot of (q_t) vs $(\ln t)$ should give a linear relationship if the Elovich model is applicable, with a slope of $(1/\beta)$ and an intercept of $[(1/\beta).\ln(\alpha\beta)]$. Some investigators have suggested that multiple linear segments in Elovich plots could indicate a changeover from one type of binding site to another; however, Sparks [23] questioned the correctness of such mechanistic interpretations.

3.4.2
Parabolic Diffusion Model

The parabolic diffusion model is used to indicate that diffusion controlled phenomena are rate limiting. It was originally derived based on radial diffusion in a cylinder where the chemical compound concentration on the cylindrical surface was constant, and initially the chemical compound concentration throughout the cylinder was uniform. It was also assumed that the diffusion of the compound of interest through the upper and lower faces of the cylinder was negligible. Following Crank [119], the parabolic diffusion model can be expressed as:

$$\left(\frac{q_t}{q_\infty}\right) = \left(\frac{4}{\pi^{1/2}}\right)\left(\frac{Dt^{1/2}}{r^2}\right) - \left(\frac{Dt}{r^2}\right) \tag{55}$$

where

- r = the average radius of the solid particle,
- q_t = as defined earlier,
- q_∞ = the corresponding quantity of sorbate at equilibrium, and
- D = the diffusion coefficient.

Equation (55) can be simply expressed as:

$$\left(\frac{q_t}{q_\infty}\right) = (R_D t^{1/2}) + \text{constant} \tag{56}$$

where R_D is the overall diffusion coefficient. If the parabolic diffusion law is valid, a plot of (q_t/q_∞) vs $(t^{1/2})$ should yield a linear relationship. The parabolic diffusion model has been applied successfully to various organic chemical reactions, especially pesticides on various solid phases [25, 120].

3.4.3
Fractional Power or Power Function Model

The Fractional Power or Power Function model can be expressed as:

$$q = kt^{\upsilon} \tag{57}$$

where

- q = the amount of sorbate per unit mass of sorbent,
- k = a constant,
- t = time, and
- υ = a positive constant (<1).

Equation (57) is empirical, except for the case where $\upsilon = 0.5$, then Eq. (57) is similar to the parabolic diffusion model. Equation (57) and various modified forms have been used by a number of researchers to describe the kinetics of solid phase sorption/desorption and chemical transformation processes [25, 121–122].

3.4.4
External Film Diffusion Model

According to Wermeulen [123] and Kuo et al. [124], if external film diffusion is the rate-controlling step, then the rate equation can be expressed by the following equation:

$$\frac{dq}{dt} = \left[\frac{K_f \cdot a}{M}(C - C^*)\right] \tag{58}$$

where

- $K_f \cdot a$ = the mass transfer coefficient,
- C = the adsorbate concentration in bulk liquid phase,
- C^* = the adsorbate concentration of the liquid that is in equilibrium with the solid phase concentration q, and
- M = the adsorbent dosage.

Assuming the adsorption isotherm can be expressed by the Langmuir model (Eqs. 3 and 4), i.e., $q = \left[\frac{QbC^*}{(1 + bC^*}\right]$ and taking advantage of the mass balance $Q = \frac{(C_0 - C)}{M}$ where C_0 is the initial adsorbate concentration, then Eq. (58) can be be changed to:

$$-\frac{dC}{dt} = K_f \cdot a\left[C - \frac{(C_0 - C)}{b \cdot [QM - (C_0 - C)]}\right] \tag{59}$$

Zogorski et al. [125] indicate that external transport is the rate-limiting step in systems having poor mixing, dilute concentration of adsorbate, small particle sizes of adsorbent, and a high affinity of adsorbate for adsorbent. Some experiments conducted at low concentrations have shown that film diffusion solely controls the adsorption kinetics of low molecular weight substances [81, 85].

3.4.5
Internal Surface Diffusion Model

In general, an adsorbate can diffuse by two mechanisms within the adsorbent, i.e., by pore and surface diffusion. For pore diffusion, the adsorbate is transported within the pore fluid. For surface diffusion, the adsorbate continues to move along the surface of the adsorbent to available adsorption sites as long as it has enough energy to leave its present site. Investigations have demonstrated that surface diffusion is the dominant mechanism, so the contribution of pore diffusion is neglected [126, 127]. Many researchers have used the surface diffusion model for describing the kinetic data or for design of adsorbents [128–130]. The partial differential equation for this model is written in spherical coordinates as:

$$\frac{\partial q(r,t)}{\partial t} = D_S \cdot \left[\frac{\partial^2 q(r,t)}{\partial r^2} + \frac{2}{r} \frac{\partial q(r,t)}{\partial t} \right] \tag{60}$$

where

- $q(r,t)$ = the solid-phase concentration along the inner particle surface,
- r = the radial coordinate with an origin at the particle center, and
- D_S = the surface diffusion coefficient.

The magnitude of D_S is a measure of how fast the molecules diffuse along the particle and therefore sets a time scale for the adsorption process. Two boundary conditions and one initial condition have to be specified in order to obtain a unique solution to Eq. (60). Initially the solid particle is free of adsorbate, which is expressed as:

$$q(r, t_{=0}) = 0 \tag{61}$$

The boundary condition at the center of the particle is:

$$\frac{\partial q(r_{=0}, t)}{\partial r} = 0 \tag{62}$$

Thus, no adsorbate fluxes across the center. Finally, the continuity of flux at the solid-liquid interface has to be satisfied:

$$\rho_p D_S \left[\frac{\partial q(r_{=d_p/2}, t)}{\partial r} \partial r \right] = K_f (C_b - C_S) \tag{63}$$

where:

- C_b and C_S = the bulk liquid and solid-liquid interface adsorbate concentrations, respectively,
- d_p = the particle diameter,
- ρ_p = the apparent density of the particle, and
- K_f = the liquid film mass transfer coefficient.

The parameter K_f is a measure of how fast the molecules diffuse across the stagnant liquid film layer. It is assumed that local equilibrium occurs at the exterior solid particle surface. The average solid phase loading, which is only a function of time, is given by:

$$q_{avg} = \left[\frac{3}{(d_p/2)^3} \right] \int_0^{d_p/2} q(r,t) \cdot r^2 \cdot dr \tag{64}$$

3.4.6
Linear-Driving-Force Approximation Model

The surface diffusion model (Eq. 60) is usually approximated by the linear-driving-force relation [124]:

$$\frac{dq}{dt} = K_p \cdot a(q^* - q) \tag{65}$$

where

- $K_p \cdot a$ = the mass transfer coefficient, and
- q = the solid-phase concentration in equilibrium with the instantaneous fluid-phase concentration outside the particle.

If the adsorption isotherm can be expressed by the Langmuir model, i.e., $q^* = \dfrac{QbC}{(1+bC)}$ and the mass balance $q = \dfrac{(C_0 - C)}{M}$ is used, Eq. (65) becomes

$$-\frac{dC}{dt} = K_p \cdot a \cdot \left[\frac{MQbC}{(1+bC)} + (C - C_0)\right] \tag{66}$$

3.4.7
Surface Reaction Model

For the case where surface reaction is the rate controlling step [124], the rate of adsorption can be expressed as:

$$q_{avg} = \left[\frac{3}{(d_p/2)^3}\right] \cdot \int_0^{d_p/2} q(r,t) \cdot r^2 \cdot dr \tag{67}$$

$$\frac{dq}{dt} = K_a \cdot \left[C \cdot (Q - q) - \frac{q}{b}\right] \tag{68}$$

where

- K_a = the surface reaction rate constant, and
- Q and b = the Langmuir adsorptive capacity and equilibrium constant, respectively.

Using the mass balance $q = \dfrac{(C_0 - C)}{M}$, Eq. (68) changes to:

$$-\frac{dC}{dt} = K_a \left[C(QM - C_0 + C) - \frac{1}{b}(C_0 - C)\right] \tag{69}$$

The adsorption process can be described as molecules leaving a solution and being held on the solid surface by chemical and physical bonding. If the bonds that form between the adsorbate and adsorbent are very strong, the process is almost always irreversible [97–99], and chemical adsorption (i.e., chemisorption) is said to have occurred. On the other hand, if the bonds that are formed are weak, as is characteristic of bonding by dispersion interactions or hydrogen bonding, then physical adsorption (i.e., physisorption) has occurred. The molecules adsorbed by physisorption are easily removed or desorbed by a change in the solution concentration of the adsorbate, and for this reason, the process is reversible. There is a difference in the activation energy for the adsorption reaction by physisorption vs chemisorption. For chemical adsorption/bonding, the activation energy is higher than 10 kcal/mole, and for dispersion interactions and hydrogen bonding, it ranges from 2 kcal/mole to 10 kcal/mole. Kuo et al. [124] showed that the adsorption rate of dissolved organics from in situ tar sand by-product waters could be described by the surface reaction kinetics (i.e., Eqs. 67–69).

3.4.8
Comparison of Kinetic Models

A number of studies reported that several kinetic models can describe rate data well, when based on correlation coefficients and standard errors of the estimates [25, 118, 131, 132]. Despite this, there often is no consistent relation between the equation which gives the best fit and the physicochemical and mineralogical properties of the adsorbent(s) being studied. Another problem with some of the kinetic equations is that they are empirical and no meaningful rate parameters can be obtained.

One of the reasons a particular kinetic model appears to be applicable may be that the study is conducted during the time range when the model is most appropriate. While sorption, for example, decreases over many orders of magnitude before equilibrium is approached, with most methods and experiments, only a portion of the entire reaction is measured and over this time range the assumptions associated with a particular equation are generally valid.

The fact that diffusion models describe a number of chemical processes in solid particles is not surprising since in most cases, mass transfer and chemical kinetics phenomena occur simultaneously and it is difficult to separate them [133–135]. Therefore, the overall kinetics of many chemical reactions in soils may often be better described by mass transfer and diffusion-based models than with simple models such as first-order kinetics. This is particularly true for slower chemical reactions in soils where a fast reaction is followed by a much slower reaction (biphasic kinetics), and is often observed in soils for many reactions involving organic and inorganic compounds.

4
Experimental Techniques and Transport Parameters

Generally, there is no simple and easy theoretical procedure which can provide exact or nearly precise quantitative predictions of what and how much will be adsorbed/desorbed by any solid phase over a period of time [9, 136–139]. Understanding sorption/desorption characteristics of any solid phase materials requires two main laboratory experimental techniques: (a) batch equilibrium testing, and (b) continuous solid phase column-leaching testing. These involve two completely different kinds of experimental tests, and the sorption characteristics determined from either one should not be confused with the other. Sorption isotherms are obtained by carrying out batch equilibrium tests and apply to solid phase suspensions [140]. The physical model which is assumed with this experiment is a system with completely dispersed solid phase particles, where all the solid particle surfaces are exposed and available for interaction with the pollutant of interest. On the other hand, column-leaching tests are performed with intact solid phase samples which have a definite matrix and solid structure. The sorption/desorption characteristics obtained from these tests are required in order to:

- Study soil sorption and desorption of pollutants in complex mixtures and/or leached from SWMs
- Estimate pore volume numbers required to achieve a specific organic pollutant breakthrough curve
- Provide information necessary for the retardation parameter calculation required in the pollutant transport equation
- Determine the transport parameters that control pollutant migration through the subsurface environment (i.e., diffusion/dispersion and diffusion coefficients)

4.1
Background and Theory

Whereas batch equilibrium tests are designed to study equilibrium sorption of solid phase particles with various pollutants, singly or in combination with other pollutants, solid phase column-leaching tests study both sorption and diffusion of organic pollutants through the subsurface environment [10, 11, 127, 141, 142].

4.1.1
Batch Equilibrium Tests

Batch equilibrium tests are conducted on solid phase suspensions, prepared with previously air-dried solids, ground to uniform powdery texture for mixing with various concentrations of the pollutants of interest in solution. The concentrations of these pollutants or the COMs leachate in the solution are designed to evaluate the capability of the suspended solids to adsorb all the pollutants possible with increasing amounts of available pollutants, consistent with interaction characteristics dictated by the surface properties of the solids and the pollutants [1, 16, 22–26, 66, 67, 71]. For a successful and proper study of solid particle sorption of pollutants, the requirement for complete dispersion of solid particles in solution is absolute [143–145]. Common practice is to use a solution to solid ratio of 10:1 [1], together with efficient sample agitation at a constant temperature (e.g., 48 h at 20 °C).

When the equilibrium concentration of the organic pollutant (C) is obtained from measurements of the liquid phase concentration of the pollutant, a sorption isotherm can be constructed. Defining the initial concentration of the pollutant without any solid particles as (C_0), the adsorption mass ratio (q) is computed for each subsample as follows:

$$q = \frac{(C_0 - C) \cdot V}{M} \tag{70}$$

where

- V = the liquid volume in the subsample, and
- M = the mass of solid particles in the subsample.

The numerator in Eq. (70) represents the mass of pollutant adsorbed onto the solid phase. This in turn is divided by the mass of the solid particles to obtain a measure of the relative mass of the constituent adsorbed on the solid phase.

The values of q are plotted as a function of the equilibrium concentration. For constituents at low or moderate concentrations, the relationship between q and C can be generated. If $n=1$, the ($q-C$) relationship will be linear (Eq. 9), and the slope of the line (i.e., K_d) defines the adsorption distribution of the pollutant. K_d is generally identified as the *distribution or partition coefficient*, and is used to describe pollutant partitioning between liquid and solids only if the reactions that cause the partitioning are fast and reversible, and if the isotherm is linear. For cases where the partitioning of the pollutants can be adequately described by the distribution coefficient (i.e., fast and reversible adsorption, with linear isotherm), the retardation factor (R) of the subsurface environment can be used as follows:

$$R = 1 + \left(\frac{P_d}{\theta}\right) \cdot K_d \tag{71}$$

where

- P_d = the dry mass density (mass of dry solids divided by the total volume of the soil/sediment specimen used in a leaching-column test) of the test specimen, and
- θ = the volumetric water content of the test specimen.

The retardation term can also be expressed as the ratio of the breakthrough time of an adsorbed pollutant relative to the elution time of a non-adsorbed tracer. In addition, parameter R can be used to estimate the number of pore volumes of flow required to achieve breakthrough, assuming that breakthrough of a non-adsorbed tracer would occur at one pore volume of flow.

The adsorption of a pollutant by solid particles does not proceed as a step function, where all the pollutant molecules are adsorbed up to a maximum capacity, with any additional amount left in solution. Hence, there will be equilibrium between the pollutants in solution/leachate and that adsorbed. For small amounts of pollutant, only a trace amount will remain in solution. The sorption isotherms must be evaluated on the basis of how much pollutant can be tolerated in the solution phase. Furthermore, part of the adsorption is reversible [97–100], and the pollutant can be desorbed with water or a salt solution.

Batch equilibrium tests used for sorption isotherm determinations involve solid suspensions (i.e., the full surface area of the solid particles is exposed to contact with the pollutant/leachate). This is to be distinguished from column-leaching tests, where the pollutant in solution/leachate travels through the solid phase sample. Because of the solid phase structure, pore geometry, and pore continuity, only a fraction of the total surface area of the solid particles comes in direct contact with the permeating pollutant in solution/leachate. The sorption isotherms obtained therefore should be identified as adsorption characteristics, to distinguish them from the sorption isotherms obtained from batch equilibrium tests. Nevertheless, batch equilibrium tests can provide valuable insight into sorption characteristics of a solid phase.

4.1.2
Continuous Column-Leaching Tests

Determination of sorption characteristics of a soil-solid phase requires simulation of passage of the pollutants in the solution/leachate being studied and the test material. To accomplish this, a soil column, sometimes called a leaching cell, is used. Figure 2 shows a schematic diagram of the continuous column-leaching test experiment. The concentration (C) of a chemical species appearing in the effluent reservoir is measured over time and the results are plotted in the form of leachate solute breakthrough curves, or relative concentration (C/C_0) vs time (t) or pore volumes (PV) of flow. A pore volume of flow for a saturated soil is the cumulative volume of flow through the soil divided by the volume of the void space in the soil. Expressing total pollutant solution/leachate flow in terms of PV as opposed to time taken for the total leachate to pass through the soil is a more convenient method for result examination. In this manner, comparison between different situations can be evaluated without complicating the problem of time-effect on sorption characteristics, i.e., one is generally interested in how much the soil can adsorb before complete exhaustion of its buffer or adsorption capacity.

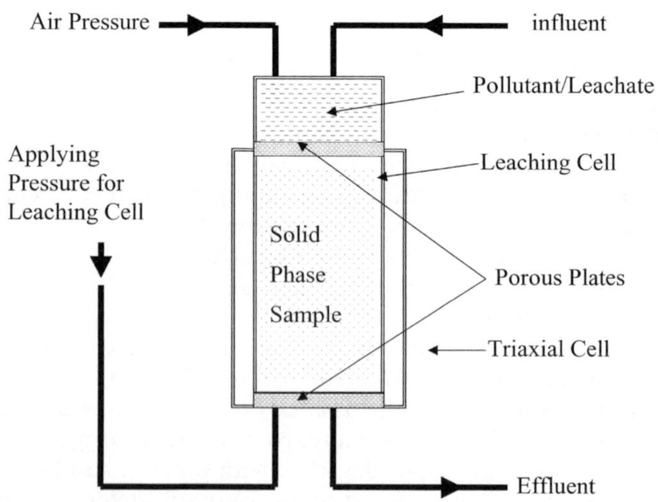

Fig. 2. Schematic diagram of the continuous column-leaching test experiment

Several workers [1, 29, 66, 67, 104, 146–149] indicated that studying pollutants and/or SWM leachate migration profiles resulting from transport of pollutants with a test soil requires that replicate samples be subjected to leaching-column tests, where various pore volumes of the same solution are applied.

4.2 Estimation of Transport Parameters

In order to predict pollutant chemodynamics of COMs and/or their leachates, the transport parameters involved in the governing sets of equations that describe the transport process need to be defined accurately [1]. In general, methods used to calculate the transport parameters fall into two broad categories, i.e., steady and transient states.

4.2.1 *Steady State Methods*

Steady state methods used to estimate transport parameters [150, 151], require the use of the general fate and transport equations, which include three different techniques: (1) decreasing source concentration, (2) time-lag method, and (3) root time method. The next sections present these methods.

4.2.1.1 *Decreasing Source Concentration*

The schematic diagram illustrating the decreasing source method for diffusion transport determination of any organic pollutant in solution or leached from COMs is shown in Fig. 3. The soil-solid sample is contained between two re-

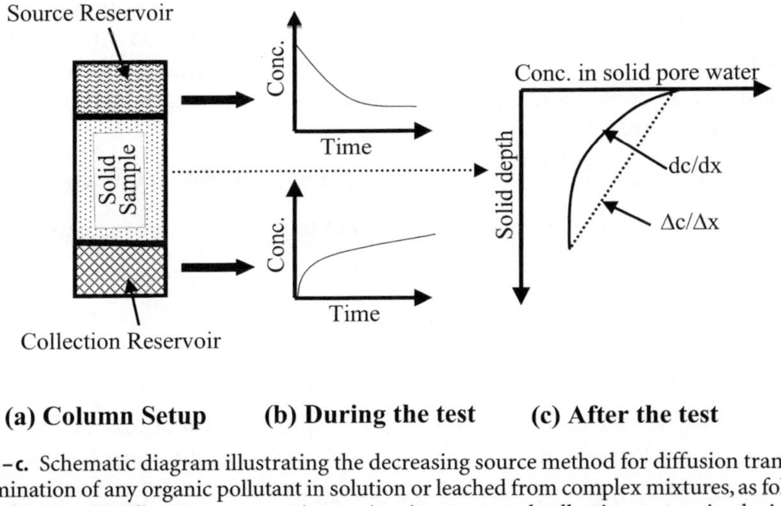

Fig. 3a–c. Schematic diagram illustrating the decreasing source method for diffusion transport determination of any organic pollutant in solution or leached from complex mixtures, as follows: **a** column setup; **b** pollutant concentration vs time in source and collection reservoirs during the test; **c** pollutant concentration in solid-pore water with depth from source after the test

servoirs, a source reservoir containing the complex mixture pollutants of interest, and a collection reservoir from which samples are withdrawn for further organic analyses (Fig. 3a). The initial test condition establishes the pollutant concentration to be higher in the source reservoir than in the collection reservoir (Fig. 3b). In this manner, this results in a chemical concentration gradient across the soil sample and pollutant diffusion across the sample (Fig. 3c). The test condition does not require replenishment of the pollutants in the source reservoir. Only the solution volumes are kept constant in both source and collection reservoirs.

When the pollutant concentration difference between the source and collection reservoir becomes smaller (i.e., when the concentration of pollutants in the collection reservoir approaches that of the source reservoir), the flux rate of pollutants decreases, and a near steady state flux (J_s) is obtained (Fig. 3c). At this time, the diffusion parameter (D) can be calculated using Fick's model as follows:

$$J_s = -D \cdot \left(\frac{\partial c}{\partial x} \right) \tag{72}$$

Hence:

$$D = -\left(\frac{\Delta x}{\Delta c} \right) \cdot J_s = -\left(\frac{L}{\Delta c} \right) \left(\frac{\Delta m}{A \cdot \Delta t} \right) = -\left(\frac{L}{A \cdot \Delta c} \right) \left(\frac{\Delta m}{\Delta t} \right) \tag{73}$$

where

- J_s = the mass flux,
- D = the diffusion parameter,
- L and A = length and cross sectional area of the soil sample, respectively, and
- Δm = change in mass of the organic pollutant in an increment of time (Δt).

For best application, the test should be conducted with initial conditions set such that the pollutant concentration differences between source and collection reservoirs are relatively small. In this manner, the difference between the curve shown in Fig. 3c and a straight line will be relatively small. Obtaining high precision and repeatability in measurements at low concentration differences and fluxes are most critical and essential. Unless that can be attained, this procedure should not be used.

Since the quantity $\left(\frac{\Delta m}{\Delta t} \right)$ in Eq. (73) can be measured or set independently of the test, only the change in mass with respect to time $\left(\frac{\Delta m}{\Delta t} \right)$ is measured during the test. At steady state (or nearly so):

$$-\left(\frac{\Delta m_1}{\Delta t} \right) = \left(\frac{\Delta m_2}{\Delta t} \right) = \left(\frac{\Delta m}{\Delta t} \right) \tag{74}$$

where

- Δ_{m_1} = the decrease in mass of the chemical species in the source reservoir, and
- Δ_{m_2} = the increase in mass of the chemical species in the collection reservoir.

The use of the difference operator in Eq. (73) implies that the concentration gradient across the sample is linear. However, due to coupled flow processes, the concentration gradient within the soil sample is non-linear. As a result, the calculated diffusion parameter using the total (across the sample length) concentration gradient may not be the same as that determined using the incremental (along the sample length) distribution of concentration, as might be directly deduced from Fig. 3c.

4.2.1.2
Time-Lag Method

This method is commonly used to obtain the diffusion coefficient through porous membranes. The schematic diagram illustrating the best technique for application of the time-lag method for determination of diffusion transport is shown in Fig. 4. As in the test setup shown in Fig. 4a, the soil is contained between the source and collection reservoirs. Using this technique for diffusion coefficient determination of pollutants requires that the following conditions are satisfied:

- The concentrations of the organic pollutant species should be higher in the source reservoir than the collection reservoir.
- The organic pollutant species diffusing from the source reservoir must be continuously replenished while the mass of the organic pollutant species diffusing into the collection reservoir is continuously removed in order to maintain a constant concentration difference across the sample. This is shown as the pollutant flushing system in Fig. 4a, b.

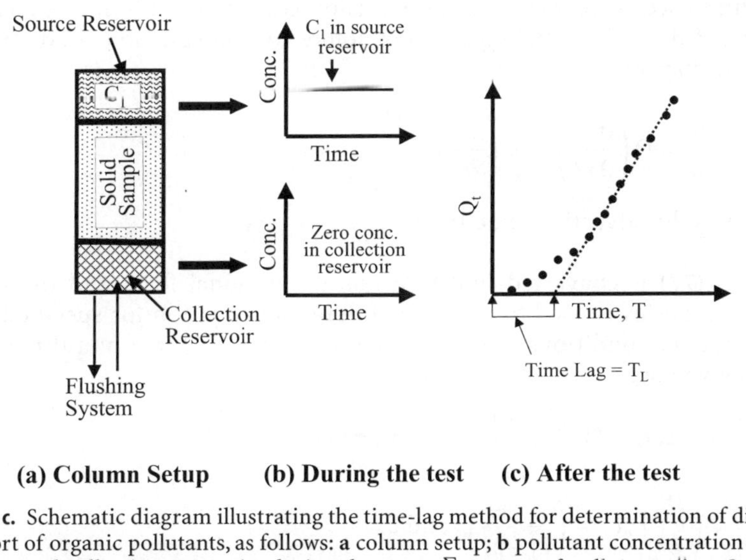

Fig. 4a–c. Schematic diagram illustrating the time-lag method for determination of diffusion transport of organic pollutants, as follows: **a** column setup; **b** pollutant concentration vs time in source and collection reservoirs during the test; **c** $\sum$ amount of pollutants (i.e., Q_t) transported through solid particles with time after the test

The total amount of diffusing substance per cross sectional area (Q_t) which has passed through the soil-solids approaches a steady state value as (t) approaches infinity [58, 119, 152]

$$Q_t = \left(\frac{Dc_1}{L}\right)\left(t - \frac{L^2}{6D}\right) \tag{75}$$

where

- L = the length of test sample, and
- C_1 = the concentration in the source reservoir, which is maintained at a constant value with time.

Equation (75) yields a straight line on a plot of Q_t vs time as shown in Fig. 4c. The intercept on the time axis is the time lag (T_L), which is given by:

$$T_L = \left(\frac{L^2}{6D}\right) \tag{76}$$

The diffusion coefficient (D) can be calculated using Eq. (76) by plotting Q_t vs time and determining the value for the intercept T_L.

4.2.1.3
Root Time Method

The root time method of data analysis for diffusion coefficient determination was developed by Mohamed and Yong [142] and Mohamed et al. [153]. The procedure used for computing the diffusion coefficient utilizes the analytical solution of the differential equation of solute transport in soil-solids (i.e., the diffusion-dispersion equation):

$$\frac{\partial c}{\partial t} = D\left(\frac{\partial^2 c}{\partial x^2}\right) - v_x\left(\frac{\partial c}{\partial x}\right) \tag{77}$$

where v_x is the advective velocity.

Equation (77) is converted first to a non-dimensional form and the Fourier Transform Series is used to solve the differential equation for specified initial and boundary conditions. The final solution of Eq. (77) in a non-dimensional form is given by:

$$c^*(\zeta, \tau) \sim (1 - \zeta) - \frac{1}{\pi} e^{-\pi^2 \tau} \sin(\pi \zeta) \tag{78}$$

where $\tau = \left(\frac{Dt}{L^2}\right)$; $c^* = \left(\frac{c - c_2}{c_1 - c_2}\right)$; and $\zeta = \frac{x}{L}$

The relative change in concentration is given by:

$$c^*_{RC} = e^{-\pi^2\tau} \tag{79}$$

where

- τ = the non-dimensional time factor,
- ζ = the non-dimensional distance,
- c^* = the non-dimensional concentration,
- c = the concentration at specific time and distance,
- c_1 = the concentration at $x = 0$,
- c_2 = the concentration at $x = L$, and
- L = the length of soil specimen.

The test, theoretical relationship between the non-dimensional relative concentration (c^*_{RC}), and the root time factor (τ) may be seen in Fig. 5. Mohamed and Yong [142] analyzed the results obtained from the diffusion experiment shown in Fig. 5a, b, using the information from solution of the equation above. The theoretical correlation in Fig. 5c shows a linear relationship up to a relative concentration of 0.2 (80% equilibrium). At a relative concentration of 0.1 (90% equilibrium), the abscissa is used to determine the point on the experimental curve corresponding to a relative concentration of 0.1 (i.e., 90% of the steady state equilibrium time).

When the data obtained from the experimental system shown in Fig. 5 are reduced in terms of relative concentrations of specific ions in the collected effluent vs square-root time, the experimental curve obtained shows a linear portion, followed by a curve (Fig. 5d). The point (D) corresponding to the initial

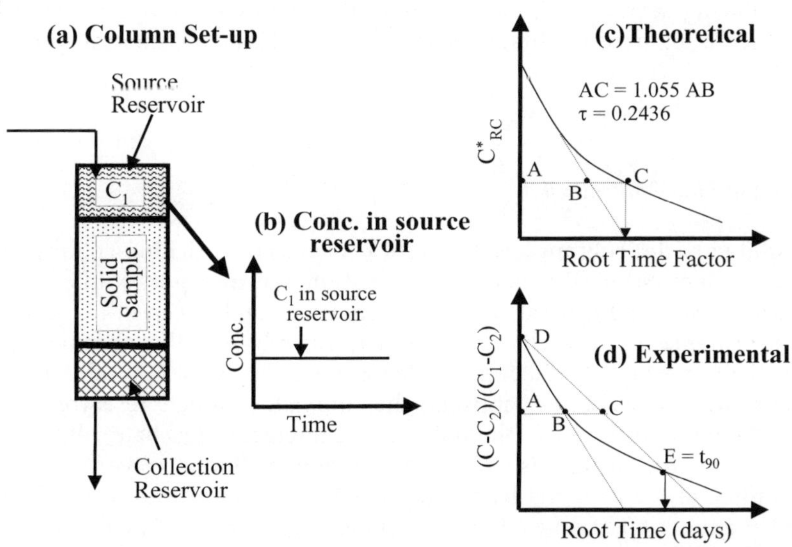

Fig. 5a–d. The theoretical relationship between the non-dimensional relative concentration (c^*_{RC}) and the root time factor (τ)

condition is obtained by projecting the linear part of the curve to zero time. A straight line (DE) is then drawn having abscissa 1.055 times the corresponding abscissa on the linear portion of the experimental plot. The intersection of the line (DE) with the experimental plot locates the point (t_{90}) corresponding to a relative concentration of 0.1 and the corresponding value t_{90}, can be obtained. The value of τ corresponding to $c^{*}_{RC} = 0.1$ is 0.2436 and the diffusion coefficient (D) is given by:

$$D = 0.2436 \left(\frac{L^2}{t_{90}} \right) \tag{80}$$

The diffusion parameter calculated by the root time method is an average parameter, and is generally considered to be operative over the range of time from initial diffusion flux to near steady state flux conditions. The method is applicable for the situation where adsorption and desorption occur, and for various pH values of the influent. The closer (DE) is to (DB) in Fig. 5 d, the greater is the accuracy of the D coefficient. It is important to note that in the case of low pH values of the influent, desorption of cations from a clay soil could produce conditions where $C_2 > C_1$. Accordingly, the experimental values for relative change in concentration would then become negative.

4.2.2
Transient Methods

In general, experiments using transient methods utilize solutions to Eq. (92) (Sect. 4.2.2.3) to obtain so-called experimentally derived diffusion coefficients. The following sections will show briefly the common transient methods of experimentation used to obtain test data for calculations of the transport coefficient.

4.2.2.1
Column-Leaching Cell Method

The solid particles column-leaching cell, known as the leaching-column test, is a common method used for pollutant sorption and transport studies through subsurface soils. The general type of system used can be seen in Fig. 2. During the performance of the leaching experiment, a steady-state flow through the soil sample will be established using distilled water as the influent fluid. Then, after steady-state flow has been established, the fluid in the influent reservoir is changed to the test solution (i.e., the pollutant of interest or leachate), with known and constant concentrations ($C_0's$) of the various pollutant constituents of the COMs to be tested as a mix leachate. The effluent concentration (C_e) is determined as a function of time and pore volumes (PV), and the data reduced in the form of breakthrough curves, of relative concentration (C_e/C_0) vs time or pore volumes of flow (Fig. 6).

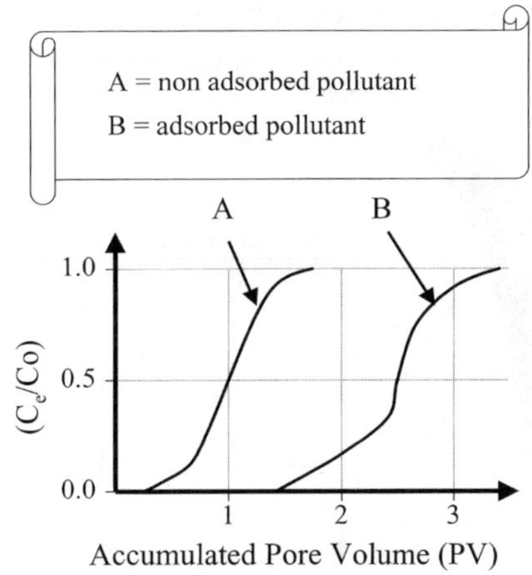

Fig. 6. The general column-leaching cell methods, with their breakthrough curves

Breakthrough curves such as those illustrated in Fig. 6 are typically analyzed using the following analytical solution for Eq. (81) [143, 154]:

$$\left(\frac{c_e}{c_0}\right) = 0.5\left[erfc\left(\frac{L - vt}{2.(Dt)^{0.5}}\right) + \exp\left(\frac{vL}{D}\right) efrc\left(\frac{L + vt}{2.(Dt)^{0.5}}\right)\right] \tag{81}$$

where

- L = the length of the soil column,
- v = the advective velocity,
- t = time, and
- $erfc$ = the complementary error function.

For any argument z, the $erfc$ is given by Eq. (82):

$$erfc(z) = 1 - erf(z) = 1 - \frac{2}{(\pi)^{0.5}}\int_0^z e^{-u^2}\,du \tag{82}$$

$$= 1 - \frac{2}{\pi}\left(z - \frac{z^3}{3X1!} - \frac{z^5}{5X2!} - \frac{z^7}{7X3!} + \ldots\right)$$

where $erf(z)$ is the error function of the argument (z). The diffusion coefficient can then be calculated once, C_e, C_0, v, L, and t are known.

The analytical technique assumes that the calculated diffusion coefficients for various individual pollutants represent average values throughout the length of the soil column. Although the interactions established between the pollutant and the soil cause continuous alteration in the transmissivity characteristics of the soil, the procedure which uses the analytical solution can only provide average values, because the values of C_e are obtained at the outlet end of the test sample. Thus, a representative diffusion coefficient should be calculated for individual layers in the soil column, and/or each pore volume passage of ef-

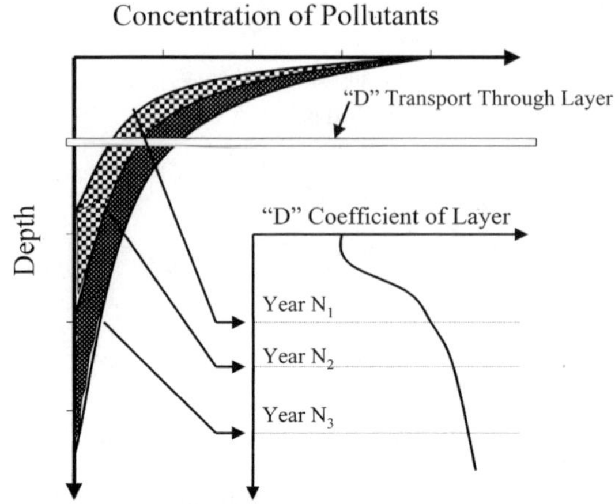

Fig. 7. Schematic diagram showing the variation of D with depth (i.e., length of the soil-solid sample) and with number of pore volumes of passage of COM solution

fluent, so long as only outlet values of pollutant concentration are the only sets of values obtained. Hence, the different values of D throughout the length of the sample cannot be calculated at any one time or pore volume passage of COM solution.

The test technique shown in Fig. 7 is used to determine the variation of D with depth (i.e., length of the soil-solid sample) and with number of pore volumes (i.e., PV) of passage of COM solution. By analyzing the various sections of a soil-solid sample, one can obtain the pollutant profile shown in Fig. 7. Casting Eq. (82) in the finite difference form yields the following:

$$\left(\frac{c_i^{j+1} - c_i^j}{\Delta t}\right) = D \cdot \left(\frac{c_{i+1}^j - 2c_i^j}{(\Delta x)^2}\right) - \upsilon \cdot \left(\frac{c_{i+1}^j - c_{i-1}^j}{2.(\Delta x)}\right) \tag{83}$$

Experimental data can be used to compute the diffusion coefficient based on Fig. 7 as a function of PV for a particular depth of soil, or as a function of depth. One should recognize the importance of the determination of D values in relation to elapsed time and distance from pollutant source. In many instances, the prediction of the advance of a pollutant plume and rate can be very sensitive to the specification of the D coefficient.

4.2.2.2
Adsorption/Desorption Function

Of the various equilibrium and non-equilibrium sorption isotherms or sorption characteristics models, the most popular are the Langmuir and Freundlich models. The correct modeling of an adsorbate undergoing both transport and adsorption through a clay soil-solid system necessitates the selection of an adsorption isotherm or characteristic model which best suits the given system. The use of an improper or inappropriate adsorption model will greatly affect the transport model, and possibly result in erroneous conclusions regarding the nature and description of both parameters and processes of the system analyzed. Thus, it is particularly useful to obtain similar conditions of adsorbate-absorbent interactions for a specific model application if adsorption is chosen for modeling. Hence, it is necessary to obtain a match both adsorbate and absorbent compositions.

The linear equilibrium isotherm adsorption relationship (Eq. 11) requires a constant rate of adsorption, and is most often not physically valid because the ability of clay solid particles to absorb pollutants decreases as the adsorbed amount of pollutant increases, contrary to expectations from the liner model. If the rate of adsorption decreases rapidly as the concentration in the pore fluid increases, the simple Freundlich type model (Eqs. 8 and 9) must be extended to properly portray the adsorption relationship. Few models can faithfully portray the adsorption relationship for multicomponent COM-pollutant systems where some of the components are adsorbed and others are desorbed. It is therefore necessary to perform initial tests with the natural system to choose the adsorption model specific to the problem at hand. From leaching-column experimental data, using field materials (soil solids and COMs solutions), and model calibration, the following general function can be successfully applied [155]:

$$S_j^{ad} = E_j - B_j \exp(-A_j C_j) \tag{84}$$

where E_j, B_j, and A_j are adsorption parameters for component j to be determined from calibration experiments. In a multicomponent solution-solid phase system, conservation of mass shows that the net desorption rate of a component species can be expressed as follows [156–160]:

$$\left(\frac{\partial S_i^{net}}{\partial t}\right) = \left(\frac{\partial S_i^{ad}}{\partial t}\right) - \left(\frac{\partial S_i^{d}}{\partial t}\right);\ i \neq j \tag{85}$$

where

- $\left(\frac{\partial S_i^{d}}{\partial t}\right)$ = the desorption rate of component i due to ion exchange, and
- $\left(\frac{\partial S_j^{ad}}{\partial t}\right)$ = the adsorption rate of component j which takes place simultaneously with desorption.

Since the desorption is a stoichiometric reaction, $\left(\frac{\partial S_j^{ad}}{\partial t}\right)$ may be expressed as:

$$\left(\frac{\partial S_j^{ad}}{\partial t}\right) = \sum_{j=1}^{m} \frac{\partial S_j^{ad}}{\partial t} \tag{86}$$

where

- S_j^{ad} = the adsorbed amount of component j, and
- m = the number of adsorbed components.

Substituting Eq. (86) into Eq. (85) gives

$$\left(\frac{\partial S_{\mathrm{i}}^{\mathrm{net}}}{\partial t}\right) = \sum_{j=1}^{m} \frac{\partial S_{\mathrm{j}}^{\mathrm{ad}}}{\partial t} - \frac{\partial S_{\mathrm{i}}^{\mathrm{d}}}{\partial t}\,; i \neq j \tag{87}$$

Substituting Eq. (84) into Eq. (85) yields

$$\left(\frac{\partial S_{\mathrm{i}}^{\mathrm{net}}}{\partial t}\right) = \sum_{j=1}^{m} A_{\mathrm{j}} B_{\mathrm{j}} \exp(-A_{\mathrm{j}} C_{\mathrm{j}}) \left(\frac{\partial C_{\mathrm{j}}}{\partial t}\right) - A_{\mathrm{ji}} B_{\mathrm{i}} \exp(-A_{\mathrm{i}} C_{\mathrm{i}}) \left(\frac{\partial C_{\mathrm{i}}}{\partial t}\right) \tag{88}$$

Equation (88) represents the general adsorption/desorption equation in a multicomponent solution system which can be used in the general transport equation

$$\frac{\partial C}{\partial t} = \frac{\partial}{\partial x}\left(D_{\pi} \frac{\partial C}{\partial x}\right) - R_{\mathrm{i}} v_{\mathrm{x}} \frac{\partial C}{\partial x} - R_{\pi} \frac{\partial^2 C^2}{\partial x^2} \pm \frac{\rho_{\mathrm{s}}}{n} \frac{\partial S}{\partial t} \tag{89}$$

In order to understand the applicability of the adsorption/desorption function in natural systems, the following hypothetical example is given [157] as follow:

- Suppose $(X_1)^{2+}$, $(X_2)^{2+}$, and $(X)^{+}$ are the organic cations of interest which are involved in the adsorption/desorption reactions experiment.
- Since $(X_1)^{2+}$ and $(X_2)^{2+}$ are divalent organic cations with a higher adsorption affinity than the monovalent organic cation $(X)^{+}$, replacement of $(X)^{+}$ originally in the clay soil-solid system should occur, i.e., desorption of $(X)^{+}$ occurs.
- Using Eq. (88) to analyze the $(X)^{+}$ desorption mechanism, we obtain:

$$\left(\frac{\partial S_{\mathrm{i}}^{\mathrm{net}}}{\partial t}\right) = A_2 B_2 e^{-A_2 C_2} \frac{\partial C_2}{\partial t} + A_3 B_3 e^{-A_3 C_3} \frac{\partial C_3}{\partial t} - A_1 B_1 e^{-A_2 C_2} \frac{\partial C_3}{\partial t} \tag{90}$$

  where the subscripts 1, 2, and 3 refer to $(X)^{+}$, $(X_1)^{2+}$, and $(X_2)^{2+}$, respectively. The A_2, B_2, A_3, and B_3 coefficients must be determined from calibration exercises, using the experimental data for $(X_1)^{2+}$ and $(X_2)^{2+}$, respectively. Also, $\frac{\partial C_2}{\partial t}$ and $\frac{\partial C_3}{\partial t}$ for all time steps can be calculated from the indicated calibration data using the finite difference technique as a method of solution.

Substituting Eq. (90) into the surface term in Eq. (89), the governing equation of $(X)^{+}$ can be given as follows:

$$[1 - A_1 B_1 e^{-A_1 C_1}] \frac{\partial C_1}{\partial t} = \frac{\partial}{\partial x}\left(D_{\pi} \frac{\partial C_1}{\partial x}\right) - R_{\mathrm{i}} v_{\mathrm{x}} \frac{\partial C_1}{\partial x} - R_{\pi} \frac{\partial^2 C_1^2}{\partial x^2} - \frac{\rho_{\mathrm{s}}}{n}\left[A_2 B_2 e^{-A_2 C_2} \frac{\partial C_2}{\partial t} + A_3 B_3 e^{-A_3 C_3} \frac{\partial C_3}{\partial t}\right] \tag{91}$$

4.2.2.3
Diffusion Function

The diffusion coefficient D_X can be expressed as a function of the concentration of the specific constituent in the pore fluid [158]. Figure 8 shows four relationships between the non-dimensional pollutant concentration and D_π (the molecular diffusion coefficient). These show the strong influence from both in characterizing the shape of the relative pollutant concentration profile. Leaching-column test information shows that case (4) in Fig. 8 is the more likely profile (i.e., $D_\pi = ae^{-bc}$). Using this function in Eq. (90), the final governing equation for $(X)^+$ will be:

$$[1 - A_1 B_1 e^{-A_1 C_1}] \frac{\partial C_1}{\partial t} = a e^{-bC_1} \left(\frac{\partial^2 C_1}{\partial x} \right) - abe^{-bC_1} \left(\frac{\partial C_1}{\partial x} \right)^2 - R_i v_x \frac{\partial C_1}{\partial x} - R_\pi \frac{\partial^2 C_1^2}{\partial x^2} - R_\pi \frac{\partial^2 C_1^2}{\partial x^2} - \frac{\rho_s}{n} \left[A_2 B_2 e^{-A_2 C_2} \frac{\partial C_2}{\partial t} + A_3 B_3 e^{-A_3 C_3} \frac{\partial C_3}{\partial t} \right] \quad (92)$$

The input data required for parameter determinations are the concentration profiles at all time steps. These data can be obtained from leaching cell experiments. Thus, for example, in the case of desorption analysis, the concentration of the absorbed pollutants considered in the hypothetical example, at all time steps, should first be obtained to determine $\frac{\partial C_j}{\partial t}$ given in Eq. (88). Concentrations of $(X_1)^{2+}$ and $(X_2)^{2+}$ should be determined at all time steps. Following this, their derivatives with respect to time $\frac{\partial C_2}{\partial t}$ and $\frac{\partial C_3}{\partial t}$ at all time steps should be calculated.

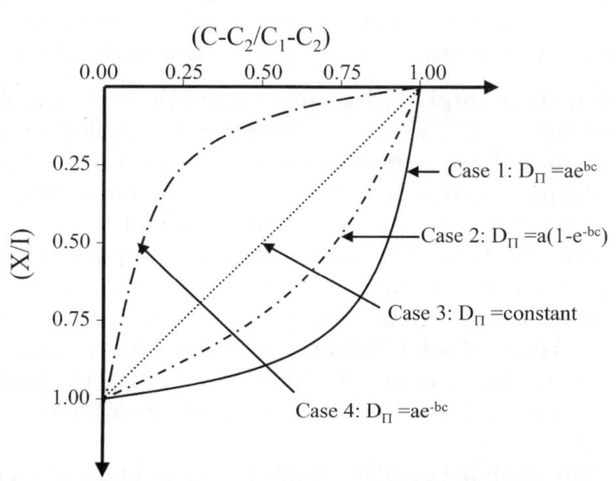

Fig. 8. Relationships between the non-dimensional pollutant concentration and molecular diffusion coefficient

If the measured and calculated concentrations are designated as $C_{\mathrm{exp}}(x,t)$ and $C_{\mathrm{calc}}(x,t)$ respectively, then the best choice of parameters (i.e., A, B, a, b, and K_{hc}) are those which minimize the following function:

$$\sigma = \sum_{i=1}^{m} ABC \,|\, C_{\mathrm{exp}}(x,t) - C_{\mathrm{calc}}(x,t) \,| \qquad (93)$$

where m is the number of measured concentrations in the experiments.

5
Slow Sorption/Desorption Process

Sorption or desorption to or from natural solid particles is an underlying process affecting the transport, degradation, and biological activity of organic compounds in the environment. Although sorption/desorption is often regarded as instantaneous for modeling purposes, it may require weeks to many months to reach equilibrium. Serious studies of sorption kinetics in soils and sediments did not begin until the mid to late 1980s, despite early circumstantial evidence going back to the 1960s that the natural degradation of certain pesticides in the field slowed or stopped after some elapsed time [107, 161–163].

All chemodynamics (i.e., fate and transport) and risk assessment models contain terms for sorption; therefore, an understanding of sorption dynamics is crucial to their success. Ignoring slow kinetics of organic contaminants can lead to an underestimation of the true extent of sorption, false predictions about the mobility, and bioavailability of contaminants, and the wrong choice of cleanup technology and engineering management. Kinetics can also be an important mechanistic tool for understanding sorption itself.

Several workers focused on updating our knowledge of the causes of slow sorption/desorption and the significance of sorption/desorption to bioavailability and the remediation of organic pollutants [107, 164–169]. The following points should be taken into consideration before discussing and evaluating the slow sorption/desorption processes of organic contaminants to solid particles:

- Because much of the sorption/desorption research been carried out in batch systems (see Sect. 4.1.1, where particles are suspended in a well-mixed aqueous solution), the present discussion is mainly for the phenomena occurring on the intraparticle scale (i.e., within individual solid soil grains or within aggregates that are stable in aqueous systems).
- Transport-related non-equilibrium behavior (e.g., physical non-equilibrium) is excluded, which plays an important role in non-ideal solute transport in the field and in some experimental column systems. Physical non-equilibrium is due to slow exchange of solute between mobile and less mobile water, such as may exist between particles or between zones of different hydraulic conductivities in the subsurface soil column, and occurs for sorbing and non-sorbing molecules alike.
- Chemisorption involving covalent bonds as well as bound residue formation is also excluded, which is defined as any organic carbon remaining after exhaustive extraction that results from degradation of parent molecules.

5.1 Equilibrium vs Non-Equilibrium Sorption

Over the last decade, some research has indicated that: (1) partition coefficients (i.e., K_d) between solid and solution phase are not measured at true equilibrium [51, 59–61], (2) the use of equilibrium rather than kinetic expressions for sorption in fate and effects models is questionable [22–24, 60, 61], and (3) sorption kinetics for some organic compounds are complex and poorly predictable [22–24, 26]. This is mainly due to what has recently been discussed as slow sorption/desorption of organic compounds to natural solid phase particles [107, 162–164, 166–182]. The following is a summary of some important points supporting this hypothesis [1, 66, 67, 170–183]:

- Bimodal behavior of the sorption/desorption of organics by various solid phase particles which occurs in fast and slow stages at a few hours to a few days.
- During sorption, the apparent sorption distribution coefficient (K_d^{app}) can increase by 30% to as much as tenfold between short contact (1–3 days) and long contact times. Desorption likewise often reveals a major slow fraction (10–96%) following a comparatively rapid release.
- Aged contaminated samples (e.g., where contact times may have been months or years) can be enriched in the slow fraction (i.e., the fraction sorbed/desorbed in the slow stage) owing to partial dissipation or degradation of more labile fractions before collection. The slow fraction of some pesticides was found to increase with contact time in the environment.
- In many studies, the slow fraction was underestimated due to incomplete contaminant recovery. This can lead to erroneous conclusions when some process of interest is being measured against the mass of contaminant believed to be present.
- Many reported K_d (a time-dependent) values represent principally the fast component rather than overall sorption. Thus, free energy correlations involving K_d are brought into question. For example, Quantitative Structure-Property Relationships (QSPRs, see Chap. 4), where molecular structure or organic contaminants are directly correlated to their K_d are based on the assumption of equilibrium or at least that all organic compounds have attained the same fractional equilibrium. However, sorption/desorption rates can depend greatly on molecular geometry and electronic properties. This is clearly evident in regard to diffusion through a viscous medium such as organic matter or a pore structure.
- A mass transfer coefficient determined from some subsurface soil column elution studies was inverse log-linearly related to the octanol-water partition coefficient (i.e., K_{OW}) for closely related compounds, and polarity in the molecule caused an additional decline in the mass transfer coefficient.

In general, this raises the question whether the sorption equilibrium assumption in fate and effects models is invalid when the fate/transport process of interest occurs over comparable or shorter time scales than sorption. The equilibrium assumption has been found to fail in a growing number of cases which were reported by several authors [162–164, 166–177, 184–190]:

- Long-term persistence was shown in subsurface soils of intrinsically biodegradable compounds even when other environmental factors were not limiting for microbial growth.
- Aging of soil-contaminant mixtures prior to the addition of microbes reduced bioavailability of such contaminants in laboratory studies and aging reduced herbicidal activity in various field studies.
- Bioremediation at contaminated sites of solid particles often levels off after an initial rapid decline (e.g., PCBs and hydrocarbons), due to the unavailability of a resistant fraction.
- Non-equilibrium sorption affects the hydrodynamic transport of contaminants by causing asymmetrical concentration vs time (i.e., elution) curves. In relatively homogeneous soil columns, this asymmetry is exhibited by early breakthrough, a decrease in peak breakthrough concentration, breakthrough front tailing, and elution-front tailing. In more heterogeneous field media, the effect of non-equilibrium sorption on transport is less distinct.
- Vadose and saturated zone studies reveal a decrease in velocity and aqueous-phase mass of the contaminant plume, relative to a non-sorbing tracer, with increasing travel time or distance.

5.2 Potential Causes

The potential causes of slow sorption are activation energy of sorptive bonds and mass-transfer limitations (molecular diffusion) [107]. Sorption can occur by physical adsorption on a surface or by partitioning into natural solid phase organic matter (SP_{OM}) or humic substances (SP_{HS}). The intermolecular interactions potentially available to neutral organic compounds, i.e., van der Waals (dispersion), dipole-dipole, dipole-induced dipole, and hydrogen bonding, are common to both adsorption and partitioning (see Chap. 2). It is noteworthy that even small, weakly polar molecules like halogenated methanes, ethanes, and ethenes exhibit slow sorption/desorption in soils [183, 191, 192]. The thermodynamic driving force for their sorption is hydrophobic expulsion from water, but their main interaction with the surface is only by dispersion and weak dipolar forces.

5.2.1 *Diffusion Limitation*

Most researchers attribute slow kinetics to some sort of diffusion limitation (e.g., diffusion is random movement under the influence of a concentration gradient [193]), because sorbing molecules are subject to diffusive constraints throughout almost the entire sorption/desorption time course due to the porous nature of particles. Particles are porous by virtue of their aggregated nature and because the lattice of individual grains in the aggregate may be fractured.

The possible diffusion processes that can be taking place during sorption mechanisms to reach all sorption sites were reported by Pignatello and Xing [107] as: (1) *film diffusion*, where diffusing molecules traverse bulk liquid; (2) *pore diffusion*, e.g., pores within the particle, and (3) *matrix diffusion*, e.g., penetrable solid phases. Diffusion coefficients of organic molecules can be expected to decrease along that same order, but few data are available for relevant natural particle systems, except for bulk aqueous diffusion. The observed kinetics in any region of the sorption vs time curve reflect one or more of these diffusive constraints, which may act in series or parallel.

The mixing that takes place in most experiments ensures that bulk liquid or vapor diffusion is not rate-limiting. Likewise, film diffusion is probably not rate-limiting. Several authors concluded that in well-mixed batch systems film resistance of Lindane and nitrobenzene on subsurface materials was insignificant compared to intraparticle diffusion, but may have been significant for nitrobenzene in columns [23, 107, 194–196]. Film diffusion is potentially rate-limiting for the initial fast stage of sorption; but it is not likely to be important in the long-term phenomena [107].

This leaves pore diffusion and matrix diffusion as likely rate-limiting steps in slow processes. Diffusion in pores can occur in pore liquids or along pore wall surfaces. Liquid and surface diffusion may act concurrently and are difficult to distinguish [197, 198]. Surface diffusion is expected to increase in relative importance: (1) in very small pores where fluids are more ordered and viscous, and where the sorbate spends a greater percentage of time on the surface, and (2) at high surface concentrations. Surface diffusion was invoked for porous resins [199] and activated carbon [200, 201] because intraparticle transport appeared to be faster than could be accounted for by liquid diffusion. A surface diffusion model was used to simulate sorption-desorption of Lindane with some success [194]. However, it has been argued that surface diffusion is insignificant on solid-soil particles because of the discontinuity of the adsorbing surface [191], if not the low mobility of the sorbate itself [202, 203].

5.2.2
Kinetic Aspects

Kinetic models proposed for sorption/desorption mechanisms including first-order, multiple first-order, Langmuir-type second-order, and various diffusion rate laws are shown in Sects. 3.2 and 3.4. All except the diffusion models conceptualize specific "sites" to or from which molecules may sorb or desorb in a first-order fashion. The following points should be taken into consideration [181, 198]:

- Most sorption/desorption kinetic models fit the data better by including an instantaneous, non-kinetic fraction described by an equilibrium sorption constant.
- None of the models are perfect, although diffusion models are more successful than first-order models when they have been compared.
- First-order kinetics is easier to apply to transport and degradation models because they do not require knowledge about particle geometry.
- Fit to a particular rate law does not by itself constitute proof of a mechanism.

To understand better the slow sorption kinetics, the following examples can show the causes of slow sorption from the perspective of kinetic behavior:

- *A single rate constant often does not apply over the entire kinetic part of the sorption/desorption curve* [107, 174, 178, 179, 181–183, 194, 201–206]:
  - In leaching field-aged residues of Atrazine and Metolachlor from a soil column, a model with a single diffusion parameter underestimated desorption at early times and overestimated desorption at late times.
  - Mass transfer coefficients obtained by modeling leaching curves depend on the contaminant residence time in the column (i.e., the flow rate).
  - In desorption studies, plots of the logarithm of the fraction remaining vs time tend to show a progressive decrease in slope, indicating increasing resistance to desorption. Hence, desorption in natural solid particles seems to be a continuum.
- *The slow fraction is inversely dependent on the initial applied concentration assuming greater importance at lower concentration* [174, 178, 179, 194, 202, 203, 206, 207]:
  - Equilibrium considerations alone may partly explain the nonlinear sorption isotherm, i.e., when n in the Freundlich model (Eq. 8) is less than unity, intraparticle retardation will increase as the concentration inside the particle declines. However, in some studies it appears that the concentration dependence is steeper than expected based on equilibrium nonlinearity.
  - In studies of trichloroethylene (TCE) vapor sorption to various porous particles at 100% relative humidity, it was reported that the slow fraction remaining after N_2 gas desorption was highly concentration dependent and not well simulated by considering only equilibrium nonlinearity.
- *Sorption is often kinetically hysteretic* [161, 164, 174, 208–211]:
  - Hysteresis means that the slow state appears to fill faster than it empties. Many examples exist of apparent "irreversible" sorption of some fraction, or at least exceedingly long times to achieve desorption, following relatively short contact times.
  - Hysteresis may be caused by experimental artifacts or degradation. Also, to assess hysteresis fairly from the desorptive direction requires that samples be at true equilibrium.

Pignatello and Xing [107] used two models, the organic matter diffusion model (OMD) and the sorption-retarded pore diffusion model (SRPD), in order to understand better the meaning of slow sorption/desorption observations and mechanisms and to explore the most likely causes of such slow process in natural solid particles. These authors reported that both OMD and SRPD mechanisms operate in the environment, often probably together in the same particle. OMD may predominate in soils that are high in natural OM and low in aggregation, while SRPD may predominate in soils where the opposite conditions exist.

5.3 Bioavailability and Remediation Technology

The bioavailability of organic compounds in soils/sediments to microbes, plants, and animals is important from the perspective of remediation and risk assessment. Cleanup technology (*ex situ* or *in situ*) of contaminated soils and bottom sediments requires mass transport of contaminants through the solid materials, which in turn depends on sorption/desorption kinetics.

Microorganisms take up substrates far more readily from the fluid than the sorbed states [212–216]. Thus, aged contaminants are resistant to degradation compared to freshly added contaminants [189, 190, 209, 217], and degradation of freshly added contaminants often tails off to leave a resistant fraction [209, 218, 219]. Bioavailability has been called a major limitation to complete bioremediation of contaminated solid sites [187]. The "solid phase-contaminant-degrader" system is dynamic and interdependent. A mechanistic-based biodegradation model must be built on the mechanism(s) governing sorption/desorption, in addition to the biological processes governing cell growth and substrate utilization in the matrix. A number of groups are now developing sorption-degradation kinetic models, where both diffusion and two-box (equilibrium and first-order kinetic compartments) sorption concepts have been explored. [185, 219–223].

The bioavailability of contaminants to wildlife and humans is also an area of critical importance, where contaminants can be taken up in pore water and by dermal contact, particle ingestion, or particle inhalation. The dynamics of sorption/desorption are not currently incorporated into exposure and risk assessment models for organic compounds, where availability, in most cases, is assumed to be 100% [224]. Recently, the following have been demonstrated and reported:

- The time between spiking and testing of solid phases affects bioavailability of the contaminant(s) of concern [163, 225].
- The kinetics of desorption control bioaccumulation of historical (e.g., aged) contamination (e.g., PAHs in benthic animals [225]) and historically contaminated soils are less toxic and/or lead to lower body burdens than equivalent amounts of spiked soils [226, 227].

In order to model contaminant bioavailability (see Chap. 4), it is crucial that we understand sorption kinetics and the factors influencing rates under the conditions of exposure. Pump-and-treat, a vapor and water extraction remediation technology widely used in environmental engineering practice, is limited in part by physical non-equilibrium and sorption non-equilibrium [228–230]. These processes both cause tailing of the contaminant plume, which increases the time invested and the volume of sparge air or water needed to achieve cleanup [228–231]. Moreover, they act to resume contamination if pumping is ceased before all the contaminant is removed [232–235]. Ways of experimentally separating out the contributions of physical and sorption non-equilibrium must be sought.

In order to achieve complete remediation, it is important that slow desorption has to be overcome [187]. Pignatello and Xing [107] considered the following conceivable approaches to promoting desorption from the slow state: (1) addition of biological agents capable of reaching remote contaminant molecules, (2) application of heat, (3) addition of chemical additives that displace the contaminant or alter the solid phase structure, and (4) physical methods that alter the soil structure.

Because molecular diffusion through natural OM and desorption from high-energy sites are expected to be strongly temperature dependent, thermal desorption is already in use in various remediation technologies for volatile contaminants. In batch application, the soil is heated to temperatures ranging from 200 °C to 500 °C in a primary chamber, and the vapors are combusted in a secondary chamber [236–238]. Steam stripping (e.g., a form of soil vapor extraction) can remove semivolatiles from the vadose zone [234, 235]. Bioremediation in a compost mode where temperatures reach 60 °C or more should prove advantageous. The success of these methods requires a fundamental understanding of sorption/desorption kinetics.

Cosolvency is another approach that could be considered in remediation technologies. In general, surfactants target specifically to the removal of slow fractions. To be effective, surfactants must penetrate the nanopore intraparticle matrix of natural OM in order to either solubilize the contaminant by micellization or alter the intraparticle properties of the sorbent in such a way as to promote desorption. The addition of surfactants gave mixed results in stimulating biodegradation [215, 239]. The use of organic cosolvents is a promising approach because cosolvents can increase desorption both thermodynamically (by enhancing solubility) and kinetically (by softening natural OM) [111, 171, 240–258].

Generally, slow sorption or desorption has made complete remediation technology difficult. However, there have recently been legitimate questions raised by some researchers [163, 187] about whether we even need to be concerned about residues that desorb so slowly and thus are apparently largely bio-unavailable. At a minimum, it is important that we understand the factors which govern slow sorption/desorption rates, their kinetics and causes at the intra-particle level of different solid phase materials (e.g., surface/subsurface and aquatic sediment particles), and how these phenomena can relate to contaminant transport, bioavailability, toxicity, remediation, and risk assessment modeling.

6 A Case Study

In the next few sections, a case study of the environmental impact of highway construction and repair materials as well as hazardous solid waste materials is presented and discussed from the view points of sorption/desorption processes.

6.1 Problem Statement

Assessment, prediction, and management of the environmental impact of solid waste material (SWM) disposal in landfills and recycled wastes mixed with asphalt that are used as highway construction and repair (C&R) materials on surface and ground waters across the USA pose many challenges. Various chemical compounds leached from such wastes can migrate from landfills/roadways to the surrounding environments, presenting a potential source of pollution. Continual SWM disposal and reuse in highway construction continue to increase, which require taking precautions for landfills planning and design, which include the selection of suitable and less harmful C&R materials and finding a means of assessing the environmental impact of their use. Both disposal of new materials and amended waste materials can pose a threat to the environment. Many wastes often contain extensive suites of potentially toxic organic [1] and inorganic compounds [66, 67]. The recent attempts at amending waste materials in mixes and fills from many industries have greatly added to the perception of highways as "linear landfills" [66, 67]. This increased landfill disposal and utilization of chemically complex C&R materials has resulted in a clear need to integrate and unify approaches towards understanding the fundamental leaching behavior and transport in the environment of these materials.

Accordingly, Eldin et al. [66, 67] proposed and developed a methodology for the evaluation of the potential environmental impact of common highway C&R materials. The project was planned in three phases, as follows:

- Phase I focused on a broad screening of common C&R material to identify the extent of the problem and to guide the succeeding phases. Phase I resulted in a comprehensive list of commonly used C&R materials, their toxicity assessment, and a preliminary description of toxicity assessment protocol, and fate and transport model.
- Phase II examined the leaching characteristics of C&R materials, full development of a predictive model, and the validation of the overall evaluation methodology. Validation of the methodology was achieved by evaluating a number of C&R materials and by broadening the evaluation criteria to include leaching kinetics, reference environments, and impact interpretation.
- Phase III focused on further validation of the methodology and modeling assumptions based on further laboratory studies, as well as on modeling enhancements and testing.

On the other hand, Aboul-Kassim [1] assessed the environmental impact of hazardous waste materials in landfills by: (1) characterizing the different organic compound fractions present in such wastes and their leachates, (2) determining the toxic effect of each fraction and individual organic compounds, and (3) studying the chemodynamics (i.e., fate and transport) of such leachates by using a battery of laboratory experiments (such as sorption/desorption, photolysis, volatilization, biodegradation).

In line with the general objectives in the present chapter we propose to discuss only the leaching and sorption/desorption experiments conducted by Aboul-Kassim [1] and Eldin et al. [66, 67]. In addition, the approach taken by these authors to predict the behavior of toxic compounds in the leachates from various SWMs/COMs is also discussed.

6.2 Types of Solid Wastes

Extensive research has been conducted on the use of the following SWMs as highway C&R materials (an alternative innovative way to recycle/reuse such wastes), soil stabilization material, roller compacted concrete, and road base stabilization materials. They include the following [1].

6.2.1 *Crumb Rubber*

More than 2 billion tires are disposed of annually in the USA. Before being recycled and/or reused, scrap tires or crumb rubber are first processed to remove any loose steel and fibers and then finely ground. Research has been conducted on the use of crumb rubber in highway construction such as in lightweight fill, subgrade insulation, and channel slope protection, as well as an additive to Portland cement concrete pavement [66, 67, 259, 260].

6.2.2 *Roofing Shingles*

Roofing shingles are a mixture of asphalt, aggregates, and reinforced fabrics which are used on top of houses as protective materials against heat, rain, or any other weathering effects. The lifetime of such roofing shingles is 10–25 years. After being removed from houses, roofing shingles are usually disposed of in landfills. One application to reuse this waste as highway construction and repair material was reported by Eldin et al. [66, 67].

6.2.3 *Coal Combustion By-Products*

There are 720 coal-fired power plants in the USA. When coal is burned in these power plants, two types of ash are produced: coal fly ash and bottom ash. Coal fly ash is the very fine particulate matter carried in the flue gas; bottom ash (or slag) is the larger, heavier particles that fall to the bottom of the hopper after combustion [261–264]. The physical and chemical characteristics of these ashes vary depending on the type of coal burned. These ashes are characterized by the following:

- *Fly ash* – the primary components are silicon dioxide, aluminum oxide, iron oxide, and calcium oxide. Fifty million tons of fly ash are produced annually in the USA. About 76% is disposed of either onsite or in state-regulated disposal areas, while the rest is reclaimed.
- *Bottom ash* has a similar chemical makeup to fly ash but has a much coarser gradation. A recent study on its use as a sub-base material showed that it had sufficient engineering properties to perform adequately.
- *Combined ash* – when fly ash and bottom ash are placed in landfills, they are generally combined. The physical properties of combined ash (including gradation, specific gravity, and loss on ignition) can vary considerably depending on the type of plant and source of coal.

6.2.4 *Municipal Solid Waste Incinerator Combustion Ash*

In 1980, 2.8 million tons of municipal solid waste was burned in the USA, yielding approximately 33% municipal waste combustion (MWC) ash. By 1990, the amount burned had increased to 32 million tons, creating approximately 25% of MWC ash or residue [265–267]. Controlled combustion of municipal solid waste produces two types of ash: fly and bottom ash. Most MWC ash (80–99%) is bottom ash; however, it usually contains a high percentage of toxic materials, and the leachates may not meet environmental standards.

6.3 Types of Solid Phases

The following sections represent the different solid phases used to study the leaching kinetics and sorption processes of different SWM leachates.

6.3.1 *Soils*

The three kinds of soils used in the present study represent a broad national geographical area. Three soils were selected from the eleven soil orders found in the US to determine the effect of soil adsorptive capacity on the reduction of contaminant toxicity in leachates prepared from COMs. The selected soils are discussed below.

6.3.1.1 *Mollisol*

The order Mollisol is distributed throughout the Ohio and Upper Mississippi Valleys. The Mollisol for this study is of the Woodburn Series and was collected from Benton County, Oregon. The soil is typically described as montmorillonitic and contains moderate quantities of organic matter. It may be slightly acidic to moderately alkaline.

6.3.1.2 *Ultisol*

The order Ultisol is distributed widely across the plains, Virginia, North Carolina, South Carolina, and Georgia, as well as other areas such as the Sierra Nevada Mountain Province and Western Oregon. The Ultisol for this study was of the Olyic series and was collected from Washington County, Oregon. The soil is typically acid, low in organic matter and high in kaolin and oxide minerals.

6.3.1.3
Aridisol

The order Aridisol is, as its name suggests, typical of arid climate conditions and found in the southwest deserts. The Aridisol for this project is of the Sagehill series and was collected from Gilliam County, Oregon. The soil is an alkaline coarse-grained soil with free $CaCO_3$. It has low infiltration rates and capacities.

6.3.2
Bottom Sediments

On the other hand, two bottom sediment samples were collected from the Willamette River (Benton County) and Yaquina Bay (Newport), Oregon. These samples represent both fresh water and estuarine environments with slightly different degrees of organic matter compositions.

6.4
Approach

Fate and transport of organic leachates from SWMs/COMs in natural environments can be approximated by a series of laboratory tests or analyses. The basic approach is to measure the mass transfer of such chemicals under controlled conditions to determine rates that can be applied to specific mathematical models.

6.4.1
Solid Waste Materials Leachate Preparations

Leaching of chemicals from complex materials or matrices is a complicated phenomenon in which many factors may influence the release of the specific organic compounds and inorganic ions. Important factors include major element chemistry, pH, redox, complexation, liquid to solid ratio, contact time, and biological activity. To describe fully the leaching of SWMs/COMs under field conditions, a battery of leaching tests was specifically designed to simulate various physical and chemical release mechanisms.

6.4.1.1
24-Hour Batch Leaching

Batch-leaching tests were designed to determine rates of desorption and equilibrium sorption relationships under conditions of high mixing, high surface areas of the solid SWMs/COMs, and continuous surface renewal. Leachate preparations for solid particle sorption were obtained from the 24-h batch-leaching test. Batch equilibrium tests were prepared in precleaned glass bottles (heated at 550 °C for 8 h in an oven, rinsed twice with methylene chloride) by adding an SWM/COM and distilled water at a ratio of 1 g to 4 ml. Sample bottles were sealed with Teflon lined caps, tumbled for 24 h, maintained at a constant room temperature of 20 °C, centrifuged for 10 min at 10,000 rpm, and filtered through a 0.45-μm filter. The filtered leachate was then placed in a glass container and stored in the dark in a 4 °C refrigerator.

6.4.1.2
Short/Long-Term Batch Leaching

Each SWM/COM was tested in a batch reactor to determine its leaching kinetics. Leaching occurs more quickly at the beginning of the test and should reach a plateau with time, signifying that the reactor has come to equilibrium. The short-term test usually lasts for 24 h with samples taken more frequently at the beginning of the test. The long-term test should continue until the solution concentrations have reached a plateau. The time and frequency of sampling may be different for each test material, depending on the leaching rate. Some common sampling times for short-term and long-term tests were 5 min, 10 min, and 20 min; 1 h, 4 h, 12 h, and 24 h; and 3 days, 5 days, and 7 days, respectively.

Short/long batch leaching tests were carried out by adding distilled water to SWMs/COMs with a solid to liquid ratio of 1 g to 4 ml. Samples were shaken until it was time to remove a sub-sample. A constant room temperature of 20 °C was maintained. Each sub-sample was then filtered through a 0.45-μm filter while the remaining samples continued tumbling until the next sub-sample time. Each filtered solution was placed in a glass container and stored in the dark in a 4 °C refrigerator for later analysis.

6.4.1.3
Column Leaching

To determine the leaching of chemical constituents from SWMs/COMs under conditions of constant surface renewal, columns (2.5 cm in diameter, 25 cm long) filled with SWMs/COMs were leached with distilled water at three different flow rates. The column tests were used to simulate leaching of highway materials under conditions of subsurface percolation of rainwater. Effluent samples from the column were taken with time for up to 80 h. The filtered solutions were measured for TOC and/or individual compound concentrations, and for toxicity.

6.4.1.4
Flat Plate Leaching

Flat plate leaching tests were used to determine the rate of leaching of contaminants from a SWM/COM surface. In these tests, the material (76 cm^2 of flat surface as a disk, 2.5 cm thick) was placed in the bottom of a beaker and the beaker then filled with 1 l of distilled water. The flux of contaminants (mg/cm^2-h) was then determined by the increase of concentration in the overlying water as a function of time.

6.4.1.5
Solid Sorption Experiments

Batch tests (i.e., tests on individual waste materials) are conducted with the provided solid suspensions (e.g., soils such as Woodburn, Sagehill, and Olyic, as well as two bottom sediment samples) prepared with previously air-dried "solids" (i.e., soils and bottom sediments), ground to a uniform powdery texture for mixing with the eluates from the 24-h batch leaching test of the different SWMs/COMs. The concentrations of eluates in solution were designed to evaluate the capability of different environmental solids to adsorb available contaminants. The solid particles were fully dispersed with the aqueous phase to achieve complete adsorption. Common practice is to use a solid:solution ratio of 1 g:4 ml [1], together with proper tumbling of the samples at a constant temperature (e.g., at least 24 h in a constant temperature environment of 20 °C).

A sorption isotherm is completed for each solid particle type and SWMs/COMs. A range of solid to solution concentrations (i.e., solid:solution) was chosen for each solid phase and waste material leachate (e.g., 50–250 mg/l), with about five data points per range. All control and test samples were performed in duplicate. The solution used in the isotherms was prepared by a 24-h batch leaching experiment with the solid test material and distilled water. The material controls consisted of the test material leachate without the solid phase particles. Chemical analyses, expressed either as TOC or as individual organic compound (e.g., aliphatic and aromatic compounds) concentrations relative to the organic carbon content of the SWM/COM, revealed the actual concentrations of various organic constituents in the leachates. Solid phase controls were also prepared for each of the test soils/sediments in order to determine the concentrations of the constituents leached from the solid phase alone.

6.5
Data Modeling

6.5.1
Batch Leaching

Batch leaching tests, conducted for different types of waste materials, resulted in data that were modeled as shown in Eq. (94):

$$C = C_a(1 - e^{kt}) \tag{94}$$

where

- C = the concentration of either individual leached organic compound,
- t = the time of leaching,
- C_a = the asymptotic concentration, and
- k = the rate coefficient (1/time).

In general, the leaching of organic compounds from SWMs/COMs involved an initial rapid release, followed by a slower release over longer time periods (Fig. 9a). For solid waste materials, disposed in landfills or used as highway construction materials and subjected to long-term environmental exposure, the rate of leaching

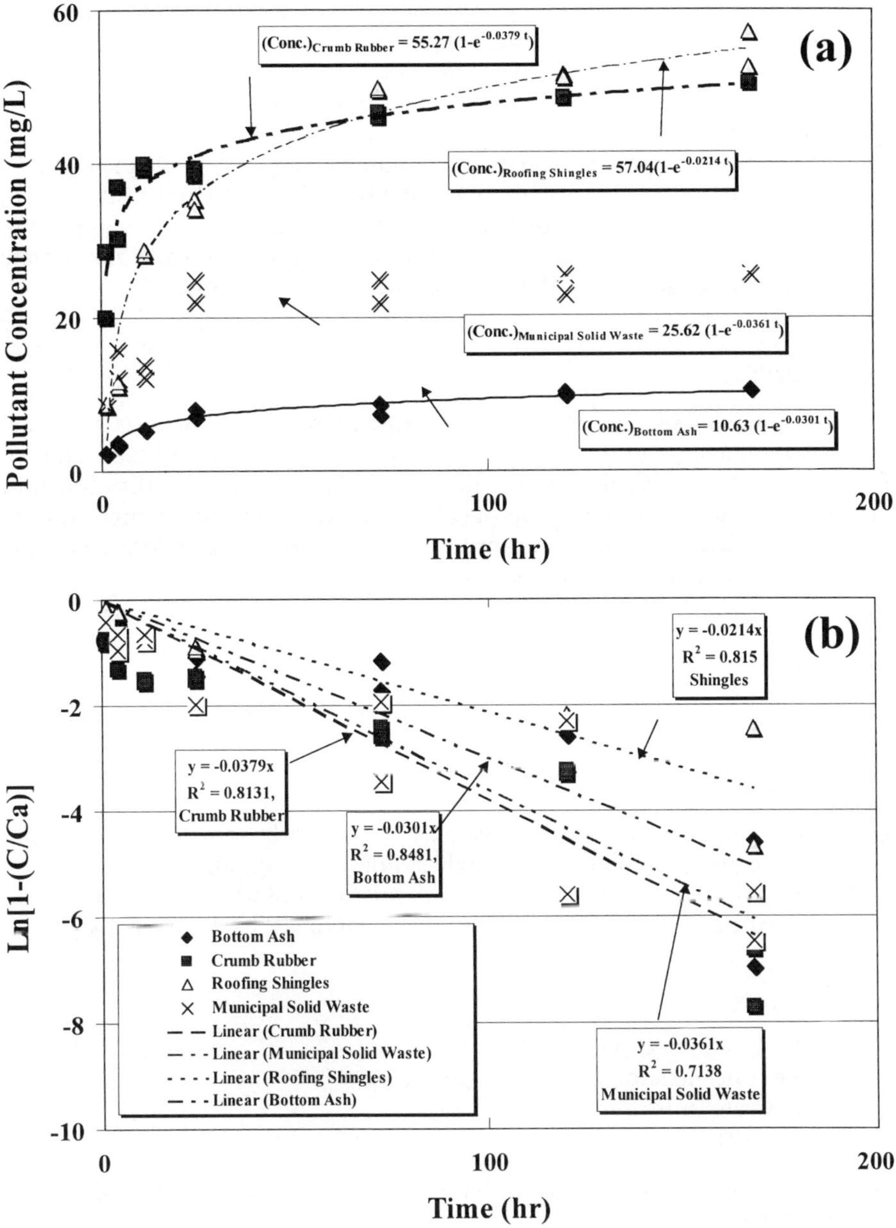

Fig. 9 a, b. Dynamic batch leaching experiment data for various solid waste material leachates: **a** before modeling; **b** after modeling

of constituents at large t values may best represent actual field conditions. As such, accurate modeling of the release near t equal to zero may not be important.

Equation (94) for SWMs leaching offers the advantage of the use of only two fitting coefficients (i.e., C_a and k) [66, 67]. The *flux* or leaching rate is proportional to the derivative (slope) of the concentration vs time, that is flux ~ dC/dt (Fig. 9a). When the C vs t formulation is nonlinear (Eq. 94), the flux is not constant and gradually decreases with increasing time (Fig. 9a). On the other hand, the modeled leaching data for all waste material leachates are shown in Fig. 9b, where a linear fit could be made to any time segment to obtain an approximate constant leaching rate during that time segment, in units of mg/l · h.

6.5.2
Column Leaching

The concentration of any contaminant(s) from highway C&R materials appearing in the effluent from the column was measured over time and the results of leachate desorption breakthrough curves [66, 67] are schematically shown in Fig. 10. The effluent concentrations of contaminants for three different flow rates were determined to follow a first-order model as shown in Eq. (95), with the coefficients fitted by the linear regressions given in Table 3:

$$C = C_0 \cdot e^{-kt} \tag{95}$$

where

- C = concentration at time t,
- C_0 = initial concentration at time 0,
- t = time, and
- k = first-order rate constant.

Leaching of contaminant(s) (expressed as TOC) clearly shows the most rapid decrease in concentration is for the highest flow rate (Fig. 10).

Additional data analyses were performed in a variety of ways, which can be used to compute the cumulative mass of a certain contaminant, as shown in Eq. (96):

$$M = \int CQ \cdot dt \tag{96}$$

where

- M = the cumulative mass leached (mg),
- C = the concentration in leachate (mg/l),
- Q = flow rate (ml/h), and
- t = time (h).

Table 3. First order regression coefficients for column leaching of TOC at ambient pH (pH ≅ 7) and pore volumes

Flow rate (ml/h)	C_0 (mg/l)	K (1/h)	R^2	Pore volume (V_p, mL)
8	8325	0.0431	0.96	265
10	8595	0.1106	0.97	256
16	6672	0.0482	0.94	243

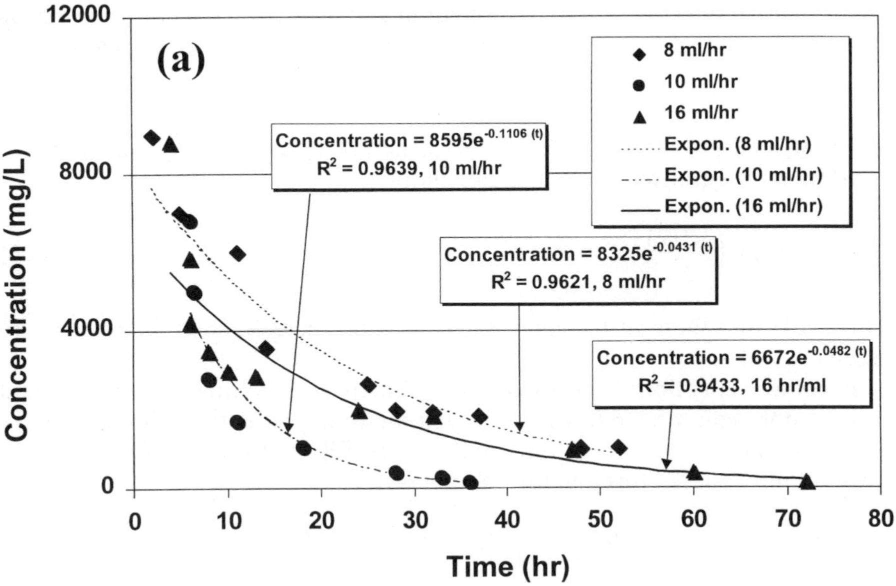

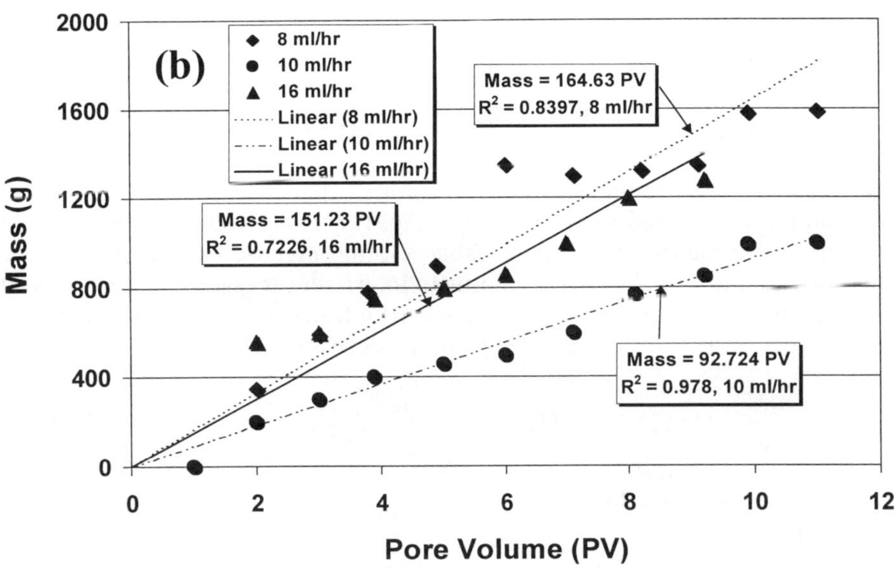

Fig. 10 a, b. Column experiments using different flow rates, first order model TOC concentration released vs: **a** time; **b** pore volume

For a constant flow rate through the column and using the concentration versus time relationship of Eq. (96), the integration yields the familiar exponential form

$$M = M_a(1 - e^{-kt}) \tag{97}$$

where M_a is the total asymptotic mass leached (mg), and may be evaluated as a constant of integration, as show in Eq. (98):

$$M_a = \left(\frac{QC_0}{k}\right) \tag{98}$$

This leads to a method for computing the total mass leached for instance to receiving water.

In column sorption/desorption tests, a dimensionless time is often used and termed pore volume [268]. One pore volume (e.g., V_p) is the volume of pores (e.g., voids) present in the column that may be filled with water. The number of pore volumes passed through the column is thus:

$$PV = \left(\frac{V}{V_p}\right) = \left(\frac{Qt}{V_p}\right) \tag{99}$$

where

- PV = the number of pore volumes,
- V_p = the volume of pores for a given column,
- V = the cumulative flow volume, and
- Q = the flow rate.

The results of total contaminant concentrations appearing in the effluent from the column are plotted vs pore volume (Fig. 10b). Pore volumes for the SWM column experiments are given in Table 3. Time may be normalized to pore volumes for additional analysis, leading to interferences regarding the trade-off between the mass leaching rate increasing with faster flow rates, but decreasing with longer times. This can be seen from the derivative of Eq. (97):

$$\frac{dM}{dt} = Q \cdot C_0 \cdot e^{-kt} = Q \cdot C_0 \cdot e^{\left[-k\left(\frac{V_P}{Q}\right)PV\right]} \tag{100}$$

where $\frac{dM}{dt}$ is the mass flux (mg/h).

The significance of cumulative mass (e.g., M) is that this may lead to a method for determining the loading which results from intermittent rainfall. A loading based strictly on time may not suffice when runoff starts, stops, and starts again.

6.5.3
Flat Plate Leaching

Results of the SWM/COM experiments reported by Eldin et al. [66, 67] are schematically shown in Fig. 11. Assuming zero order kinetics, the increase of

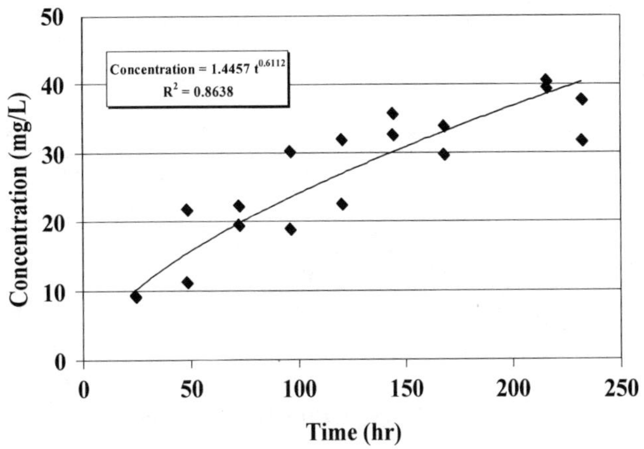

Fig. 11. Flat plate leaching test: TOC concentration in leachate as a function of time

contaminant concentration is given from the fitted line on the figure as:

$$Y = a \cdot t \tag{101}$$

where

- Y = the contaminant concentration (mg/l),
- a = the intercept, and
- t = the leaching time (h).

Knowing the average surface area of the SWM/COM flat plate and the volume of the leaching solution, the constant flux, F (for zero-order kinetics) of organic compounds from the SWM/COM, is calculated as:

$$F = \left(\frac{V}{A}\right) \cdot \frac{dC}{dt} = h \cdot \frac{dC}{dt} \tag{102}$$

where

- F = the flux ($mg/cm^2 \cdot h$),
- V = the eluate volume (cm^3),
- A = the surface area (cm^2), and
- $\frac{V}{A}$ = the depth (cm).

6.5.4
Solid Phase Sorption

Since TOC, for some solid wastes, was used as a criterion to measure leachate sorption for organic compounds, TOC by itself is considered as a single component system (i.e., SCS, see Sect. 2.1). To represent the SCS equilibrium system for various waste materials, the sorption characteristics of different soils and sediments were analyzed and evaluated using three different sorption iso-

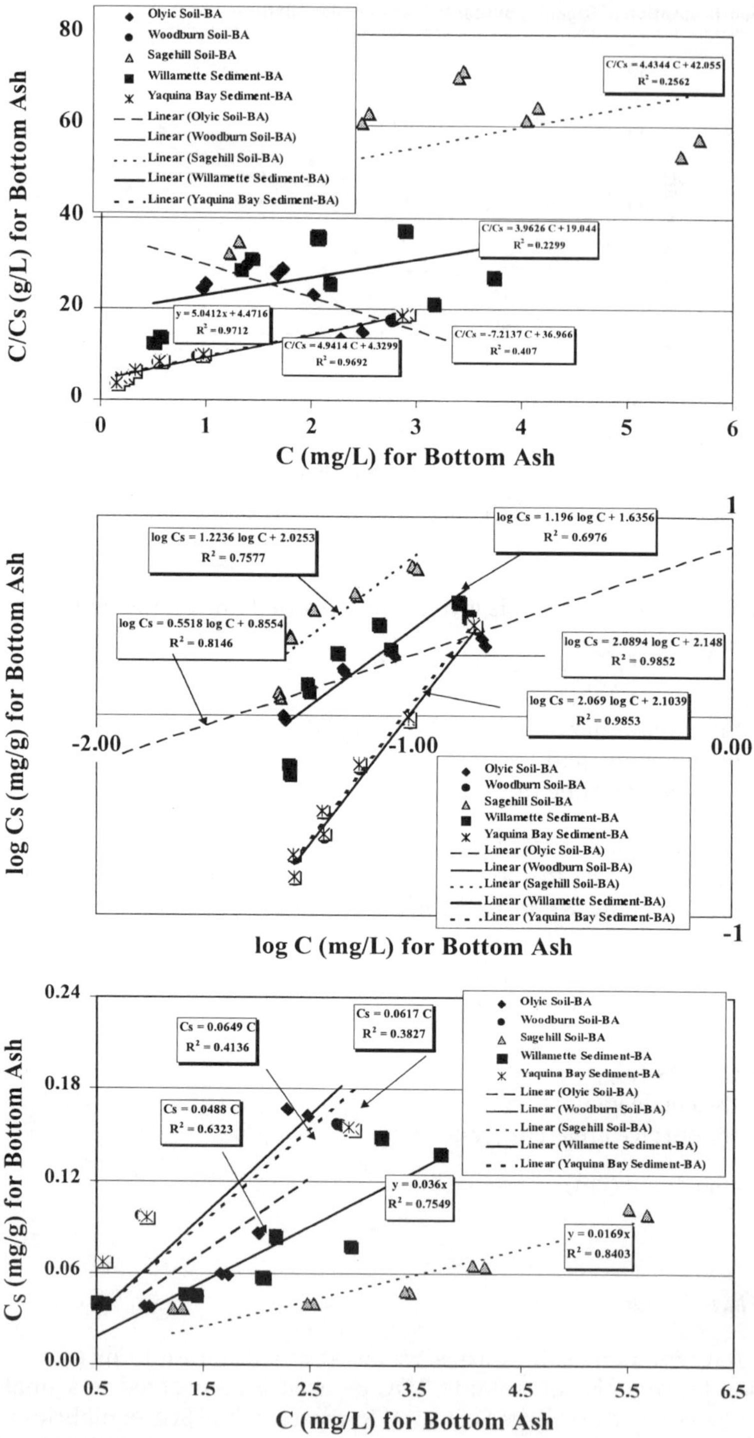

Fig. 12. Isotherm sorption models for bottom ash solid waste, representing Langmuir (C/C_s vs C), Freundlich ($\log C_s$ vs $\log C$), and Linear (C_s vs C) models

Table 4. Summary regression equation constants for various solid waste material leachates on different solid phases (Note: base 10 logs)

Waste type	Solid phase type	Model isotherm	Y axis	Intercept	Slope	X axis	R^2
Bottom ash	Olyic soil	Linear	C_s	0.0000	0.0488	C	0.6323
		Langmuir	C/C_s	36.9660	−7.2137	C	0.4070
		Freundlich	$\log C_s$	0.8554	0.5518	$\log C$	0.8146
	Woodburn soil	Linear	C_s	0.0000	0.0649	C	0.4136
		Langmuir	C/C_s	4.3299	4.9414	C	0.9692
		Freundlich	$\log C_s$	2.1039	2.0690	$\log C$	0.9853
	Sagehill soil	Linear	C_s	0.0000	0.0169	C	0.8403
		Langmuir	C/C_s	42.0550	4.4344	C	0.2562
		Freundlich	$\log C_s$	2.0253	1.2236	$\log C$	0.7577
	Willamette river sediment	Linear	C_s	0.0000	0.0360	C	0.7549
		Langmuir	C/C_s	19.0440	3.9626	C	0.2299
		Freundlich	$\log C_s$	1.6356	1.1960	$\log C$	0.6976
	Yaquina bay sediment	Linear	C_s	0.0000	0.0617	C	0.3827
		Langmuir	C/C_s	4.4716	5.0412	C	0.9712
		Freundlich	$\log C_s$	2.1480	2.0894	$\log C$	0.9852
Crumb rubber	Olyic soil	Linear	C_s	0.0000	0.4710	C	0.7120
		Langmuir	C/C_s	38.4770	−1.4355	C	0.6030
		Freundlich	$\log C_s$	1.2053	0.5852	$\log C$	0.8933
	Woodburn soil	Linear	C_s	0.0000	0.0413	C	0.2690
		Langmuir	C/C_s	4.8269	1.3492	C	0.8398
		Freundlich	$\log C_s$	1.725	2.5408	$\log C$	0.8963
	Sagehill soil	Linear	C_s	0.0000	0.0161	C	0.8618
		Langmuir	C/C_s	41.3570	0.9107	C	0.4591
		Freundlich	$\log C_s$	1.9436	1.3515	$\log C$	0.8564
	Willamette river sediment	Linear	C_s	0.0000	0.0124	C	−0.0881
		Langmuir	C/C_s	33.2820	1.7783	C	0.8899
		Freundlich	$\log C_s$	2.2068	1.6759	$\log C$	0.8155
	Yaquina bay sediment	Linear	C_s	0.0000	0.0106	C	−1.1394
		Langmuir	C/C_s	23.631	2.5079	C	0.8909
		Freundlich	$\log C_s$	2.9569	2.8739	$\log C$	0.7882

Table 4 (continued)

Waste type	Solid phase type	Model isotherm	Y axis	Intercept	Slope	X axis	R^2
Roofing shingles	Olyic soil	Linear	C_s	0.0000	0.0490	C	0.7104
		Langmuir	C/C_s	37.2260	–1.3999	C	0.6096
		Freundlich	$\log C_s$	1.1947	0.5828	$\log C$	0.8938
	Woodburn soil	Linear	C_s	0.0000	0.0431	C	0.3031
		Langmuir	C/C_s	4.7090	1.2861	C	0.8356
		Freundlich	$\log C_s$	1.6713	2.5099	$\log C$	0.8972
	Sagehill soil	Linear	C_s	0.0000	0.0171	C	0.8865
		Langmuir	C/C_s	40.7510	0.7882	C	0.3998
		Freundlich	$\log C_s$	1.8922	1.3079	$\log C$	0.8577
	Willamette river sediment	Linear	C_s	0.0000	0.0133	C	0.1806
		Langmuir	C/C_s	34.024	1.5448	C	0.8637
		Freundlich	$\log C_s$	2.1254	1.5989	$\log C$	0.8372
	Yaquina bay sediment	Linear	C_s	0.0000	0.0114	C	–0.5408
		Langmuir	C/C_s	24.6480	2.2350	C	0.8594
		Freundlich	$\log C_s$	2.7341	2.6092	$\log C$	0.7893
Municipal solid waste incinerator bottom ash	Olyic soil	Linear	C_s	0.0000	0.0247	C	0.9402
		Langmuir	C/C_s	34.073	0.8461	C	0.1295
		Freundlich	$\log C_s$	1.6939	1.1269	$\log C$	0.8448
	Woodburn soil	Linear	C_s	0.0000	0.0254	C	ND
		Langmuir	C/C_s	7.8078	3.5670	C	0.8358
		Freundlich	$\log C_s$	2.7351	2.7742	$\log C$	0.8439
	Sagehill soil	Linear	C_s	0.0000	0.0067	C	ND
		Langmuir	C/C_s	13.614	9.1697	C	0.9771
		Freundlich	$\log C_s$	6.2770	5.1310	$\log C$	0.5725
	Willamette river sediment	Linear	C_s	0.0000	0.0288	C	0.1626
		Langmuir	C/C_s	8.2438	3.1175	C	0.8872
		Freundlich	$\log C_s$	2.5497	2.6051	$\log C$	0.9128
	Yaquina bay sediment	Linear	C_s	0.0000	0.0298	C	0.5329
		Langmuir	C/C_s	14.553	2.2335	C	0.9769
		Freundlich	$\log C_s$	1.9516	1.7056	$\log C$	0.9827

ND = not determined

therms: Langmuir (Eqs. 3 and 4), Freundlich (Eqs. 8 and 9) and linear (Eq. 11) models.

Isotherm plots of TOC data for only Bottom Ash Solid waste and isotherm equations for the different solid phases are shown in Fig. 12, and the isotherm parameters determined from statistical regression analyses with their coefficients are given in Table 4.

Both the Langmuir and Freundlich equations automatically pass through the origin, but the linear model is forced through the origin. For the three soils and two sediment samples, most of the three isotherm models yielded statistically significant regressions, with the Freundlich isotherm giving the "best" model, based on a criterion of maximum coefficient of determination (R^2). However, even the linear isotherm model would be reasonable for some of these solid samples.

In summary, the present case study involved sorption/desorption processes with distilled water of a variety of hazardous solid wastes and highway C&R materials which are complex organic mixtures. The following are some of the findings:

- The water quality of the leachates was quantified in terms of both chemical constituents (this chapter) and toxicity (see Chap. 4).
- Sorption and/or desorption processes, a part of the removal/reduction/retardation (RRR) processes for chemical constituents in leachates, were determined by a testing methodology using a series of laboratory simulations.
- Leachate chemical constituents (expressed as either individual organic compound or TOC content) were specifically identified and determined by laboratory instrumental methods.
- The results from this case study can be used as input to the general comprehensive RRR fate and transport model (i.e., which includes volatilization, photolysis, biodegradation, and sorption/desorption modules) in order to predict organic leachate-generated chemical loads and concentrations at highway boundary or landfill sites.
- The potential impacts of organic leachates from complex mixtures on surface and ground waters appear to be of environmental concern, thus the testing methodology provides a systematic approach for such evaluations.

7 Conclusions

Sorption and desorption of contaminants into, onto, or from subsurface soils, bottom sediments, and suspended solids constitute a consideration in the characterization of the nature of both solid phases and contaminants. There is no simple and easy theoretical procedure that provides an exact quantitative prediction of what and how much of what will be sorbed/desorbed by a certain solid phase over a period of time, and to predict the sorption/desorption-time relationship and the fate of contaminants once they are released into the environment.

It is important to differentiate between the two different types of sorption/desorption tests (i.e., batch and column-leaching), and the sorption characteristics determined from one should not be confused with the other. Sorption isotherms obtained with batch equilibrium tests are applied mainly to solid suspensions. The physical model, assumed with this situation, is one of a completely dispersed solid particle system, where all solid particle surfaces are exposed and available for interactions with the contaminants of concern. In contrast, column-leaching tests are performed with intact solid samples, and the sorption characteristics obtained from them are the results of contaminant interactions with a structured system where not all-solid particle surfaces are exposed or available for interactions with the contaminants.

The purpose of laboratory testing to obtain contaminant-solid phase relationships is not only to obtain some insight into the accumulation and transmission characteristics of the solid materials with specific regard to the contaminant(s) of interest, but also to obtain physical input for transport modeling and chemodynamic purposes. It is also most important to conduct tests with the actual contaminant leachate or chemical species and also with the solid particle samples representative of the field matrix.

There are a good number of sorption/desorption isotherm models which were developed in order to reflect the actual sorption/desorption processes occurring in the natural environment. Some models have a sound theoretical basis; however, they may have only limited experimental utility because the assumptions involved in the development of the relationship apply only to a limited number of sorption/desorption processes. Other models are more empirical in their derivation, but tend to be more generally applicable.

In the present chapter, two main groups of models have been discussed, namely single component system (SCS) and multicomponent system models. SCS adsorption models actually deal with one pollutant component in an aqueous system or in an SWM leachate. To represent the equilibrium relation for SCS adsorption, a number of isotherm models reported in the literature were reviewed and comprise the following: double-reciprocal Langmuir, BET, Freundlich, Langmuir-Freundlich, Linear, and Toth models. Multicomponent pollutants in an aqueous environment and/or leachate of SWMs usually consist of more than one compound in the exposed environment. Multicomponent adsorption involves competition among pollutants to occupy the limited adsorbent surface available and the interactions between different adsorbates. A number of models have been developed to predict multicomponent adsorption equilibria using data from SCS adsorption isotherms. Multicomponent equilibria models include multicomponent Langmuir, modified multicomponent Langmuir, multicomponent Langmuir-Freundlich, Ideal Adsorbed Solution, and Simplified Competitive Equilibrium models.

For simple systems considerable success has been achieved but there is still no established method with universal proven applicability, and this problem remains as one of the more challenging obstacles to the development of improved methods of process design.

Sorption/desorption on solid particles can be, in some cases, exceedingly slow. The rate-limiting nature of sorption/desorption has widespread implications but is poorly understood and predicted. Its importance is appreciated by considering if sorption/desorption occurs on time scales of months or longer, and true equilibrium may exist in only limited environments. Understanding the causes of slow sorption/desorption has been hampered by the heterogeneity of natural particles as a sorptive and diffusive medium. The rate parameters of this slow sorption/desorption mechanism depend on the solid phase itself and history of exposure.

Kinetics plays an important role in understanding the reaction rate between pollutant and solid phases. In general, it is incorrect to conclude that a particular reaction order fits the data based simply on data conformity to an integrated equation. Multiple integrated equations should also be tested in order to show that the reaction rate is not affected by species whose concentrations do not change considerably during an experiment.

A number of kinetic models reported in the literature could describe rate data very well when based on correlation coefficients and standard errors of the estimates. Despite this, there often is no consistent relation between the equation, which gives the best fit and the physico-chemical and mineralogical properties of the adsorbent(s) being studied. Another problem with some of the kinetic models is that they are empirical and no meaningful rate parameters can be obtained. In general, the overall kinetics of many pollutant-solid phase chemical interactions may often be better described by mass transfer and diffusion-based models than with simple models such as first-order kinetics. This is particularly true for slower chemical reactions where a fast reaction is followed by a much slower reaction (biphasic kinetics), and is often observed in various solid phases involving organic and inorganic compounds.

Simulation and predictive modeling of contaminant transport in the environment are only as good as the data input used in these models. Field methods differ from laboratory methods in that an increase in the scale of measurement relative to most laboratory methods is involved. Determination of transport parameters (i.e., transmission coefficients) must also use actual contaminant chemical species and field solid phase samples if realistic values are to be specified for the transport models. The choice of type of test, e.g., leaching cells and diffusion tests, depends on personal preference and availability of material. No test is significantly better than another. Most of the tests for diffusion evaluation are flawed to a certain extent.

Because of the possible wide differences among properties and characteristics of solid phases and the varied chemical compositions of contaminants or a contaminant leachate, field measurement variables present average properties over a large volume/area. The problem which complicates the picture is that ideal models are applied to a material or space which is highly non-ideal, non-uniform, and does not permit easy specification or identification of both initial and boundary conditions. To avoid this discrepancy, field and laboratory methods should be developed or modified to complement one another. Thus, ideal theory needs to be supported with physical evidence if rational applications to field studies and environmental simulation are desired.

References

1. Aboul-Kassim TAT (1998) Ph.D. Dissertation, Department of Civil, Construction and Environmental Engineering, College of Engineering, Oregon State University, Corvallis, Oregon, USA
2. Anderson DC (1981) MS Thesis, Soil and Crop Sciences Dept, Texas A & M University, College Station, TX, p 223
3. Bergen BJ, Rahn KA, Nelson WG (1998) Environ Sci Technol 32:3496
4. Cheremisinoff PN, Gigliello KA, O'Neill TK (1989) Groundwater leachate. Technomic Publishing, p 145
5. Cherry JA (1983) J Hydrology, Special Issue, Elsevier, p 197
6. Abriola LM, Pinder GF (1985) Water Resour Res 21:11
7. Abriola LM, Pinder GF (1985) Water Resour Res 21:19
8. Aiken GR, McKnight RL, MacCarthy P (1985) Humic substances in soil, sediments, and water. Wiley, NY, p 645
9. Alvarez J, Carton A, Isla T, Herguedas A (1995) Third International Conference on Water Pollution: Modeling, Measuring and Prediction, Computational Mechanics, Billerica, MA, p 389
10. Aminabhavi TM, Naik HG (1998) J Hazard Mater 60:175
11. Aminabhavi TM, Naik HG (1999) J Hazard Mater 64:251
12. Daniel DE (1984) Predicting hydraulic conductivity of clay liners. Proc ASCE, Geotech Eng Div 110:285
13. Elliott HA, Liberati MR, Huang CP (1986) J Environ Qual 15:214
14. Farquhar GJ, Sykes JF (1980) Landfill leachate migration in soil. Proc. of Leachate Management Seminar, University of Toronto, Toronto, p 157
15. Buffle J, Stumm W (1994) In: Buffle J, Devitre RR (eds) Chemical and biological regulation of aquatic systems. CRC Press, Boca Raton, FL, p 42
16. Celis R, Hermosín MC, Cox L, Cornejo J (1999) Environ Sci Technol 33:1200
17. Chiou CT, Peter LJ, Freed VH (1979) Science 206:831
18. Chiou CT, Schmedding DW, Manes M (1982) Environ Sci Tech 16:4
19. Fernandez F, Quigley RM (1988) Can Geotech J 25:582
20. Sparks DL, Suarez DL (eds) (1991) Rates of soil chemical processes. SSSA Spec Publ No. 27, Soil Sci. Soc. Am., Madison, Wl.
21. Sparks DL (1985) Adv Agron 38:231
22. Sparks DL (1986) Soil physical chemistry. CRC Press, Inc, Boca Raton, FL p 308
23. Sparks DL (1989) Kinetics of soil chemical processes. Academic Press, San Diego, CA
24. Sparks DL (1992) In: Nirenberg WA (ed) Encyclopedia of earth systems science, vol 4. Academic Press, San Diego, CA, pp 219–229
25. Sparks DL (1995) Environmental soil chemistry. Academic Press, p 267
26. Sposito G (1984) The surface chemistry of soils. Oxford University Press, NY, p 234
27. Stumm W, Morgan JJ (1981) Aquatic chemistry, 2nd edn. Wiley, New York, p 463
28. Stumm W (ed) (1990) Aquatic chemical kinetics. Wiley, New York
29. Huang E, Weber WJ Jr (1997) Environ Sci Technol 31:2562
30. Huang W, Young TM, Schlautman MA, Yu H, Weber WJ Jr (1997) Environ Sci Technol 31:1703
31. Schlebaum W, Schraa G, van Riemsdijk WH (1999) Environ Sci Technol 33:1413
32. Schwarzenbach RP, Gschwend PM, Imboden M (1993) Environmental organic chemistry. Wiley, p 681
33. Thibaud-Erkey C, Guo Y, Erkey C, Akgerman A (1996) Environ Sci Technol 30:2127
34. Yu JW, Neretnieks I (1990) Ind Eng Chem Res 29:220
35. Langmuir I (1918) J Am Chem Soc 40:1362
36. Brunauer S, Emmet PH, Teller E (1938) J Am Chem Soc 60:309
37. Freundlich H (1926) Colloid and capillary chemistry. Methuen, London
38. Sips R (1948) J Chem Phys 16:490
39. Banerjee S, Howard PH (1988) Environ Sci Technol 22:839

40. Banerjee S, Yalkowsky SH, Valvani SC (1980) Environ Sci Technol 14:1227
41. Bowman BT, Sans WW (1993) J Environ Sci Health B18:667
42. Bruggeman WA, van der Steen J, Hutzinger O (1982) J Chromatogr 238:335
43. Burkhard LP, Kuehl DW, Veith GD (1985) Chemosphere 14:1551
44. Chiou CT (1981) In: Hazard assessment of chemicals current developments. Academic Press, New York, pp 117–153
45. Chiou CT (1985) Environ Sci Technol 19:57
46. Chiou CT, Freed VH, Schmedding DW (1977) Environ Sci Technol 11:475
47. Doucette WJ, Andren AW (1987) Environ Sci Technol 21:521
48. Doucette WJ, Andren AW (1988) Chemosphere 17:345
49. Hawker DW (1989) Chemosphere 19:1586
50. Hawker DW, Connell DW (1988) Environ Sci Technol 22:382
51. Isnard P, Lambert S (1989) Chemosphere 18:1837
52. Jury WA, Spencer WF, Farmer WJ (1983) J Environ Qual 12:558
53. Jury WA, Farmer WJ, Spencer WF (1984) J Environ Qual 13:567
54. Jury WA, Farmer WJ, Spencer WF (1984) J Environ Qual 13:573
55. Karickhoff SW (1981) Chemosphere 10:833
56. Karickhoff SW (1988) J Hydraul Eng 110:707
57. Karickhoff SW, Morris KR (1985) Environ Toxicol Chem 4:469
58. Karickhoff SW, Brown DS, Scott TA (1979) Water Res 13:241
59. Mackay D (1991) Multimedia environmental models. The fugacity approach. Lewis Publishers, Chelsea, MI, p 553
60. Mackay D, Paterson S (1991) Environ Sci Technol 25:427
61. Mackay D, Stiver WH (1991) In: Grover R, Lessna AJ (eds) Environmental chemistry of herbicides, vol II. CRC Press, Boca Raton, FL, pp 281–297
62. McDuffie B (1981) Chemosphere 10:73
63. Paterson S, Mackay D (1985) In: Hutzinger O (ed) The Handbook of Environmental Chemistry. vol 2, part C. Springer, Berlin Heidelberg New York, pp 121–140
64. Toth J (1971) Acta Chim Acad Sci Hung 69:311
65. Jossens L, Prausnitz JM (1978) Chem Eng Sci 33:1097
66. Eldin NE, Greene JC, Williamson KJ, Quigley MM, Lundy JR, Azizian M, Aboul-Kassim TAT, Frey K, Huber WC, Thayumanavan P, Edwards K, Kwon P, Martinez MA (1997) Environmental impact of highway construction and repair materials on surface and ground waters. Interim Progress Report-Phase II, National Cooperative Highway Research Project (NCHRP) 25–9, Department of Civil, Construction, and Environmental Engineering, Oregon State University, Corvallis, OR
67. Eldin NE, Huber WC, Williamson KJ, Nelson PO, Lundy JR, Quigley MM, Azizian M, Frey K, Thayumanavan P, Aboul-Kassim TAT (1998) Environmental impact of highway construction and repair materials on surface and ground waters. Final Progress Report-Phase II, National Cooperative Highway Research Project (NCHRP) 25–9, Department of Civil, Construction, and Environmental Engineering, Oregon State University, Corvallis, Oregon
68. Dugenest S, Combrisson J, Casabianca H, Grenier-Loustalot MF (1999) Environ Sci Technol 33:1110
69. Hitchon B, Trudell M (1985) Hazardous wastes in groundwater, a soluble dilemma. Proc Second Canadian/American Conference on Hydrogeology, p 255
70. Sykes JF, Soyupak S, Farquhar GJ (1982) Water Resour Res 18:135
71. Crittenden JC, Luft P, Hand DW (1985) Water Res 19:1537
72. Crittenden JC, Luft P, Hand DW, Oravitz JL, Loper SW, Ari M (1985) Environ Sci Technol 19:1037
73. Kibbey TCG, Hayes KF (1997) Environ Sci Technol 31:1171
74. Ruthven DM (1984) Principles of adsorption and adsorption processes. Wiley, New York, p 445
75. Yon CM, Turnock PH (1971) Am Inst Chem Eng Symp Ser 67:75
76. Yonge DM, Keinath TM (1986) J Water Pollut Control Fed 58:77

77. Butler JAV, Ockrent C (1930) J Phys Chem 34:2841
78. Ali MA (1996) Environ Sci Technol 30:1061
79. Crittenden JC, Wong BWC, Thacker WE, Snoeyink VL, Hinrichs RL (1980) J Water Pollut Control Fed 52:2780
80. DiGiano FA, Baldauf G, Sontheimer H (1978) Chem Eng Sci 33:1667
81. Fritz W, Schlunder EU (1981) Chem Eng Sci 36:721
82. Hsieh JSC, Turian RM, Tien C (1977) AlChE J 23:263
83. Kilduff JE, Karanfil T, Chin Y-P, Weber WJ Jr (1996) Environ Sci Technol 30:1336
84. Kilduff JE, Karanfil T, Weber WJ Jr (1996) Environ Sci Technol 30:1344
85. Merk W, Fritz W, Schlunder EU (1981) Chem Eng Sci 36:743
86. Murin CJ, Snoeyink VL (1979) Environ Sci Technol 13:305
87. Thayumanavan P (1997) MS Thesis, College of Engineering, Department of Civil, Construction and Environmental Engineering, Oregon State University, Corvallis, OR
88. Wilmanski K, Breemen AN (1990) Water Res 24:773
89. Xing B, Pignatello JJ (1997) Environ Sci Technol 31:792
90. Xing B, Pignatello JJ (1998) Environ Sci Technol 32:614
91. Xing B, Pignatello JJ, Gigliotti B (1996) Environ Sci Technol 30:2432
92. Yen CY, Singer PC (1984) J Environ Eng 110:976
93. Jain JS, Snoeyink VL (1973) J Water Pollut Control Fed 45:2463
94. Radke CJ, Prausnitz JM (1972) AlChE J 18:761
95. Smith EH, Weber WJ Jr (1988) Environ Sci Technol 22:213
96. Weber WJ Jr, Smith EH (1987) Environ Sci Technol 21:1040
97. Grant TM, King CJ (1990) Ind Eng Chem Res 29:264
98. Kan AK, Fu G, Hunter MA, Tomson MB (1997) Environ Sci Technol 31:2176
99. Kan AT, Fu G, Hunter M, Chen W, Ward CH, Tomson MB (1998) Environ Sci Technol 32:892
100. Yong RN, Elmonayeri DS, Chong TS (1985) Eng Geology 21:279
101. Thacker WE, Crittenden JC, Snoeyink VL (1984) J Water Pollut Control Fed 56:243
102. Speitel GE Jr, Lu CJ, Turakhia M, Zhu XJ (1989) Environ Sci Technol 23:68
103. Arocha MA, Jackman AP, McCoy BJ (1996) Environ Sci Technol 30:1500
104. Gevao B, Jones KC (1998) Environ Sci Technol 32:640
105. Huang W, Weber WJ Jr (1998) Environ Sci Technol 32:3549
106. Piatt JJ, Brusseau ML (1998) Environ Sci Technol 32:1604
107. Pignatello JJ, Xing B (1996) Environ Sci Technol 30:1
108. Bunnett JF (1986) In: Bernasconi CF (ed) Investigtions of rates and mechanisms of reactions, 4th edn. Wiley, New York, pp 171–250
109. Skopp J (1986) J Environ Qual 15:205
110. Tang L, Sparks DL (1993) Soil Sci Soc Am J 57:42
111. Ko S-O, Schlautman MA (1998) Environ Sci Technol 32:2776
112. Ko S-O, Schlautman MA, Carraway ER (1998) Environ Sci Technol 32:2769
113. Cornelissen G, van Noort PCM, Parsons JR, Govers HAJ (1997) Environ Sci Technol 31:454
114. Ren FY, Harris JM (1996) Anal Chem 68:1651
115. Werth CJ, Reinhard M (1997) Environ Sci Technol 31:689
116. Werth CJ, Reinhard M (1997) Environ Sci Technol 31:697
117. Low MJD (1960) Chem Rev 60:267
118. Chien SH, Clayton WR (1980) Soil Sci Soc Am J 44:265
119. Crank J (ed) (1976) The mathematics of diffusion. 2nd edn. Oxford University Press, Clarendon, London
120. Weber WJ Jr, Gould JP (1966) Adv Chem Ser 60:280
121. Kuo S, Lotse EG (1974) Soil Sci Soc Am Proc 38:50
122. Havlin JL, Westfall DG (1985) Soil Sci Soc Am J 49:366
123. Wermeulen T (1958) Adv Chem Eng 2:163
124. Kuo JF, Fedram EO, Hines AL, McTeman WF (1987) Chem Eng Comm 50:201
125. Zogorski JS, Faust SD, Haas JH Jr (1976) J Colloid Interface Sci 55:329

126. Brecher LE, Frantz DC, Kostecki JA (1967) Am Inst Chem Eng Symp Ser 63:2530
127. Komiyama H, Smith JM (1974) AlChE J 20:1110
128. van Vleet BM, Weber WJ Jr (1981) J Water Pollut Control Fed 53:1585
129. Wu SC, Gschwend PM (1986) Environ Sci Technol 20:717
130. Traegner UK, Suidan MT (1989) Water Res 23:267
131. Onken AB, Matheson RL (1982) Soil Sci Soc Am J 46:276
132. Sparks DL, Jardine PM (1984) Soil Sci 138:115
133. Mukherji S, Peters CA, Weber WJ Jr (1997) Environ Sci Technol 31:416
134. Ramaswami A, Ghoshal S, Luthy RG (1997) Environ Sci Technol 31:2268
135. Ramaswami A, Luthy RG (1997) Environ Sci Technol 31:2260
136. Beckett GD, Huntley D (1998) Environ Sci Technol 32:287
137. Bouchard DC, Enfield CG, Piwoni MD (1989) In: Sawhney BL and Brown K (eds) Reactions and movement of organic chemicals in soils. Soil Science Society of America: American Society of Agronomy, Madison, WI, USA Series: SSSA special publication, no 22, p 349
138. Schwarzenbach RP, Westall J (1981) Environ Sci Technol 15:1360
139. van de Weerd H, van Riemsdijk WH, Leijnse A (1999) Environ Sci Technol 33:1675
140. Alammawi AM (1988) PhD thesis, McGill University, p 453
141. Khandelwal A, Rabideau AJ, Shen P (1998) Environ Sci Technol 32:1333
142. Mohamed AMO, Yong RN (1993) Proc ASCE/CSCE, 3rd National Conf Environ Eng, Montreal, Canada
143. Ogata A (1970) Theory of dispersion in a Granular Medium. US Geological Survey, Professional Paper No 411-1
144. Yong RN, Warith MA (1990) Groundwater contamination by industrial and domestic wastes: a case study. Proc CSCE, Conf Environ Eng 1:648
145. Yong RN, Warith MA (1990) Contaminant migration effect on dispersion coefficients. ASTM, STP 1095:69
146. Carmichael LM, Christman RF, Pfaender FK (1997) Environ Sci Technol 31:126
147. Cornelissen G, Rigterink H, Ferdinandy MMA, van Noort PCM (1998) Environ Sci Technol 32:966
148. Culver TB, Hallisey SB, Sahoo D, Deitsch JJ, Smith JA (1997) Environ Sci Technol 31:1581
149. Fukushima M, Oba K, Tanaka S, Nakayasu K, Nakamura H, Hasebe K (1997) Environ Sci Technol 31:2218
150. Bennett DH, Kastenberg WE, McKone TE (1999) Environ Sci Technol 33:503
151. Bennett DH, McKone TE, Matthies M, Kastenberg WE (1998) Environ Sci Technol 32:4023
152. Iwata S, Tabuchi T, Warkentin BP (1988) Environ Sci Technol 22:112
153. Mohamed AMO, Yong RN, Tan BK (1992) Mitigation of acidic mine drainage: engineered soil barriers for reactive tailings. ASCE National Conf Environmental Engineering, Water Forum 92, Baltimore, Maryland
154. Ogata A, Banks RB (1961) J Environ Qual 2:29
155. Yong RN, Samani HMV (1989) Analysis of two-dimensional solute transport in clay soils using irreversible thermodynamics. Proc CANCAM 23:54
156. Yong RN, Sethi AJ, Ludwig HP, Jorgensen MA (1979) ASCE, Geotech Eng Div 105:1193
157. Yong RN, Mohamed AMO, Warkentin BP (1992) In: Development in geotechnical engineering. Prentice Hall, Englewood Cliffs, NJ
158. Connaughton DF, Stedinger JR, Lion LW, Shuler ML (1993) Description of time varying desorption kinetics: release of naphthalene from contaminated soils. Environ Sci Technol 27:2397
159. DiToro DM, Horzempa LM (1982) Reversible and resistant components of PCB adsorption-desorption: isotherms. Environ Sci Technol 16:594
160. Scheidegger AM, Sparks DL (1996) A critical assessment of sorption-desorption mechanisms at the soil mineral/water interface. Soil Sci 161:813
161. Pignatello JJ (1989) In: Sawhney BL, Brown K (eds) Reactions and movement of organic chemicals in soil. Soil Science Society of America, Madison, WI, pp 45-97

162. Alexander M (1994) Biodegradation and bioremediation. Academic Press, New York
163. Alexander M (1995) Draft report environmentally acceptable endpoints in soil. Gas Research Institute, Environment & Safety Research Group
164. Brusseau ML, Rao PSC (1989) Crit Rev Environ Control 19:33
165. Gaber HM, Inskeep WP, Comfort SD, Wraith JM (1995) Soil Sci Soc Am J 59:60
166. Goltz MN, Roberts PV (1988) J Contam Hydrol 3:37
167. Roberts PV, Goltz MN, Mackay DM (1986) Water Resour Res 22:2047
168. van Genuchten MT, Wagenet RJ (1989) Soil Sci Soc Am J 53:1303
169. Velocchi A (1989) J Water Resour Res 25:273
170. Brusseau ML, Rao PSC (1991) Environ Sci Technol 25:1501
171. Brusseau ML, Wood AL, Rao PSC (1991) Environ Sci Technol 25:903
172. Eischenback A, Kaestner M, Bierl R, Schaefer G, Mahro B (1994) Chemosphere 28: 683
173. Huang LQ, Pignatello JJJ (1990) Assoc Off Anal Chem 73:443
174. Kan AT, Fu G, Tomson MB (1994) Environ Sci Technol 28:859
175. Karickhoff SW (1980) In: Baker RA (ed) Contaminants and sediments. Ann Arbor Science, Ann Arbor, MI, pp 193
176. Koskinen WC, Cheng HH, Jarvis LEJ, Sorenson BA (1994) Int J Environ Anal Chem 58:379
177. Laitinen A, Michaux A, Aaltonen O (1994) Environ Technol 15:715
178. Pignatello JJ (1990) Environ Toxicol Chem 9:1107
179. Pignatello JJ (1990) Environ Toxicol Chem 9:1117
180. Pignatello JJ, Huang LQ (1991) J Environ Qual 20:222
181. Pignatello JJ, Ferrandino FJ, Huang LQ (1993) Environ Sci Technol 27:1563
182. Sawhney BL, Pignatello JJ, Steinberg SM (1988) J Environ Qual 17:149
183. Steinberg S (1992) Chemosphere 24:1301
184. Carroll KM, Harkness MR, Bracco AA, Balcarcel RR (1994) Environ Sci Technol 28:253
185. Hatzinger PB, Alexander M (1995) Environ Sci Technol 29:537
186. Jaynes DB (1991) Soil Sci Soc Am J 55:658
187. Loehr RC, Webster MT (1995) In: Draft report environmentally acceptable endpoints in soil. Gas Research Institute Environment & Safety Research Group, chap 2
188. Pignatello JJ, Frink CR, Marin PA, Droste EX (1990) J Contam Hydrol 5:195
189. Scribner SL, Benzing TR, Sun S, Boyd S (1992) J Environ Qual 21:115
190. Steinberg SM, Pignatello JJ, Sawhney BL (1987) Environ Sci Technol 21:1201
191. Ball WP, Roberts PV (1991) Environ Sci Technol 25:1223
192. Pavlostathis SG, Jaglal KJ (1991) Environ Sci Technol 25:274
193. Crank J (1975) The mathematics of diffusion. Oxford University Press, Oxford, United Kingdom
194. Miller CT, Pedit JA (1992) Environ Sci Technol 26:1417
195. Miller CT, Weber WJ (1988) Water Res 22:465
196. Weber WJ, Miller CT (1988) Water Res 22:457
197. Weber WJJ, McGinley PM, Katz LE (1991) Water Res 25:499
198. Pedit JA, Miller CT (1994) Environ Sci Technol 28:2094
199. Komiyama H, Smith JM (1974) Am Inst Chem Eng J 20:728
200. Critenden JC, Weber WJ Jr (1978) J Environ Eng Div (Am Soc Civ Eng) 104:185
201. Critenden JC, Hand DW, Arora H, Lykins BW (1987) J Am Water Works Assoc 79:74
202. Farrell J, Reinhard M (1994) Environ Sci Technol 28:53
203. Farrell J, Reinhard M (1994) Environ Sci Technol 28:63
204. Brusseau ML (1992) J Contam Hydrol 9:353
205. Connaughton DF, Stedinger JR, Lion LW, Shuler ML (1993) Environ Sci Technol 27:2397
206. Harmon TC, Roberts PV (1994) Environ Sci Technol 28:1650
207. Xing B, Pignatello JJ (1995) Environ Chem Div Proc, ACS 35:432
208. Di Toro DM, Horzempa LM (1982) Environ Sci Technol 16:594
209. Fu G, Kan AT, Tomson M (1994) Environ Toxicol Chem 13:1559
210. Pignatello JJ (1991) Environ Toxicol Chem 10:1399

211. Rao PSC, Davidson JM (1980) In: Overcash MR, Davidson JM (eds) Environmental impact of nonpoint source pollution. Ann Arbor Science Publishers, Ann Arbor, MI, pp 23
212. Alvarez-Cohen L, McCarthy PL, Roberts PV (1993) Environ Sci Technol 27:2141
213. Siantar DP, Feinberg BA, Fripiat JJ (1994) Clays Clay Miner 42:187
214. Miller ME, Alexander M (1991) Environ Sci Technol 25:240
215. Aronstein BN, Calvillo YM, Alexander M (1991) Environ Sci Technol 25:1728
216. Ogram AV, Jessup RE, Ou LT, Rao PSC (1985) Appl Environ Microbiol 49:582
217. Allard A, Hynning P, Remberger M, Neilson AH (1994) Appl Environ Microbiol 60:777
218. Robinson KG, Farmer WS, Novak JT (1990) Water Res 24:345
219. Guerin WF, Boyd SA (1992) Appl Environ Microbiol 58:1142
220. Rijnaarts HHM, Bachmann A, Jumelet JC, Zehnder AJB (1990) Environ Sci Technol 24:1349
221. Rao PSC, Bellin CA, Brusseau ML (1993) In: Linn DM (ed) Sorption and degradation of pesticides and organic chemicals in soil. Soil Science Society of America, Madison, WI, chap 1, p 1
222. Fry VA, Istok JD (1994) Water Resour Res 30:2413
223. Scow KM, Hutson J (1992) Soil Sci Soc Am J 56:119
224. Heiger-Bernays W, Menzie C, Montgomery C, Edwards D, Pauwels S (1995) In: Draft report environmentally acceptable endpoints in soil. Gas Research Institute Environment & Safety Research Group, chap 3
225. Landrum PF (1989) Environ Sci Technol 23:588
226. Umbreit TH, Hesse EJ, Gallo MA (1986) Science 232:497
227. Varanasi U, Richnert WL, Stein JW, Brown DW, Sanborn HR (1985) Environ Sci Technol 19:836
228. MacKay DM, Cherry JA (1989) Environ Sci Technol 23:630
229. MacDonald JA, Kavanaugh MC (1994) Environ Sci Technol 28:362 A
230. Travis CC, MacInnis JM (1992) Environ Sci Technol 26:1885
231. Bolick JJ Jr, Wilson DJ (1994) Sep Sci Technol 29:701
232. Whiffin RB, Bahr JM (1985) Proceedings of the 4th National Symposium on Aquifer Restoration and Ground Water Monitoring. National Water Well Association, Worthington, OH, pp 75
233. Wilson DJ, Rodríguez-Maroto JM, Gómez-Lahoz C (1994) Sep Sci Technol 29:1645
234. Rodríguez-Maroto JM, Gómez-Lahoz C, Wilson DJ, Clarke AN (1995) Sep Sci Technol 30:317
235. Rodríguez-Maroto JM, Wilson DJ, Gómez-Lahoz C, Clarke AN (1995) Sep Sci Technol 30:521
236. Lightly JS, Silcox GD, Pershing DW, Cundy VA, Linz DG (1990) Environ Sci Technol 24:750
237. DeCicco SG, Troxler WL (1989) In: Freeman HM (ed) Standard handbook of hazardous waste treatment and disposal. McGraw-Hill, New York
238. Bucalá V, Saito H, Howard JB, Peters WA (1994) Environ Sci Technol 28:1801
239. Laha S, Luthy RG (1991) Environ Sci Technol 25:1920
240. Banerjee S, Castrogivanni MA (1987) J Chromatography 396:169
241. Banerjee S, Yalkowsy SH (1988) Anal Chem 60:2153
242. Ji W, Brusseau ML (1998) Water Resour Res 34:1635
243. Johnson JC, Sun S, Jaffé PR (1999) Environ Sci Technol 33:1286
244. Kilduff JE, Wigton A (1999) Environ Sci Technol 33:250
245. Kile DE, Chiou CT (1989) In: Suffet IH MacCarthy P (eds) Aquatic humic substances: influence on fate and treatment of pollutants. Adv Chem Ser 219:131
246. Kile DE, Chiou CT, Helburn RS (1990) Environ Sci Technol 24:205
247. Li Z, Bowman RS (1998) Environ Sci Technol 32:2278
248. Nkedi-Kizza P, Rao PSC, Hornsby AG (1985) Environ Sci Technol 19:975–979
249. Nzengung VA (1996) Environ Sci Technol 30:89
250. Nzengung VA, Nkedi-Kizza P, Jessup RE, Voudrias EA (1997) Environ Sci Technol 31:1470
251. Rao P, Suresh C, Lee LS, Pinal R (1990) Environ Sci Technol 24:647

252. Sahoo D, Smith JA, Imbrigiotta TE, McLellan HM (1998) Environ Sci Technol 32:1686
253. Smith JA, Sahoo D, McLellan HM, Imbrigiotta TE (1997) Environ Sci Technol 31:3565
254. Takimoto K, Ito K, Mukai T, Okada M (1998) Environ Sci Technol 32:3907
255. Tiehm A, Stieber M, Werner P, Frimmel FH (1997) Environ Sci Technol 31:2570
256. Walters RW, Ostazeski SA, Guiseppi-Elie A (1989) Environ Sci Technol 23:480
257. Woodburn KB, Lee LS, Rao PSC, Delfino JJ (1989) Environ Sci Technol 23:407
258. Zachara JM, Ainsworth CC, Schmidt RL, Resch CT (1988) J Contaminant Hydrol 2:343
259. Humphrey DN, Eaton RA (1993) Tire chips as a subgrade insulation – field trial. Symposium proceedings: Recovery and Effective Reuse of Discarded Materials and Byproducts for Construction of Highway Facilities. Federal Highway Administration, FHA, 11
260. North Carolina Department of Transportation Materials and Tests Unit (NC-DOT-MAT) (1993) A laboratory evaluation on crumb rubber on strength performance of concrete. Symposium proceedings: Recovery and Effective Reuse of Discarded Materials and By-products for Construction of Highway Facilities. Federal Highway Administration, FHA, 34
261. Collins RJ, Ciesielski SK (1993) Recycling and use of waste materials and byproducts in highway construction. Federal Highway Administration, FHA 1–2:356
262. Ahmed I (1991) Use of waste materials in highway construction. Rep FHWA/IN/JHRP91/3
263. Hunsucker DQ (1993) Evaluating the use of ponded fly ash in roadway base course. Symposium proceedings: Recovery and Effective Reuse of Discarded Materials and By-products for Construction of Highway Facilities. Federal Highway Administration, FHA 61
264. Vassiladou EE (1993) Utilization of fly and bottom ash as a partial fine aggregate replacement in asphalt concrete mixtures. Symposium proceedings: Recovery and Effective Reuse of Discarded Materials and By-products for Construction of Highway Facilities. Federal Highway Administration, FHA 101
265. Dewey G (1993) Municipal waste combustion ash as an aggregate substitute in bituminous mixture. Symposium proceedings: Recovery and Effective Reuse of Discarded Materials and By-products for Construction of Highway Facilities. Federal Highway Administration, FHA 64
266. Martin WP, Gast RG, Meyer GW (1976) Land application of waste materials: unresolved problems and future outlook. Soil Conserv Soc Am 25:300
267. Pojasek RB (1980) Toxic hazardous waste disposal. Ann Arbor Science Publishers
268. Mercer JW, Waddell RK (1993) In: Maidment DR (ed) Handbook of hydrology, chap 16. McGraw-Hill, New York, NY

CHAPTER 4

QSAR/QSPR and Multicomponent Joint Toxic Effect Modeling of Organic Pollutants at Aqueous-Solid Phase Interfaces

Tarek A. T. Aboul-Kassim[1], Bernd R. T. Simoneit[2]

[1] Department of Civil, Construction and Environmental Engineering, College of Engineering, Oregon State University, 202 Apperson Hall, Corvallis, OR 97331, USA
e-mail: Tarek.Kassim@orst.edu

[2] Environmental and Petroleum Geochemistry Group, College of Oceanic and Atmospheric Sciences, Oregon State University, Corvallis, OR 97331, USA
e-mail: simoneit@oce.orst.edu

Information about environmental chemodynamics of organic pollutants is a basic need in environmental planning, restoration, and engineering management. Sorption/desorption, an important chemodynamic behavior of various pollutants, can greatly influence the mobility and bioavailability of these compounds in different environmental compartments. Accordingly, aqueous-solid phase interfaces are significant in determining: (1) the route and rates by which organic pollutants can transfer to and from these interfaces, (2) the ultimate behavior and fate of pollutants, and (3) their toxicity, genotoxicity, and bioavailability to microorganisms.

When the rates of sorption or desorption processes are known, environmental fate modeling can provide an educated estimate and prediction on the accessibility and bioavailability of a target pollutant to a specific transport mechanism in the environment. Hence, the present chapter is an attempt to assess fate (i.e., in terms of pollutant mobility using predictive sorption or desorption coefficients) as well as effects (i.e., in terms of bioavailability) of various pollutants and to correlate these observations for development of predictive relationships.

In order to fulfill this general objective in the present chapter, the following interdisciplinary approaches are covered: (1) an overview of some physical and chemical properties of organic pollutants in complex mixtures which can affect their sorption/desorption chemodynamics, (2) a discussion of the fundamentals of both quantitative structure-activity and structure-property relationships (QSARs and QSPRs, respectively), with special emphasis on using molecular connectivity indices as useful properties to predict pollutant mobility and bioavailability, and (3) a review of the multicomponent (i.e., multipollutant) joint toxic/genotoxic effect models (i.e., additivity, synergism, antagonism) to predict the bioavailable fraction and action of organic pollutants at aqueous-solid phase interfaces.

The applicability of using these interdisciplinary approaches, which include incorporation of various physical and chemical properties of the pollutants, QSARs/QSPRs and multicomponent joint action modeling are discussed and evaluated using a group of toxic and carcinogenic pollutants, i.e., polychlorinated biphenyls (PCBs) and polycyclic aromatic hydrocarbons (PAHs).

Keywords. Organic pollutants, PAHs, PCBs, Aqueous-solid phase environment, QSAR, QSPR, multicomponent joint effect, Modeling

The Handbook of Environmental Chemistry Vol. 5 Part E
Pollutant-Solid Phase Interactions: Mechanism, Chemistry and Modeling
(by T. A. T. Aboul-Kassim, B. R. T. Simoneit)

Abbreviations

AI	Additivity index
COMs	Complex organic mixtures
DP_{HS}	Dissolved phase humic substances
EC_{50}	Ecological concentration at which 50% growth inhibition of *Selenastrum capricornutum* occurs
FA	Factor analysis
HOMO	The highest occupied molecular orbital
$K_{d\text{-oil}}$	PCB partition coefficient for residual transformer oil and water
$K_{d\text{-PC}}$	Partition coefficient for PCB dielectric fluid-water
K_{OW}	Octanol-water partitioning coefficient
LFER	Linear Free Energy Relationships
LUMO	The lowest unoccupied molecular orbital
MCI	Molecular connectivity index
MOLCONN	Molecular connectivity
MTI	Mixture toxicity index
PAHs	Polycyclic aromatic hydrocarbons
PCA	Principal component analysis
PCBs	Polychlorinated biphenyls
QSAR	Quantitative structure-activity relationship
QSPR	Quantitative structure-property relationship
REG	Regression procedure
SAR	Structure-activity relationship
TU	Toxic unit
VP	Vapor pressure

1
Introduction

The behavior of organic pollutants in the aqueous-solid phase environment is governed mainly by both physical and chemical properties of such compounds and a variety of complex processes, the most important of which is sorption. Sorption of organic pollutants can influence the mobility and biological activity (i.e., bioavailability) of many pollutants. This is the direct result of sorption, which determines the distribution of the pollutant in question between the aqueous phase and the solid phase. For instance, adsorption is defined as the excess of solute concentration at the solid-liquid interface over the concentration in the bulk solution regardless of the nature of the interface region or of the interaction between the pollutant of interest and the solid surface causing the excess (see Chap. 2). Any process which proceeds at a more rapid rate in the solid solution than in the sorbed state (such as transport), will be retarded as a result of sorption. Conversely, reactions such as degradation can either be enhanced or impeded by sorption depending on the exact nature of the degradation process. Surface catalyzed reactions, such as the degradation of Parathion on kaolinite surfaces, will be enhanced by sorption whereas solution phase reactions will be slowed down due to sorption of one of the reactants [1–9].

Accordingly, sorption has received a tremendous amount of attention and any method or modeling technique which can reliably predict the sorption of a solute will be of great importance to scientists, environmental engineers, and decision makers (references herein and in Chaps. 2 and 3). The present chapter is an attempt to introduce an advanced modeling approach which combines the physical and chemical properties of pollutants, quantitative structure-activity, and structure-property relationships (i.e., QSARs and QSPRs, respectively), and the multicomponent joint toxic effect in order to predict the sorption/desorption coefficients, and to determine the bioavailable fraction and the action of various organic pollutants at the aqueous-solid phase interface.

The goals of the present chapter are to: (1) provide an overview of some physical and chemical properties of pollutants that can affect their sorption/desorption behavior, (2) discuss the fundamentals of QSARs and QSPRs, with special emphasis on using molecular connectivity indices as useful properties to predict pollutant mobility and bioavailability, (3) review different multicomponent joint action models which cover additivity, synergism, and antagonism that help predict the bioavailable fraction and action of organic pollutants at aqueous-solid phase interfaces, and (4) apply, use, and evaluate these interdisciplinary approaches on a group of toxic and carcinogenic organic contaminants, i.e., PCBs and PAHs.

2
Mobility and Bioavailability Prediction at Aqueous-Solid Interfaces: Approach

The present section presents an advanced modeling approach which can be used and applied to predict and determine both the mobility and bioavailability of organic pollutants at aqueous-solid phase interfaces. Hence, a review of some of the physical and chemical properties of organic pollutants is necessary in discussing the relationships between pollutant chemical structures and their properties (i.e., QSPR) or their toxic/genotoxic effect (i.e., QSAR). Various joint action models of pollutants are presented as well, in order to predict the combined toxic effects of such compounds at interfaces.

2.1
Properties of Organic Pollutants in Complex Mixtures

Predicting sorption coefficients and hence the mobility of organic pollutants in aqueous-solid systems requires complete knowledge and analysis of various physical and chemical properties of such pollutants. This includes properties such as solubility, equilibrium vapor pressure, Henry's law constant, partition coefficient, as well as pK_a and pK_b values. Such properties can initially help determine the sorption-desorption behavior of organic pollutants once they are released, directly and/or indirectly, to the aqueous environment and then are in direct contact with solid phases. The following sections briefly summarize these properties.

2.1.1
Solubility

The tendency of any organic pollutant to move from complex organic mixtures (COMs) into the surrounding aqueous medium is expressed as the concentration of a saturated solution in equilibrium with excess solid. This equilibrium process is dependent on the balance between those forces holding the molecules or ions in the COM and the solvating ability of the solvent [10, 11]. The measurement of this parameter is called the solubility. A solubility measurement does not usually impose excessive demands on chemical techniques; however, its assessment for very sparingly soluble compounds requires specialized procedures and introduces some conceptual problems. This situation happens to be of some consequence because many organic compounds, which are known to be significant environmental pollutants, have very low water solubilities.

The main problem is well represented by the variability in the values given in the literature for the solubility of many organic compounds [12–29]. Using different techniques to determine the solubilities of organic chemicals sometimes yields values varying by a factor of 2 to 4 [10–29] (Table 1). Since many of the chemicals of environmental significance have low water solubilities, one needs to be aware of the problems involved in measuring this parameter. Thus, in searching the literature data one should note the procedures used for obtaining solubilities. It is advantageous if more than one investigator has determined the solubility for a given pollutant and the values are the same and/or similar.

A review of the commonly used experimental methods for solubility determinations is presented in Table 1. Briefly, batch equilibration is the conventional method of preparing saturated solutions for solubility determinations, where an excess amount of solute chemical is added to water and equilibrium is achieved

Table 1. Review of the methods used for the determination of some physical and chemical properties of organic pollutants

Properties	Methods	References
Solubility	Gravimetric or volumetric methods – An excess amount of chemical compound is added to a flask containing water to achieve saturation solution by shaking, stirring, centrifuging until the water is saturated with solute and undissolved solid or liquid residue appears, often as a cloudy phase – For liquids, successive known amounts of solute may be added to water and allowed to reach equilibrium, and the volume of excess undissolved solute is measured.	Abramowitz and Yalkowsky [10], Bohon and Claussen [12], Booth and Everson [13]
	Instrumental methods – UV spectrometry	Andrews and Keffer [14], Bohon and Claussen [12], Yalkowsky et al. [15, 16]
	– Gas chromatographic analysis with FID, ECD or other detectors	Chiou et al. [17], McAuliffe [18], Mackay et al. [19]
	– Fluorescence spectrophotometer	Mackay and Shiu [20]
	– High-pressure liquid chromatography (HPLC) with R.I., UV or fluorescence detection	Doucette and Andren [21], May et al. [22, 23], Shiu et al. [24], Wasik et al. [25]
	– Nephelometric methods	Davis and Parke [26], Davis et al. [27], Hollifield [28]
	– Equilibrium batch stripping	Dunnivant et al. [11], Mackay et al. [29]
Vapor pressure	– Comparative ebulliometry	Ambrose [32]
	– Effusion methods, torsion and weight-loss	Balson [34], Bradley and Cleasby [35], Hamaker and Kerlinger [36]
	– Gas saturation or transpiration methods	Spencer and Cliath [37–39], Sinke [40], Macknick and Prausnitz [41], Westcott et al. [42], Rordorf [43–45]
	– Dynamic coupled-column liquid chromatographic method	Sonnefeld et al. [46]
	– Calculation from evaporation rates and vapor pressures of reference compound – Calculation from GC retention time data	Guckel et al. [47, 48], Dobbs and Grant [49], Dobbs and Cull [50], Bidleman [51], Burkhard et al. [52], Foreman and Bidleman [53], Hamilton [54], Hinckley et al. [55], Kim et al. [56], Westcott and Bidleman [57]
K_{OW}	– EPICS (Equilibrium Partitioning In Closed Systems) method – Wetted-wall column – Headspace analyses – Calculation from vapor pressure and solubility – Direct measurement by use of pressure gauges: – Diaphragm gauge – Rodebush gauge – Inclined–piston gauge	Fujita et al. [60], Leo et al. [61], Hansch and Leo [62], Rekker [63], Bowman and Sans [64], Chiou [65–67], Chiou et al. [68, 70–77], Chiou and Kile [69], Howard [78, 79], Hansch et al. [80], De Bruijn et al. [81], De Bruijn and Hermens [82], Doucette and Andren [83], Isnard and Lambert [84], McDuffie [85], Miller et al. [86], Woodburn et al. [87]

by shaking gently or slow stirring [11, 29]. The aim is to prevent emulsion or suspension formation and thus avoid additional procedures such as filtration or centrifugation. However, experimental difficulties can still occur because of emulsion formation or microcrystal suspension with sparingly soluble compounds such as higher molecular weight *n*-alkanes and polycyclic aromatic hydrocarbons (PAHs). Thus, an alternative approach is to coat a thin layer of the compound on the surface of the equilibration flask before water is added. An accurate "generator column" method has also been developed [22, 23, 30] where a column is packed with an inert solid support (e.g., glass beads or Chromosorb) and then coated with the solute chemical. Water is pumped through the column at a controlled, known flow rate to achieve saturation. The method of concentration measurement of the saturated solution depends mainly on the solute solubility and its chemical properties. In general, solubility of organic compounds is reported at a defined temperature in distilled water. On the other hand, the pH of any compound which can ionize (e.g., phenols) must be reported because the extent of ionization affects the solubility.

2.1.2
Equilibrium Vapor Pressure

The equilibrium vapor pressure of organic compounds is comparable to solubility in that it is a measure of the volatilization tendency from liquid or solid phases. The equilibrium vapor pressure of a gas can be conceived as its solubility in air. The vapor pressure of a liquid or solid is the pressure of the gas in equilibrium with the liquid or solid at a given temperature. The thermodynamic "Clausius-Clapeyron" expression (Eq. 1, [31]) describing this equilibrium is

$$\frac{d \ln p}{d(1/T)} = \left(\frac{-\Delta H}{R}\right) \tag{1}$$

where ΔH is the heat of vaporization, T is the absolute temperature, and R is the universal gas constant.

The previous equation can be also expressed in an integral form as

$$\log p = A - BT \tag{2}$$

in which $B = \dfrac{-\Delta H}{2.303\,R}$, where $-\Delta H$ is assumed to be constant. Since Eq. (2) is linear only over a relatively narrow temperature range, other equations have been suggested, such as the Antoine expression (Eq. 3, [31]):

$$\log p = \left[A - \frac{B}{(t + C)}\right] \tag{3}$$

where A, B, and C are constants characteristic of the substance and temperature range, and t the temperature in °C.

It is interesting to mention that partitioning into the vapor phase is generally significant only for those pollutants with high vapor pressures; however, even though very small, the vapor pressure of solids can be of major consequence under certain circumstances in defining the organic pollutant distribution and chemodynamics in the environment. Thus, the concentration term *vapor density* is often used in discussions of vapor phase systems [32]. Vapor density is related to the equilibrium vapor pressure through the equation of state for a gas:

$$PV = nRT \tag{4}$$

where n is the number of moles, m is the mass in grams, and M is the gram molecular weight.

Substituting $\left(\frac{m}{M}\right)$ for the number of moles (n) gives the following equation:

$$PV = \left(\frac{m}{M}\right) \cdot RT \tag{5}$$

Since density is mass/unit volume,

$$\left(\frac{m}{M}\right) = \left(\frac{PM}{RT}\right) \tag{6}$$

If the volume, $V = 1$ l,

$$(d_0) = \left(\frac{PM}{RT}\right) \tag{7}$$

where (d_0) is the vapor density, P is the equilibrium vapor pressure in atmospheres, and R is 0.082 l atm/mol · °K.

The vapor pressure (P_A) above a solution where the mole fraction of component (A) is X_A is defined by Raoult's Law [31, 33–36]:

$$P_A = X_A \cdot P_A^0 \tag{8}$$

where P_A^0 is the vapor pressure of the pure substance at that temperature. If more than one component in the solution is volatile, the total pressure above the solution is the sum of the partial pressures of the components is

$$P_{Total} = P_A + P_B + \ldots = X_A P_A^0 + X_B P_B^0 \tag{9}$$

Solutions obey Raoult's Law when interactions between like and unlike molecules are identical. Positive ($P_A > X_A P_A^0$) and, on rare occasions, negative ($P_A > X_A P_A^0$) deviations from Raoult's Law are observed depending on the nature of the components in the solution and are accounted for by the activity coefficient (γ):

$$P_A = X_A \cdot \gamma_A \cdot P_A^0 \tag{10}$$

The activity coefficient is unity under ideal conditions.

Basically, the vapor pressure determination involves the measurement of the saturation concentration or pressure of the solute in a gas phase [37–45]. It can be determined directly from the actual concentrations and/or indirectly based on an evaporation rate measurement or chromatographic retention time [46–57]. Vapor pressures are strongly temperature dependent. Some methods and approaches for vapor pressure determinations are listed in Table 1 [32, 34–57].

2.1.3
Henry's Law Constant

Generally, the higher the pressure, the higher is the solubility of a gas in a liquid. This relationship is expressed quantitatively by Henry's Law which states that the mass of gas (m) dissolved by a given volume of solvent at a constant temperature is proportional to the gas pressure (p) with which it is in equilibrium:

$$m = k \cdot p \tag{11}$$

If the mass of gas dissolved by the given volume is converted to a concentration term, the pressure to vapor density, the Henry's Law relation may be expressed as

$$\frac{C_V}{C_L} = \text{constant}\,(H) \tag{12}$$

where C_V and C_L are the concentrations of gas in both vapor and liquid phases, respectively. The Henry's Law Constant (H) is thus a distribution coefficient indicating the tendency of an organic pollutant to distribute between a solvent and the vapor phase.

Henry's Law is obeyed with organic pollutants of low solubility provided the pressures are not high or temperatures too low – conditions under which one might expect deviations from ideal behavior. Experimental values for Henry's Law constant may be obtained by equilibrating a pollutant between the solvent and vapor phase and measuring its concentration in those two phases. Providing the solubility is low ($P_A < 0.1$) Henry's Law constant can be calculated from the equilibrium vapor pressure (P_A) and solubility (S):

$$H = \left(\frac{P^0}{S}\right) \tag{13}$$

Generally, pollutants with low vapor pressures may often have significant Henry's Law constants because of low water solubilities. In a simplified sense, the aqueous environment is so unfavorable that distribution into the vapor phase becomes a favorable transition.

The Henry's law constant is an air-water partition coefficient, which can be determined by measurement of solute concentrations in both phases [11, 58, 59]. Some effort has been devoted to devising techniques in which concentrations are measured in only one phase and the other concentration is deduced by a mass balance. These methods are generally more accurate. The principal difficulty arises with hydrophobic, low volatility compounds which have only small concentrations in both phases. Henry's law constant can also be regarded as a ratio of vapor pressure to solubility (Eq. 13); thus it is subject to the same effects, which electrolytes have on solubility and temperature has on both properties.

2.1.4
Partition Coefficient

The concentrations of any single molecular species in two phases, which are in equilibrium, have a constant ratio to each other and this is defined as follows:

$$P = K = \frac{C_2}{C_1} \tag{14}$$

It assumes that there are no significant solute-solute interactions and no strong solute-solvent interactions which would influence the distribution process. Concentrations are expressed as mass/unit volume, and usually C_1 refers to an aqueous phase and C_2 to a non-aqueous phase. The equilibrium constant (*P* or *K*) defining this system is referred to as the partition coefficient or distribution ratio. The thermodynamic partition coefficient (P') is given by the ratio of the respective mole fractions as follows:

$$P' = \frac{X_0}{X_w} \tag{15}$$

It must be noted that the partition coefficient is not the ratio of the pollutant solubilities in the two pure liquids. This change can result in significant differences, particularly with compounds of low aqueous solubility. The measurement of partition coefficients may be complicated by the involvement of other equilibrium processes such as pK_a and pH values. For example, the following reaction shows the dissociation of a monoprotic organic acid:

$$HA \Leftrightarrow H^+ + A^- \tag{16}$$

Thus, on measuring a partition coefficient of *HA*, it is imperative to know which species is being measured, i.e., neutral (undissociated, *HA*) or charged species (A^-). Mathematical procedures can be used to take into account the complicating equilibria, and partition coefficients can be calculated for both the nonionized and ionized species of organic acids. The difference in partition coefficient between the two species is approximately

$$\Delta \log P = (\log P_{ion}) - (\log P_{neutral}) \tag{17}$$

Another approach to the same type of situation is simply to measure the distribution of total solute in both phases to provide a partition ratio that is sometimes referred to as an apparent partition coefficient. Obviously, for COM materials containing aliphatic acids or bases, this ratio can vary drastically with changes in pH.

As an example of a partition coefficient, the octanol-water partition coefficient (K_{OW}) is determined by similar experimental procedures as those for solubility (Table 1), employing shake flask or generator-column techniques [60–87]. Concentrations in both the water and octanol phases may be determined and analyzed after equilibration and the partition coefficient is calculated from the concentration ratio C_0/C_w. This is actually the ratio of solute concentration in octanol saturated with water to that in water saturated with octanol.

Values of K_{OW} have been successfully calculated from molecular structure; thus there has been a tendency to calculate K_{OW} rather than measure it, especially for difficult hydrophobic chemicals [65–85]. These calculations are, in some cases, extrapolations and can be seriously in error. Any calculated $\log K_{OW}$ value above 7 should be regarded as suspect, while a value above 8 should be treated with extreme caution [78, 79, 81, 82, 86, 87].

2.1.4.1
Empirical vs Predictive Measurements

Recently, extensive research on partition coefficients has been developed in the field of medicinal chemistry because it has been observed that the action of drugs may be correlated with their partition coefficients. This parameter is an important component of structure-activity relationships (Sect. 2.2) for different series of biologically active compounds as well as for predicting environmental behavior and chemodynamics of complex mixtures [21, 62, 80–85, 88–90]. The octanol/water (K_{OW}) system is used almost exclusively in such comparisons.

Using predictive models for measuring environmental chemodynamics of organic pollutants in complex mixtures requires literature data on partition coefficient values. In some cases the values cited are not strictly experimental, being derived from linear free energy relations, while in others wide variations are reported in experimental values. The main problem is how one should evaluate which values are correct. Thus, Table 2 provides some basis to discriminate between reported values of partition coefficients, as well as predictive equations for partition coefficient calculations [21, 62, 65–85].

2.1.4.2
Relationship with Water Solubility

A number of empirical relationships have been published which could be used to predict partition coefficients from solubility data [19–29, 65, 72, 78–97]. Comparisons among these relationships may be confusing since different sets of compounds and different solubility terms are used. A theoretical analysis of partition coefficient with reference to aqueous solubility is important because it illustrates the thermodynamic principles underlying the partitioning process. The objective of that relationship is its utility for both predicting and validating reported values for partition coefficients.

A single equation can represent with some precision the relation between partition coefficient and solubility for a diverse group of organic liquids. Partition coefficients for solids do not correlate well with relations established

Table 2. Some basis to discriminate between reported values of partition coefficients

Methods		Approach	Advantages/Disadvantages
Empirical [65–79]	Equilibration technique	The most direct approach is to equilibrate the organic pollutant in the octanol/water system and measure its concentration in both phases	Concentrations derived from mass balance calculation, though less time-consuming, can introduce considerably more uncertainty
		On occasion, the concentration is measured in only one phase, with concentration in the other being derived from a mass balance calculation	Other experimental considerations in obtaining accurate values by this approach have been discussed by several workers
	HPLC retention times	Partition coefficients can also be derived from retention times in high-pressure liquid chromatography (HPLC) analyses	This approach provides some experimental advantages that simplify the analytical procedures and allow the handling of mixtures
		The retention times of test solutes are correlated with reference compounds whose partition coefficients in octanol/water (K_{OW}) are known	The reliability of this technique depends on the extent to which the stationary and mobile phases simulate the octanol/water system
			Abnormally low K_{OW} values have been obtained with sparingly soluble compounds, presumably because they do not achieve true equilibrium during the separation
Predictive [21, 62, 80–85]	π Values	Partition coefficient can be treated as an additive constitutive property, and for a given molecule can be considered an additive function of its component parts	π Values can provide an estimate of the partition coefficient of some organic compounds, providing an experimental value is available for a structurally related analogue
		This is based on the fact that the energetics of transferring a -CH_3 group from one environment to another is relatively constant from compound to compound – hence the term ***linear free energy*** relations	For example, if one needs to know K_{OW} for 2,3-dimethylphenanthrene, and **log *P*** for phenanthrene is known to be 4.09 and $\boldsymbol{\pi}\mathbf{CH_3} = 0.71$ for an aromatic ring substituent, the following relation could be used: $\log P_{(\text{dimethyl phenanthrene})} = P_{(\text{phenanthrene})} + \pi \cdot CH_3 = 4.09 + 2(0.71) = 5.51$

		Given this relation, a quantity is defined as follows for different radicals or functional groups: $\pi = \log P_X - \log P_H$ This type of analysis has been used to derive a series of π values.	This value agrees well with an experimental value of 5.58 This approach becomes less accurate with a greater difference between the unknown and the reference compound. More deviation would be expected with polar substituents (i.e. -OH, -COOH, -NO_2) than with the less polar groups (-CH_3, -NH_2, and -Cl)
	Fragment constant	The partition coefficient is expressed as the sum of its component fragments: $\log P = \left(\frac{n}{1}\right) \cdot a_n \cdot f_n$ where: (**a**) is the number of fragment (**f**) of type (**n**) in the molecule Adjustment for steric effects, bond type and different interactions gives a complex calculation usually accomplished with computer software	Fragments may be as fundamental as certain types of carbon atoms or hydrogen atoms, or may refer to multiple atom groupings such as -OH or -C-NH_2 Such a procedure is based on numerous assumptions and the accuracy with which it will predict the partition coefficient for a given compound will depend on how well it confirms to those assumptions

with liquids. However, this inconsistency can be overcome by incorporating a melting point correction (M) in the solubility term for solids. This disparity between liquids and solids is because dissolving a solid involves an additional step of breaking down the highly ordered structure, which has already been overcome in a liquid. This distinction is not a factor for partition coefficients since the solution process is equivalent in both phases for any compound. The melting point correction converts the solubility of the solid [$S_{(S)}$] to the solubility of the super-cooled liquid [$S_{(S.C.L)}$]:

$$\log S_{(S.C.L)} = \log S_{(S)} + \log M \tag{18}$$

and can be rearranged as

$$\log M = \left(\frac{\Delta H_f}{2.303\,R}\right) \cdot \left(\frac{T_m - T}{T \cdot T_m}\right) \tag{19}$$

where H_f is the molar heat of fusion, R is the universal gas constant (1.9865 cal/mol · °K), T_m is the melting point of the solid (°K), and T is the temperature under consideration (°K).

Since heats of fusion are not always available, the following approximation can be used to calculate the melting point correction:

$$\log M = \frac{K \cdot (T_m - T)}{2.303} \tag{20}$$

where $K = 0.02273$ °C. This approximation is based on the observation that the entropy change on melting $\left(\frac{\Delta H}{T_m}\right)$ is relatively constant at 13.46 cal/mole · °K. Thus

$$K = \left(\frac{\Delta H_f}{T_m}\right) \cdot \left(\frac{1}{RT}\right) \tag{21}$$

which is an expression defining the relation between solubility and partition coefficient for both liquids and solids, providing appropriate corrections are made for the latter. This relation deviates more from the ideal line at lower solubilities which is expected because departure from ideal behavior is more pronounced with lower solubilities. If solubility/partition coefficient combinations deviate significantly from the regression line, there is a good possibility that either value, or perhaps both, could be in error [19–29, 65, 72, 78–97]. It is often quite a challenge to decide which of several cited values for the partition coefficient is most accurate. Assuming the solubility data is accurate, this relationship can provide a basis for making such a discrimination.

2.1.5
pK

Whether a toxic pollutant in a COM or a solid waste material (SWM) leachate carries a charge or exists as a neutral species will have a dramatic effect on its environmental chemodynamics. This is a possibility with weak organic acids and bases, and is a function of the *pK* of the particular organic compound and pH of the surrounding environment. For instance, the dissociation of any weak organic acid (proton donor) may be represented as

$$HA + H_2O \Leftrightarrow H_3O^+ + A^- \tag{22}$$

and the equilibrium constant K_a defined as

$$K_a = \frac{[H^+] \cdot [A^-]}{[HA]} \tag{23}$$

where $[H_2O]$ is not considered and $[H^+] = [H_3O^+]$. The logarithmic form of Eq. (23) is as follows:

$$pK_a = -\log K_a \tag{24}$$

which is known as the Henderson-Hasselbach Equation relating Eqs. (22) and (23) as follows:

$$\mathrm{pH} = pKa + \log \frac{[A^-]}{[HA]} \tag{25}$$

Equation (25) can be used to calculate the composition of buffer solutions where pH is the dependent variable and $[A^-]$ and $[HA]$ are variables which can be controlled experimentally. In environmental chemodynamics studies of complex organic mixtures, a relation expressing $[A^-]$ and $[HA]$ as a function of pH and *pK* is needed. Providing the total concentration of the A containing species is C_T:

$$C_T = [HA] + [A^-] \tag{26}$$

and it follows that:

$$[HA] = \frac{C_T \cdot [H^+]}{K_a + [H^+]} \quad \text{and} \quad [A^-] = \frac{C_T \cdot K_a}{K_a + [H^+]} \tag{27}$$

On the other hand, the general case for an organic base (proton acceptor) can be given as

$$B + H_2O \Leftrightarrow BH^+ + OH^- \tag{28}$$

where

$$K_b = \frac{[BH^+] \cdot [OH^-]}{[B]} \tag{29}$$

Since pH rather than pOH is most widely used in environmental chemistry equations, it is most common to use an acidity constant for the conjugate acid of the base. In this case the equilibrium is expressed as

$$BH^+ + H_2O \Leftrightarrow H_3O^+ + B \tag{30}$$

and

$$K_a = \frac{[H^+] \cdot [B]}{[BH^+]} \tag{31}$$

In this situation K_a and K_b are related, where

$$K_W = K_a \cdot K_b = 1 \times 10^{-4} \quad \text{or} \quad pK_a + pK_b = 14 \tag{32}$$

Extensive collections of pK values are available in the literature, e.g., [98–101]. It is also possible to predict pK values for a broad range of organic acids and bases using linear free energy relationships based on a systematic treatment of electronic (inductive, electrostatic, etc.) effects of substituents which modify the charge on the acidic and basic center. Quantitative treatment of these effects involves the use of the Hammett Equation which has been a real landmark in mechanistic organic chemistry. A Hammett parameter (σ), defined as follows:

$$\sigma = \log K_X - \log K_H \tag{33}$$

or

$$\sigma = (pK_H - pK_X) \tag{34}$$

was introduced, where K_H is the dissociation constant for an organic acid (e.g., benzoic acid) in water at 25 °C, and K_x is the dissociation constant under the same experimental conditions of the benzoic acid derivative with a substituent in the meta or para position.

Positive values of σ indicate electron withdrawing by the substituent, while negative values indicate electron release to the benzene ring of the acid. A listing of some σ values is provided in the literature [98–101]. Quantitative predictions of pK values use the Hammett equation as follows:

$$\log K_X = \rho\sigma + \log K_H \tag{35}$$

or

$$pK_X = pK_H - \rho\sigma \tag{36}$$

The slope (ρ) is an indication of the sensitivity to the electronic effects from the substituents. Calculating the pK of a given organic acid or base involves selecting the correct equation and incorporating the σ values for the appropriate substituents:

$$pK_X = pK_H - \rho \cdot (\textstyle\sum \sigma) \tag{37}$$

In addition, it is possible to extend the analysis to include an ortho substituent and the associated steric effects [98–101]. Thus it is possible by this procedure to predict with some accuracy the pK_a and pK_b of organic acids and bases leached from COMs.

In summary, understanding environmental partitioning at aqueous-solid phase interfaces of organic pollutants in complex mixtures requires the complete knowledge and analysis of most of the important physical and chemical properties of such compounds. These properties can initially determine the behavior and ultimate partitioning of such pollutants once they are released to the environment. Definitive experimental values for these parameters are required before any organic compound can be used and applied in environmental modeling; however, partitioning of COMs will result in an inadvertent release of some intermediates or by-products into the environment. Chances are that no experimental values are available for these intermediates or by-products and decisions concerning their environmental behavior and partitioning are required before the necessary data could be generated. Even through predicted values may be less accurate than experimental values in this situation, they are better than no values at all.

2.2 Quantitative Structure-Activity and Structure-Property Relationships

The second modeling approach discussed in this section presents an overview of the fundamentals of quantitative structure-activity relationships (i.e., QSARs [102–130]) and quantitative structure-property relationships (i.e., QSPRs [131–139]). It will show how such an approach can be used in order to estimate and predict sorption/desorption coefficients of various organic pollutants in environmental systems.

QSARs are defined as the systematic categorization of atoms or molecules according to common features called structure, and to relate these assignments to the values of measured properties [140–165]. A property or activity of a molecule is a characteristic which can be determined or measured. By subjecting a target compound to a form of energy, numerical values can be obtained. Repeated subjection of a molecule to such an assault yields numerical measurements which are highly reproducible. By defining the physical events underway in such a process, we can define the observations as a property. A profile of measured properties is characteristic to that atom or molecule under investigation. Thus, every organic compound has a boiling point, molar refraction, partition coefficient, density, etc. Information about its form or structure is not self evident from physical property measurements. The structure is inferred from these measurements because it is known that properties are a consequence of structure [166–169].

Although QSAR/QSPR has been used almost exclusively and extensively in drug design and pharmaceutical research [151, 170–172], several studies have shown that they can be used as effectively in modeling environmental fate processes [173–191]. This may be explained by the similarity of the underlying processes that give drugs their beneficial effects and environmental pollutants their adverse effects. However, there are some important differences in characteristics and approaches between using QSAR in pharmaceutical vs environmental research, and some of these are summarized in Table 3.

QSAR/QSPR analyses describe the dependence of activity on structure and typically include several physical-chemical parameters, such as electronic

Table 3. Some important differences in characteristics and approaches between using QSAR in pharmaceutical and environmental research

QSAR in drug design research	QSAR in environmental sciences
Objectives	
Optimize biological activity of drugs	Estimate rates of fate processes
Find new active lead compounds	Analyze Processes
Characteristics	
Response in isolated systems	Whole organism response
Effects are specific and well defined	Net effects (mortality growth, etc.)
Specific mechanism of action	Specific & nonspecific mechanisms
Receptor is known in most cases	Receptor unknown in most cases
Techniques	
Hansch Approach	Hansch Approach
Multivariate Analysis	Multivariate Analysis
Computerized molecular modeling	Molecular modeling not applied

(σ, pKa), hydrophobic (π, P_{ow}, K_{ow}), and steric (E_s, MR) properties [141, 175, 176, 181, 192–199]. Since the properties of a molecule are dependent on the nature of the independent atoms and their chemical bonds, a fixed relationship exists between topological indices conveying information on bond types and bond characteristics and properties exhibited by a molecule [134–136, 200–203]. These topological or structural indices may be defined as a count of selected topological features such as the number of skeletal atoms or bonds, the number of bonds or atoms of a given type, the number of double bonds, the number of rings, and other structural parameters. Molecular topology provides a rationale for correlating interactions between a molecule and its environment through molecular connectivity indices, which are based on the graphical depiction of molecular structure and may be described by a set of numerical values [103, 204–217].

In line with the main objective of the present chapter, the next section discusses structure-activity relationships (i.e., SAR), such as molecular connectivity indices, and how these can be used to predict pollutant mobility and bioavailability.

2.2.1
Molecular Connectivity

At the molecular level, the structure of an organic pollutant is defined by a few characteristics:

1. The total number of atoms
2. The number of different kinds of atoms
3. The linking pattern or bonding scheme of the atoms

These three elements of structural information depict a molecule as a graphic structural formula [218–237]. There are two general approaches to structure description. In the first, the identities of atoms and their connections form one set of information about molecular structure called the "topology" of the molecule [102–105, 154, 166, 167, 235–237]. The second includes various three-dimensional aspects called "molecular topography". Characteristics such as size, shape, volume, surface area, etc., can be directly explained by three-dimensional molecular topography [238–240].

Generally the properties of a molecule are dependent upon the three-dimensional topography of the molecule, and the geometry which in turn depends on molecular topology (nature of the individual atoms and the bonded connections between them). Because of the relationship between bond types and characteristics such as bond strength, length, and polarity, there are relationships between topology and properties. Hence, it is most useful to express molecular structure in terms of its molecular topology [103, 221–226, 241–248]. The starting point in representing molecular structure is the molecular skeleton that in chemical graph theory is defined as the hydrogen suppressed graph.

The most basic element in the molecular structure is the existence of a connection or a chemical bond between a pair of adjacent atoms. The whole set of connections can be represented in a matrix form called the connectivity matrix [249–253]. Once all the information is written in the matrix form, relevant information can be extracted. The number of connected atoms to a skeletal atom in a molecule, called the vertex degree or valence, is equal to the number of σ bonds involving that atom, after hydrogen bonds have been suppressed.

2.2.2
Nomenclature of Molecular Connectivity Indices

The most successful of all topological indices at present is the molecular connectivity index (MCI) or a system of molecular connectivity indices. Their numerous applications in various areas of physics, chemistry, biology, pharmacology (drug design), and environmental sciences outnumber all other existing topological indices, the number of which is approaching 100 [108, 221, 222, 224–226, 254–261]. There are two major reasons for this:

1. These indices are based on sound chemical, structural (topologic and geometrical), and mathematical grounds.
2. They were developed with the idea of paralleling important physico-chemical properties such as boiling point, mobility on chromatographic columns, enthalpies of formation, and total molecular surface areas.

The following nomenclature is used to designate molecular connectivity indices [262–265]. The Greek letter chi (χ) is used to represent the index itself. Two superscripts and one subscript are used to specify a particular index. The left-side superscript (zero or a positive integer) is used to designate the order of index. The right-side superscript (letter **v**) differentiates between valence- and nonvalence-type indices. The right-side subscript (P, C, PC, or CH) specifies the subclass of molecular connectivity index, which may be a path, cluster, path/cluster, or chain-type index. If no subscript is indicated, a path-type index is assumed.

2.2.2.1
The Path-Type MCIs

The concept of the molecular connectivity index (originally called branching index) was introduced by Randic [266]. The information used in the calculation of molecular connectivity indices is the number and type of atoms and bonds as well as the numbers of total and valence electrons [176, 178, 181, 267, 268]. These data are readily available for all compounds, synthetic or hypothetical, from their structural formulas. All molecular connectivity indices are calculated only for the non-hydrogen part of the molecule [269–271]. Each non-hydrogen atom is described by its atomic δ value, which is equal to the number of adjacent non-hydrogen atoms. For example, the first-order (${}^1\chi$) molecular connectivity index is calculated from the atomic δ values using Eq. (38):

$$ {}^1\chi = \sum(\delta_\mathrm{i} * \delta_\mathrm{j})^{-0.5} \qquad (38) $$

where i and j correspond to the pairs of adjacent non-hydrogen atoms and summation is over all bonds between non-hydrogen atoms.

The first-order molecular connectivity index has been used very extensively in various QSPR and QSAR studies [269, 272, 273]. Thus, the question of its physical meaning has been raised many times. It has been found, in several studies [103, 178–180, 266, 274, 275], that this particular index correlates extremely well with the molecular surface area. It seems this index is a simple and very accurate measure of molecular surface for various classes of compounds and consequently correlates nicely with the majority of molecular surface dependent properties and processes.

Its counterpart, the first-order (${}^1\chi^v$) valence molecular connectivity index, is also calculated from the non-hydrogen part of the molecule and was suggested by several authors [103, 276, 277]. In the valence approximation, non-hydrogen atoms are described by their atomic valence δ^v values, which are calculated from their electron configuration by the following equation:

$$ \delta^v = \left(\frac{Z^v - h}{Z - Z^v - 1}\right) \qquad (39) $$

where Z^v is the number of valence electrons in the atom, Z is its atomic number, and h is the number of hydrogen atoms bound to the same atom.

By analogy with Eq. (38), the ${}^1\chi^v$ index is then calculated from the atomic δ^v values using Eq. (40):

$$ {}^1\chi^v = \sum(\delta_\mathrm{i}^v * \delta_\mathrm{j}^v)^{-0.5} \qquad (40) $$

A system of molecular connectivity indices was developed and extensively exploited by Kier and Hall [102–104, 113], Hall and Kier [108, 109, 115, 120, 125] and Kier [107]. The zero-order (${}^0\chi$) and second-order (${}^2\chi$) molecular connectivity indices are the closest members to the ${}^1\chi$ index described above. The

${}^0\chi$ and ${}^2\chi$ indices are calculated from the same input information (atomic δ values) using Eqs. (41) and (42), respectively:

$${}^0\chi = \sum (\delta_i)^{-0.5} \quad (41)$$

$${}^2\chi = \sum (\delta_i * \delta_j * \delta_k)^{-0.5} \quad (42)$$

where *i*, *j*, and *k* correspond to three consecutive non-hydrogen atoms and summations are over all non-hydrogen atoms and over all pairs of adjacent bonds between non-hydrogen atoms, respectively. Their valence analogs are defined identically as for the first-order valence molecular connectivity index. The zero-order valence and the second-order valence molecular connectivity indices are useful in modeling and estimation of acute and chronic toxicity [278–280] and of fish bioconcentration factors [179–181], respectively, for many classes of commercial organic compounds. It was suggested that the ${}^0\chi^v$ index is a simple and sound approximation for the molecular volume, thus correlating strongly with many molecular properties where molecular bulk plays an important role [280].

For molecular connectivity indices with orders higher than 2, it is also necessary to specify the subclass of index. There are four subclasses of higher order indices: path, cluster, path/cluster, and chain. These subclasses are defined by the type of structural subunits they are describing, a subunit over which the summation is to be taken when the respective indices are calculated. Naturally, the valence counterparts of all four subclasses of higher order indices can be easily defined by analogy, described above for the first-order valence molecular connectivity index.

From a chemical structural point of view, the path-type indices [102, 103, 106–109, 111–113] can be divided into two subgroups:

- The first subgroup contains the zero-, first-, and second-order indices.
- The second subgroup all other higher order indices.

The first subgroup best describes global molecular properties such as size, surface, volume, while the second subgroup describes more and more (as the order of index increases) local structural properties and possibly long-range interactions.

2.2.2.2
The Cluster and Path/Cluster MCIs

The main characteristic of cluster-type indices is that all bonds are connected to the common, central atom (star-type structure). The third-order cluster molecular connectivity index (${}^3\chi_c$) is the first, simplest member of the cluster-type indices where three bonds are joined to the common central atom [102–104, 111–113, 152–154, 166, 167, 269]. The simplest chemical structure it refers to is the non-hydrogen part of *tert*-butane. This index is then calculated using Eq. (43):

$${}^2\chi = \sum (\delta_i * \delta_j * \delta_k)^{-0,5} \quad (43)$$

where *i*, *j*, *k*, and *l* correspond to the individual non-hydrogen atoms that form the subgraph, and the summation is over all *tert*-butane-type subgraphs in a molecule. For cluster-type indices, orders higher than four do not have much chemical and structural sense for organic compounds.

The fourth-order path/cluster molecular connectivity index ($^4\chi_{pc}$) is the first, simplest member of the path/cluster-type indices. It refers to subgraphs consisting of four adjacent bonds between non-hydrogen atoms, three of which are joined to the same non-hydrogen atom [169, 221, 281–285]. Structurally (chemically) this subgraph corresponds to the non-hydrogen part of iso-pentane. This index is then calculated using Eq. (44):

$$^4\chi_{pc} = \sum (\delta_i * \delta_j * \delta_k * \delta_l \delta_m)^{-0.5} \tag{44}$$

where *i*, *j*, *k*, *l*, and *m* correspond to the individual non-hydrogen atoms that form the subgraph, and the summation is over all iso-pentane-type subgraphs in a molecule. For path/cluster-type indices, orders higher than six do not have much chemical and structural sense either. In addition, it becomes very difficult to understand what the structural and physical meaning of higher order path/cluster-type indices is.

The cluster and path/cluster indices describe mainly local structural properties, such as the extent or degree of branching in a molecule. They are highly sensitive to changes in branching, and their value rapidly increases with the degree of branching. As such they may be useful as steric descriptors. From these two classes of molecular connectivity indices the most interesting and commonly used are the third-order cluster and fourth-order path/cluster indices.

The second structural property described by the $^4\chi_{pc}$ index is the substitution pattern on the benzene ring. The value of the $^4\chi_{pc}$ index increases sharply with the degree of substitution, while in the isomeric classes of substituted benzenes it increases with the proximity of substituents. Thus, this structural parameter has also been found to be very useful in describing activities and properties of polysubstituted benzenes [103], chlorinated benzenes [279], and polychlorinated biphenyls [286].

2.2.2.3
The Chain-Type MCIs

The chain-type molecular connectivity indices describe the type of rings that are present in a molecule as well as the substitution patterns on those rings. Thus, chain-type indices also describe more local-type properties [204–208, 221, 224–226]. Their specificity is that they describe the same number of non-hydrogen atoms and bonds. For all other classes of molecular connectivity indices the corresponding subgraphs always contain more atoms than bonds. The lowest order for the chain-type index is third-order and increases up to the largest ring in any particular molecule. In this class of molecular connectivity indices the most interesting and commonly used are the sixth-order ($^6\chi_{CH}$) and seventh-order ($^7\chi_{CH}$) chain-type indices since they are related to benzene rings. The $^7\chi_{CH}$ index corresponds to monosubstituted benzene rings. The latter index was found to be very useful in describing the chromatographic behavior of chlorinated benzenes [103, 204–208, 231–235, 279].

In summary, molecular structure and topological indices aid in identifying structural features responsible for toxic organic compound chemodynamics at the molecular level which has influenced their use in developing relationships that accurately predict a broad range of physico-chemical [123–130, 162, 163, 209–213, 228–230, 241–244, 252, 253, 272–277] and biological [111–115, 155–157, 168, 169, 204–208, 231–240, 267, 268, 270, 271, 281–285] responses, resulting recently in more consistent statistically relevant and reliable models [177, 179, 180, 287, 288]. The molecular connectivity indices have been shown to be rich in structural information related to topological, geometric, and spatial attributes [103, 214–217, 224–226]. Information about different topological and geometric properties of a chemical structure is encoded in different molecular connectivity indices [227, 245–251]. The relative degree of branching of a molecule is encoded in the $^1\chi$ index when compared to other structural isomers. This translates into encoding molecular bulk or volume and surface area. The $^0\chi$ index encodes information about atoms, the $^2\chi$ index carries information about three atom fragments which are the minimum number necessary to describe a plane, while the $^3\chi_p$ index encodes information about three dimensional attributes such as conformation. The $^3\chi_{pc}$ index encodes information useful to the structural analysis of substituted rings. Information such as degree of substitution, length and heteroatom content of these groups is contained in $^4\chi_p$ and $^4\chi^{\upsilon}_{pc}$ indices.

2.2.3
Modeling Techniques

The molecular shape of organic compounds influences biological activity, especially where enzymes and receptors are involved. Several research studies have been conducted to address the problem of finding a mathematical means to express differences in geometric features such as those evidenced in the measurement of both size (a bulk measure) and shape (vectorial quantity) of molecules. The first has been to find parameters suitable for use in the Hansch equation. Taft's E_s parameter or its variants derived from the acid and base hydrolysis rates of aliphatic esters has been most widely used [102–104, 289]. Kier and Hall [103] have adapted the molecular connectivity index χ for QSAR correlations, a number derived originally by Randic [266] from graph theoretical principles to express the relative topology of variously branched hydrocarbon isomers. Many χ terms can be calculated for a given molecule, differing in the number of atoms taken together ($^n\chi$), and these may include or ignore the valence weighted indices ($^n\chi_v$) for the specific atoms or bond types present. The various terms for the molecules of a series may be tested as parameters in the usual multiple regression correlation model [103, 105–108, 266].

Other approaches to expressing topological differences include treating the problem of directionality of steric effects by the direct expedient of modeling a substituent and calculating its extension in five orthogonal directions (e.g., the minimal steric difference method, [289]). Other approaches [111–115, 290–295] include the use of quantum mechanical methods and molecular modeling techniques. A brief discussion about different modeling techniques commonly used is presented here. The various aspects of statistical analysis associated with multivariate data analysis for model development is also discussed briefly.

2.2.3.1
Free Energy Models

Among the first models proposed using QSAR methods is the one by Hansch and co-workers [60–62, 80, 102–110, 152, 195, 296–298]. They proposed that the early observations of the importance of relative lipophilicity to biological potency into the useful formalism of Linear Free Energy Relationships (LFER) to provide a general QSAR model in biological contexts. As a suitable measure of lipophilicity, the partition coefficient ($\log K_{OW}$) between *l*-octanol and water was proposed, and it was further demonstrated that this was roughly an additive and constitutive property and hence calculable in principle from molecular structure. Using a probabilistic model for transport across biological membranes, Hansch proposed the following equations (also called the Hansch Equation):

$$\log\left(\frac{1}{C}\right) = -k\pi^2 + k'\pi + \rho\sigma + k'' \qquad (45)$$

$$\log\left(\frac{1}{C}\right) = -k(\log K_{OW})^2 + k'(\log K_{OW}) + \rho\sigma + k'' \qquad (46)$$

where C is the molar concentration (or dose) for a constant biological response (EC_{50}, LC_{50}, genotoxic induction value, etc.), π is the substituent lipophilicity, $\log K_{OW}$ is the partition coefficient, $k\sigma$ is the Hammett value for substituent electronic effect, and k, k', ρ and k'' are regression coefficients derived from statistical curve fitting.

The reciprocal of the concentration reflects that higher potency is associated with lower dosage, and the negative sign for the π^2 or $(\log K_{OW})^2$ term reflects the expectation of an optimum lipophilicity. Multiple linear regression techniques may be used to determine these coefficients. A number of statistics are derived from such a calculation, which allow the statistical significance of the resulting correlation to be assessed. The most important of these are:

- The standard error of the estimate, also called standard deviation.
- r^2, the coefficient of determination or percentage of data variance accounted for by the model.
- F, a statistic for assessing the overall significance of the derived equation (statistical tables list critical values for the appropriate number of degrees of freedom and confidence level).
- t values (also compared with statistical tables) and confidence intervals (usually 95%) for the individual regression coefficients in the equation.

Also the cross-correlation coefficients between the independent variables in the equation are very important in multiparameter equations. These must be low to assure true-independence or orthogonality of the variables, a necessary condition for meaningful results in multivariate linear regression models.

The applicability of Eq. (45) to a broad range of biological (i.e., toxic, genotoxic) structure-activity relationships has been demonstrated convincingly by Hansch and associates and many others in the years since 1964 [60–62, 80, 120–122, 160, 161, 195, 204–208, 281–285, 289, 296–298]. The success of this model led to its generalization to include additional parameters in attempts to minimize residual variance in such correlations, a wide variety of physicochemical parameters and properties, structural and topological features, molecular orbital indices, and for constant but for theoretically unaccountable features, indicator or "dummy" variables (1 or 0) have been employed. A widespread use of Eq. (45) has provided an important stimulus for the review and extension of established scales of substituent effects, and even for the development of new ones. It should be cautioned here, however, that the general validity or indeed the need for these latter scales has not been established.

Lipophilicity in particular, as reflected in partition coefficients between aqueous and non-aqueous media most commonly water (or aqueous buffer) and *l*-octanol, has received much attention [105, 141, 152, 153, 176, 199, 232, 233]. LogK_{OW} for the octanol-water system has been shown to be approximately additive and constitutive, and hence, schemes for its a priori calculation from molecular structure have been devised using either substituent π values or substructural fragment constants [289, 299]. The approximate nature of any partition coefficient has been frequently emphasized and, indeed, some of the structural features that cause unreliability have been identified and accommodated. Other complications such as steric effects, conformational effects, and substitution at the active positions of hetero-aromatic rings have been observed but cannot as yet be accounted for completely and systematically. Theoretical statistical and topological methods to approach some of these problems have been reported [116–119, 175, 289, 300]. The observations of linear relationships among partition coefficients between water and various organic solvents have been extended and qualified to include other dose-response relationships [120–122, 160, 161, 299–302].

The success of the Hansch model in demonstrating that free energy correlations can be successfully applied to biological processes has prompted many researchers to reexamine the derivation of the Hansch equation. Using the principles of theoretical pharmacology or pharmacokinetics, improved theoretical models have been sought to accommodate more complex relationships between biological activity and chemical structure or properties, or to broaden the scope of Eq. (45) to include, for example, ionizable compounds. The free energy model of Hansch and its elaboration has been by far the most widely used. This has been due not only to its many successful applications, but also to its simplicity, its direct conceptual lineage to establish physical organic chemical properties, and the ready availability of a database of substituent parameters.

2.2.3.2
Free Wilson Mathematical Model

The idea that substituents should contribute constant increments or decrements to biological response in a related series of compounds has probably been a long held intuition of medicinal chemists trained in organic chemistry. However, in the recent past are a few demonstrations of this reported in the literature. The same time that the Hansch model was proposed, Free and Wilson demonstrated a general mathematical method for assessing both the occurrence of additive substituent effects and quantitatively estimating their magnitude [116–119, 158, 159, 289, 298]. According to their method, the molecules of a drug series can be structurally partitioned into a common moiety or core which has various substituents in multiple positions. In this approach, a series of linear equations in the form of Eq. (45) are constructed:

$$B'A_{\mathrm{j}} = \sum_{j} a_{\mathrm{j}} X_{\mathrm{ij}} + \mu \qquad (47)$$

where *BA* is the biological activity, X_{j} is the j-th substituent with a value of 1 if present and 0 if not, a_{j} is the contribution of the j-th substituent to *BA*, and μ is the average overall activity.

All contributions at each position of substitution should sum to zero. The series of linear equations thus generated is solved by the method of least squares for terms a_{j} and μ. There must be several more equations than unknowns and each substituent should appear more than once at a position in different combinations with substituents at other positions. The attractiveness of this model, also referred to as the de novo method, is as follows:

- Any set of quantitative biological data may be employed as the dependent variable.
- No independently measured substituent constants are required.
- The molecules of a series may be structurally partitioned in any convenient manner.
- Multiple sites of variable substitution are easily accommodated.

There are also several limitations [298] which include the following points:

- A substantial number of compounds with varying substituent combinations is required for a meaningful analysis.
- The derived substituent contributions give no reasonable basis for extrapolating predictions beyond the substituent matrix analyzed.
- The model will break down if non-linear dependence on substituent properties is important or if there are interactions between the substituents.

2.2.3.3
Discriminant Analysis

In many cases of interest the biological measurements available are semiquantitative or qualitative in nature, and activity assessments must be evaluated. Such data may arise from measurements with inherent imprecision, subjective evaluation of behavioral or response observations, or a combination of several criteria of interest into a single index. Neglecting the question to what extent this type of data is suitable for correlation in free energy models, it is nevertheless interesting to try to obtain some insight into the operative properties or structural parameters responsible for the variations in such data. Discriminant analysis has been proposed to deal with this type of a problem [1, 303–307]. This method seeks a linear combination of parameters called a linear discriminant function that will successfully classify the observations into their observed or assigned categories. Parameters are added or deleted to improve discrimination and the results are judged by the number of observations correctly classified.

2.2.3.4 Cluster Analysis

Cluster analysis is simply a method to group entities, for which a number of properties or parameters exist, by similarity [292, 308–313]. Various distance measurements are used, and the analysis is performed in a sequential manner, reducing the number of clusters at each step. Such a procedure has been described for use in drug design and environmental engineering research as a way to group substituents that have the most similarity when various combinations of the electronic, steric, and statistically derived parameters are considered.

2.2.3.5 Principal Components and Factor Analysis

Principal Component Analysis (PCA) is the most popular technique of multivariate analysis used in environmental chemistry and toxicology [313–316]. Both PCA and factor analysis (FA) aim to reduce the dimensionality of a set of data but the approaches to do so are different for the two techniques. Each provides a different insight into the data structure, with PCA concentrating on explaining the diagonal elements of the covariance matrix, while FA the off-diagonal elements [313, 316–319]. Theoretically, PCA corresponds to a mathematical decomposition of the descriptor matrix, X, into means (x_k), scores (t_{ia}), loadings (p_{ak}), and residuals (e_{ik}), which can be expressed as

$$x_{ik} = x_k + \sum_{a=1}^{A} t_{ia} \cdot p_{ak} + e_{ik} \qquad (48)$$

where x_{ik} are data elements used to describe the structural variation within the class of compounds, t_{ia} is the location of the i-th compound along the a-th principal component (PC), and p_{ak} loadings describe how much and in what way the k-th chemical descriptor contributes to a certain PC.

In the case of PCA, the following points should be considered:

- Principal Components (i.e, PCs) are linear combinations of random or statistical variables, which have special properties in terms of variances.
- The central idea of PCA is to reduce the dimensionality of a data set that may consist of a large number of interrelated variables while retaining as much as possible of the variation present in the data set [317–320].

- One of the statistical concerns in PCA is cross correlation between independent variables under consideration. This can simply be assessed by examination of the correlation matrix of the parameters responsible for variations of such data. Further manipulations can be performed on this matrix or on the variance-covariance matrix including the dependent variable. By methods of linear algebra such a matrix may be transformed by prescribed methods into one containing non-zero elements only on the diagonal. These are called eigen values of the matrix and associated with each of these is an eigen vector that is a linear combination of the original set of variables. Eigen vectors, unlike the original set of variables, have the property of being exactly orthogonal, that is the correlation coefficient between any two of them is zero.
- If a set of variables has substantial covariance, it will turn out that most of the total variance will be accounted for by a number of eigen vectors equal to a fraction of the original number of variables. A reduced set containing only the major eigen vectors or principal components may then be examined or used in various ways. This method is often used as a preprocessing tool. If only the principal components are considered, new orthogonal variables can be constructed from the eigen vectors and hence the dimensionality of the parameter space can be reduced, while most of the information in the original variable set is retained. This is particularly useful in the multidimensional methods that may be used as a preliminary step for series design in multiple regression analysis of the Hansch variety and pattern recognition.

On the other hand, factor analysis involves other manipulations of the eigen vectors and aims to gain insight into the structure of a multidimensional data set. The use of this technique was first proposed in biological structure-activity relationship (i.e., SAR) and illustrated with an analysis of the activities of 21 diphenylaminopropanol derivatives in 11 biological tests [116–119, 289]. This method has been more commonly used to determine the intrinsic dimensionality of certain experimentally determined chemical properties which are the number of fundamental factors required to account for the variance. One of the best FA techniques is the Q-mode, which is based on grouping a multivariate data set based on the data structure defined by the similarity between samples [1, 313–316]. It is devoted exclusively to the interpretation of the inter-object relationships in a data set, rather than to the inter-variable (or covariance) relationships explored with R-mode factor analysis. The measure of similarity used is the cosine theta matrix, i.e., the matrix whose elements are the cosine of the angles between all sample pairs [1, 313–316].

The goal of Q-mode FA is to determine the absolute abundance of the dominant components (i.e., physical or chemical properties) for environmental contaminants. It provides a description of the multivariate data set in terms of a few end members (associations or factors, usually orthogonal) that account for the variance within the data set. A factor score represents the importance of each variable in each end member. The set of scores for all factors makes up the factor score matrix. The importance of each variable in each end member is represented by a factor score, which is a unit vector in n (number of variables) dimensional space, with each element having a value between -1 and 1 and the sum of the squared elements equal to 1.00. The relative importance of each end member factor in each sample (i.e., a pollutant) is its factor loading value. The complete set of factor loadings describing each SWM/COM sample in terms of its end members is the factor-loading matrix.

2.2.3.6
Pattern Recognition

Pattern recognition is an ensemble of techniques that utilizes artificial intelligence to predict biological response [321–327] or chemical characteristics [295, 328–332]. As they have been applied to QSAR these methods comprise yet another approach for examining structural features and/or chemical properties for underlying patterns which are associated with different biological effects [333–337]. Accurate classification of untested compounds is again the primary goal. This is carried out in two stages. First, a set of compounds, designated the training set, is chosen for which the correct classification is known. A set of molecular or property descriptors is generated for each compound. A suitable classification algorithm is then applied to find some combination and weight of the descriptors that allows perfect classification [338]. Many different statistical and geometric techniques have been used and compared for this purpose [339–342]. The derived classification is then applied in the second step to compounds not included in the training set to test predictability. Performance is judged by the percentage of correct predictions. Repeating the training procedure several times with slightly altered but randomly varied training sets usually tests the robustness of the classifications. The two-pattern recognition systems that were used earliest in QSAR work are called ARTHUR [343] and ADAPT [102–105, 289].

In summary, the QSAR and QSPR approaches, as well as their modeling techniques, are important and a basic need for environmental planning and engineering management. Molecular connectivity indices (MCIs) are a sensitive property for many organic pollutants. Such MCIs can be used to predict the partitioning of pollutants at interfaces as will be seen in Sect. 3.

2.3
Joint Toxic Effect of Multicomponent Pollutant Mixtures

The third approach described here presents how and why a mixture of toxic and/or carcinogenic compounds can exhibit greater impacts in the environment than the individual constituents themselves. Such an impact, called the joint toxic effect of multiple chemicals, has been recognized as an important consideration in environmental chemodynamics. An understanding of and ability to predict joint effects of chemical mixtures is beneficial to provide meaningful inputs in managing the environmental hazards of synthetic compounds. This prediction of mixture toxicity/carcinogenicity can provide an insight about the bioavailable fraction of pollutants at aqueous-solid phase interfaces, and greatly enhance the decision-making processes in optimizing, limiting or preventing the disposal and/or recycling of solid wastes until they meet certain environmental criteria.

The toxic effects of chemical mixtures on different aquatic biota have been extensively studied; however, very few studies have evaluated such effects on fresh water algae [344–346]. Because of the important role of fresh water algae in determining the toxicity of various pollutants derived from municipal and industrial wastewater runoff and solid waste leachates, and their widespread distribution in the aquatic system, we will illustrate this by analyzing and predicting the joint toxicity of PAH mixtures using the fresh water alga *Selenastrum capricornutum* (as described in Sect. 3.2).

The study of joint toxic effects originated with the analysis of the effect of two compounds in binary mixtures. Plackett and Hewlett [344] identified four types of joint effects as follows:

- *Similar vs dissimilar*, depending on whether the sites of action and modes of primary action of the two compounds are the same or different.
- *Interactive vs noninteractive*, depending on whether one compound does or does not influence the biological action of the other.

If the response of the organism is produced by a combination of the two compounds, then they are said to exert joint action. This joint action can be further classified into simply additive, more than additive (i.e., synergistic), and less than additive (i.e., antagonistic). When this scheme is applied to multicomponent mixtures present in leachates of solid wastes, the analysis becomes more complex because the joint actions of different compound pairs may fall into different types of joint action. In the next section, three different modeling schemes are presented.

2.3.1
Toxic Unit Concept

In quantifying the joint actions of PAHs in mixtures, for instance, the concept of toxic unit (TU) is used. It is defined as

$$TU_i = \left(\frac{z_i}{Z_i}\right) \tag{49}$$

where z_i is the concentration of compound i in a mixture that causes a certain response, and Z_i is its concentration causing the same response when acting singly.

In fresh water algal toxicity this response could be 50% inhibition of the algal growth. If the TUs of all PAHs in a mixture are equal, then the PAH mixture is referred to as an equitoxic or a uniform mixture.

Using the TU concept, alternative schemes have been proposed to characterize the degree of joint action of multiple compounds acting together. In the first scheme, the sum of the TUs of the components M (i.e., $M = \sum TU_i$) is used as an index to categorize the type of joint action as follows:

- If $M = 1$, the components are simply additive (also referred to as concentration addition).
- If $M < 1$, they are more than additive (also referred to as synergism).
- If $M > 1$, they are less than additive (also referred to as antagonism).

Hermens et al. [345] evaluated literature toxicity data on fish and found average $M=\sum TU_i=0.9$ in mixtures of 50 nonreactive compounds, and average $M=\sum TU_i=1.1$ in 17-component mixtures. They concluded that the compounds acted together by simple addition since M values were very close to 1.

2.3.2 *Additive Index*

In the second scheme proposed by Marking [346], an additive index (AI) is used as the index where

$$AI=\left(\frac{1}{M}\right)-1\,; \quad M=\leq 1 \tag{50}$$

$$AI=1-M\,; \quad \text{if} \quad M=>1 \tag{51}$$

According to this scheme, when $AI=0$, components are simply additive; if $AI>0$, then they are more than additive, and if $AI<0$, they are less than additive. Lewis and Perry [347] applied this scheme to analyze the joint effects of equitoxic mixtures of three compounds on bluegills and found that AI value ranged from 0.30 to –1.23. Even though several AI values in that study deviated significantly from 0, they concluded that the compounds acted by simple addition, based on the average AI of 0.05.

2.3.3 *Mixture Toxicity Index*

The third scheme proposed by Konemann [348] uses a mixture toxicity index (MTI) defined as

$$MTI-1\left(\frac{\log M}{\log M_0}\right) \tag{52}$$

where

$$M_0=\left(\frac{M}{\text{the largest } TU_i \text{ in the mixture}}\right) \tag{53}$$

In this scheme, $MTI=1$ implies simply additive, $MTI=0$ implies independent action, $MTI<0$ implies antagonism, $MTI>1$ implies supra-addition, and $1>MTI>0$ implies partial addition. Broderius and Kahl [349] used this scheme to analyze joint effects of several equitoxic 7-, 14-, and 21-component mixtures, and concluded simple additivity with MTI values ranging from 0.93 to 1.06. Hermens et al. [350] evaluated the joint effects of 14 miscellaneous compounds to *Daphnia magna* and concluded simple addition, with an average MTI of 0.95.

In summary, the different joint effect models of multicomponent pollutant mixtures (i.e., the toxic unit, additive and mixture toxicity indices) were presented. Using such models to analyze the joint effect of a group of toxic and carcinogenic organic compounds such as polycyclic aromatic hydrocarbons will be presented and evaluated in Sect. 3.2.

3
Mobility and Bioavailability of Organic Pollutants: Applications

This section represents different case studies to explain how physical and chemical properties, QSAR and QSPR approaches, and multicomponent toxic effect models can be used to predict the mobility and bioavailability of organic pollutants at aqueous-solid phase interfaces. Such interdisciplinary approaches are applied here to two groups of toxic and carcinogenic compounds.

3.1
Polychlorinated Biphenyls

Polychlorinated biphenyls (PCBs) are a family of compounds, manufactured in the United States from 1930–1975, which were used in a number of discard applications and extensively as an electrical insulating fluid (see Chap. 1). Environmental concerns have led to strict controls on the use of PCBs and standards for cleanup of PCB discharges. One of the purposes of this section is to present information on the chemical and physical characteristics of these compounds. Based on this, the mechanisms of their movement in the surface/subsurface environment can be explained.

PCBs are relatively insoluble, viscous, and display a strong tendency toward sorption on solid particles. Their transport in the surface and movement through the subsurface is limited by their chemical and physical characteristics. Manufacturers normally marketed PCBs as mixtures of biphenyls. The combination of the various biphenyls in the mixture controlled the properties of the mixture.

PCBs are attractive for industrial applications because of their stability and dielectric properties [351–354]. Figure 1 shows the structure of the biphenyl molecule along with examples of chlorination that can occur at any of the positions on the rings. The physical and chemical properties of both isomers and mixtures used in industrial applications depend upon the degree and position of the chlorine atoms [355–358]. There are 209 possible chlorobiphenyl isomers and Table 4 lists the number of isomers for various degrees of substitution. However, many of these isomers do not occur in significant amounts in commercial products, and isomers with four or five chlorine atoms on one ring but none on the other are not detectable in PCB mixtures [359–362].

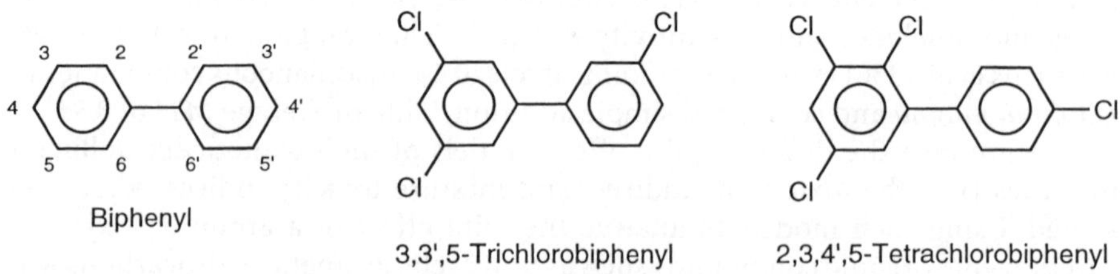

Fig. 1. The biphenyl molecule and its numbering system

Table 4. The numbers of possible substitution isomers of PCBs

Degree of substitution	Number of isomers
Mono	3
Di	12
Tri	24
Tetra	42
Penta	46
Hexa	42
Hepta	24
Octa	12
Nona	3
Deca	1
Total	209

The five largest uses for PCBs prior to 1970 were dielectric fluids in capacitors, plasticizers, lubricants, transformer fluids, and hydraulic fluids. They were also used widely in protective coatings, sealers, putty, grinding fluids, printing inks, pattern waxes, carbonless paper, etc. (see Chap. 1). Because of this widespread PCB use they are found throughout the environment [363–365]. A number of important properties of PCBs are discussed below along with information on their distribution and persistence in the environment.

3.1.1 *PCB Compositions*

Monsanto Chemical Company was the sole producer of PCBs in the United States, marketing them under the trade name Aroclor. A four-digit number identified the mixture of biphenyls found in a particular product. The first two digits (usually "12") indicated that the mixture contained polychlorinated biphenyls. The second two numbers indicated the percentage of chlorine in the mixture. For example, the name Aroclor 1254 indicates a PCB mixture with 54% chlorine. The only exception to this numbering system was Aroclor 1016 which contained 41% chlorine. This Aroclor, although similar to Aroclor 1242, contained lower chlorinated biphenyls than Aroclor 1242 [363, 366, 367]. PCBs were also marketed as Kanechlor and Santotherm in Japan, as Phenoclor and Pyralene in France, as Fenclor in Italy, as Clophen in Germany, as Chemko in Czechoslovakia, and as Sovol in Russia [363, 368].

Transformer fluids containing PCBs are of two types:

1. Oil filled transformers with a relatively low concentration of PCBs.
2. Transformers filled with Askarel which contained a significant percentage of PCBs combined with other fluidizers.

ASTM standard method D2283–86 defines the Askarel mixtures used by the utility industry (Table 5). The result of retrofilling older Askarel transformers is the presence of trace PCBs in refurbished oil filled equipment. McGraw [369]

Table 5. Askarel components in weight percent (after ASTM [367])

Askarel Formulation		Type						
Component	Description	A	B	C	D	E	F	H[a]
Hexachlorobiphenyl	Biphenyl chlorinated to a chlorine content of 60 weight percent	60	45					
Pentachlorobiphenyl	Biphenyl chlorinated to a chlorine content of 54 weight percent				70	45	60	
Trichlorobiphenyl	Biphenyl chlorinated to a chlorine content of 42 weight percent			80		10		
Trichlorobenzene	A mixture of isomers of trichlorobenzene	40			30	40		
Tri-tetra blend	A mixture of isomers of tri- and tetrachlorobenzene		55	20		5	40	100

[a] Non-PCB contains no PCB.

notes that about 2–4% of the oil originally placed in such a transformer remains within the coil and core structure after draining. This residual PCB can contaminate the mineral oil after retrofilling.

The Aroclor mixtures that were commonly in commercial use are listed in Table 6, with PCB isomers, molecular weights, and percentages of chlorine in each [368, 370–373]. Table 7 lists the specific isomers found in three of the major Aroclors used by the utility industry. This table also provides a listing of key environmental parameters used to evaluate the fate and transport of these PCBs.

Several workers noted that the patterns of biphenyls detected in various environmental media have different characteristics [368, 375, 376]. The composition

Table 6. Compositions of Aroclors manufactured for commercial use [368, 374]

Number of Cl atoms	MW (g/mol)	Cl (wt%)	Aroclor 1221	1232	1242	1248	1254	1260	1016
0	154	0	11	<0.1	<0.1	–	<0.1	–	<0.1
1	189	18.8	51	31	1	–	<0.1	–	1
2	223	31.8	32	24	16	2	0.5	–	20
3	258	41.3	4	28	49	18	1	–	57
4	292	48.6	2	12	25	40	21	–	21
5	326	54.3	<0.5	4	8	36	48	12	<0.1
6	361	58.9	–	<0.1	–	4	23	38	–
7	395	62.8	–	–	<0.1	–	6	41	–
8	430	66.0	–	–	–	–	–	8	–
9	464	68.7	–	–	–	–	–	1	–
Average MW of mixtures			201	232	267	300	328	376	258

Table 7. Physical and chemical properties of selected PCB isomers

Chlorine pattern	# Cl atoms	MW (g/mol)	Wt% of isomers			Solubility (μg/l)	Vapor pressure (mm Hg)	Henry's law constant (atm-m^3/mol)	Log K_{OW}	Log K_{OC}
			1242	1254	1260					
2-	1	188.7	0.0	0.0	0.0	5900.0	1.51E–02	6.35E–04[a]	3.9	3.2[a]
3-	1	188.7	0.0	0.0	0.0	3500.0	7.14E–03	5.07E–04[a]	4.4	3.7[a]
4-	1	188.7	0.0	0.0	0.0	1910.0	1.73E–03	2.25E–04[a]	4.6	3.9[a]
2,2-	2	223.1	0.0	0.0	0.0	1500.0	1.32E–03	2.30E–04[a]	4.9	4.3
2,4′	2	223.1	10.7	0.0	0.0	637.0	9.57E–04[a]	3.52E–04	5.1	4.5
2,2′,3-	3	257.5	6.5	0.0	0.0	231.0[a]	2.05E–04[a]	2.00E–04	5.6	5.0
2,2′,4-	3	257.5	7.6	0.0	0.0	231.0[a]	2.05E–04[a]	3.01E–04[a]	5.6	5.0
2,2′,5-	3	257.5	11.9	0.0	0.0	248.0	2.05E–04[a]	2.50E–04	5.6	5.0
2,3,4′-	3	257.5	3.1	0.0	0.0	231.0[a]	2.05E–04[a]	3.01E–04[a]	5.6	4.9
2,4,4′-	3	257.5	10.3	0.0	0.0	258.0	2.05E–04[a]	2.00E–04	5.8	5.2
2,4′,5-	3	257.5	10.1	0.0	0.0	231.0[a]	2.05E–04[a]	1.90E–04	5.7	5.1
2,3,4-	3	257.5	7.6	0.0	0.0	78.0	4.00E–04	5.76E–03[a]	5.8	5.2
2,2′,3,3′-	4	292.0	0.5	0.0	0.0	34.0	1.36E–04	1.00E–04	5.6	5.0
2,2′,3,4-	4	292.0	3.6	0.0	0.0	70.0[a]	4.38E–05[a]	1.40E–04	6.0	5.4
2,2′,3,4′-	4	292.0	3.1	0.0	0.0	70.0[a]	4.38E–05[a]	1.40E–04	5.8	5.3
2,2′,3,5′-	4	292.0	3.9	0.0	0.0	170.0	4.38E–05[a]	9.90E–05	6.0	5.4
2,2′,4,4′-	4	292.0	1.9	0.0	0.0	68.0	4.38E–05[a]	1.90E–04	5.9	5.3
2,2′,4,5′-	4	292.0	2.9	0.0	0.0	70.0[a]	4.38E–05[a]	2.10E–04	6.1	5.5
2,2′,5,5′-	4	292.0	4.2	3.2	0.0	26.5	4.90E–05	3.25E–03[a]	6.1	5.5
2,3,4,4′-	4	292.0	3.9	2.1	0.0	58.0	4.38E–05[a]	2.10E–04[a]	5.9	5.3
2,3′,4,4′-	4	292.0	3.9	7.0	3.4	70.0[a]	4.38E–05[a]	2.40E–04[a]	5.8	5.2
2,3′,4′,5-	4	292.0	1.8	7.6	0.0	41.0	4.38E–05[a]	1.00E–04	5.9	5.3
2,4,4′,5-	4	292.0	0.0	1.6	0.0	70.0[a]	4.38E–05[a]	1.00E–04	6.1	5.5
2,2′,3,3′,4-	5	326.4	0.0	0.6	4.3	21.0[a]	9.35E–06[a]	1.91E–04[a]	6.2	5.8
2,2′,3,4,4′-	5	326.4	0.0	1.4	0.0	21.0[a]	9.35E–06[a]	6.60E–05	6.2	5.7
2,2′,3,4,5′-	5	326.4	0.0	4.4	0.0	22.0	9.35E–06[a]	7.40E–05	6.5	6.0
2,2′,3′,4,5-	5	326.4	0.0	2.4	0.0	21.0[a]	9.35E–06[a]	7.40E–05	6.6	6.1
2,2′,4,4′,5-	5	326.4	0.0	4.3	0.0	21.0[a]	9.35E–06[a]	7.80E–05	6.4	5.8

Table 7 (continued)

Chlorine pattern	# Cl atoms	MW (g/mol)	Wt% of isomers			Solubility (µg/l)	Vapor pressure (mm Hg)	Henry's law constant (atm-m^3/mol)	Log K_{OW}	Log K_{OC}
			1242	1254	1260					
2,2′,4,5,5′-	5	326.4	0.0	11.3	4.5	10.3	1.10E–05	4.59E–04[a]	6.4	5.7
2,3,3′,4′,6-	5	326.4	0.0	11.9	0.0	21.0[a]	9.35E–06[a]	1.91E–04[a]	6.5	5.8
2,3′,4,4′,5-	5	326.4	0.0	15.6	0.0	21.0[a]	9.35E–06	1.91E–04[a]	6.4	5.7
2,2′,3,3′,4,4′-	6	360.9	0.0	1.3	0.0	6.0[a]	2.00E–06[a]	1.30E–05	7.0	6.5
2,2′,3,4,4′,5′-	6	360.9	0.0	9.5	11.7	6.0[a]	2.00E–06[a]	2.10E–05	7.0	6.5
2,2′,3,4,5,5′-	6	360.9	0.0	0.0	2.2	6.0[a]	2.00E–06[a]	2.30E–05	6.8	6.2
2,2′,3,4,5′,6-	6	360.9	0.0	0.0	14.5	6.0[a]	2.00E–06[a]	1.58E–04[a]	6.7	6.2
2,2′,3,4′,5,5′-	6	360.9	0.0	4.9	0.0	6.0[a]	2.00E–06[a]	2.50E–05	6.9	6.2
2,2′,3,5,5′,6-	6	360.9	0.0	0.0	1.6	6.0[a]	2.00E–06[a]	5.90E–05	6.6	6.2
2,2′,4,4′,5,5′-	6	360.9	0.0	8.1	19.0	8.8[a]	2.00E–06[a]	1.30E–04	6.9	6.4
2,2′,3,3′,4,4′,5-	7	395.3	0.0	0.0	3.8	2.0[a]	4.27E–07[a]	9.00E–06	7.3	6.6
2,2′,3,3′,4,5,6′-	7	395.3	0.0	0.0	2.1	2.0[a]	4.27E–07[a]	1.40E–05	7.1[a]	6.6
2,2′,3,3′,4,5′,6-	7	395.3	0.0	0.0	7.7	2.0[a]	4.27E–07[a]	1.11E–04[a]	7.2	6.6
2,2′,3,3′,4′,5,6-	7	395.3	0.0	0.5	0.3	2.0[a]	4.27E–07[a]	1.11E–04[a]	7.1	6.6
2,2′,3,3′,5,6,6′-	7	395.3	0.0	0.0	2.6	2.0[a]	4.27E–07[a]	2.40E–05	6.7	6.6
2,2′,3,4,4′,5,5′-	7	395.3	0.0	0.0	14.5	2.0[a]	4.27E–07[a]	1.00E–05	7.4	6.6
2,2′,3,4,5,5′,6-	7	395.3	0.0	0.0	5.8	2.0[a]	4.27E–07[a]	1.60E–05	7.0	6.6
2,2′,3,4′,5,5′,6-	7	395.3	0.0	0.0	2.0	2.0[a]	4.27E–07[a]	1.11E–04[a]	7.2	6.6
2,2′,3,3′,4,4′,5,5′-	8	429.8	0.0	0.0	0.8	7.0	9.14E–08[a]	1.00E–05	7.8	7.3
2,2′,3,3′,4,4′,5′,6-	8	429.8	0.0	0.0	1.4	0.5[a]	9.14E–08[a]	1.10E–05	7.1[a]	7.3
2,2′,3,3′,4′,5,5′,6-	8	429.8	0.0	0.0	1.5	0.5[a]	9.14E–08[a]	9.01E–05[a]	7.1[a]	6.6[a]
2,2′,3,3′,4,4′,5,5′,6,6′-	10	498.8	0.0	0.0	0.0	0.0	5.23E–10	4.60E–05[a]	8.2	7.8[a]

[a] These numbers are estimated using regression equations developed in this report.

of atmospheric samples contains lighter weight chlorobiphenyls than are found in water or soil. The reason for this phenomenon is a direct result of the chemical and partitioning characteristics of the individual chlorinated biphenyl compounds.

3.1.2
Property-Property Relationships

An effective assessment of the environmental impact of PCBs should consider the individual isomers that make up the PCB mixtures. This opinion is supported by several authors [358, 363, 368, 377–380], indicating that:

- The lower chlorinated isomers are more water soluble, readily vaporized, and rapidly biodegraded than the highly chlorinated isomers.
- Partitioning is stronger with the highly chlorinated isomers.
- The composition of PCBs in the atmosphere is similar to that of Aroclor 1242, while PCBs in surface waters approach the composition of Aroclor 1254. PCBs in the terrestrial environment are expected to be heavier still, approximating Aroclor 1260.

The molecular weight, solubility, vapor pressure, Henry's Law constants, log K_{OW}, and log K_{OC} of the various biphenyls and Aroclors at 25 °C are listed in Table 7. Because it is not practical to include all 209 isomers in Table 7, only isomers present at significant percentages in the Aroclors and used by the industry are included. Decachlorobiphenyl is included to provide an example for the highest weight isomer (also used as internal standard for quantitation). The selection of isomers is based on information presented in Griffin and Chian [363] and Girvin et al. [358]. Several properties listed in Table 7 are not readily available in the literature, so we estimated them for the particular isomer based upon the property-property regression equations shown in various figures provided in this chapter (Figs. 2–8). These equations can be used to estimate the property as long as the user understands that measured values are likely to be slightly different from the estimates.

3.1.2.1
Partition Coefficients

Partitioning of PCBs into other organic compound mixtures or phases found in the environment alters environmental parameters used to estimate their fate and transport. For example, dissolved phase humic substances (i.e., DP_{HS}) can increase the apparent solubility of organic pollutants [381–390] (see Chap. 2).

The most common partition coefficient encountered in environmental work (Sect. 2.1.4) is the octanol water partition coefficient (K_{OW}) and the solid phase carbon-water partition coefficient (K_{OC}). A partition coefficient for dissolved organic matter-water (i.e., $K_{d\text{-}OM}$) or dissolved organic carbon-water (i.e., $K_{d\text{-}OC}$) occasionally appears in the literature. In the case of PCBs, Boyd and Sun [378] defined a partition coefficient for residual transformer oil and water as $K_{d\text{-}oil}$, while Sun and Boyd [379] defined a coefficient for PCB dielectric fluid-water as $K_{d\text{-}PCB}$. These authors [378, 379] identified a total partition coefficient that combines coefficients for several components of the soil-water system. They defined this as follows:

$$K_p = \sum f_m \cdot K_m \tag{54}$$

where K_p is the overall partition coefficient, f_m is the fraction of material in medium, and K_m is the partition coefficient for medium.

Any assessment of PCBs, leaching from electric utility equipments, in the environment must first consider partitioning into the various media involved in the electric equipment themselves before partitioning into other environmental solid phases. For example, most PCBs research used either pure PCB isomers or Aroclors without the fluidizers normally found in utility equipment. These fluidizers, such as mineral oil and chlorinated solvents, used in the equipment all act as partitioning media for PCB isomers. In general, lower molecular weight PCB isomers partition into higher molecular weight isomer mixtures along with partitioning into the fluidizers. Combining the work of several workers in the field [65–77, 378, 379, 382] the following relationship can be defined:

$$K_p = f_{OC} \cdot K_{OC} + f_{mo} \cdot K_{mo} + f_{PCB} \cdot K_{PCB} \tag{55}$$

where K_p is the total partition coefficient, f_{OC} is the fraction of solid particle organic carbon, K_{OC} is the partition coefficient for organic carbon-water, f_{mo} is the fraction of mineral oil in solid phase, K_{mo} is the partition coefficient for mineral oil-water, f_{PCB} is the fraction of Aroclor in solid phase, and K_{PCB} is the partition coefficient for Aroclor-water.

Hydrophobic pollutants such as PCBs often partition into lipid rather than into water. The K_{OW} measures this partitioning. This coefficient provides an indication of the degree to which a pollutant accumulates into fatty tissues and any organic phase. This coefficient is especially useful for determining the release of PCBs from mineral oil transformer fluids, and Hawker and Connell [391] pro-

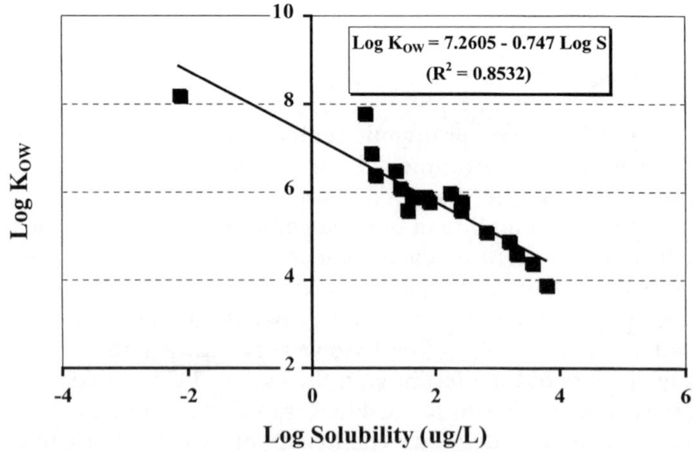

Fig. 2. PCB solubility-K_{OW} relationship (based on data presented in Table 7)

vided a listing of the K_{OW} for 180 PCB isomers. In general, PCB isomers partition into an oil phase rather than a water phase and residual oil in the solid phase is approximately ten times more effective for retaining PCBs than solid phase organic matter (i.e., SP_{OM}). Partitioning into an oil phase significantly reduces the mobility of PCBs and other hydrophobic pollutants [378]. K_{OW} can be estimated from the solubility of the PCBs themselves. The regression equation shown in Fig. 2 provides an estimate of K_{OW} for the PCB isomers, and the coefficient is also highly correlated with the degree of chlorination of the biphenyl (Fig. 3).

Solid phase organic carbon (i.e., K_{OC}) controls partitioning of hydrophobic contaminants such as PCB isomers [392–402]. K_{OC} is a measure of this partitioning. K_{OC} can be estimated from either solubility or K_{OW} as derived in this chapter and shown in Fig. 4.

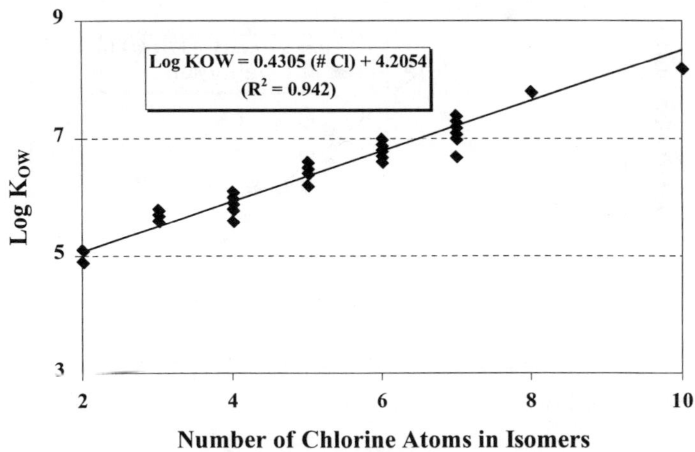

Fig. 3. The log K_{OW} of PCBs vs the number of chlorine atoms in isomers

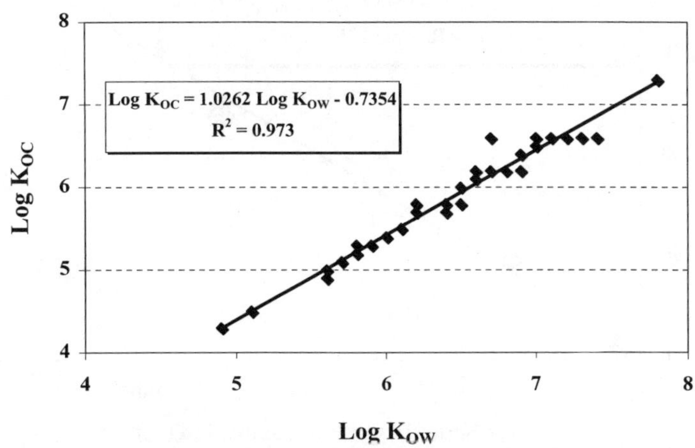

Fig. 4. The log K_{OW}-K_{OC} relationship (based on data presented in Table 7)

3.1.2.2
Solubility

Aqueous solubility (see Sect. 2.1.1) controls the loss of PCBs via surface and groundwater migration and transport, and is a major factor in understanding the environmental fate of PCB contaminants. The solubility of PCB isomers decreases as the degree of chlorination increases, as shown in Fig. 5. It should be noted that solubility data included in Table 7 and shown in Fig. 5 are based upon pure isomers. When an isomer is part of a mixture such as the Aroclors, solubility is reduced. Figure 6 shows the relationship between the solubility of the pure

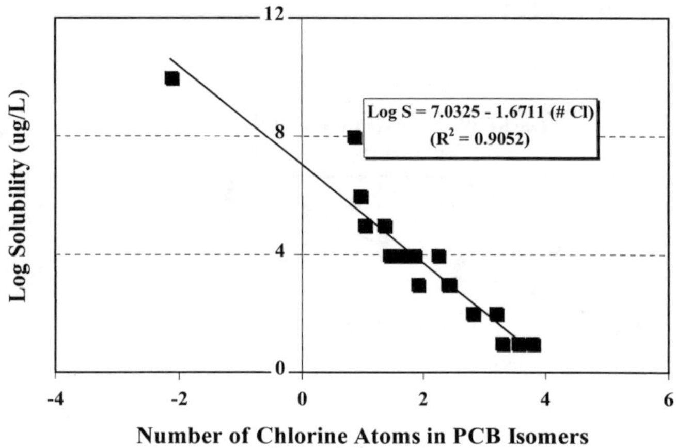

Fig. 5. Solubility of PCB isomers (based on data presented in Table 7)

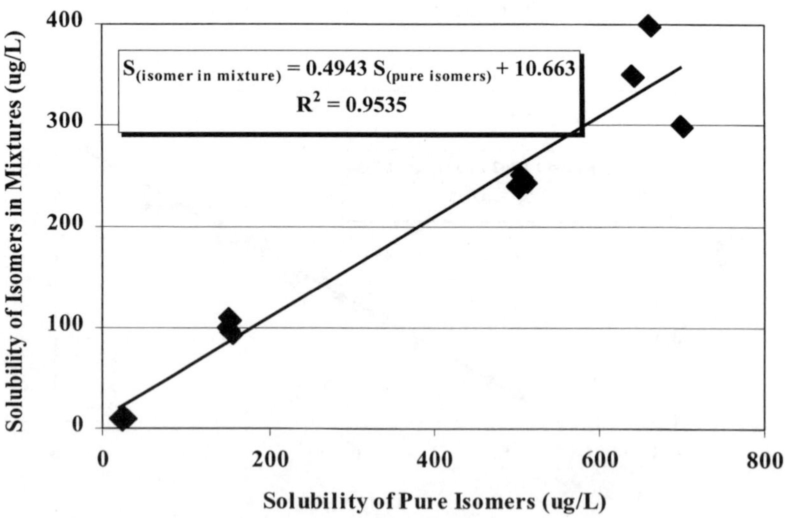

Fig. 6. Effects of PCB mixtures on solubility of individual PCB isomers

isomer and the same isomer when it is part of Aroclor 1241 or 1254. This difference is due to the partitioning of the isomers among the other biphenyls found in the mixture [378, 379]. Boyd and Sun [378] noted that 2-chlorophenyl partitioned into an Aroclor mixture by a factor of approximately five times more than into octanol.

Environmental releases of PCBs often accompany releases of carriers from utility equipment. An example would be mineral oil released from oil filled transformers. When PCBs are present in a mineral oil-PCB mixture the aqueous solubility of the PCBs is reduced significantly. Two factors play a role in this reduction: partitioning of the lipophilic (oil-loving) PCBs into the oil phase, and the reduced interaction of the PCBs with precipitation or groundwater caused by the hydrophobic nature of the oil matrix. Interpretation of aqueous PCB concentrations in the field must consider the presence of dissolved organic carbon (DOC) [382, 386, 397, 403].

3.1.2.3 Vapor Pressure

Vapor pressure (i.e., VP) is a measure of the amount of contaminant present in the air at a particular temperature (see Sect. 2.1.2). VP is one of the main factors controlling the vaporization of PCBs from aqueous or solid phase environments into the atmosphere. Figure 7 shows the vapor pressure at 25 °C for PCB isomers, indicating that VP decreases as the degree of chlorination increases.

Many of the PCBs found in the aquatic environment (e.g., in lakes and in the Arctic and Antarctic) have migrated via atmospheric dispersion of vapors [404–410]. Vaporization of PCBs from soil decreases as the amount of humic material in the solid phase increases due to mainly partitioning processes [381–390]. Griffin and Chian [363] note that vaporization of PCBs from suspensions of solids or humic acids is reduced by the presence of these materials.

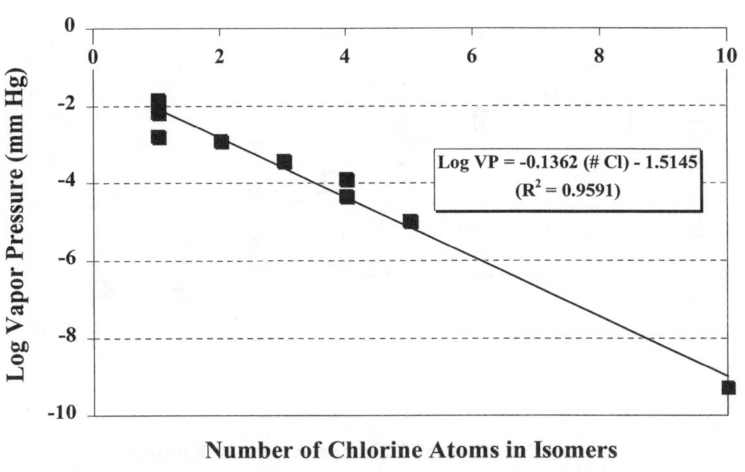

Fig. 7. Vapor pressure of PCB isomers (based on data presented in Table 7)

Humic acids at 500 mg/l reduced the volatilization of Aroclor 1242 from 3.5% to 2.6%. A solid particle suspension at a concentration of 6400 mg/l reduced the loss to 0.74%.

3.1.2.4
Henry's Law Constant

Henry's Law constant (i.e., H, see Sect. 2.1.3) expresses the equilibrium relationship between solution concentration of a PCB isomer and air concentration. This H constant is a major factor used in estimating the loss of PCBs from solid and water phases. Several workers measured H constants for various PCB isomers [411, 412]. Burkhard et al. [52] estimated H by calculating the ratio of the vapor pressure of the pure compound to its aqueous solubility (Eq. 13, Sect. 2.1.3). Henry's Law constant is temperature dependent and must be corrected for environmental conditions. The data and estimates presented in Table 7 are for 25 °C. Nicholson et al. [413] outlined procedures for adjusting the constants for temperature effects.

Burkhard et al. [52] noted that H constants for pure isomers were reduced by two- to threefold when the isomers were in Aroclor mixtures. The reduction may be due to changes in both solubility and vapor pressure resulting from interactions of the isomers with other isomers found in the Aroclors. The literature suggests that H does not change with the degree of chlorination alone, but there is variability within the isomers of any chlorination group [411]. Figure 8 shows the effects of chlorination on H constants and shows the degree of spread in the values reported for each level of chlorination. Brunner et al. [411] indicated that a more sensitive estimation method is based not only on chlorination of the biphenyl, but also on the degree of chlorination of the ortho-positions on the biphenyl molecule. The H constant increases as the degree of substitution on the ortho-position increases.

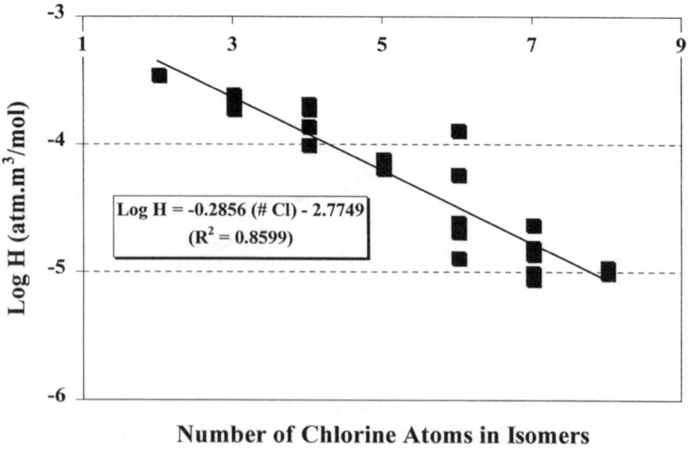

Fig. 8. Henry's law constant for PCB isomers

3.1.3 *Environmental Fate*

Each of the properties of the PCB isomers, listed above (Sect. 3.1.2) and either measured or calculated using various equations presented in Sect. 2.1, plays a role in the environmental distribution of these contaminants, especially at air-solid and water-solid interfaces. From the physical and chemical properties specific for PCBs and their isomers (Table 7, Figs. 2–8), the following information evaluates routes by which PCBs are lost from a particular source, spill or environmental compartment, that includes air-solid or aqueous-solid phase interfaces. These include vaporization (i.e., solid $\rightarrow$ air process), sorption/desorption and partitioning (i.e., water $\leftrightarrow$ solid processes) and biodegradation (i.e., water $\leftrightarrow$ biosolid interactions).

3.1.3.1 *Loss Due to Vaporization*

As mentioned in Sect. 3.1.2.2, vapor transport is believed to be one of the major routes of movement of PCBs through the environment. In general, low-molecular weight PCBs volatilize more readily than high-molecular weight species. Because of this tendency, there is an atmospheric enrichment of low molecular weight isomers, while high-molecular weight species tend to be enriched in the solid phase environment [404–407, 410].

The partitioning exhibited through the Henry's Law constant can be used to estimate the vaporization of various PCB contaminants from solid surfaces. In the presence of water, organic compounds volatilize more rapidly than would be expected based upon vaporization of the pure compound. This tendency accounts for the presence of low vapor pressure contaminants, such as the PCBs, in the atmosphere at higher concentrations than one would estimate from the chemistry of the pure compounds [403, 408, 409]

Contaminants migrate from surfaces via diffusion. This effect plays a role in the migration of PCB contaminants from and through soil particles. The less soluble a substance is in a liquid or air, the slower its absolute rate of diffusion into previously pure liquid or air [414].

Lewis et al. [415] measured PCB emissions from several contaminant waste landfills. At an uncontrolled site, air concentrations ranged up to 18 $\mu g/m^3$. However, at a landfill designed to meet regulatory standards, the levels were below the detection limit of 0.006 $\mu g/m^3$. Chromatograms of the air samples indicated that the PCB pattern resembled that of Aroclor 1242 with a preponderance of the peaks in the low molecular weight region. In contrast, Murphy et al. [416] noted emissions of PCBs from a number of landfills in the Great Lakes Region; however, the air concentrations were lower by several orders of magnitude compared to those seen by Lewis et al. [415]. Murphy et al. [416] also measured the PCB concentrations in the stack gases released from sewage sludge and municipal refuse incinerators and found concentrations ranging up to 2000 $\mu g/m^3$.

Pal et al. [368] gave volatilization half-lives reported in the literature for a number of Aroclors, ranging from 10–12 days for pure water and up to 52 days

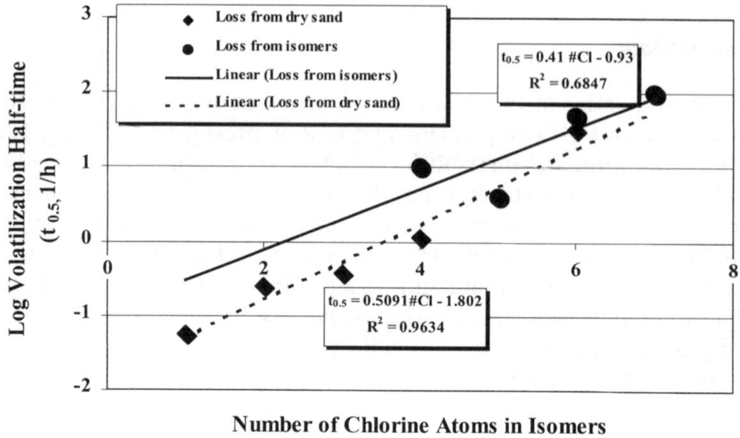

Fig. 9. Vapor loss of PCB isomers

for Aroclor 1260 from river water. Mackay and Leinonen [417] estimated volatilization half-lives for the Aroclors as follows: 12.1 h for Aroclor 1242, 10.3 h for Aroclor 1254, and 10.2 h for Aroclor 1260.

Vaporization of the PCB isomers from pure contaminant plated onto a surface generally is dependent upon the degree of chlorination. Figure 9 shows the effect of chlorination on the rate of PCB volatilization from pure isomer and from dry sand (data were taken from [359, 417–419]).

3.1.3.2
Sorption, Partitioning, and Retardation

PCBs in any solid phase system do not move at the same rate, for instance as groundwater, because of sorption/desorption mechanisms onto/from solid particle surfaces and partitioning into the solid phase organic matter. Chiou [397] noted that several organic contaminants preferentially bind more strongly to the humin than to the humic and fulvic acid fractions of any solid phase. Garbarini and Lion [420] showed that toluene and trichloroethylene partitioned the strongest in the most resistant fraction of the solid environment. This disparity between partitioning into the various fractions may account in part for the observation by Di Toro and Horzempa [421] of the reversible and resistant component of sorption-desorption of PCBs.

Girvin et al. [358] evaluated the release of PCBs from electrical substation soils contaminated with transformer fluids. They observed that there are two phases to the uptake and release of PCBs with these soils. The initial phase is a rapid, labile phase that is followed by a slower, nonlabile phase. The labile phase occurs at a scale of hours to days while the nonlabile phase releases over weeks and months. Girvin et al. [422] also reviewed the effects of adsorption on the mobility of PCBs and their transport. In an example presented for a hexachlorobiphenyl, these authors noted that the PCB isomer would have a retardation factor R_f of 1400 for the particular case given. This means that the groundwater would migrate at a rate 1400 times faster than the PCB isomer. The retardation factor depends primarily upon the partitioning of the isomer between the soil organic carbon and the aqueous phase. Retardation of any PCB isomers can be estimated by the following equation [358]:

$$R_f = 1 + \left(\frac{bd}{n}\right) \cdot K_p = \upsilon/x \tag{56}$$

where R_f is the retardation factor, bd is the bulk density of solid phase, n is the soil solid porosity, K_p is the soil water partition coefficient, υ is the average velocity of groundwater, and x is the rate of advance of the PCB front. Thus the larger the K_p, the more the retardation of the PCB.

3.1.3.3
Biodegradation

Microorganisms (i.e., biosolids) have been shown to degrade PCBs to various degrees depending upon the solid particle type and other environmental parameters [363, 368, 423]. Figure 10 shows the degradation of the Aroclors by microorganisms in activated sewage sludge [368, 424]. The less chlorinated isomers are degraded more readily by biosolid phases thus contributing to an enrichment of the higher molecular weight compounds [423].

In summary, property-property relationships of environmental contaminates and their isomers are useful in order to estimate other isomer properties which have never been measured or are not readily available in the literature. This can be done by developing property-property regression equations for the particular isomers of interests. In addition, environmental fate and behavior of such contaminants and their isomers could also be predicted using such relation-

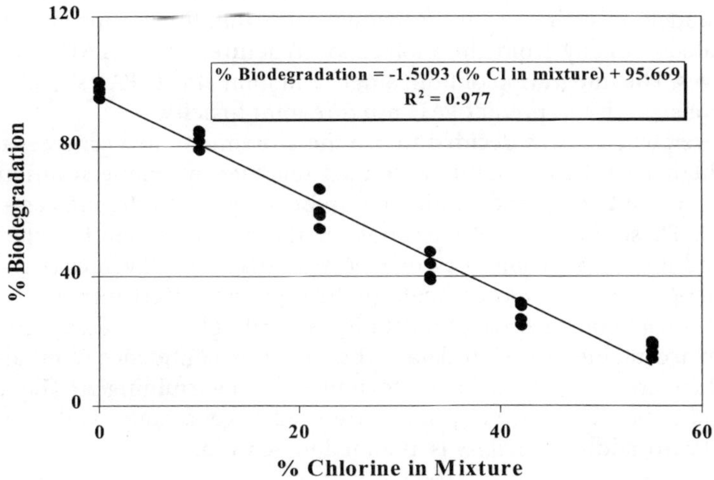

Fig. 10. Biodegradation of PCB by activated sludge (based on data from [368, 424])

ships. It should be noted that these equations and relationships could give slightly different estimates than the measured ones.

3.2 Modeling Multicomponent Toxic Effects of PAHs

In this part, a case study representing PAH-containing leachates from solid waste materials (SWMs) reported by Aboul-Kassim [1] is presented in terms of the joint toxic/carcinogenic actions of such PAHs in mixtures. Thus, different schemes discussed in Sect. 2.3 (i.e., the toxic unit *TU*, the additivity index *AI*, and the mixture toxicity index *MTI*) for analyzing joint effects of multipollutants on the fresh water alga *Selenastrum capricornutum* chronic 96-h toxicity due to PAH mixtures are presented and discussed. MOLecular CONNectivity-Quantitative Structure-Activity Relationship (i.e., MOLCONN-QSAR) techniques are then used to develop a predictive model to estimate the concentrations of PAH components, in organic mixtures in an aqueous system and/or derived from SWM leachate, that would jointly cause 50% inhibition of the *Selenastrum capricornutum* toxicity. The application of this multicomponent mixture chronic toxicity approach is demonstrated based on the experimental ecotoxicity data of 11 "8-component" PAH mixtures on alga *Selenastrum capricornutum* which was reported by Aboul-Kassim [1].

3.2.1 *Model Development*

If the joint effects of a set of organic compounds in a mixture can be accepted to be simply additive, then their concentrations in any mixture that would result in a certain response can be readily estimated from their respective individual concentrations causing the same response when acting singly. The practical utility of this deduction was further enhanced by Aboul-Kassim [1] by incorporating QSAR models to estimate the individual 50% inhibition concentration (i.e., EC_{50}) values directly from the molecular structures of the PAH components themselves. The integration of both single and joint effects PAHs-QSAR models can be constructive to predict PAH mixture joint toxicity.

However, when it was decided to use the aforementioned schemes to determine whether PAH compounds would act together by simple addition or not, statistically valid acceptance limits had to be assigned to the indices – *TU*, *AI*, and *MTI*. These limits should account for the variances due to experimental errors and the reproducibility associated with the z_i and Z values (Eq. 49). This would help to analyze and estimate multicomponent PAH mixture-combined toxicity with a known degree of reliability. Accordingly, the main approach used to assign acceptable ranges of data is that the 95% confidence intervals for the EC_{50} values are substituted in the formulas for determining *AI* (Eq. 50). The lower and upper limits of EC_{50} values are used to get a range, and if that range included zero, additive toxicity is assumed to be valid.

3.2.2
PAHs and Algal Toxicity Testing

A total of 11 polycyclic aromatic hydrocarbons (PAHs, Table 8) were assayed in non-equitoxic ratios in 11 "8-component" mixtures. These PAHs were selected based on their presence in most of the waste materials studied by Aboul-Kassim [1], covering a whole range of PAH chemical and physical properties. The organic pollutants assayed and their respective EC_{50} are listed in Table 8. The different PAHs, prepared singly or in mixtures, and the fresh water alga *Selenastrum capricornutum* culture assay were determined according to Aboul-Kassim [1].

Table 8. Toxicity values (EC_{50}) and molecular connectivity indices for various PAH compounds

Compound tested		CAS #	Chemical formula	MW (g/mol)	EC_{50} (mg/l)	$^3\chi^p$	$^6\chi^v_p$	$^2\chi^v$
Name	Symbol							
Naphthalene	NP	91-20-3	$C_{10}H_8$	128.2	19.536	3.47	0.30	2.35
1-Methylnaphthalene	1-MN	90-12-0	$C_{11}H_{10}$	142.2	19.000	4.10	0.41	2.84
2-Methylnaphthalene	2-MN	91-57-6	$C_{11}H_{10}$	142.2	12.000	4.30	0.42	2.54
2,6-Dimethylnaphthalene	DMN	581-40-2	$C_{12}H_{12}$	156.2	14.122	4.50	0.55	2.84
Acenaphthylene	ACY	208-96-8	$C_{12}H_8$	150.2	9.7016	4.84	0.67	3.13
Phenanthrene	PH	85-01-8	$C_{14}H_{10}$	178.2	6.0000	5.39	0.74	3.51
Anthracene	AN	120-12-7	$C_{14}H_{10}$	178.2	2.5000	5.34	0.73	3.55
Fluoranthene	FLU	206-44-0	$C_{16}H_{10}$	202.3	0.0644	6.73	1.23	4.25
Benzo[*a*]pyrene	BaP	50-32-8	$C_{20}H_{12}$	252.3	0.0008	8.62	1.74	5.45
Benzo[*e*]pyrene	BeP	192-97-2	$C_{20}H_{12}$	252.3	0.0028	8.74	1.78	5.74
Perylene	PER	198-55-0	$C_{20}H_{12}$	252.3	0.0001	10.0	2.38	6.20

3.2.3
Chronic 96-h Toxicity Measurement

For each PAH mixture, two reactors were used as controls and the remaining reactors were dosed with PAH mixtures. Chronic 96-h algal toxicity of the dosed reactors was compared against those of the control reactors to determine the percent inhibition. The EC_{50} values were obtained from percent inhibition vs PAH concentration plots. In the study by Aboul-Kassim [1], non-uniform PAH mixtures (i.e., 11 different "8-PAH" mixtures) were assayed. For each PAH component test mixture, two PAH components were added at 0.08 TU (i.e., $TU_1 = 0.08$, $TU_2 = 0.08$); two more at 0.09 TU (i.e., $TU_3 = 0.09$, $TU_4 = 0.09$); three more at 0.1 TU (i.e., $TU_5 = 0.1$, $TU_6 = 0.1$, $TU_7 = 0.1$); and the eighth PAH component was added at various TUs (i.e., 0.1, 0.2, 0.3, 0.4, and 0.5) to determine the TU_8 which would induce 50% growth inhibition. If all the eight PAHs in that mixture acted by simple addition, then 50% growth inhibition would occur at a TU_8 of 0.36 since the other seven PAH components together add up to $\sum TU_i$ of 0.64. The experimental TU that would cause 50% growth inhibition due to different PAHs could be found from a plot of percent inhibition vs TU of the eighth PAH compound, and compared against the expected value of TU_8 of 0.36 to verify simple addition.

3.2.4
Molecular Connectivity-QSAR Model for PAH Chronic Toxicity Prediction

A QSAR (see Sect. 2.2) is a mathematical model between a property (activity) of a certain compound, in this case the chronic toxicity value of a certain PAH compound, and the descriptors of that PAH. The descriptors are chemical or physical characteristics obtained experimentally or from the structure of the compound itself. In order to develop the MOLCONN-QSAR model, Aboul-Kassim [1]:

1. Prepared a training data set of chronic toxicity measurements to statistically establish the relationship between chronic toxicity and PAH descriptors of interest.
2. Used a QSAR modeling technique to predict the chronic toxicity of untested PAH compounds for which the descriptors are known.

Here, molecular connectivity (i.e., MOLCONN) models were used as descriptors of PAHs. It is a method of describing molecular structure based solely on the molecule's bonding and branching patterns (see Sect. 2.2.1). Using a simple algorithm, a series of indices called zero order (${}^0\chi$), first order (${}^1\chi$) and so forth, based on increasingly larger molecular fragments (called subgraphs) were computed for PAH compounds, as follows:

- Simple indices, which encode information on sigma bonded electrons that can be observed directly from bonding patterns in structural formulae of PAHs.
- Valence indices, (denoted with a υ superscript) which encode sigma, pi, and lone electrons and thus include more information about the specific elements included in the PAH compound.
- Indices of order greater than two which can be computed as either path, cluster, path/cluster, or chain depending upon the configuration of the molecular fragments (Eqs. 41–47).

Simple and valence indices up to sixth order were computed for all the PAHs used in the present study database. The program MOLCONN2 [133, 152, 154, 156] performed these calculations using the chemical structural formula as input. SAS [425] was used on a mainframe computer to perform statistical analyses. First, indices were selected which explained the greatest amount of variance in the data (i.e., R^2 procedure). These indices were then used in a multiple linear regression analysis (REG procedure).

Using the 96-h chronic toxicity data of the different PAHs and their molecular connectivity indices (i.e., MCIs), the following MOLCONN-QSAR models were developed:

$$\log EC_{50}\,(\mathrm{mg/l}) = 4.9861 - 0.888\,({}^3\chi^{\mathrm{p}}) \tag{57}$$

$$\log EC_{50}\,(\mathrm{mg/l}) = 2.4784 - 2.8352({}^6\chi_{\mathrm{p}}^{\upsilon}) \tag{58}$$

$$\log EC_{50}\,(\mathrm{mg/l}) = 5.1341 - 1.4212({}^2\chi^{\upsilon}) \tag{59}$$

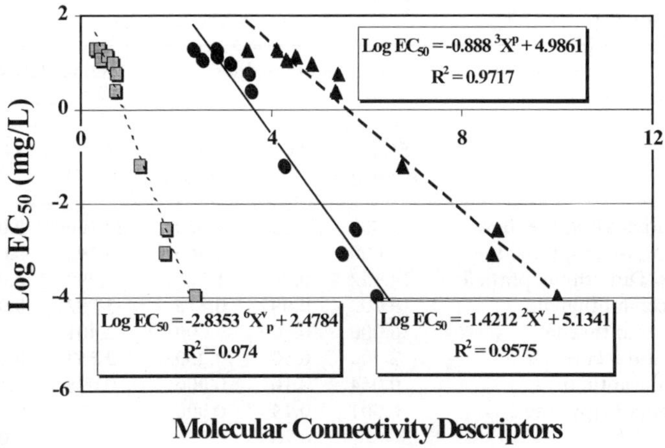

Fig. 11. The molecular connectivity-QSAR toxicity model for PAHs

where ${}^3\chi^{\mathrm{p}}$, ${}^6\chi^v_{\mathrm{p}}$, and ${}^2\chi^v$ are MCIs, listed in Table 8 and shown in Fig. 11. All the relationships were significantly correlated; however Eq. (58) (i.e., EC_{50} vs ${}^6\chi^v_{\mathrm{p}}$) was used to develop the MOLCONN-QSAR model. The rationale behind that is:

- It has the highest correlation coefficient ($R^2 = 0.9740$) among the other MCIs
- The inclusion of higher order indices, such as the sixth order index used here, indicates that a critical dimension or number of atoms in a chain is influential

As shown in Fig. 11, the negative coefficient on ${}^6\chi^v_{\mathrm{p}}$ reflects the fact that, beyond a critical dimension, the increasing size, particularly increasing chain size, reflected by ${}^6\chi^v_{\mathrm{p}}$ decreases a molecule's EC_{50} value (i.e., increases its chronic toxicity).

The MOLCONN-QSAR model represented by Eq. (61) (Fig. 11) was used to predict concentrations of the components in the PAH mixtures that would jointly cause 50% growth inhibition. The individual concentrations of the com-

Table 9. Experimental and predicted toxicity values for eleven PAH mixtures

PAH mixture ID	PAH compounds in mixtures	Experimental values			MOLCONN-QSAR predicted values		
		EC_{50} (mg/l)	TU_i	Conc (mg/l)	EC_{50} (mg/l)	TU_i	Conc (mg/l)
8-C1	Naphthalene	19.537	0.08	1.563	42.446	0.08	3.396
	1-Methylnaphthalene	19.000	0.08	1.520	20.700	0.08	1.656
	2-Methylnaphthalene	12.000	0.09	1.080	19.392	0.09	1.745
	2,6-Dimethylnaphthalene	14.122	0.09	1.271	8.299	0.09	0.747
	Acenaphthylene	9.702	0.10	0.970	3.792	0.10	0.379
	Phenanthrene	6.000	0.10	0.600	2.401	0.10	0.240
	Anthracene	2.500	0.10	0.250	2.563	0.10	0.256
	Fluoranthene	0.064	0.21	0.009	0.098	0.36	0.035
	Total TU_i		0.85			1.00	

Table 9 (continued)

PAH mixture ID	PAH compounds in mixtures	Experimental values			MOLCONN-QSAR predicted values		
		EC_{50} (mg/l)	TU_i	Conc (mg/l)	EC_{50} (mg/l)	TU_i	Conc (mg/l)
8-C2	1-Methylnaphthalene	19.000	0.08	1.520	20.700	0.08	1.656
	2-Methylnaphthalene	12.000	0.08	0.960	19.392	0.08	1.551
	2,6-Dimethylnaphthalene	14.122	0.09	1.271	8.299	0.09	0.747
	Acenaphthylene	9.702	0.09	0.873	3.792	0.09	0.341
	Phenanthrene	6.000	0.10	0.600	2.401	0.10	0.240
	Anthracene	2.500	0.10	0.250	2.563	0.10	0.256
	Fluoranthene	0.064	0.10	0.006	0.098	0.10	0.010
	Benzo[*a*]pyrene	0.001	0.14	0.000	0.004	0.36	0.001
	Total TU_i		0.78			1.00	
8-C3	2-Methylnaphthalene	12.000	0.08	0.960	19.392	0.08	1.551
	2,6-Dimethylnaphthalene	14.122	0.08	1.130	8.299	0.08	0.664
	Acenaphthylene	9.702	0.09	0.873	3.792	0.09	0.341
	Phenanthrene	6.000	0.09	0.540	2.401	0.09	0.216
	Anthracene	2.500	0.10	0.250	2.563	0.10	0.256
	Fluoranthene	0.064	0.10	0.006	0.098	0.10	0.010
	Benzo[*a*]pyrene	0.001	0.10	0.000	0.004	0.10	0.000
	Benzo[*e*]pyrene	0.003	0.17	0.000	0.003	0.36	0.001
	Total TU_i		0.83			1.00	
8-C4	2,6-Dimethylnaphthalene	14.122	0.08	1.130	8.299	0.08	0.664
	Acenaphthylene	9.702	0.08	0.776	3.792	0.08	0.303
	Phenanthrene	6.000	0.09	0.540	2.401	0.09	0.216
	Anthracene	2.500	0.09	0.225	2.563	0.09	0.231
	Fluoranthene	0.064	0.10	0.006	0.098	0.10	0.010
	Benzo[*a*]pyrene	0.001	0.10	0.000	0.004	0.10	0.000
	Benzo[*e*]pyrene	0.003	0.10	0.000	0.003	0.10	0.000
	Perylene	0.000	0.07	0.000	0.000	0.36	0.000
	Total TU_i		0.71			1.00	
8-C5	Acenaphthylene	9.702	0.08	0.776	3.792	0.08	0.303
	Phenanthrene	6.000	0.08	0.480	2.401	0.08	0.192
	Anthracene	2.500	0.09	0.225	2.563	0.09	0.231
	Fluoranthene	0.064	0.09	0.006	0.098	0.09	0.009
	Benzo[*a*]pyrene	0.001	0.10	0.000	0.004	0.10	0.000
	Benzo[*e*]pyrene	0.003	0.10	0.000	0.003	0.10	0.000
	Perylene	0.000	0.10	0.000	0.000	0.10	0.000
	Naphthalene	19.537	0.21	4.103	42.446	0.36	15.281
	Total TU_i		0.85			1.00	
8-C6	Phenanthrene	6.000	0.08	0.480	2.401	0.08	0.192
	Anthracene	2.500	0.08	0.200	2.563	0.08	0.205
	Fluoranthene	0.064	0.09	0.006	0.098	0.09	0.009
	Benzo[*a*]pyrene	0.001	0.09	0.000	0.004	0.09	0.000
	Benzo[*e*]pyrene	0.003	0.10	0.000	0.003	0.10	0.000
	Perylene	0.000	0.10	0.000	0.000	0.10	0.000
	Naphthalene	19.537	0.10	1.954	42.446	0.10	4.245
	1-Methylnaphthalene	19.000	0.24	5.320	20.700	0.36	7.452
	Total TU_i		0.88			1.00	

Table 9 (continued)

PAH mixture ID	PAH compounds in mixtures	Experimental values			MOLCONN-QSAR predicted values		
		EC_{50} (mg/l)	TU_i	Conc (mg/l)	EC_{50} (mg/l)	TU_i	Conc (mg/l)
8-C7	Anthracene	2.500	0.08	0.200	2.563	0.08	0.205
	Fluoranthene	0.064	0.08	0.005	0.098	0.08	0.008
	Benzo[*a*]pyrene	0.001	0.09	0.000	0.004	0.09	0.000
	Benzo[*e*]pyrene	0.003	0.09	0.000	0.003	0.09	0.000
	Perylene	0.000	0.10	0.000	0.000	0.10	0.000
	Naphthalene	19.537	0.10	1.954	42.446	0.10	4.245
	1-Methylnaphthalene	19.000	0.10	1.900	20.700	0.10	2.070
	2-Methylnaphthalene	12.000	0.23	2.880	19.392	0.36	6.981
	Total TU_i		0.87			1.00	
8-C8	Fluoranthene	0.064	0.08	0.005	0.098	0.08	0.008
	Benzo[*a*]pyrene	0.001	0.08	0.000	0.004	0.08	0.000
	Benzo[*e*]pyrene	0.003	0.09	0.000	0.003	0.09	0.000
	Perylene	0.000	0.09	0.000	0.000	0.09	0.000
	Naphthalene	19.537	0.10	1.954	42.446	0.10	4.245
	1-Methylnaphthalene	19.000	0.10	1.900	20.700	0.10	2.070
	2-Methylnaphthalene	12.000	0.10	1.200	19.392	0.10	1.939
	2,6-Dimethylnaphthalene	14.122	0.18	2.542	8.299	0.36	2.988
	Total TU_i		0.82			1.00	
8-C9	Benzo[*a*]pyrene	0.001	0.08	0.000	0.004	0.08	0.000
	Benzo[*e*]pyrene	0.003	0.08	0.000	0.003	0.08	0.000
	Perylene	0.000	0.09	0.000	0.000	0.09	0.000
	Naphthalene	19.537	0.09	1.758	42.446	0.09	3.820
	1-Methylnaphthalene	19.000	0.10	1.900	20.700	0.10	2.070
	2-Methylnaphthalene	12.000	0.10	1.200	19.392	0.10	1.939
	2,6-Dimethylnaphthalene	14.122	0.10	1.412	8.299	0.10	0.830
	Acenaphthylene	9.702	0.23	2.134	3.792	0.36	1.365
	Total TU_i		0.87			1.00	
8-C10	Benzo[*e*]pyrene	0.003	0.08	0.000	0.003	0.08	0.000
	Perylene	0.000	0.08	0.000	0.000	0.08	0.000
	Naphthalene	19.537	0.09	1.758	42.446	0.09	3.820
	1-Methylnaphthalene	19.000	0.09	1.710	20.700	0.09	1.863
	2-Methylnaphthalene	12.000	0.10	1.200	19.392	0.10	1.939
	2,6-Dimethylnaphthalene	14.122	0.10	1.412	8.299	0.10	0.830
	Acenaphthylene	9.702	0.10	0.970	3.792	0.10	0.379
	Phenanthrene	6.000	0.29	1.740	2.401	0.36	0.864
	Total TU_i		0.93			1.00	
8-C11	Perylene	0.000	0.08	0.000	0.000	0.08	0.000
	Naphthalene	19.537	0.08	1.563	42.446	0.08	3.396
	1-Methylnaphthalene	19.000	0.09	1.710	20.700	0.09	1.863
	2-Methylnaphthalene	12.000	0.09	1.080	19.392	0.09	1.745
	2,6-Dimethylnaphthalene	14.122	0.10	1.412	8.299	0.10	0.830
	Acenaphthylene	9.702	0.10	0.970	3.792	0.10	0.379
	Phenanthrene	6.000	0.10	0.600	2.401	0.10	0.240
	Anthracene	2.500	0.30	0.750	2.563	0.36	0.923
	Total TU_i		0.94			1.00	

ponents in the PAH mixture C_i were determined by multiplying the EC_{50} values by their respective TU_i values. The TU_i value for the eighth component (i.e., TU_8) is taken as 0.36 assuming the simple additivity model (i.e., $\sum TU_i = 1$). The simple additivity model was then verified, and these calculations are illustrated in Table 9 for the 11 "8-component" PAH mixtures tested.

3.2.5
Data Interpretation

The experimental results and the procedure used in determining the *TU* of the eighth PAH compound (i.e., TU_8) that would induce 50% growth inhibition for the fresh water alga *Selenastrum capricornutum* are detailed in Fig. 12 for only a single test on the eighth component PAH mixture (i.e., 8-C1, see Table 9 for the mixture composition).

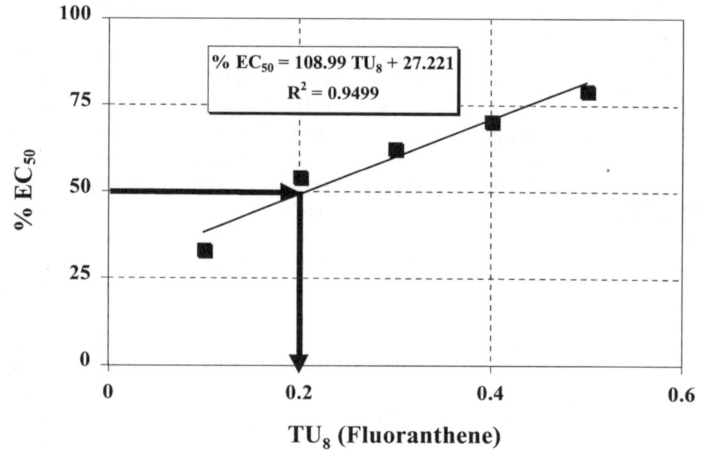

Fig. 12. Procedure for analyzing and determining the *TU* of the eighth PAH compound in mixture 8-C1 (i.e., TU_8 for fluoranthene, Table 9) that would induce 50% growth inhibition for the fresh water alga *Selenastrum capricornutum*

In general, three separate runs were conducted for each PAH mixture. A plot of % growth inhibition (i.e., % EC_{50}) vs TU_8, generated from three different runs on PAH mixture 8-CI, is shown in Fig. 13. The correlation coefficient (R^2) in such plots for all the eleven PAH mixtures assayed exceeded 0.8.

The $\sum TU_i$, *AI*, and *MTI* values found for the 11 "8-PAH" mixtures are summarized in Table 10 along with the experimentally determined values for TU_8 (see Fig. 13). In the case of $\sum TU_i$ (Table 10) joint toxic effect model, all the observed values were below 1 (i.e., indicating synergism rather than a simple additivity model), with a low value of 0.71 for PAH mixture 8-C4 (i.e., mixture causing highest synergism) and a high value of 0.94 for PAH mixture 8-C11 (i.e., close to simple addition). For the *AI* joint toxicity model, all the TU_8 values of the different PAH mixtures did not record a zero value (Table 10), with a high value of 0.29 (for mixture 8-C2) and a low of 0.07 (for mixture 8-C11). On the other

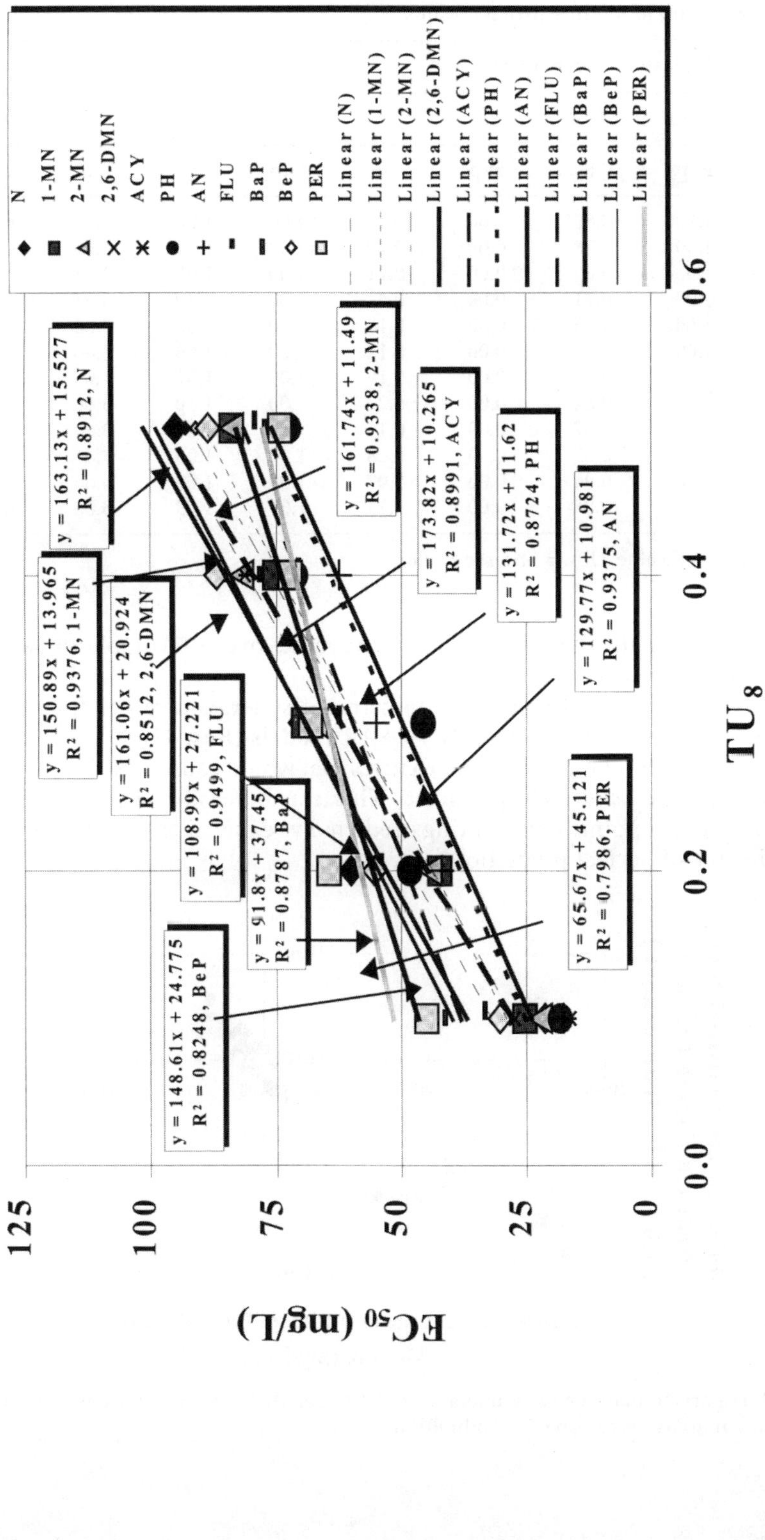

Fig. 13. Determination of TU_8 in non-uniform PAH mixture tests

Table 10. Summary of mixture toxicity results

PAH mixtures	Indicators of synergism[a]							
	$\sum TU_i$		AI		MTI		TU_8	
	Exp.	Obs.	Exp.	Obs.	Exp.	Obs.	Exp.	Obs.
8-C1	1.00	0.85	0.00	0.18	1.00	1.08	3.66	0.21
8-C2	1.00	0.78	0.00	0.29	1.00	1.13	3.66	0.14
8-C3	1.00	0.83	0.00	0.21	1.00	1.09	3.66	0.17
8-C4	1.00	0.71	0.00	0.41	1.00	1.20	3.66	0.07
8-C5	1.00	0.85	0.00	0.18	1.00	1.08	3.66	0.21
8-C6	1.00	0.88	0.00	0.14	1.00	1.06	3.66	0.24
8-C7	1.00	0.87	0.00	0.15	1.00	1.07	3.66	0.23
8-C8	1.00	0.82	0.00	0.22	1.00	1.10	3.66	0.18
8-C9	1.00	0.87	0.00	0.15	1.00	1.07	3.66	0.23
8-C10	1.00	0.93	0.00	0.08	1.00	1.03	3.66	0.29
8-C11	1.00	0.94	0.00	0.07	1.00	1.03	3.66	0.30
Mean	1.00	0.85	0.00	0.19	1.00	1.09	3.66	0.21

[a] Exp. = expected value, Obs. = observed value.

hand, for the *MTI* all PAH mixtures recorded values over 1.00 (Table 10), with an average of 1.09.

Thus, when Aboul-Kassim [1] demonstrated synergism rather than simple additivity using the PAH MOLCONN-QSAR models (Eqs. 57–59), the concentrations of the PAH components in mixtures that would cause 50% inhibition by joint action were accurately predicted. This can be easily seen from the associations of data points representing predicted vs experimental concentrations along the line of perfect prediction (Fig. 14).

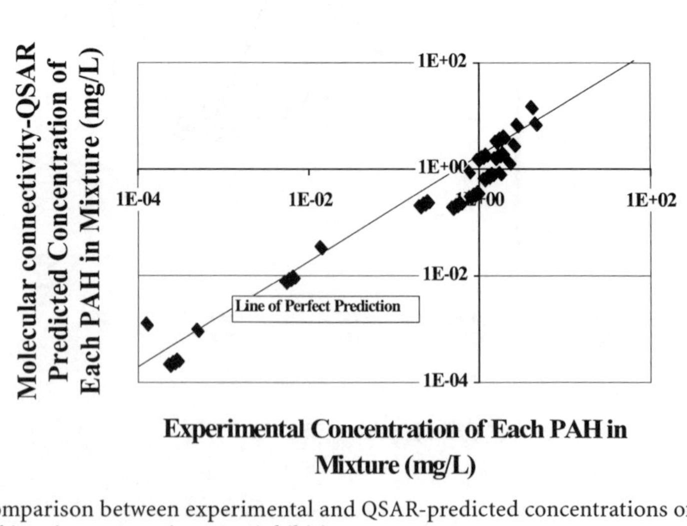

Fig. 14. Comparison between experimental and QSAR-predicted concentrations of each PAH compound in mixtures causing 50% inhibition

In summary, all the values observed by the use of different joint toxic effect modeling procedures indicated synergism rather than simple additive effects of the different PAH mixtures. Such a finding can have a big effect when measuring the bioavailable fraction of organic compounds at aqueous-solid phase interfaces. This is due to the fact that measured bioavailable fractions might be either overestimated because of synergism, or underestimated because of antagonism.

3.3
Predictive QSPR Model for Estimating Sorption Coefficients

Sorption/desorption is the key property for estimating the mobility of organic pollutants in solid phases. There is a real need to predict such mobility at different aqueous-solid phase interfaces. Solid phase sorption influences the extent of pollutant volatilization from the solid phase surface, its lateral or vertical transport, and biotic or abiotic processes (e.g., biodegradation, bioavailability, hydrolysis, and photolysis). For instance, transport through a soil phase includes several processes such as bulk flow, dispersive flow, diffusion through macropores, and molecular diffusion. The transport rate of an organic pollutant depends mainly on the partitioning between the vapor, liquid, and solid phase of an aqueous-solid phase system.

In order to understand the complex interactions of organic pollutants at aqueous-solid phase interfaces and to predict their mobility, which can be determined from their sorption coefficients, it is necessary to consider:

- The variation of molecular and structural properties of the pollutants concerned.
- The heterogeneous solid phase chemistry and physics.

A solid phase, as discussed in detail in Chap. 2, is composed of varying amounts of mineral and organic matter which influence the crumb structure and the binding capacity, by the association of clay minerals with organic matter of the solid. The ability of a solid phase to sorb organic pollutants is also influenced by variable system conditions and differing environmental conditions.

For nonpolar pollutants, sorption to the organic matter of the solid phase can be regarded as a distribution process between a polar aqueous phase and a nonpolar organic phase, i.e., the organic matter. For this type of sorption, several significant correlations have been published between the sorption coefficient and compound-specific properties describing partitioning between hydrophilic/hydrophobic systems such as water solubility, 1-octanol/water distribution coefficient, and capacity factors in reversed-phase chromatography [426–430]. In the present section, sorption coefficients of various PAH compounds determined with five different sorbents are shown to be predicted accurately using a quantitative structure-property relationship (i.e., QSPR) model.

3.3.1
Model Development

The following shows the development of the predictive QSPR model. The descriptors considered for the model include geometric and topological descriptors, electronic properties, charge-dependent properties, physico-chemical properties, and accumulation factors. The solid phases studied include three soil types (mollisol, ultisol, and aridisol) and two aquatic sediments.

3.3.1.1
Determination of Sorption Coefficients

The sorption behavior of 11 PAH compounds (a training set, Table 11) on various solid phases (e.g., three soils and two sediments) with different properties to relevant sorption (e.g., organic carbon content, clay content, pH, cation exchange capacity "CEC"; Table 12), was determined by batch equilibrium studies [1]. Batch equilibrium tests were designed to determine rates of equilibrium sorption under conditions of high mixing and high surface areas of the solid particles (see Chap. 3).

Table 11. Difference between predicted and observed $\log K_{OC}$ values for the training and validation data sets

Data set	Compound name	Difference between predicted and observed $\log K_{OC}$ values	
		Soils	Sediments
Training data set	Naphthalene	4.39	–3.86
	1-Methylnaphthalene	1.05	–0.84
	2-Methylnaphthalene	1.05	–0.82
	2,6-Dimethylnaphthalene	–0.89	1.47
	Acenaphthylene	–1.81	0.55
	Phenanthrene	–3.09	3.23
	Anthracene	–3.59	3.69
	Fluoranthene	–0.21	–1.00
	Benzo[*a*]pyrene	0.30	–0.05
	Benzo[*e*]pyrene	0.30	–0.05
	Perylene	2.50	–2.35
Validation data set	Indane	11.25	–6.5
	1,2-Dimethylnaphthalene	0.41	1.04
	1,3-Dimethylnaphthalene	0.31	0.62
	1,4-Dimethylnaphthalene	–0.39	1.05
	1,5-Dimethylnaphthalene	–0.69	1.14
	2,3-Dimethylnaphthalene	–0.49	1.00
	1-Ethylnaphthalene	–0.29	0.92
	2-Ethylnaphthalene	–0.69	1.02
	1,4,5-Trimethylnaphthalene	1.20	–0.18
	Biphenyl	12.80	2.08
	4-Methylbiphenyl	11.76	2.43

Table 11 (continued)

Data set	Compound name	Difference between predicted and observed log K_{OC} values	
		Soils	Sediments
	4,4′-Dimethylbiphenyl	12.81	0.46
	Diphenylmethane	13.46	3.32
	Bibenzyl	–2.69	4.85
	trans-Stilbene	–1.89	3.46
	Acenaphthene	–3.00	1.86
	Fluorene	–4.35	3.81
	1-Methylfluorene	0.51	–1.27
	1-Methylphenanthrene	1.85	–2.14
	9-Methylanthracene	1.85	–2.24
	2-Methylanthracene	1.85	–2.24
	9,10-Dimethylanthracene	5.41	–4.79
	Pyrene	–1.11	–0.20
	Benzo[*a*]fluorene	–0.85	–0.06
	Benzo[*b*]fluorene	0.85	–2.26
	Chrysene	7.80	21.69
	Triphenylene	–4.70	4.08
	p-Terphenyl	7.05	–7.97
	Naphthacene	–0.10	–0.82
	Benz[*a*]anthracene	7.30	–8.12
	Benzo[*b*]fluoranthene	2.50	–2.35
	Benzo[*j*]fluoranthene	–3.90	4.25
	Benzo[*k*]fluoranthene	–0.60	0.85
	7,12-Dimethylbenz[*a*]anthracene	–1.59	2.07
	9,10-Dimethylbenz[*a*]anthracene	–1.59	2.07
	3-Methylcholanthrene	0.86	0.11
	Benzo[*ghi*]perylene	5.16	–4.09
	Dibenz[*a,c*]anthracene	–0.59	2.14
	Dibenz[*a,h*]anthracene	0.71	0.74
	Dibenz[*a,j*]anthracene	0.51	0.94
	Pentacene	–0.49	2.04
	Coronene	–3.60	6.40

Table 12. Characteristics of the different solid phase particles

Solid phases		$\%C_{org}$	pH	CEC (mval/100 g)	$CaCO_3$ (%)	Grain size analysis		
Type	Name					Sand (%)	Silt (%)	Clay (%)
Soils	Olyic	6.18	6.8	25.3	14.2	64.2	21.3	14.5
	Woodburn	6.44	6.9	18.8	10.3	55.3	28.7	16.0
	Sagehill	1.91	6.7	11.7	8.7	66.2	12.3	21.5
Sediments	Willamette River, OR	1.81	7.1	14.7	12.6	26.9	49.5	23.6
	Yaquina Bay, OR	2.58	7.3	13.4	13.9	30.5	55.3	14.2

Table 13. Summary of average values of the regression equation constants for the training data set (11 PAHs) on different solid phases (Note: base 10 logs)

Solid phase type		Model isotherm	Y Axis	Intercept	Slope	X axis	R^2
Soils	Olyic	Linear	C_s	0.0000	0.05422	C	0.7665
		Langmuir	C/C_s	36.1966	−7.6737	C	0.5437
		Freundlich	$\log C_s$	0.8944	0.65068	$\log C$	0.8996
	Woodburn	Linear	C_s	0.0000	0.0649	C	0.3536
		Langmuir	C/C_s	5.3679	4.0567	C	0.8862
		Freundlich	$\log C_s$	2.6734	2.7456	$\log C$	0.9735
	Sagehill	Linear	C_s	0.0000	0.0754	C	0.6777
		Langmuir	C/C_s	38.3664	3.2456	C	0.2867
		Freundlich	$\log C_s$	2.6935	1.7546	$\log C$	0.8999
Sediments	Willamette River, OR	Linear	C_s	0.0000	0.0664	C	0.7696
		Langmuir	C/C_s	11.5671	3.7566	C	0.2699
		Freundlich	$\log C_s$	3.6454	1.5765	$\log C$	0.8866
	Yaquina Bay, OR	Linear	C_s	0.0000	0.0764	C	0.5388
		Langmuir	C/C_s	4.8451	5.7472	C	0.8986
		Freundlich	$\log C_s$	3.1467	2.7382	$\log C$	0.9895

Solutions with a defined solid/solution ratio and containing one of five initial concentrations of the PAH of interest in the training set (Table 11) were tumbled for 24 h until equilibrium was reached. After centrifugation of the samples, the PAH concentration was determined in the liquid phase. The determined PAH concentrations in the liquid and solid phases (by difference) were used to calculate distribution coefficients (k_d) and to obtain k_f values using the Freundlich equation as shown in Chap. 3. In order to reduce the variance in sorption coefficients (i.e., K_{OC}), the distribution coefficient (k_d or k_f) was frequently normalized to the organic carbon content (%OC) of the solid phase particles. For the training data set of 11 PAH compounds (Table 11), different sorption isotherm models were tested, namely linear, Langmuir, and Freundlich isotherms (see Chap. 3 for the corresponding equations). In general, the Freundlich isotherm model showed high correlation coefficients (Table 13).

3.3.1.2
Descriptor Calculations

The parameters used for regression analysis in the present case study were calculated based on the Quantum Chemical Programs Exchange program according to the AMI algorithm [102–107, 133, 154, 166]. The quantum mechanical parameters calculated were the highest occupied and lowest unoccupied molecular orbital (HOMO and LUMO) and the difference between them, the ionization potential, electronegativity, dipole moment, and charge distribution [111, 112, 431]. Depending on the charge distribution obtained, some further electronic descriptors, e.g., self-polarizability and the probability of a nucleophilic/electrophilic attack, were calculated according to Schuurmann [432]. Geometrical descriptors calculated were the molar refraction, the molecular volume, and the minimum and maximum diameter of the molecule [433–436]. The Randic indices used for analysis were calculated according to Kier and Hall [102–104, 107, 108].

3.3.2
Model Testing and Validation

For the training data set (Table 11), few parameters of the 22 physical-chemical properties of the PAHs showed high significance vs log K_{OC}. These are the log K_{OW}, molecular weight, and molecular connectivity indices (${}^3X_p$, ${}^6X_\upsilon^p$, ${}^2X_\upsilon$). Table 14

Table 14. Correlation coefficients between log K_{OC} and molecular descriptors of PAHs

Molecular descriptors		log K_{OC} Soils	log K_{OC} Sediments
Log K_{OW}		0.8392	0.8563
Molecular weight		0.8637	0.7837
Geometric and topological parameters	Molar refraction	0.5796	0.5736
	Molar volume	0.6886	0.5716
Molecular connectivity indices	${}^3X_p$	0.8950	0.9494
	${}^6X_\upsilon^p$	0.9131	0.9010
	${}^2X_\upsilon$	0.8634	0.8914
Electronic descriptors	Self polarizability	0.6495	0.7174
	D_N^a	0.5895	0.4673
	Q_{tot}^b	0.6568	0.6811
	Q_{ave}^c	0.4862	0.5636

[a] Probability of nucleophilic attack. [b] Total charge of molecule. [c] Average charge of molecule.

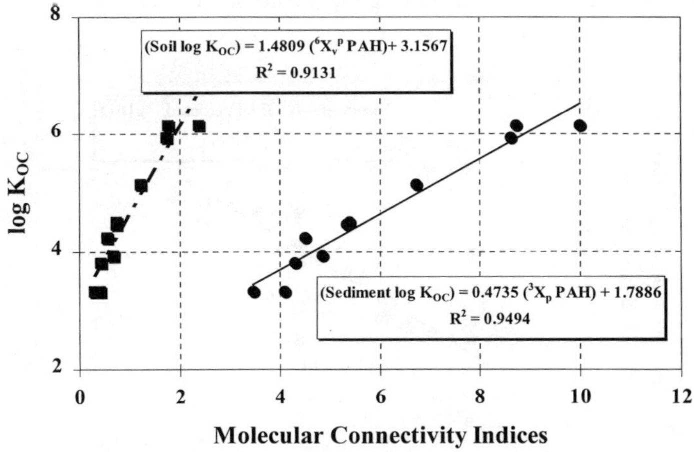

Fig. 15. Sorption coefficient (log K_{OC}) values vs molecular connectivity indices in the PAH training set for both soil and sediment solids

shows the correlation coefficients between $\log K_{OC}$ and various molecular descriptors, while Fig. 15 illustrates the regression equation models used for $\log K_{OC}$ prediction for both soil and sediment solids from their molecular connectivity indices.

The validation data set constitutes 42 PAHs (Table 11) comprising both unsubstituted and substituted compounds with a wide range of physical and chemical properties. Predictive models developed for PAH compounds in the training data set (Fig. 15) were used to predict values of sorption coefficients. All predicted and observed values were regressed, and recorded significant R^2 values as shown in Figs. 16 and 17, while the difference between such values are presented in Table 11.

In summary, it can be stated that the characterization of sorption of hydrophobic compounds to the organic matter of solid phase particles by K_{OC} values is a useful model for solids with a high organic carbon content and negligible

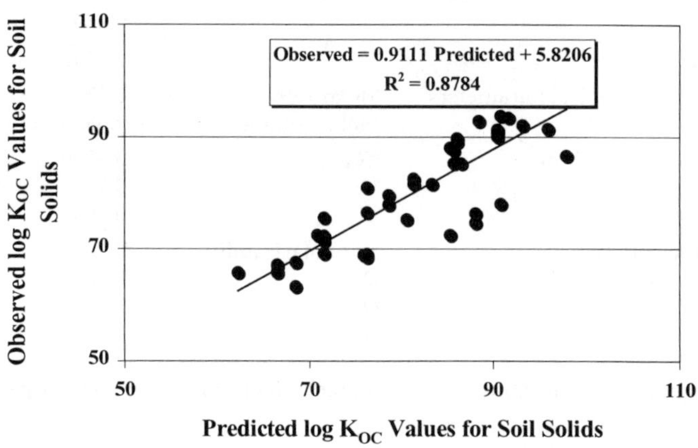

Fig. 16. Predicted vs observed values of sorption coefficients for soil solids

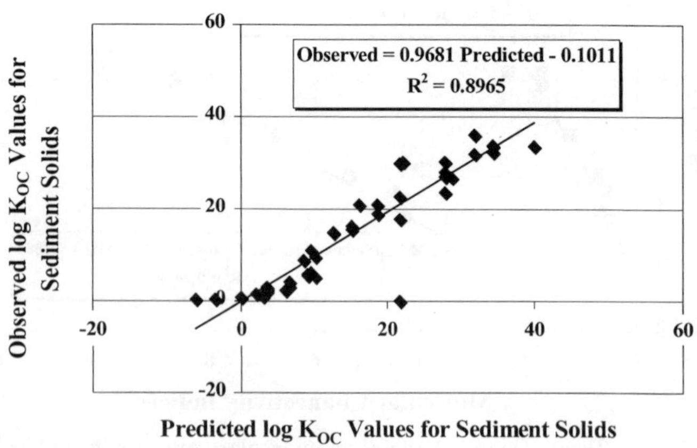

Fig. 17. Predicted vs observed values of sorption coefficients for sediment solids

clay content. For solids with a high clay content this is not applicable. Furthermore, this fact impedes the applicability to solids of other regions and composition.

4 Conclusions

Understanding environmental partitioning mechanisms of organic pollutants at aqueous-solid phase interfaces (i.e., water-soil, water-sediment, water-suspended solids, water-biosolids) requires the complete knowledge and analysis of most of the important physical and chemical properties of such pollutants. These properties, in some degrees of precision and accuracy, can initially determine the chemodynamics of the pollutants once they are released to the environment. Even through the predicted values may be slightly more or less accurate than experimental values, they are considered to be better than no values at all. Comparisons between predicted and experimental values of various physical and chemical properties of pollutants can provide an insight about the accuracy and precision of the developed models.

The applications of quantitative structure-activity and quantitative structure-property relationships (i.e., QSARs and QSPRs, respectively) as well as their various modeling techniques for environmental planning and engineering management are important. Generally speaking, the molecular connectivity indices (i.e., *MCI*s) are still a sensitive property for many organic pollutants. *MCI*s can be used to predict effectively the partitioning of pollutants at interfaces. The predictive ability of the QSAR and QSPR models generated for various chemical compounds depends strongly on the composition of the selected training sets themselves. Most often the standard way of selecting this training set is either to take all available data in a database, or to start with a lead compound and generate the training set by changing one substituent at a time. Usually these approaches lead to training sets with low information content and thereby to QSARs with low predictive power. Because of the complex mechanism of biological activity, the models used for QSAR/QSPR must necessarily be statistical in nature. Furthermore, due to the variation in biological mechanisms it is necessary to have separate models for different classes of compounds.

A mixture of toxic and/or carcinogenic compounds can exhibit a greater impact at aqueous-solid phase interfaces than the individual constituents themselves. Such an impact (i.e., the joint toxic effect of multiple chemicals) must be studied and modeled where it has an importance in environmental chemodynamics studies. An understanding and ability to predict joint effects of chemical mixtures is useful in order to assess, predict, and manage the environmental hazards of synthetic compounds. This prediction of mixture toxicity/carcinogenicity can provide an insight about the bioavailable fraction of pollutants at aqueous-solid phase interfaces, and greatly enhance the decision-making processes in optimizing, limiting, or preventing the disposal and/or recycling of solid wastes and synthetic chemicals until they meet certain environmental criteria.

References

1. Aboul-Kassim TAT (1998) Ph.D. Dissertation. Department of Civil, Construction and Environmental Engineering, Oregon State University, Corvallis, Oregon, USA
2. Ruthven DM (1984) Principles of adsorption and adsorption processes. Wiley, New York, p 445
3. Scheidegger AM, Sparks DL (1996) A critical assessment of sorption-desorption mechanisms at the soil mineral/water interface. Soil Science 161:813
4. Schlebaum W, Schraa G, van Riemsdijk WH (1999) Environ Sci Technol 33:1413
5. Schwarzenbach RP, Westall J (1981) Environ Sci Technol 15:1360
6. Thibaud-Erkey C, Guo Y, Erkey C, Akgerman A (1996) Environ Sci Technol 30:2127
7. Xing B, Pignatello JJ (1997) Environ Sci Technol 31:792
8. Xing B, Pignatello JJ (1998) Environ Sci Technol 32:614
9. Xing B, Pignatello JJ, Gigliotti B (1996) Environ Sci Technol 30:2432
10. Abramowitz R, Yalkowsky SH (1990) Chemosphere 21:1221
11. Dunnivant FM, Coate JT, Elzerman AW (1988) Environ Sci Technol 22:448
12. Bohon RL, Claussen WF (1951) J Am Chem Soc 73:1571
13. Booth HS, Everson HE (1948) Ind Eng Chem 40:1491
14. Andrews LJ, Keffer RM (1950) J Am Chem Soc 72:3644
15. Yalkowsky SH, Orr RJ, Valvani SC (1979) I&EC Fundam 18:351
16. Yalkowsky SH, Valvani SS, Mackay D (1983) Res Rev 85:43
17. Chiou CT, Freed VH, Schmedding DW (1977) Environ Sci Technol 11:475
18. McAuliffe C (1968) J Phys Chem 76:1267
19. Mackay D, Bobra AM, Shiu W-Y, Yalkowsky SH (1980) Chemosphere 9:701
20. Mackay D, Shiu WY (1981) J Phys Chem Ref Data 11:1175
21. Doucette WJ, Andren AW (1988) Chemosphere 17:243
22. May WE, Wasik SP, Freeman DH (1978) Anal Chem 50:175
23. May WE, Wasik SP, Freeman DH (1978) Anal Chem 50:997
24. Shiu WY, Doucette W, Gobas FAP, Mackay D, Andren AW (1988) Environ Sci Technol 22:651
25. Wasik SP, Miller MM, Tewari YB, May WE, Sonnefeld WJ, DeVoe H, Zoller WH (1983) Res Rev 85:29
26. Davis WW, Parke TV Jr (1942) J Am Chem Soc 64:101
27. Davis WW, Krahl ME, Clowes GH (1942) J Am Chem Soc 64:108
28. Hollifield HC (1979) Bull Environ Contam Toxicol 23:579
29. Mackay D, Shiu WY, Wolkoff AW (1975) ASTM STP, American Society for Testing and Materials, Philadelphia, Pa, 573:251
30. Weil L, Dure G, Quentin KL (1974) Z Wasser Abwasser Forsch 7:169
31. Stumm W, Morgan JJ (1981) Aquatic chemistry, 2nd edn. Wiley, New York, p 463
32. Ambrose D (1981) J Chem Thermodyn 13:1161
33. Dearden JC (1990) In: Karcher W, Devillers J (eds) Practical applications of quantitative structure-activity relationships (QSARs) in environmental chemistry and toxicology. Kluwer Academic Publishers, Dordrecht, Netherlands, p 25
34. Balson EW (1947) Trans Farad Soc 43:54
35. Bradley RS, Cleasby TG (1953) J Chem Soc 34:1690
36. Hamaker JW, Kerlinger HO (1969) Adv Chem Ser 86:39
37. Spencer WF, Cliath MM (1969) Environ Sci Technol 3:670
38. Spencer WF, Cliath MM (1970) J Agric Food Chem 18:529
39. Spencer WF, Cliath MM (1972) J Agric Food Chem 20:645
40. Sinke GC (1974) J Chem Thermodyn 6:311
41. Macknick AB, Prausnitz JM (1979) J Chem Eng Data 24:175
42. Westcott JW, Bidleman TF (1982) J Chromatogr 210:331
43. Rordorf BF (1985) Chemosphere 14:885
44. Rordorf BF (1985) Thermochimica Acta 85:435
45. Rordorf BF (1986) Chemosphere 15:1325

46. Sonnefeld WJ, Zoller WH, May WE (1983) Anal Chem 55: 275
47. Guckel W, Rittig R, Synnatschke G (1974) Pestic Sci 5: 393
48. Macknick AB, Prausnitz JM (1979) J Chem Eng Data 24: 175
49. Dobbs AJ, Grant C (1980) Pestic Sci 11: 29
50. Dobbs AJ, Cull MR (1982) Environ Pollut 3: 289
51. Bidleman TF (1984) Anal Chem 56: 2490
52. Burkhard LP, Kuehl DW, Veith GD (1985) Chemosphere 14: 1551
53. Foreman WT, Bidleman TF (1985) J Chromatogr 330: 203
54. Hamilton DJ (1980) J Chromatogr 195: 75
55. Hinckley DA, Bidleman TF, Foreman WT (1990) J Chem Eng Data 35: 232
56. Kim Y-H, Woodrow JE, Seiber JN (1984) J Chromotagr 314: 37
57. Westcott JW, Bidleman TF (1982) J Chromatogr 210: 331
58. Fendinger NJ, Glotfelty DE (1990) Environ Toxicol Chem 9: 731
59. Fendinger NJ, Glotfelty DE (1988) Environ Sci Technol 22: 1289
60. Fujita T, Iwasa J, Hansch C (1964) J Am Chem Soc 86: 5175
61. Leo A, Hansch C, Elkins D (1971) Chem Rev 71: 525
62. Hansch C, Leo AJ (1979) Substituent constants for correlation analysis in chemistry and biology. Wiley, NY
63. Rekker RF (1977) The hydrophobic fragmental constant. Elsevier, Amsterdam/New York, NY, 220 pp
64. Bowman BT, Sans WW (1983) J Environ Sci Health B18: 667
65. Chiou CT (1981) In: Hazard assessment of chemicals current developments. Academic Press, New York, p 117
66. Chiou CT (1985) Environ Sci Technol 19: 57
67. Chiou CT (1989) In: MacCarthy P, Malcolm RL, Clapp E, Bloom P (eds) Humic substances in soil and crop sciences. American Society of Agronomy, Madison, Wisconsin, p 214
68. Chiou CT, Freed VH, Schmedding DW (1977) Environ Sci Technol 11: 475
69. Chiou CT, Kile DE (1998) Environ Sci Technol 32: 338
70. Chiou CT, Kile DE, Brinton TI, Malcolm RL, Leenheer JA, MacCarthy P (1987) Environ Sci Technol 21: 1231
71. Chiou CT, Malcolm RL, Brinton TI, Kile DE (1986) Environ Sci Technol 20: 502
72. Chiou CT, McGroddy SE, Kile DE (1998) Environ Sci Technol 32: 264
73. Chiou CT, Peter LJ, Freed VH (1979) Science 206: 831
74. Chiou CT, Porter PE, Schmedding DW (1983) Environ Sci Technol 17: 227
75. Chiou CT, Porter PE, Shoup TD (1984) Environ Sci Technol 18: 295
76. Chiou CT, Schmedding DW, Manes M (1982) Environ Sci Technol 16: 4
77. Chiou CT, Shoup TD, Porter PE (1985) Org Geochem 8: 9
78. Howard PH (1989) Handbook of fate and exposure data for organic chemicals. I. Large production and priority pollutants. Lewis Publishers, Chelsea, Michigan, 351 pp
79. Howard PH (1990) Handbook of fate and exposure data for organic chemicals. II. Solvents. Lewis Publishers, Chelsea, MI, 429 pp
80. Hansch C, Quinlan JE, Lawrence CL (1968) J Org Chem 33: 347
81. De Bruijn J, Busser G, Seinen W, Hermens J (1989) Environ Toxicol Chem 8: 499
82. De Bruijn J, Hermens J (1990) Quant Struct Act Relat 9: 11
83. Doucette WJ, Andren AW (1988) Chemosphere 17: 345
84. Isnard P, Lambert S (1989) Chemosphere 18: 1837
85. McDuffie B (1981) Chemosphere 10: 73
86. Miller MM, Ghodbane S, Wasik SP, Tewari YB, Martire DE (1984) J Chem Eng Data 29: 184
87. Woodburn KB, Doucette WJ, Andren AW (1984) Environ Sci Technol 18: 457
88. Doucette WJ, Andren AW (1987) Environ Sci Technol 21: 521
89. Hawker DW, Connell DW (1988) Environ Sci Technol 22: 382
90. Howard PH, Boethling RS, Jarvis WF, Meylan WM, Michalenlco EM (1991) Handbook of environmental degradation rates. Lewis Publishers, Chelsea, Michigan, pp 556
91. Al-Sahhaf TA (1989) J Environ Sci Health A24: 49
92. Banerjee S (1985) Environ Sci Technol 19: 369

93. Hawker DW (1989) Chemosphere 19:1586
94. Kenaga EE, Goring CA (1980) In: Eaton JE (ed) Aquatic toxicology. Vol 707, ASTM, Philadelphia, PA, p 78
95. Mailhot H, Peters RH (1988) Environ Sci Technol 22:1479
96. Miller MM, Wasik SP, Huang GL, Shiu WY, Mackay D (1985) Environ Sci Technol 19:522
97. Tomlinson E, Hafkenscheid TL (1986) In: Dunn WJ III, Block JH, Pearlman RS (eds) Partition coefficient, determination and estimation. Pergamon Press, New York, p 101
98. Burkhard LP (1984) PhD Thesis, University of Wisconsin-Madison, Wisconsin
99. Dean JD (ed) (1985) Lange's handbook of chemistry, 13th edn. McGraw-Hill, New York, NY, pp 876
100. Hutzinger O (1980) The handbook of environmental chemistry. Vol 2, Part A. Springer, Berlin Heidelberg New York
101. Schwarzenbach RP, Gschwend PM, Imboden M (1993) Environmental organic chemistry. Wiley, pp 681
102. Kier LB, Hall LH (1999) Molecular structure description: the electrotopological state. Academic Press
103. Kier LB, Hall LH (1986) Molecular connectivity in structure-activity analysis. Wiley, Chichester, UK
104. Kier LB, Hall LH (1976) Molecular connectivity in chemistry and drug research. Research Studies Press
105. Bicerano J (1996) Prediction of polymer properties Marcel Dekker
106. Hall LH (1990) In: Rouvray DH (ed) Computational chemical graph theory. Nova Science Publishers, NY, chap 8, p 213
107. Kier LB (1990) In: Rouvray DH (ed) Computational chemical graph theory. Nova Press, NY, chap 6, p 152
108. Hall LH, Kier LB (1991) In: Boyd D, Lipkowitz K (eds) Reviews of computational chemistry. VCH Publishers, chap 9, pp 367
109. Hall LH, Kier LB (1992) In: Patei S, Rappoport Z (eds) Chemistry of the functional groups, vol 22. Wiley, Chichester, England, chap 5, p 186
110. Kier LB, Hall LH (1992) In: Testa B (ed) Advances in drug research, vol 22. Academic Press, p 1
111. Hall LH, Mohney B, Kier LB (1991) Quant Struct-Act Relat 10:43
112. Hall LH, Mohney B, Kier LB (1991) J Chem Inf Comput Sci 31:76
113. Kier LB, Hall LH (1992) In: Biaggi A (ed) Design of bioactive compounds, a telesymposium. Prous Publishers
114. Pogliani L (1992) J Pharm Sci 81:334
115. Hall LH, Kier LB (1992) Med Chem Res 2:497
116. Hall LH, Kier LB, Brown BB (1995) J Chem Inf Comput Sci 35:1074
117. Cummins DJ, Andrews CW, Bentley JA, Cory M (1996) J Chem Inf Comput Sci 36:750
118. Cheng C, Maggiora G, Lajiness M, Johnson M (1996) J Chem Inf Comput Sci 36:909
119. Lewis RA, Mason JS, McLay IM (1997) J Chem Inf Comput Sci 37:599
120. Hall LH, Kier LB (1988) J Environ Tox Chem 8:19
121. Hall LH, Maynard EL, Kier LB (1989) J Environ Tox Chem 8:431
122. Hall LH, Maynard EL, Kier LB (1989) J Environ Tox Chem 8:783
123. Hall LH, Aaserud D (1989) Quant Struct-Act Relat 8:296
124. Hall LH, Mohney B, Kier LB (1991) J Chem Inf Comput Sci 31:76
125. Hall LH, Kier LB (1991) J Math Chem 7:229
126. Hall LH, Mohney B, Kier LB (1991) Quant Struct-Act Relat 10:43
127. Pogliani L (1993) Comput Chem 17:283
128. Pogliani L (1993) J Phys Chem 97:6731
129. Pogliani L (1994) J Phys Chem 98:1494
130. Amic D, Davidovic-Amic D, Trinajstic N (1995) J Chem Inf Comput Sci 35:136
131. Kier LB, Hall LH (1976) Molecular connectivity in chemistry and drug research. Academic Press, New York, NY

132. Kier LB, Hall LH (1986) Molecular connectivity in structure-activity analysis. Wiley, New York, NY
133. Hall LH (1990) In: Rouvray DH (ed) Computational aspects of molecular connectivity and its role in structure-property modeling computational chemical graph theory. Nova Science Publishers, New York, NY, chap 8, p 202
134. Katritzky AR, Lobanov VS, Karelson M (1995) Chem Soc Rev 24:279
135. Murugan R, Grendze MP, Toomey JE, Katritzky AR, Karelson M, Lobanov V, Rachwal P (1994) Chemtech 24:17
136. Stanton DT, Jurs PC (1990) Anal Chem 62:2323
137. Katritzky AR, Ignatchenko ES, Barcock RA, Lobanov VS (1994) Anal Chem 66:1799
138. Katritzky AR, Lobanov VS, Karelson M (1994) CODESSA software. Copyright 1994, University of Florida. SemiChem, Inc, Shawnee, KS
139. Katritzky AR, Lobanov VS, Karelson M (1994) CODESSA Version 2.0 Users Manual
140. Charton M, Ciszewska GR, Ginos J, Standifer KM, Brooks AI, Brown GP, Ryan H, Moro JP, Pasternak GW (1998) Quant Struct-Act Relat 17:109
141. Clare BW (1990) J Med Chem 33:687
142. Fujimura KI, Ota A, Kawashima Y (1996) Chemical Pharm Bull 44:542
143. Fujita T (1997) Quant Struct-Act Relat 16:107
144. Gandhe BR, Purnanand P, Prasad R, Danikhel RK, Shinde SK, Srivastava RK, Batra BS, Rao KM (1990) Pestic Sci 29:379
145. Gupta SP (1987) Chem Rev 97:1183
146. Kawashima Y, Yamada Y, Asaka T, Misawa Y, Kashimura M, Morimoto S, Ono T, Nagate T, Hatayama K (1994) Chem Pharm Bull 42:1088
147. Kawashima Y, Sato M, Yamamoto S, Shimazaki Y, Chiba Y, Satake M, Iwata C, Hatayama K (1995) Chem Pharm Bull 43:1132
148. Miyoshi H, Tsujishita H, Tokutake N, Fujita T (1990) Biochim Biophys Acta 1016:99
149. Nezu Y, Wada N, Yoshida F, Miyazawa T, Shimizu T, Fujita T (1998) Pestic Sci 52:343
150. Paleti A, Gupta SP (1997) Quant Struct-Act Rel 16:367
151. Zahradnik P, Konecny V, Loos D, Zuziova J, Leska J (1989) Chem Pap 43:537
152. Hall LH (1995) In: van der Waterbeemd H (ed) Chemometric methods in molecular design. VCH Publishers, Weinheim, Germany
153. Pogliani L (1994) Curr Top Peptide Protein Res 1:119
154. Hall LH, Kier LB (1999) In: Devillers J, Balaban AT (eds) Topological indices and related descriptors in QSAR and QSPR. Gordon and Breach, Reading, UK
155. Pogliani L (1992) J Pharm Sci 81:967
156. Hall LH, Mohney BK, Kier LB (1993) Quant Struct-Act Relat 12:44
157. Pogliani L (1993) J Phys Chem 97:6731
158. Kier LB (1997) Med Chem Res 7:394
159. Brown RD, Martin YC (1998) SAR and QSAR Environ Res 8:23
160. Basak SC, Grunwald GD (1994) SAR and QSAR Environ Res 2:289
161. Cronin MTD (1996) SAR and QSAR Environ Res 5:167
162. Kier LB, Hall LH (1995) SAR and QSAR Environ Res 3:97
163. Hall LH, Kier LB (1995) J Chem Inf Comput Sci 35:1039
164. Hall LH, Vaughn TA (1997) Med Chem Res 7:407
165. Gough J, Hall LH (1999) Environm Tox Chem 36:65
166. Kier LB, Hall LH (1999) In: Devillers J, Balaban AT (eds) Topological indices and related descriptors in QSAR and QSPR. Gordon and Breach, Reading, UK
167. Kier LB, Hall LH (1999) In: Devillers J, Balaban AT (eds) Topological indices and related descriptors in QSAR and QSPR. Gordon and Breach, Reading, UK
168. Dang P, Madan AK (1994) J Chem Inf Comput Sci 34:1162
169. Luco L, Sosa M, Cesco J, Tonn C, Giodano G (1994) Pestic Sci 41:1
170. King RD, Muggleton S, Lewis RA, Sternberg MJE (1992) Proc Nat Acad Sci USA 89:11,322
171. Pastor M, Alvarez Builla J (1991) Quant Struct-Act Relat 10:350
172. Rekker RF (1992) Quant Struct-Act Relat 11:195
173. Benigni R, Richard AM (1998) Chem Toxicity Meth 14:264

174. Benigni R, Andreoli C, Giuliani A (1994) Enviro Molecular Mutagenesis 24 : 208
175. Ertl P (1997) Quant Struct-Act Relat 16 : 377
176. Mannhold R, Rekker RF, Sonntag C, Ter Laak AM, Dross K, Polymeropoulos EE (1995) J Pharm Sci 84 : 1410
177. Nirmalakhandan NN, Speece RE (1989) Environ Sci Technol 33 : 708
178. Petrauskas AA, Svedas VK (1991) J Chromatogr 585 : 3
179. Sabljic A (1987) Environ Sci Technol 21 : 358
180. Sabljic A (1987) Z Gesamt Hyg 33 : 493
181. Shapiro S, Guggenheim B (1998) Quant Struct-Act Relat 17 : 338
182. Egolf LM, Wessel MD, Jurs PC (1994) J Chem Inf Comput Sci 34 : 947
183. Bonchev D (1983) Information of theoretic indices for characterization of chemical structure. Wiley, NY
184. Kier LB (1990) In: Rouvray DH (ed) Computational chemical graph theory. Nova Press, NY, chap 6, pp 152
185. Weiner H (1947) J Am Chem Soc 69 : 17
186. Zefirov NS, Kirpichenok MA, Izmailov FF, Trofimov MI (1987) Dokl Akad Nauk SSSR 296 : 883
187. Stankevich MI, Stankevich IV, Zefirov NS (1988) Russ Chem Rev 57 : 191
188. Huibers PDT, Lobanov VS, Katritzky AR, Shah DO, Karelson M (1996) Langmuir 12 : 1462
189. Huibers PDT, Lobanov VS, Katritzky AR, Shah DO, Karelson M (1997) J Colloid Interface Sci 187 : 113
190. Huibers PDT, Shah DO, Katritzky AR (1997) J Colloid Interface Sci 193 : 132
191. Huibers PDT, Katritzky AR (1998) J Chem Inf Comput Sci 38 : 283
192. Rouvray DH (1986) Sci Am 255 : 40
193. Rouvray DH (1987) J Comput Chem 8 : 470
194. Rouvray DH (1989) J Mol Struct 185 : 187
195. Agarwal A, Pearson PP, Taylor EW, Li HB, Dahlgren T, Herslof M, Yang Y, Lambert G, Nelson DL, Regon JW, Martin AR (1993) J Med Chem 36 : 4006
196. Moore AS, Pope JD, Barnett JT, Sqarez LA (1989) EPA/600/3-89/080 Environmental Research Library, US-EPA, Athens, GA
197. Nevalainen T, Kolehmainen E (1994) Environ Toxicol Chem 13 : 1699
198. Okey RW, Stensel HD (1996) Water Res 30 : 2206
199. Siatra-Papastaikoudi T, Papadaki-Valiraki A, Tsantili-Kakoulidou A, Tzouvelekis L, Mentis A (1994) Chem Pharm Bull 42 : 392
200. Katritzky AR, Mu L, Karelson MA (1996) J Chem Inf Comput Sci 36 : 1162
201. Ivanciuc O (1997) J Chem Inf Comput Sci 37 : 405
202. Stanton DT, Jurs PC (1992) J Chem Inf Comput Sci 32 : 109
203. Nelson TM, Jurs PC (1994) J Chem Inf Comput Sci 34 : 601
204. Goel A, Madan AK (1995) J Chem Inf Comput Sci 35 : 504
205. Goel A, Madan AK (1995) J Chem Inf Comput Sci 35 : 510
206. Galvez J, Garcia-Domenech R, de Julian-Ortiz JV, Soler R (1995) J Chem Inf Comput Sci 35 : 272
207. Garcia-Domenech R, Garcia-March FG, Soler RM, Galvez J, Anton-Fos GM, Julian-Ortiz JV (1996) Quant Struct-Act Relat 15 : 201
208. Pogliani L (1996) J Phys Chem 100 : 18,065
209. Corbella R, Rodriguez M, Sanchez M, Montelongo M (1995) Chromatographia 40 : 532
210. Hu Q, Wang X, Brusseau M (1995) J Environm Tox Chem 14 : 1133
211. Saxena AK (1995) Quant Struct-Act Relat 14 : 31
212. Hall LH, Kier LB, Brown BB (1995) J Chem Inf Comput Sci 35 : 1074
213. Hall LH, Story CT (1996) J Chem Inf Comput Sci 36 : 1004
214. Kier LB, Hall LH (1989) Pharm Res 6 : 497
215. Hall LH, Kier LB (1990) Quant Struct-Act Relat 9 : 115
216. Kier LB, Hall LH (1990) Pharm Res 7 : 801
217. Kier LB, Hall LH (1990) Reports Theoret Chem 1 : 121
218. Blum DJW, Suffet IH, Duguet JP (1994) Water Res 28 : 687

219. Govers H, Ruepert C, Aiking H (1984) Chemosphere 13:227
220. Hall LH, Kier LB (1984) Bull Environ Contamin Toxicol 32:354
221. Malacarne D, Pesenti R, Paolucci M, Parodi S (1993) Environ Health Perspect 101:332
222. Hall LH, Maynard EL, Kier LB (1989) J Environ Tox Chem 8:783
223. Hall LH, Maynard EL, Kier LB (1989) J Environ Tox Chem 8:431
224. Kier LB, Hall LH (1991) Quant Struc-Act Relat 10:134
225. Kier LB, Hall LH, Frazer JW (1993) J Chem Inf Comput Sci 33:143
226. Hall LH, Kier LB, Frazer JW (1993) J Chem Inf Comput Sci 33:148
227. Hall LH, Kier LB (1993) J Chem Inf Comput Sci 33:598
228. Huibers PD, Lobanov VS, Katritkzy AR, Shah DO, Karelson M (1996) Langmuir 34:76
229. Basak SC, Gute BD, Grunwald G (1996) J Chem Inf Comput Sci 36:1054
230. Gombar VK, Enslein K (1996) J Chem Inf Comput Sci 36:1127
231. Pogliani L (1996) J Chem Inf Comput Sci 36:1082
232. Galvez J, Gomez-Icehon MJ, Garcia-Domenech R, Castell JV (1996) Bioorg Med Chem Lett 6:2301
233. Kellogg GE, Kier LB, Gaillard P, Hall LH (1996) J Comp Aid Molec Des 10:513
234. Soskic M, Plavsic D, Trinajstic N (1996) J Chem Inf Comput Sci 36:146
235. Amic D, Davidovic-Amic D, Beslo D, Lucic B, Trinajstic N (1997) J Chem Inf Comput Sci 37:581
236. Testa B (1997) Med Chem Res 7:340
237. Pogliani L (1997) Med Chem Res 7:380
238. Kellogg G (1997) Med Chem Res 7:417
239. Luco J, Ferretti F (1997) J Chem Inf Comput Sci 37:392
240. Garcia-Domenech R, de Julian-Ortiz JV (1998) J Chem Inf Comput Sci 38:445
241. Bicerano J (1996) Prediction of polymer properties. Marcel Drekker Publishing
242. Huibers PDT, Lobanov VS, Katritkzy AR, Shah DO, Karelson M (1997) J Colloid Interface Sci 187:113
243. Hall LH, Story CT (1997) SAR and QSAR Environ Res 6:139
244. Huuskonen J, Salo M, Taskinen J (1997) J Pharm Sci 86:450
245. Katritzky AR, Gordeeva E (1993) J Chem Inf Comput Sci 33:835
246. Skvortsova M, Baskin II, Slovkhotova O, Palyulin AA, Zefirov N (1994) J Chem Inf Comput Sci 33:630
247. Kier LB, Hall LH (1994) Quant Struct-Act Relat 12:383
248. Hall LH, Stewart D (1994) SAR and QSAR Environ Res 2:181
249. Pogliani L (1995) J Phys Chem 99:925
250. Hall LH, Fisk JB (1994) J Chem Inf Comput Sci 34:1184
251. Hall LH, Kier LB(1995) J Chem Inf Comput Sci 35:1039
252. Huuskonen J, Salo M, Taskinen J (1998) J Chem Inf Comput Sci 38:450
253. Finizio A, DiGuardo A, Vighi M (1994) SAR and QSAR Environ Res 2:249
254. Dowdy DL (1996) The 17th Annual Meeting of the Society of Environmental Toxicology and Chemistry, Washington, DC (USA), 17–21 Nov
255. Dowdy DL, McKone TE, Hsieh DPH (1994) The Society of Environmental Toxicology and Chemistry 15th Annual Meeting: Ecological Risk – Science, Law and Policy, Denver, CO (USA), 30 Oct–3 Nov
256. Leegwater DC (1989) Aquat Toxicol 15:157
257. Dowdy DL, McKone TE, Hsieh DPH (1995) The 2nd World Congress of the Society of Environmental Toxicology and Chemistry, Vancouver, British Columbia (Canada), 5–9 Nov
258. Liao YY, Wang ZT, Chen JW, Han SK, Wang LS, Lu GY, Zhao TN (1996) Bull Environ Contam Toxicol 56:711
259. Schramke JA, Murphy SF, Doucette WJ, Hintze WD (1996) The 17th Annual Meeting of the Society of Environmental Toxicology and Chemistry, Washington, DC (USA), 17–21 Nov
260. Pogliani L (1995) J Phys Chem 99:925
261. von der Lieth C-W, Stumpf-Nothof K, Prior U (1996) J Chem Inf Comput Sci 36:711

262. Pogliani L (1996) J Phys Chem 100:18,065
263. Kellogg GE, Kier LB, Gaillard P, Hall LH (1996) J Comp Aid Molec Des 10:513
264. Kier LB, Hall LH (1997) J Chem Inf Comput Sci 37:548
265. Milne GWA (1997) J Chem Inf Comput Sci 37:639
266. Randic M (1975) J Am Chem Soc, 97:6609
267. de Gregorio C, Kier LB, Hall LH (1998) J Comp Aid Molec Des 12:557
268. Gough JD, Hall LH (1999) J Chem Inf Comput Sci 39:23
269. Fraaije JG, Van Vlimmeren BAC (1996) Biophys J 70:A258
270. Stanton DT (1999) J Chem Inf Comput Sci 39:11
271. Gao H, Williams C, Labute P, Bajorath J (1999) J Chem Inf Comput Sci 39:164
272. Finizio A, Sicbaldi F, Vighi M (1995) SAR and QSAR Environ Res 3:71
273. Huuskonen JJ, Villa AEP, Tetko IV (1999) J Pharm Sci 54:67
274. Nikolic S, Trinajstic N (1999) SAR and QSAR Environ Res 9:117
275. Stanton DT (1999) J Chem Inf Comput Sci 39:11
276. Pompe M, Novic M (1999) J Chem Inf Comput Sci 39:59
277. Pogliani L (1999) J Chem Inf Comput Sci 39:104
278. Sabljic A (1983) Bull Environ Contam Toxicol 30:80–83
279. Sabljic A (1985) J Chromatogr 319:1
280. Protic M, Sabljic A (1989) Aquat Toxicol 14:47
281. Galvez J, Garcia-Domenech R, de Julian-Ortiz JV, Soler R (1994) J Chem Inf Comput Sci 34:1198
282. Kier LK, Hall LH (1994) Quant Struct-Act Relat 12:383
283. Hall LH, Stewart D (1994) SAR and QSAR Environ Res 2:181
284. Pogliani L (1995) J Phys Chem 99:925
285. Amic D, Davidovic-Amic D, Juric A, Luric B, Trinajstic N (1995) J Chem Inf Comput Sci 35:1034
286. Sabljic A, Gusten H (1989) Chemosphere 19:1503
287. Sabljic A, Piver WT (1992) Environ Toxicol Chem 11:961
288. Sabljic A, Gusten H, Verhaar H, Hermens J (1995) Chemosphere 31:4489
289. Topliss JG (1983) Quantitative structure-activity relationships of drugs. Academic Press, New York, pp 253
290. Damborsky J, Schultz TW (1997) Chemosphere 34:429
291. Enslein K (1993) In Vitro Toxicol 6:163
292. Eriksson L, Jonsson J, Hellberg S, Lindgren F, Skagerberg B, Sjostrom M, Wold S, Berglind R (1990) Enivron Toxicol Chem 9:1339
293. Nendza M (1991) Chemosphere 23:497
294. Roghair CJ, Buijze AB, Yedema ESE, Hermens JLM (1994) Chemosphere 28:989
295. Verhaar HJM, Morroni JR, Reardon KF, Hays SM, Gaver DP Jr, Carpenter RL, Yang RSH (1997) Environ Health Perspect 105:179
296. Dove S, Franke R (1991) Quant Struct-Act Relat 10:23
297. Roy NK, Nidiry ESJ, Vasu K, Bedi S, Lalljee B, Singh B (1996) J Agric Food Chem 44:3971
298. Tmej C, Chiba P, Huber M, Richter E, Hitzler M, Schaper KJ, Ecker G (1998) Archiv der Pharmazie 331:233
299. Reddy KN, Locke MA (1994) Chemosphere 28:1929
300. Catana C (1995) Toxicol Model 1:181
301. Bintein S, Devillers J (1994) Chemosphere 28:1171
302. Furay VJ, Smith S (1995) Bull Environ Contam Toxicol 54:36
303. Huang Q, Wang X, Liao Y, Kong L, Han S, Wang L (1995) Bull Environ Contam Toxicol 55:796
304. McFadden WH, Bradford DC, Eglinton G, HajIbrahim S, Nicolaides N (1979) J Chromatog Sci 17:518
305. McSharry C, Anderson K, Speekenbrink A, Lewis C, Boyd G (1993) J Med Chem 48:496
306. Olubuyide IO, Festing MF, Chapman C, Higginson J, Whicher JT (1997) Gastroenterol 18:15
307. Sanchez ML, Sanz J (1994) Atmos Environ 28:1147

308. Anderson TW (1984) An introduction to multivariate statistical analysis, 2nd edn. Wiley, New York, 244 pp
309. Andreyev AD, Yevtushenko NY (1996) Hydrobiol J 32 : 9
310. Dorling SR, Davies TD, Pierce CE (1992) Atmos Environ 26 : 2575
311. Eriksson L, Jonsson J, Sjostrom M, Wold S (1989) Chemometrics and Intelligent Lab Sys 7 : 131
312. Jonsson J, Eriksson L, Sjostrom M, Wold S, Tosato M (1989) Chemometrics and Intelligent Lab Sys 5 : 169
313. Aboul-Kassim TAT, Simoneit BRT (1995) Environ Sci Technol 29 : 2473
314. Aboul-Kassim TAT, Simoneit BRT (1995) Mar Pollut Bull 30 : 63
315. Aboul-Kassim TAT, Simoneit BRT (1996) Mar Chem 54 : 135
316. Jolliffe IT (1986) Principal component analysis. Springer, Berlin Heidelberg New York, 155 pp
317. Meglen RR (1992) In: Farrington JW (ed) Marine organic geochemistry: review and challenges for the future. Mar Chem 39 : 217
318. Zitko V (1989) Mar Pollut Bull 20 : 26
319. Zitko V (1994) Mar Pollut Bull 28 : 718
320. Holden H, LeDrew E (1998) Remote Sens Environ 65 : 217
321. Anthony ML, Sweatman BC, Beddell CR, Lindon JC, Nicholson JK (1994) Mol Pharmacol 46 : 199
322. Burse VW, Groce DF, Caudill SP, Korver MP, Phillips DL, McClure PC, Lapeza CR Jr, Head SL, Miller DT, Buckley DJ (1994) Sci Total Environ 144 : 153
323. Chopicka J, Zagrodzki P, Zachwieja Z, Krosniak M, Fota M (1995) Analyst 120 : 943
324. Gingeras TR, Ghandour G, Wang E, Berno A, Small PM, Drobniewski F, Alland D, Desmond E, Holodniy M, Drenkow J (1998) Genome Res 8 : 435
325. Guckert JB, Nold SC, Boston HL, White DC (1995) Can J Fish Aquat Sci 49 : 2579
326. Holmes E, Nicholls AW, Lindon JC, Ramos S, Spraul M, Neidig P, Connor SC, Connelly J, Damment SJ, Haselden J, Nicholson JK (1998) Biomed 11 : 235
327. Naf C, Broman D, Pettersen H, Rolff C, Zebuhr Y (1992) Environ Sci Technol 26 : 1444
328. Salt DW, Yildiz N, Livingstone DJ, Tinsley CJ (1992) Pestic Sci 36 : 161
329. Sloof JE (1995) Atmos Environ 29 : 333
330. Wenning RJ, Harris MA, Finley B, Paustenbach DJ, Bedbury H (1995) Ecotoxicol Environ Saf 25 : 103
331. Witzmann FA (1994) Two-dimensional protein pattern recognition in chemical toxicity. Government Reports Announcements & Index (GRA&I) Issue 19 NTIS/AD-A280 359/1, p 47
332. Yatsenko V (1996) J Chromatogr 722 : 233
333. Katsnelson BA, Polzik EV, Kazantsev VS, Privalova LI, Kuz'min SV (1997) Med Tr Prom Ekol 4 : 7
334. McDonald CR, Norstrom RJ, Turle R (1991) Chemosphere 25 : 129
335. Simmons BP, Spear RC (1993) Environ Sci Technol 27 : 2430
336. Simpson R, Williams R, Ellis R, Culverhouse PF (1992) Mar Ecol Prog Ser 79 : 303
337. Truquet P, Lassus P, Honsell G, Le Dean L (1996) Aquat Living Resour 9 : 273
338. Armanino C, Roda A, Ius A, Casolino MC, Bacigalupo MA (1996) Environmetrics 7 : 537
339. Landis WG, Matthews RA (1993) Development of pattern recognition techniques for the evaluation of toxicant impacts to multispecies systems. Government Reports Announcements & Index (GRA&I) Issue 21 NTIS/AD-A267 197/2, p 255
340. Lavine BK (1992) Chemometrics Intell Lab Syst 15 : 219
341. Lavine BK, Stine A, Mayfield HT (1993) Anal Chim Acta 277 : 357
342. Lavine BK, Stine AB, Qin XH (1997) Application of pattern recognition techniques to problems in advanced pollution monitoring. Government Reports Announcements & Index (GRA&I) Issue 02 NTIS/AD-A313 960/7, p 238
343. Kirschner GL, Kowalski BR (1979) In: Arriens EJ (ed) Medicinal chemistry. Academic Press, New York, vol 8, p 34
344. Plackett RL, Hewlett PS (1967) Biometrics 23 : 27

345. Hermens J, Canton H, Steyger N, Wegman R (1984) Aquat Toxicol 5:315
346. Marking LL (1977) In: Aquatic toxicology and hazard assessment. ASTM STP Publ 634, ASTM, West Conshohocken, PA, p 99
347. Lewis MA, Perry RL (1981). In: Aquatic toxicology and hazard assessment. ASTM STP Publ 737, ASTM, West Conshohocken, PA, p 402
348. Konemann H (1981) Toxicology 19:229
349. Broderius S, Kahl M (1985) Aquat Toxicol 6:307
350. Hermens J, Konemann H, Leeuwangh P, Musch A (1985) Environ Toxicol Chem 4:273
351. Mackay D (1982) US-EPA, the utility solid waste activities group. The Edison Electric Institute and The National Rural Electric Cooperative Association
352. De Filippis PS, Carsella M, Pochetti F (1999) Ind Eng Chem Res 38:380
353. Gustafsson O, Gschwend PM, Buesseler KO (1997) Environ Sci Technol 31:3544
354. Jackson LJ, Schindler DE (1996) Environ Sci Technol 30:1861
355. Amaro AR, Oakley GG, Bauer U, Spielmann HP, Robertson LW (1996) Chem Res Toxicol 9:623
356. Bandh C, Ishaq R, Broman D, Naf C, Ronquist-Nii Y, Zebuhr Y (1996) Environ Sci Technol 30:214
357. Kawahara FK, Michalakos PM (1997) Ind Eng Chem Res 36:1580
358. Girvin DC, Sklarew DS, Scott AJ, Zipperer JP (1990) Release and attenuation of PCB congeners, measurements of desorption kinetics and equilibrium sorption partition coefficients. GS6875, Electric Power Research Institute, Palo Alto, CA 94304
359. Lawruk TS, Lachman CE, Jourdan SW, Fleeker JR, Hayes MC, Herzog DP, Rubio FM (1996) Environ Sci Technol 30:695
360. Macdonald RW, Ikonomou MG, Paton DW (1998) Environ Sci Technol 32:331
361. Morrison HA, Gobas FAP, Lazar R, Whittle DM, Haffner GD (1998) Environ Sci Technol 32:3862
362. Oakley GG, Devanaboyina U, Robertson LW, Gupta RC (1996) Chem Res Toxicol 9:1285
363. Griffin RA, Chian ESK (1980) Attenuation of water soluble PCBs by earth materials. EPA600/280027, US-EPA, MERL, Cincinatti, OH 45268
364. Madenjian CP, Hesselberg RJ, DeSorcie TJ, Schmidt LJ, Stedman RM, Quintal RT, Begnoche LJ, Passino-Reader DR (1998) Environ Sci Technol 32:886
365. Pascoe GA, Riley MJ, Floyd TA, Gould CL (1998) Environ Sci Technol 32:813
366. Simcik MF, Franz TP, Zhang H, Eisenreich SJ (1998) Environ Sci Technol 32:251
367. ASTM (1986) Chlorinated aromatic hydrocarbons (Askarels) for transformers. D228386. Annual book of ASTM standards, vol 10.03. American Society for Testing of Materials, Philadelphia, PA, 19103
368. Pal D, Weber JB, Overcash MR (1980) Residue Rev 74:46
369. McGraw MG (1983) The PCB problem, separation fact from fiction. Electric World 49
370. DeWeerd KA, Bedard DL (1999) Environ Sci Technol 33:2057
371. Gevao B, Hamilton-Taylor J, Murdoch C, Jones KC, Kelly M, Tabner BJ (1997) Environ Sci Technol 31:3274
372. Hofelt CS, Shea D (1997) Environ Sci Technol 31:154
373. Huckins JN, Petty JD, Prest HF, Lebo JA, Orazio CE (1997) Environ Sci Technol 31:3732
374. Mieure JP, Hicks O, Kaley RG, Saeger VW (1976) National conference on PCBs. Electric Power Research Institute, Palo Alto, CA 94304
375. Lead WA, Steinnes E, Jones KC (1996) Environ Sci Technol 30:524
376. Lee RGM, Hung H, Mackay D, Jones KC (1998) Environ Sci Technol 32:2172
377. Brown JF, Wagner RE, Bedard DL, Carnahan JC, Unterman R (1988) Proc 1987 EPRI PCB Seminar: EA/EL5612. Electric Power Research Institute, Palo Alto, CA94304
378. Boyd SA, Sun S (1990) Environ Sci Technol 24:142
379. Sun S, Boyd SA (1991) J Environ Qualit 20:557
380. Tucker ES, Litschgi WJ, Mees WM (1975) Bull Environ Contam Technol 17:513
381. Aminabhavi TM, Naik HG (1998) J Haz Mater 60:175
382. Chiou CT, Malcolm RL, Brinton TI, Kile DE (1986) Environ Sci Technol 20:502
383. Pignatello JJ, Xing B (1996) Environ Sci Technol 30:1

384. O'Connor DJ, Connolly JP (1980) Water Res 14:1517
385. Voice TC, Rice CP, Weber WJ (1983) Environ Sci Technol 17:513
386. Gschwend PM, Wu S (1985) Environ Sci Technol 19:90
387. Walters RW, Ostazeski SA, Guiseppi-Elie A (1989) Environ Sci Technol 23:480
388. Hassett JP, Anderson MA (1979) Environ Sci Technol 13:1526
389. Hassett JP, Anderson MA (1982) Water Res 16:681
390. Whitehouse B (1985) Estuarine Coastal Shelf Sci 20:393
391. Hawker DW, Connell DW (1988) Environ Sci Technol 22:382
392. Engebretson RR, Amos T, von Wandruszka R (1996) Environ Sci Technol 30:990
393. Chang S, Berner RA (1998) Environ Sci Technol 32:2883
394. Banerjee S, Yalkowsky SH, Valvani SC (1980) Environ Sci Technol 14:1227
395. Celis R, Hermosín MC, Cox L, Cornejo J (1999) Environ Sci Technol 33:1200
396. Lee CM, Meyers SL, Wright CL Jr, Coates JT, Haskell PA, Falta RW Jr (1998) Environ Sci Technol 32:3574
397. Chiou CT (1989) In: MacCarthy P, Malcolm RL, Clapp E, Bloom P (eds) Humic substances in soil and crop sciences. American Society of Agronomy, Madison, Wisconsin, p 214
398. De Bruijn J, Hermens J (1990) Quant Struct Act Relat 9:11
399. De Bruijn J, Busser G, Seinen W, Hermens J (1989) Environ Toxicol Chem 8:499
400. Doucette WJ, Andren AW (1988) Chemosphere 17:345
401. Woodrow BN, Dorsey JG (1997) Environ Sci Technol 31:2812
402. Zimmerman JB, Kibbey TCG, Cowell MA, Hayes KF (1999) Environ Sci Technol 33:169
403. Voice TC, Weber WJ Jr (1985) Environ Sci Technol 19:789
404. Skoglund RS, Stange K, Swackhamer DL (1996) Environ Sci Technol 30:2113
405. Thomas G, Sweetman AJ, Ockenden WA, Mackay D, Jones KC (1998) Environ Sci Technol 32:936
406. Wiberg K, Letcher R, Sandau C, Duffe J, Norstrom R, Haglund P, Bidleman T (1998) Anal Chem 70:3845
407. Shea D, Hofelt CS (1997) Environ Sci Technol 31:3734
408. Jeremiason JD, Eisenreich SJ, Baker JE, Eadie BJ (1998) Environ Sci Technol 32:3249
409. Pearson RF, Hornbuckle KC, Eisenreich SJ, Swackhamer DL (1996) Environ Sci Technol 30:1429
410. Madenjian CP, Schmidt LJ, Desorcie TJ, Hesselberg RJ, Quintal RT, Begnoche LJ, Elliott RF, Bouchard PM, Holey ME (1998) Environ Sci Technol 32:3063
411. Brunner S, Hornung E, Santi H, Wolff E, Piringer OG, Altschuh H, Bruggemann R (1990) Environ Sci Technol 24:1751
412. Dunnivant FM, Coates JT, Elzerman AW (1988) Environ Sci Technol 22:448
413. Nicholson BC, Maquire BP, Bursill DB (1984) Environ Sci Technol 18:518
414. Springer C, Thibodeaux LJ, Chatrathi S (1983) In: Francis CW, Auerbach SI (eds) Environmental and solid wastes, characterization, treatment, and disposal. Butterworths, Boston, MA, 209
415. Lewis RG, Martin BE, Sgontz DL, Howes JE Jr (1985) Environ Sci Technol 19:986
416. Murphy TJ, Formanski LJ, Brownawell B, Meyer JA (1985) Environ Sci Technol 19:942
417. Mackay D, Leinonen PJ (1975) Environ Sci Technol 9:1178
418. Haque R, Schmedding DW, Freed VH (1974) Environ Sci Technol 8:139
419. Haque R, Kohnert R (1976) J Environ Sci Health B11:253
420. Garbarini DR, Lion LW (1986) Environ Sci Technol 12:1263
421. Di Toro DM, Horzempa LM (1982) Environ Sci Technol 16:594
422. Girvin DC, Sklarew DS, Scott AJ (1988) Proc 1987 EPRI PCB Seminar. EA/EL 5612. Electric Power Research Institute, Palo Alto, CA 94304
423. Hankin L, Sawhney BL (1984) Soil Sci 137:401
424. Baxter RM, Sutherland DA (1984) Environ Sci Technol 18:608
425. SAS Institute (1997) SAS Statistical Package, Version 6.12, Cary, NC
426. Kenaga EE, Goring CA (1980) In: Eaton JG (ed) Aquatic toxicology. ASTM, Philadelphia, p 707
427. Karickhoff SW (1981) Chemosphere 10:833

428. Karickhoff SW, Brown DS, Scott TA (1979) Water Res 13:241
429. Karickhoff SW (1980) In: Bakerw A (ed) Contaminants and sediments: 2. Ann Arbor Science Publishers, Ann Arbor, Ml, p 193
430. Hodson J, Williams NA (1988) Chemosphere 17:67
431. Pearson RG (1986) Proc Natl Acad Sci USA 83:8440
432. Schuurmann G (1990) Environ Toxicol Chem 9:417
433. Pearlman RS (1980) Molecular surface area and volumes and their use in structure/activity relationships. Marcel Dekker, New York, p 412
434. Pearlman RS (1981) QCPE Bull 1:16
435. ADAPT Program (1986) Version 2.0, Molecular Design Limited, San Leandro, CA
436. Schuurmann G (1990) Quant Struc-Act Relat 9:326

CHAPTER 5

Microbial Transformations at Aqueous-Solid Phase Interfaces: A Bioremediation Approach

Tarek A. T. Aboul-Kassim[1], Bernd R. T. Simoneit[2]

[1] Department of Civil, Construction and Environmental Engineering, College of Engineering, Oregon State University, 202 Apperson Hall, Corvallis, OR 97331, USA
e-mail: Tarek.Kassim@orst.edu

[2] Environmental and Petroleum Geochemistry Group, College of Oceanic and Atmospheric Sciences, Oregon State University, Corvallis, OR 97331, USA
e-mail: simoneit@oce.orst.edu

The application of state-of-the-science analytical techniques to environmental chemodynamics of organic contaminants has provided society with various information of concern. The air we breathe, the water we drink, the soil in which our crops are grown, and the environments where populations of humans, animals, and plants grow are contaminated with a variety of synthetic organic chemicals. Many of these contaminants are industrial chemicals that have been, deliberately or inadvertently, discharged into surface and ground waters, or onto soils and bottom sediments following their intended use. Others are by-products of manufacturing operations that do not utilize waste-treatment facilities or by-products that were inadequately treated.

Biodegradation is the most important fate and biotransformation mechanism for various organic compounds at aqueous-solid phase interfaces, compared to other abiotic chemodynamic processes (e.g., photolysis, volatilization, and hydrolysis). It frequently, although not necessarily, leads to the conversion of much of the organic C, N, P, S, and halogens in the original contaminant to inorganic products. Such a conversion of an organic substrate to inorganic products is known as *mineralization* or *ultimate biodegradation*. Thus, in the mineralization/biodegradation of organic compounds, CO_2 and inorganic forms of N, P, and S are released by the microorganisms in aqueous-solid phase environments. This biotransformation process appears to result largely, or entirely in some interfacial environments, from microbial activity. Indeed, microorganisms are the dominant means of converting synthetic chemicals to inorganic products in the ambient environment.

Accordingly, the present chapter is designed to present the basic principles of microbial associations at aqueous-solid phase interfaces, the types and mechanisms of biodegradation and biotransformation, and to show how those principles relate to bioremediation engineering technologies. It considers the microbiological, chemical, environmental, engineering, and technological aspects of biodegradation. However, it does not cover all facets because the information is too extensive and diverse, and the knowledge base is expanding too rapidly to be covered in a single chapter. Nevertheless, there are general key principles that underlie the science and engineering technology. Thus, the present chapter focuses mainly on state-of-the-knowledge about the major groups of microorganisms, the biodegradation processes and factors affecting them, and the microbial transformations of various toxic and carcinogenic organic contaminants at interfaces. In addition, several case studies showing the application of biodegradation concepts in bioremediation technology of contaminated environments are also presented and discussed.

Keywords. Organic pollutants, Microbial transformations, Biodegradation, Bioremediation, Microorganisms, Aqueous-solid phase systems, Contaminated Sediments

The Handbook of Environmental Chemistry Vol. 5 Part E
Pollutant-Solid Phase Interactions: Mechanism, Chemistry and Modeling
(by T. A. T Aboul-Kassim, B. R. T. Simoneit)

List of Abbreviations

ABS	Alkylbenzene sulfonate
AHA	Aldrich humic acid
APEC	Alkylphenol ethoxycarboxylate
APEO	Alkylphenol ethoxylate
BB	Bromobiphenyl
BETX	Benzene, ethylbenzene, toluene, and xylene
CBs	Chlorobiphenyls
CDD	Chlorinated dibenzo-*p*-dioxin
CF	Chloroform
CM	Chloromethane
CO_2	Carbon dioxide
COMs	Complex organic mixtures
CSIA	Compound-specific isotope analysis
CT	Carbon tetrachloride
DCA	Dichloroethane
DCB	Dichlorobenzidine
DCE	Dichloroethylene
DCM	Dichloromethane
DOC	Dissolved organic matter
Eh	Redox potential
GC	Gas chromatography
GC-C-IRMS	Gas chromatography-combustion-isotope ratio mass spectrometry
GC-MS	Gas chromatography-mass spectrometry
HMW	High molecular weight
HpCDD	Heptachlorodibenzo-*p*-dioxin
LAS	Linear alkylbenzene sulfonate
LMW	Low molecular weight
LSC	Liquid scintilation count
N	Nitrogen
NPEOs	Nonylphenol ethoxylates
OPEC	Octylphenol ethoxycarboxylate

P	Phosphorus
PAHs	Polycyclic aromatic hydrocarbons
PCBs	Polychlorinated biphenyls
PCDD	Polychlorinated dibenzo-*p*-dioxin
PCE	Perchloroethylene
PMA	Polymaleic acid
S	Sulfur
SOM	Solid organic matter
TBOS	Tetrabutoxysilane
TCA	Trichloroethane
TCDD	Tetrachlorodibenzo-*p*-dioxin
TCE	Trichloroethylene
TCP	Trichlorophenol
TEAP	Terminal electron-accepting processes
TeCA	Tetrachloroethane
TKEBS	Tetrakis(2-ethylbutoxy)silane
TNT	Trinitrotoluene
TOC	Total organic carbon
VC	Vinyl chloride

1 Introduction

Synthetic organic compounds are found in simple or complex mixtures in various environmental multimedia, and especially at aqueous-solid phase interfaces. These complex mixtures may be associated with the release, storage, or transport of many compounds in surface or ground waters, waste-treatment systems, soils, or sediments. The number of compounds found to date is enormous, and the types of mixtures are similarly numerous [1]. Moreover, the concentrations of individual organic compounds vary appreciably, and they may be higher than 1.0 g/l of water or µg/kg of soil/sediment at sites subject to spills from tank cars, trucks, or oil/cargo ships, to discharge of industrial waste, or to leakages from storage or disposal facilities for industrial chemicals [2–11]. In contrast, the concentrations may be lower than 1.0 µg/l of water or µg/kg of soil/sediment at some distance from the point of release, marine/terrestrial spill, or storage [12–15]. Even at these low concentrations some organic compounds are toxic, or risk analyses suggest that exposure of large populations to the low levels will result in deleterious effects to a few individuals. In addition, some chemicals at low concentrations are subject to biomagnification and may reach levels that have deleterious effects on humans, animals, or plants.

Synthetic organic compounds are, in general, present in the human environment (e.g., in areas used for food and feed production, and in environments supporting natural populations of animals and plants). Modern society relies on a striking array of organic chemicals, and the quantities used are staggering. Values for the annual production of organic compounds in the United States alone show the large tonnages that are part of human activities in industry and agriculture (Fig. 1). Although many of these chemicals are consumed or destroyed, a significant amount is released into the air, water, soil, and sediment compartments of the global environment. The quantity released varies with the compound and its particular use, but regulatory agencies in industrialized countries have found that significant percentages of the total quantities consumed by industry, agriculture, and domestic pursuits do, indeed, find their way into water-soil and water-sediment interfaces [1, 16–19].

Predicting the hazards of an organic compound to humans, animals, or plants requires information not only on its toxicity to living microorganisms but also on the degree of exposure of the microorganisms to the compound (see Chap. 4). The mere discharge of an organic compound does not, in itself, constitute a hazard; however, the individual human, animal, or plant must be exposed to it. In evaluating exposure, the transport of a compound and its fate must be con-

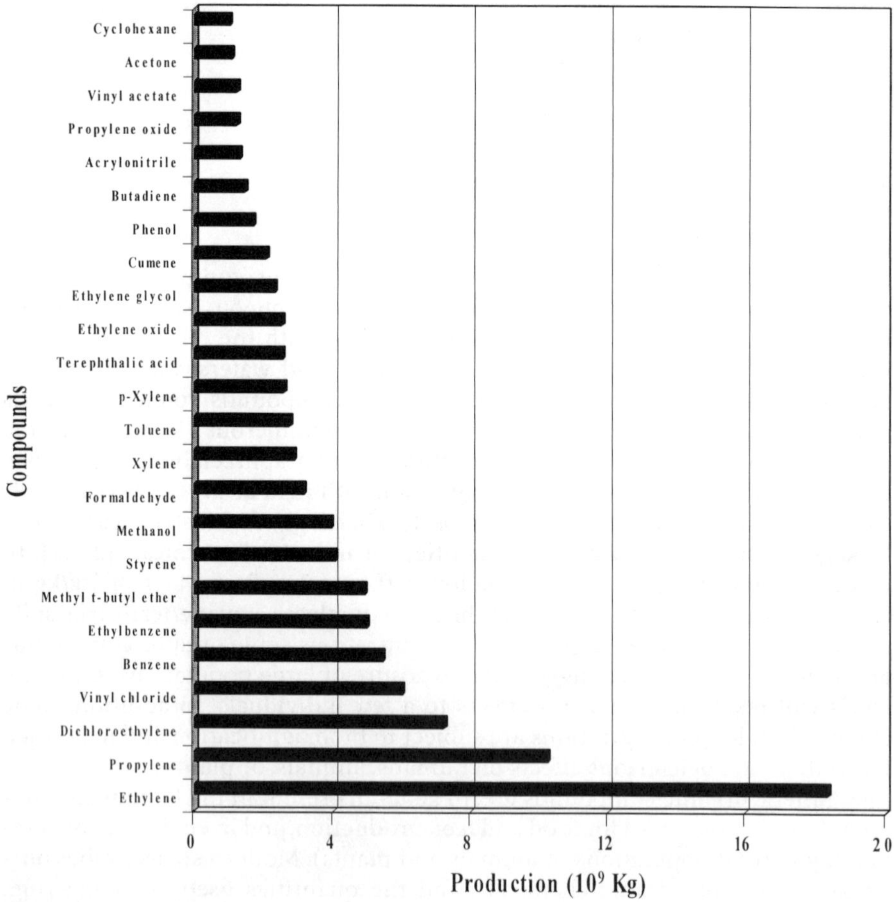

Fig. 1. Annual production of synthetic organic chemical feedstock in the United States during 1992

sidered. A molecule that is not subject to environmental transport is not considered as a health or environmental problem except to species at the specific point of release. Thus, information on dissemination of a compound from the point of its release to the point where it could have an effect is of great relevance [20–22]. However, the compound may be structurally modified or totally destroyed during transport, and its fate during transport, i.e., modification or destruction, is crucial to defining the exposure. A compound which is modified to yield products that are more or less toxic, is totally degraded or biomagnified, can represent a greater or lesser hazard to species potentially subject to injury.

At the specific site of discharge or during its transport, an organic molecule may be acted on by various abiotic mechanisms, such as volatilization, photolysis, hydrolysis, and sorption/desorption. However, the biotic (i.e., microbial attack) mechanism is considered to be the most effective and destructive mechanism bringing about significant changes to organic compounds. Such microbial transformations, which involve enzymes as catalysts, frequently bring about extensive modification in the structure and toxicological properties of pollutants or potential pollutants. These biotic processes may result in the complete conversion of the organic molecule to inorganic products, cause major changes which result in new organic products, or occasionally result in only minor modifications.

Accordingly, biodegradation (i.e., microbial transformation) can be defined as the biologically catalyzed reduction in complexity of organic compounds [22–32]. In the case of organic matter at aqueous-solid phase interfaces, biodegradation frequently, although not necessarily, leads to the conversion of much of the C, N, P, and S in the original organic compounds to inorganic products. Such a conversion is known as *mineralization* or *ultimate biodegradation*. Because biodegradation results in the total destruction of the parent organic compounds and their conversion to inorganic products, such a process is beneficial. In contrast, non-biological and many biological processes degrade organic compounds, but convert them to other organic products. Some of these products are toxic, but others are not. Nevertheless, the environmental accumulation of an organic product is still a cause for some concern because a material not presently known to be harmful may, with new techniques or measurements of new toxicological manifestations, become undesirable. Thus, biodegradation is especially important in removing actual or possible chemical hazards for humans, animals, or plants from natural environments.

The main objectives of the present chapter are to present the basic principles of the microbial associations at aqueous-solid phase interfaces, the types and mechanisms of biodegradation and biotransformation, and to show how those principles relate to bioremediation technologies. The multidisciplinary information in the present chapter considers some of the microbiological, chemical, environmental, engineering, and technological aspects of biodegradation. However, it does not cover all facets because the information is too large and diverse, and the knowledge base is expanding too extensively to be covered in a single chapter. Nevertheless, there are key general principles that underlie the science and engineering technology. Hence, the current chapter portrays the state-of-the-knowledge about the major groups of microorganisms present at interfaces, the biodegradation processes and factors affecting them, and the microbial transformations of various toxic and carcinogenic organic contaminants occurring at interfaces. Moreover, several case studies showing the application of biodegradation concepts in bioremediation technology of various contaminated environments are also presented and discussed.

2 Microbial Associations at Aqueous-Solid Phase Interfaces

The abiotic characteristics of aqueous-solid phase interfaces strongly influence chemical/biochemical reactions in the interface microenvironment of aqueous-solid phases. These reactions at interfaces are controlled mainly by biotic activity. Specifically, all aqueous-solid phase microenvironments contain living microorganisms that mediate biochemical transformations. Solid phases (e.g., soil and sediment particles) usually contain billions of microorganisms, with the aqueous phase containing smaller, but still significant, populations [22, 33–39].

The present section discusses the different types and classifications of microbial associations present at aqueous-solid phase interfaces, their energy generations, and factors affecting their growth and activity. The following is a summary.

2.1 Types and Classifications

The major groups of microorganisms at aqueous-solid phase interfaces include viruses, bacteria, fungi, algae, and macro fauna (e.g., protozoa and arthropods). All of these microorganisms have specific ecological niches and functions, and each contributes to the overall biotic activity of this microenvironment.

Both bacteria and fungi are particularly important with respect to biochemical transformations and have a critical role in influencing the fate and mitigation of many organic pollutants. Accordingly, a large subdivision of bacteria includes the actinomycetes, which are often treated as a separate group of microorganisms because of their unique characteristics. In the following discussion, a broad overview of bacteria, actinomycetes, and fungi is presented in order to examine their significance with respect to detoxification and biodegradation. The importance of solid phase microflora is illustrated by their numbers and biomass. The relative estimates of the abundances of solid phase microbes in the aqueous-solid phase environment of bacteria, actinomycetes and fungi are 10^8, 10^7, and 10^6 number/g solid phase, respectively [2, 40–45]. It is obvious that very large populations can be sustained in/on solid particles. In addition, the groups are very diverse, so that large numbers of different microorganisms can mediate an almost infinite number of biochemical transformations. The following is a brief description of these three groups.

2.1.1
Bacteria

Bacteria, which are the most numerous microorganisms in aqueous and solid phases, are prokaryotic microorganisms lacking a nuclear membrane. They are characterized by a complex cell envelope, which contains cytoplasm but no cell organelles [42, 43, 46–48]. Bacteria are capable of rapid growth and reproduction which occur by binary fission. Genetic exchange occurs predominantly by conjugation (i.e., cell-to-cell contact) or transduction (i.e., exchange via viruses), although transformation (i.e., transfer of naked DNA) also occurs [41, 42, 47, 48]. The size of bacteria generally ranges from 0.1 μm to 2.0 μm.

In general, bacteria are the most abundant soil phase microorganisms, with a biomass of about 500 kg/ha to the depth of the root zone [48]. Generally, *aerobes* are more prevalent than *anaerobes*. As the depth decreases in the terrestrial environment, the number of anaerobes increases relative to aerobes. Bacteria are critically involved in almost all aqueous-solid phase interface biochemical/ microbial transformations, including the metabolism of both organic and inorganic chemicals. The importance of bacteria in the fate and mitigation of organic pollutants cannot be overestimated. Because of their prevalence and diversity, as well as fast growth rates and adaptability, they have an almost unlimited ability to degrade most natural products and many xenobiotics.

Bacteria can be classified according to several characteristics, including their mode of nutrition, type of electron acceptor, ecological conditions, and dominance [41–43, 45, 47, 49–53]. The following is a brief summary.

2.1.1.1
Mode of Nutrition

According to the mode of nutrition, that is either specific to various groups of bacteria or characteristic to certain environments, bacteria can be classified into:

- Autotrophic mode: strict solid autotrophs obtain energy from inorganic sources and carbon from carbon dioxide. These kinds of microorganisms generally have few growth-factor requirements. Autotrophic bacteria can obtain energy from the oxidation of inorganic substances (i.e., chemoautotrophs) or obtain energy from photosynthesis (i.e., photoautotrophs).
- Heterotrophic mode: heterotrophs obtain energy and carbon from organic substances. Thus, chemoheterotrophs obtain energy from oxidations, whereas photoheterotrophs obtain energy from photosynthesis with an organic electron donor requirement.

2.1.1.2
Type of Electron Acceptor

According to the type of electron acceptor, bacteria can be classified into:

- Aerobic: aerobic microorganisms utilize oxygen as a terminal electron acceptor and possess superoxide dismutase or catalase enzymes which are capable of degrading peroxide radicals.
- Anaerobic: anaerobic microorganisms do not utilize oxygen as a terminal electron acceptor. *Strict anaerobes* do not possess superoxide dismutase or catalase enzymes and are thus poisoned by the presence of oxygen. Although other kinds of anaerobes do possess these enzymes, they utilize terminal electron acceptors other than oxygen, such as nitrate or sulfate. *Facultative anaerobes* can use oxygen or combined forms of oxygen as terminal electron acceptors.

2.1.1.3
Ecological Status

Indigenous bacteria can be *autochthonous* or *zymogenous*. The former metabolize slowly in soil and sediments, utilizing their organic matter as a substrate. The latter are adapted to intervals of dormant and rapid growth, depending on substrate availability. *Allochthonous* microorganisms usually survive only for short periods of time. However, the most recent theory of classification is founded on the concept of *r* and *K selection*. Microorganisms adapted to living under conditions in which substrate is plentiful are designated as *K-selected*. Microorganisms that are *r-selected* live in environments in which substrate is the limiting factor, except for occasional flushes of substrate. *r-Selected* microorganisms rely on rapid growth rates when substrate is available, and generally occur in uncrowded environments. In contrast, *K-selected* microorganisms exist in crowded environments and are highly competitive.

2.1.1.4
Dominance

According to the dominance of bacteria at aqueous-solid phase interfaces, they can be classified into the following groups:

- *Arthrobacter*: the most numerous bacteria on/in solid phases are *Arthrobacter*, as determined by plating procedures. They represent as much as 40% of the culturable solid phase bacteria. These autochthonous microorganisms are pleomorphic and Gram-variable. Young cells are Gram-negative rods, which later become Gram-positive cocci.
- *Streptomyces*: *Streptomyces* microorganisms are actually actinomycetes. They are Gram-positive, chemoheterotrophic microorganisms that can comprise 5–20% of the bacterial count in solid phases. These microorganisms produce antibiotics, including streptomycin.
- *Pseudomonas*: *Pseudomonas* are Gram-negative microorganisms ubiquitous and diverse in nature. They are generally heterotrophic and aerobic, but some

are facultative autotrophs. As a group, they possess many different enzyme systems and are capable of degrading a wide variety of organic compounds. These microorganisms can comprise 10–20% of the bacterial population.
- *Bacillus: Bacilli* are characterized as Gram-positive aerobic microorganisms which produce endospores. This genus is heterotrophic and diverse, comprising 10% of the bacterial population.

2.1.2
Actinomycetes

Actinomycetes are microorganisms that are technically classified as bacteria, but are unique enough to be discussed here as an individual group. They have some characteristics in common with bacteria, but are also similar in some respects to fungi. For the most part, they are aerobic, chemoheterotrophic microorganisms consisting of elongated single cells [41, 42, 46, 48]. They display a tendency to branch into filaments, or hyphae, that resemble fungal mycelia. These hyphae are morphologically similar to those of fungi, but are smaller in diameter (about 0.5–2 μm [41, 42, 48, 49]). The total number of actinomycetes is often about 10^7 per gram of a solid phase. Generally, the population of actinomycetes is 1–2 orders of magnitude less than that of other bacteria in solid phases. The genus *Streptomyces* dominates the actinomycetes, and these Gram-positive microorganisms may represent 90% of the total actinomycetes population.

Like all bacteria, actinomycetes are prokaryotic microorganisms. In addition, the adenine-thymine and guanine-cytosine contents of bacteria and actinomycetes are similar, as are the cell wall constituents of both types of microorganisms. Actinomycetes filaments are also about the same size as those of bacteria.

Like fungi, however, actinomycetes display extensive mycelial branching, and both types of microorganisms form aerial mycelia and conidia. Moreover, growth of actinomycetes in liquid culture tends to produce fungus-like clumps or pellets rather than the turbidity produced by bacteria. Finally, growth rates in fungi and actinomycetes are not exponential as they are in bacteria; rather, they are cubic [35, 42].

Actinomycetes can metabolize a wide variety of organic substrates, including organic compounds that are generally not metabolized, such as phenols and steroids. They are also important in the metabolism of heterocyclic compounds such as complex nitrogen compounds and pyrimidines [42, 49]. The breakdown products of their metabolites are frequently aromatic, and these metabolites are important in the formation of humic substances and soil humus [42, 49].

2.1.3
Fungi

The fungi (e.g., molds, mildews, rusts, yeasts, or mushrooms) are the third major group of solid phase microorganisms. However, they differ from bacteria and actinomycetes in that they are eukaryotic. They are all heterotrophic, and most are aerobic, with the exception of yeasts, which are fermenting microorganisms.

Although fungi are eukaryotic, they contain no chlorophyll; therefore, they do not photosynthesize. Among the fungi in solid phases, the filamentous molds are most critically involved in the degradation of organic substrates and hence in the fate and mitigation of organic pollution. These fungi are characterized by extensive branching and mycelial growth, as well as by the production of sexual and asexual spores [36, 42, 49]. Some of the most common genera of fungi are *Penicillium, Aspergillus, Fusarium, Rhizactonia, Alternaria,* and *Rhizopus.*

Based on plate counts, the populations of fungi are on the order of 10^6 per gram of soil and sediments, although such estimates are biased toward sporulating species. The diameter of fungal hyphae can be 10–50 μm. This size, which helps to distinguish them morphologically from the smaller actinomycetes, results in a total biomass of about 1500 kg/ha of soil. Thus, their biomass is greater than that of the bacteria and actinomycetes, even though they are numerically less prevalent in most soils [41, 42, 50, 51].

Fungi are heavily involved in the degradation of organic matter. As a group, they contain extremely diverse enzyme systems and efficiently degrade sugars, organic acids, and complex compounds (e.g., cellulose or lignin [36, 41–43]). Fungi are very important in controlling the ultimate fate of organic pollutants. Because they are more tolerant of acidic solid systems ($pH < 5.5$) than bacteria or actinomycetes, they are more actively involved in the degradation of organics in acidic solid particles.

In general, microorganisms on/in solid phases can be viewed as a vast biological entity whose parts live in unison, with diverse capabilities for the degradation of all natural organics and many xenobiotics. The major characteristics and differences of bacteria, actinomycetes, and fungi are compared in Table 1.

Table 1. The major characteristics and differences of bacteria, actinomycetes, and fungi

Parameter	Bacteria	Actinomycetes	Fungi
Population	Numerous	Intermediate	Least numerous
Biomass	Both have similar biomass		Largest biomass
Degree of branching	Slight	Filamentous	Extensive filamentous
Growth in liquid cultures	Yes, forming turbidity	Yes, forming pellets	Yes, forming pellets
Growth rate	Exponential	Cubic	Cubic
Cell wall	Murein, teichoic acid and lipopolysaccharide		Chitin or cellulose
Competitiveness for simple organics	Most competitive	Least competitive	Intermediate competitive
Fix nitrogen	Yes	Yes	No
Aerobic	Aerobic, anaerobic	Mostly aerobic	Aerobic, except yeast
Moisture stress	Least tolerant	Intermediate tolerant	Most tolerant
Optimal pH	6–8	6–8	6–8
Competitiveness in solid phases	All solid phases	Dominate dry and high-pH phases	Dominate low-pH phases

2.2
Energy Generation

Microbial activity requires energy, and all microorganisms generate energy. This energy is subsequently stored as adenosine triphosphate (ATP), which can then be utilized for growth and metabolism as needed, subject to the second law of thermodynamics [2, 23, 35, 41, 42, 51, 54].

2.2.1
A Theoretical Approach

The second law of thermodynamics states: "*In a chemical reaction, only part of the energy is used to do work, while the rest of the energy is lost as entropy*". The Gibbs free energy (ΔG) is the amount of energy available for work for any chemical reaction. For the reaction

$$A + B \Leftrightarrow C + D \tag{1}$$

the thermodynamic equilibrium constant is defined as

$$K_{eq} = \frac{[C][D]}{[A][B]} \tag{2}$$

where $[C]$ and $[D]$ are the product concentrations, and $[A]$ and $[B]$ are the reactant concentrations. Two cases should be considered here. When the product formation is favored, that is if

$$[C][D] \geq [A][B] \tag{3}$$

then $K_{eq} > 1$ and $\ln K_{eq}$ is positive (e.g., if $K_{eq} = 2.0$, then $\ln K_{eq} = 0.69$). If the product formation is not favored (Eq. 1):

$$[C][D] \leq [A][B] \tag{4}$$

then $K_{eq} < 1$ and $\ln K_{eq}$ is negative (e.g., if $K_{eq} = 0.2$, then $\ln K_{eq} = -1.61$).

In general, the relationship between the equilibrium constant K_{eq} and the free energy ΔG can be given by

$$\Delta G = -RT \ln K_{eq} \tag{5}$$

where R is the universal gas constant, and T is the absolute temperature (°K).

Table 2 illustrates the effect of the Gibbs free energy on the spontaneity of a chemical/biochemical reaction and the resulting release of energy. Thus, it is useful to use ΔG values for any biochemical reaction mediated by microbes to determine whether energy is liberated for work, and how much energy is liberated.

Table 2. The effect of the Gibbs free energy on the spontaneity of a chemical reaction

ΔG	K_{eq}	$\ln K_{eq}$	Energy status
Negative	$K_{eq} > 1$	Positive	Energy is released and the reaction proceeds spontaneously
Positive	$0 < K_{eq} < 1$	Negative	Energy must be added to promote the reaction

2.2.2
Mechanisms

Microbial associations which live at aqueous-solid phase interfaces can in general generate energy via several mechanisms that can be divided into two main categories, as follows.

2.2.2.1
Photosynthesis

Energy supplied by sunlight is necessary for photosynthesis (Fig. 2). The fixed organic carbon is then used to generate energy via respiration. Examples of microorganisms on/in solid phases which carry out photosynthesis are *Rhodospirillum*, *Chromatium*, and *Chlorobium* [36, 41, 42, 46, 49, 50].

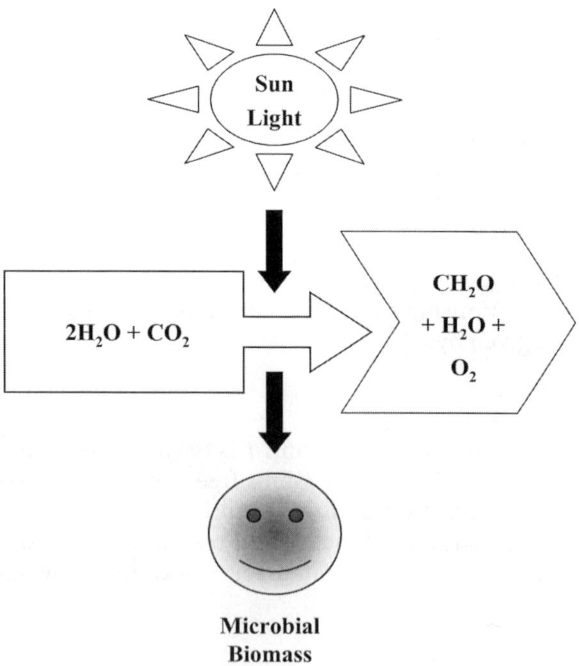

Fig. 2. The photosynthesis process of microorganisms

2.2.2.2 Respiration

Microorganisms at aqueous-solid phase interfaces have different respiration modes which include: (a) aerobic heterotrophic, (b) aerobic autotrophic, (c) facultative anaerobic heterotrophic, (d) facultative anaerobic autotrophic, and (e) anaerobic heterotrophic respiration modes [36, 41, 43, 47, 55]. Table 3 shows the main differences between these different respiration modes.

Overall, there are many ways in which microorganisms at aqueous-solid phase interfaces can generate energy. The mechanisms listed above illustrate the diversity of soil/sediment microorganisms and explain their ability to break down or transform almost any natural organic substance. In addition, enzyme systems have evolved to metabolize complex organic molecules. These enzymes can also be used to degrade xenobiotics with similar chemical structures. Xenobiotics, which do not degrade easily, are normally chemically different

Table 3. Main differences between various respiration modes of microorganisms

Type	Microbial respiration processes
Aerobic heterotrophic	Many microorganisms on solid phases undergo aerobic heterotrophic respiration (e.g., *Pseudomonas* and *Bacillus*), as follows: (a) $C_6H_{12}O_6 + 6O_2 \rightarrow 6CO_2 + 6H_2O$ ($\Delta G = -2870$ kJ/mol)
Aerobic autotrophic	The reactions carried out by *Nitrosomonas* and *Nitrobacter* (reactions b and c, respectively) are known as nitrification, while those carried out by *Beggiatoa* and *Thiobacillus thiooxidans* (reactions d and e, respectively) are examples of sulfur oxidation: (b) $NH_3 + 1.5O_2 \rightarrow HNO_2 + H_2O$ ($\Delta G = -280$ kJ/mol) (c) $KNO_2 + 0.5O_2 \rightarrow KNO_3$ ($\Delta G = -73.2$ kJ/mol) (d) $2H_2S + O_2 \rightarrow 2H_2O + 2S$ ($\Delta G = -350$ kJ/mol) (e) $2S + 3O_2 + 2H_2O \rightarrow 2H_2SO_4$ ($\Delta G = -992$ kJ/mol) All of the reactions above (b–e) illustrate how microorganisms on solid phases mediate reactions that can cause or negate pollution. For example, nitrification (reactions b–c) and sulfur oxidation (reactions d–e) can result in the production of specific pollutants, i.e., nitrate and sulfuric acid
Facultative anaerobic	The bacterium *Pseudomonas denitrificans* is capable of this kind of metabolism utilizing nitrate as a terminal electron acceptor rather than oxygen. This bacterium can use oxygen as a terminal electron acceptor if it is available, and aerobic respiration is more efficient than anaerobic respiration. (f) $5C_6H_{12}O_6 + 24KNO_3 \rightarrow 30CO_2 + 18H_2O + 24KOH + 12N_2$ ($\Delta G = -150$ kJ/mol)
Facultative anaerobic autotrophic	The best example for facultative anaerobic autotrophic respiration is represented by *Thiobacillus denitrificans*, as shown in the denitrification reaction g: (g) $S + 2KNO_3 \rightarrow K_2SO_4 + N_2 + O_2$ ($\Delta G = -280$ kJ/mol)
Anaerobic heterotrophic	The conversion of lactic acid to acetic acid, mediated by *Desulfovibrio*, is shown in reaction h: (h) $2CH_3CHOHCOOH + SO_4^{2-} \rightarrow 2CH_3COOH + HS^- + H_2CO_3 + HCO_3^-$ (lactic acid) (acetic acid) ($\Delta G = -170$ kJ/mol)

from any known natural organic substance. Hence, microorganisms have not evolved enzyme systems capable of metabolizing such compounds.

2.3
Factors Affecting Growth and Activity

In order to understand the factors that limit microbial activity in the micro-environments of aqueous-solid phase interfaces, it is necessary that biotic and abiotic factors be discussed. These factors include the following.

2.3.1
Biotic Stress

Since indigenous soil/sediment microbes are in competition with one another, the presence of large numbers of microorganisms results in: (a) biotic stress factors, such as competition for substrate, water, or growth factors; (b) secretion of inhibitory or toxic substances, including antibiotics, that harm neighboring microorganisms; and (c) predation or parasitism on neighboring microbes (e.g., phages infect both bacteria and fungi). On the other hand, because of biotic stress, non-indigenous microorganisms that are introduced into a solid phase environment often survive for fairly short periods of time. This effect has important consequences for other microorganisms introduced to aid biodegradation [40, 42, 50, 51, 56, 57].

2.3.2
Abiotic Stress

The abiotic stress affecting microbial activity and growth in an interfacial microenvironment include factors such as light, moisture, temperature, pH, soil/sediment grain size, carbon/nitrogen content, and redox potential [40–43, 46, 47, 49–51, 56–58].

2.3.2.1
Light

Generally, solid phases are impermeable to light, that is, no sunlight penetrates beyond the top few centimeters of a soil surface or even top few millimeters of a bottom sediment in a shallow aquatic environment. Phototrophic microorganisms are therefore very limited on/in these solid phases. However, at the soil surface physical parameters such as temperature and moisture fluctuate significantly throughout the day and also seasonally. Hence most soils tend to provide a harsh environment for photosynthesizing microorganisms. A few phototrophic microorganisms, including algae, have the ability to switch to a heterotrophic respiratory mode of nutrition in the absence of light. Such a change can be found at significant depths within soils. Normally, these microorganisms are not competitive with other indigenous heterotrophic microorganisms for organic substrates.

2.3.2.2
Moisture

The moisture content varies in any soil environment, and microorganisms must be adaptable to a wide range of moisture contents. Soil aeration is dependent on soil moisture, saturated soils tend to be anaerobic, whereas dry soils are usually aerobic. But soil is a heterogeneous environment; even saturated soils contain pockets of aerobic regimes, and dry soils harbor anaerobic microsites which exist within the centers of secondary aggregates. Although bacteria are the least tolerant of low soil moisture, as a group they are the most flexible with respect to soil aeration. They include aerobes, anaerobes, and facultative anaerobes, whereas the actinomycetes and fungi are predominantly aerobic.

2.3.2.3
Temperature

Temperatures vary widely and most soil/sediment populations are resistant to wide fluctuations in temperatures although solid phase populations can be psychrophilic, mesophilic, or thermophilic, depending on the geographic location of the solid phase environment.

2.3.2.4
pH

Undisturbed soils and bottom sediments usually have fairly stable pH values within the range of 6–8, and most microorganisms have pH tolerance optima within this range. There are exceptions to this rule, as exemplified by *Thiobacillus thiooxidans*, a microorganism that oxidizes sulfur to sulfuric acid (e.g., in mine tailings) and has a pH optimum of 2–3. Within a solid phase microenvironment, pH variation can also occur due to local decomposition of an organic residue to organic acids. Thus, the solid phase behaves as a heterogeneous or discontinuous environment, allowing microorganisms with differing pH optima to coexist in close proximity.

2.3.2.5
Grain Size

Almost all soil/sediment particles contain populations of microorganisms regardless of their grain sizes. Most nutrients are associated with clay or silt particles, which also retain solid phase moisture efficiently. Thus, solid particles with at least some silt or clay particles offer a more favorable habitat for microorganisms than do particles without these materials.

2.3.2.6
Carbon/Nitrogen Content

Carbon and nitrogen are both nutrients that are found in solid particles. Since they are present in low concentrations, the growth and activity of microorganisms is limited. In fact, many microorganisms exist in solid particles under limited starvation conditions and hence are dormant. Without added substrate or amendment, these microorganisms generally metabolize at low rates. Solid phase humus (see Chap. 2) represents a source of organic nutrients that is mineralized slowly by autochthonous microorganisms. Similarly, specific microbial populations can utilize xenobiotics as a substrate, even though the rate of degradation is generally quite slow.

2.3.2.7
Redox Potential

Redox potential (Eh) is the measurement of the tendency of an environment to oxidize or reduce substrate, i.e., the availability of different terminal electron acceptors that are necessary for specific microorganisms. Such electron acceptors exist only at specific redox potentials, which are measured in millivolts (mV). An aerobic (i.e., oxidizing) solid phase environment has a redox potential or Eh of +800 mV, while an anaerobic (reducing) solid phase environment has an Eh of about 0 to –300 mV [59,60,62]. Oxygen is found in solid particles at a redox potential of about +800 mV [52]. When solid particles are placed in a closed container, oxygen is used by aerobic microorganisms as a terminal electron acceptor until all of it is depleted. As this process occurs, the redox potential of the solid phase decreases, and other compounds can be used as terminal electron acceptors. The fact that different terminal electron acceptors are available for various microorganisms having diverse pH requirements means that some solid phase environments are more suitable than others for various groups of microorganisms.

3
Microbial Transformations of Organic Pollutants

Besides several physical and chemical factors, biological factors (e.g., biodegradation) can affect the fate and transport of organic pollutants at aqueous-solid phase interfaces. Although less well-studied than the physical and chemical factors, the biological factors are receiving increasing attention due to growing interest in the use of biological approaches to bioremediation of contaminated sites. The presence of microorganisms at aqueous-solid interfaces can affect the distribution, movement, and concentration of pollutants through a process called *biodegradation*. Indeed, some organic pollutants have very short lifetimes under normal environmental conditions because they serve as nutrient sources for actively growing microorganisms. For other pollutants, the effect of microorganisms may be limited for a variety of reasons, such as low numbers of degrading microorganisms, microbe-resistant pollutant structure, and adverse environmental conditions.

This section focuses on: (1) a discussion of the overall process of biodegradation, (2) a review of the different types, aspects and phases of biodegradation of several classes of organic pollutants, (3) an examination of the environmental factors affecting biodegradation and biotransformation mechanisms, and (4) a description of the different biodegradation and biotransformation pathways.

3.1 Overview

Biodegradation is the breakdown of organic compounds through microbial activity. Biodegradable organic compounds serve as the food source, or substrate for microbes, and the availability of an organic compound to such microbes is its bioavailability. Bioavailability (see Chap. 4), which is one important aspect in the biodegradation of any substrate, depends largely on water. Microbial cells are 70–90% water, and the food they obtain comes from the water surrounding the cell [2, 41, 43, 49, 61–75]. Thus, the bioavailability of a substrate refers to the amount of substrate in the water solution around the cell. One important factor that reduces bioavailability is sorption of a substrate by solid phases (see Chap. 2).

Biodegradation of organic pollutants can be explained in terms of a series of biological degradation steps or a pathway, which ultimately results in the oxidation of the parent compound. Often, the degradation of these compounds generates energy. Complete biodegradation (i.e., mineralization) involves oxidation of the parent compound to form carbon dioxide and water, a process that provides both carbon and energy for growth and reproduction of microbial cells. Figure 3 illustrates the mineralization of any organic compound under aerobic conditions.

The series of degradation steps comprising mineralization is similar, whether the carbon source is a simple sugar (e.g., glucose), a plant polymer (e.g., cellulose), or a pollutant molecule [49, 50, 62–64, 72, 73]. Each degradation step in the pathway is facilitated by a specific catalyst (i.e., an enzyme) made by the degrading cell. Enzymes are found mostly within a cell (i.e., internal enzymes),

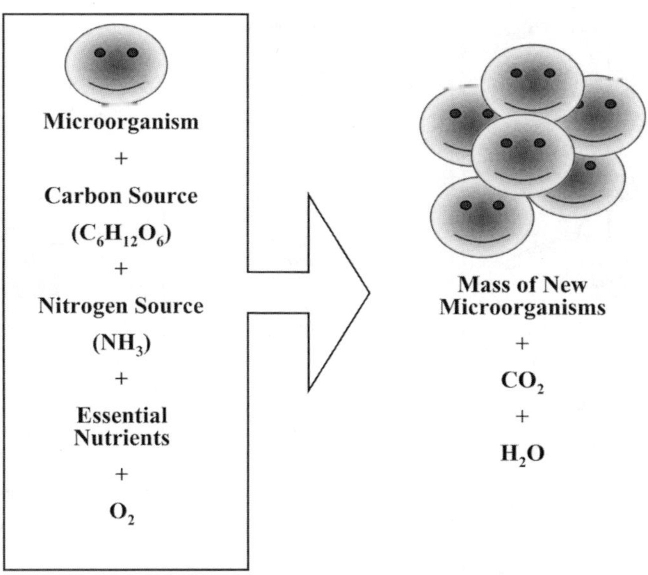

Fig. 3. Aerobic mineralization of an organic compound

but they are also made and released from a cell to help initiate degradation reactions. Enzymes found external to cells (i.e., exoenzymes) are important in the degradation of macromolecules such as the plant polymer cellulose because macromolecules must be broken down into smaller subunits to allow transport into a microbial cell. Both internal enzymes and exoenzymes are essential to the degradation process, where degradation will stop at any step if the appropriate enzyme is not present (Fig. 4). Thus, a different enzyme catalyzes each step of the

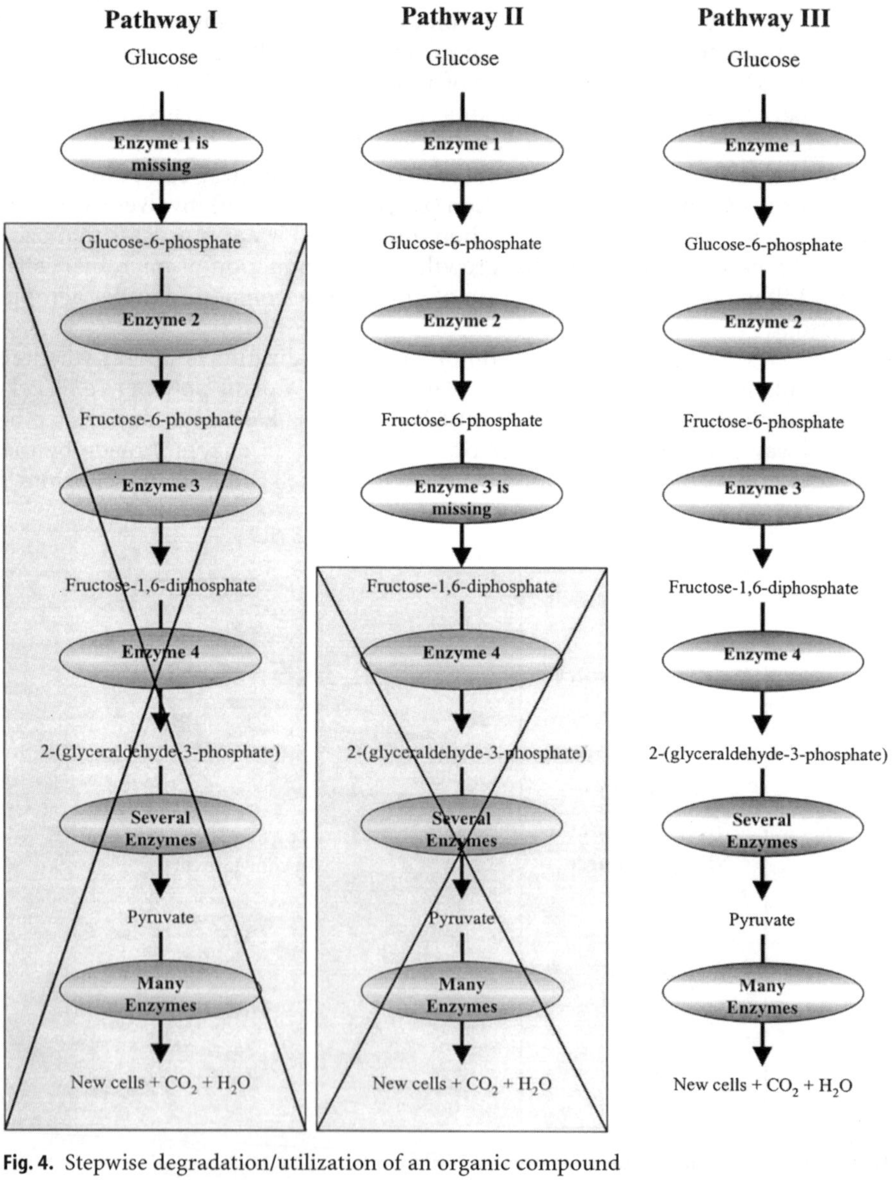

Fig. 4. Stepwise degradation/utilization of an organic compound

biodegradation pathway. If any one enzyme is missing, the product of the reaction it catalyzes is not formed as shown by the shaded boxes in Fig. 4. The reaction stops at that point and no further product is made. Lack of appropriate biodegrading enzymes is one common reason for the persistence of some pollutants, particularly those with unusual chemical structures which the existing enzymes do not recognize. Thus, degradation depends mainly on chemical structure. Pollutants that are structurally similar to natural substrates usually

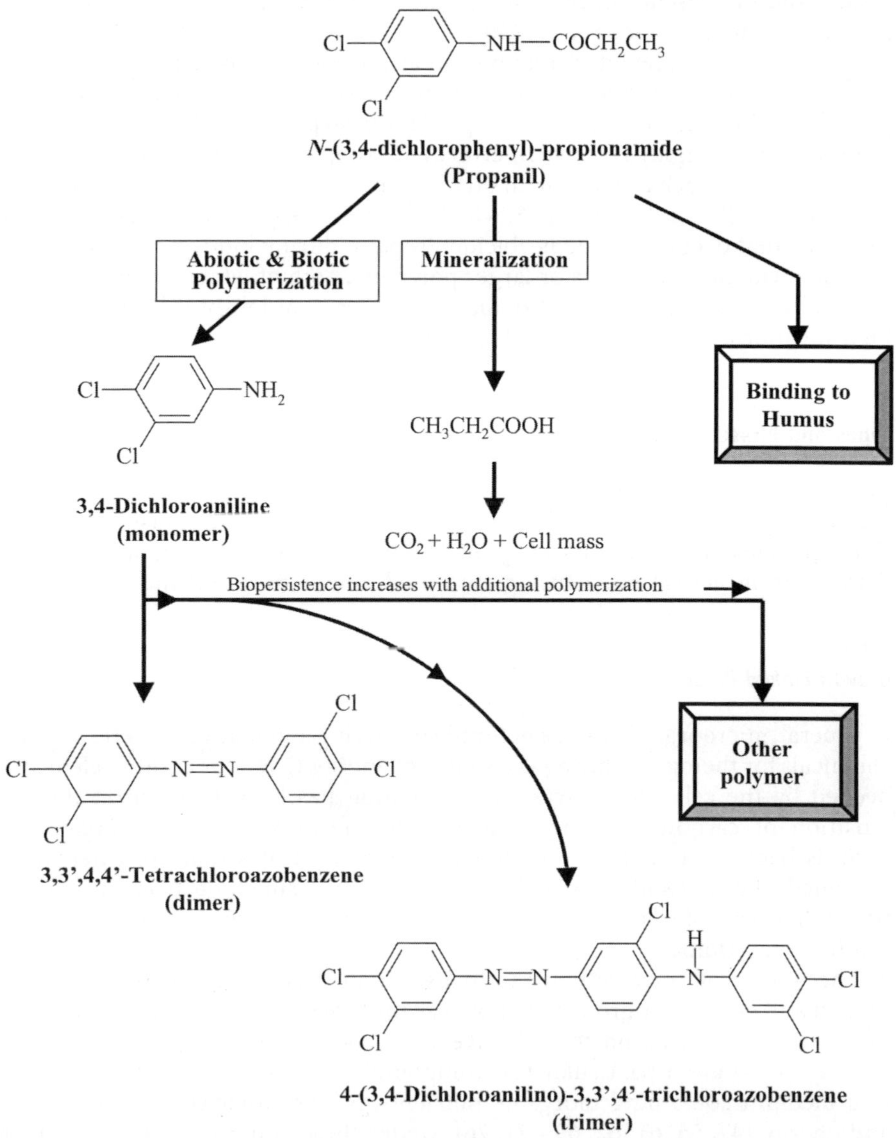

Fig. 5. Polymerization reactions that occur with the herbicide Propanil during biodegradation

degrade easily while pollutants that are dissimilar to natural substrates often degrade slowly, or not at all.

Biodegradation can also be described as a chemical reaction. As Fig. 3 shows, in the presence of oxygen and a nitrogen source (such as ammonia, NH_3), glucose is converted to new cell mass, carbon dioxide, and water. Like glucose, many pollutant molecules (e.g., most gasoline components and many of the herbicides and pesticides) can be biodegraded under the correct conditions. Some organic compounds are only partially degraded. Incomplete degradation can result from the absence of the appropriate degrading enzyme, or it may result from co-metabolism (see Sect. 3.2.7).

Partial or incomplete degradation can also result in polymerization, that is, the synthesis of compounds more complex and stable than the parent compound. This occurs when initial degradation steps, often catalyzed by exoenzymes, create highly reactive intermediate compounds, which can then combine either with each other or with other organic matter present in the environment. This is illustrated in Fig. 5, which shows some possible polymerization reactions that occur with the herbicide Propanil during biodegradation. These include formation of dimers or larger polymers, both of which are quite stable in the environment. Such stability may be the result of low bioavailability (i.e., high sorption), or the absence of degrading enzymes.

3.2
Types and Phases

The following section discusses the different types and phases of microbial degradation of organic pollutants present at aqueous-solid phase interfaces. This includes a discussion of growth-linked biodegradation, acclimation, detoxification, activation, defusing, threshold, and co-metabolism.

3.2.1
Growth-Linked Biodegradation

In general, microorganisms use naturally occurring and many synthetic organic chemicals for their growth, i.e., as a source of energy, C, N, P, or another elements needed by the cells themselves. Most attention has been focused on the acquisition of carbon and energy to sustain the growth of microorganisms such as bacteria and fungi. For the synthetic substrates that are extensively degraded, the molecule is simply another organic substrate from which the population can obtain the needed elements or the energy required for biosynthetic reactions.

The ability of microorganisms to use organic compounds as sources of carbon and energy for growth is known as the "enrichment-culture technique". This method is based on the selective advantage gained by a microorganism that is able to use a particular test compound as a carbon and energy source in a medium containing inorganic nutrients, but no other sources of carbon and energy [47, 55, 61, 62, 69–71, 76]. Under these conditions, a species that is able to grow by utilizing that organic compound will multiply. Few other bacteria and fungi will proliferate in this medium. However, species which use products excreted by the populations acting on the added organic nutrient will also flourish, and thus the final isolation of a microorganism in a pure culture requires plating on an agar medium so that individual colonies can be selected. That agar medium is also made selective by having a single source of carbon and energy. Repeated transfer of the enrichment through solutions that contain the test compound and inorganic nutrients further increases the degree of selectivity before plating, because organic materials and unwanted species from the original environmental sample are diluted by the serial transfers.

The enrichment-culture technique has been the basis for the isolation of pure cultures of bacteria and fungi that are able to use a large number of organic molecules as carbon and energy sources [41, 43, 49]. However, attempts to obtain microorganisms which are able to grow on a variety of other organic compounds have met with failure. Large numbers of bacteria and fungi have been isolated which grow on one or more synthetic compounds. Much of the early literature deals with sugars, amino acids, organic acids, and other cellular or tissue constituents of living microorganisms. However, a variety of pesticides have also been shown to support the growth of one or another bacterium or fungus [2, 46, 47, 50, 56, 77, 78]. Under these conditions, bacteria increase in numbers and fungi increase in biomass in culture media. At the same time, the organic compound disappears, typically at a rate that parallels the increase in cell numbers or biomass. As the concentration of the carbon source declines, the rate of cell or biomass increase diminishes until, when all the substrate is consumed, the population rise ends.

As a rule, biodegradation (i.e., mineralization) of organic compounds is characteristic of growth-linked biodegradation, in which the microorganism converts the substrate to CO_2, cell components, and products typical of the usual catabolic pathways. It is likely, however, that mineralization in nature occasionally may not be linked to growth but instead results from non-proliferating populations. Conversely, some species growing at the expense of a carbon compound may still not mineralize and produce CO_2 from the substrate [40, 43, 50, 51, 79]. However, if O_2 is present, the organic products excreted by one species probably will be converted to CO_2 by another species, so that even if the initial population does not produce CO_2, the second species will. The net effect is still one of mineralization.

An organic pollutant (see Chap. 1) that represents a novel carbon and energy source for a particular microbial population still is transformed by the metabolic pathways that are characteristic of heterotrophic microorganisms. For the microorganism to grow on that pollutant molecule, it must be converted to the intermediates which characterize these major metabolic sequences. If the pollutant cannot be modified enzymatically to yield such intermediates, it will not serve as a carbon and energy source because the energy-yielding and biosynthetic processes cannot function. Thus, the initial phases of biodegradation involve modification of the novel substrate to yield a product that is itself an intermediate or following further metabolism is converted to an intermediate in these ubiquitous metabolic sequences [49, 56, 57, 78]. This need to convert the synthetic molecule to intermediates is characteristic of both aerobes and anaerobes as they derive carbon and energy from the substrate.

It should be stressed, however, that an organic compound need not be a substrate for growth in order to be metabolized by microorganisms. Two categories of transformations exist, either: (1) the biodegradation provides carbon and energy to support growth, and the process therefore is growth-linked, or (2) the biodegradation is not linked to replication. The following examples illustrate the main differences between these categories [43, 47, 49–51, 80–83]:

- The number of microbial cells or the biomass of the species acting on the organic compound of interest increases as degradation proceeds. During a typical growth-linked mineralization brought about by bacteria, the cells use some of the energy and carbon of their organic substrate to make new cells, and this increasingly large population causes a progressively more rapid mineralization. In these instances, the mineralization reflects the population changes.
- During the decomposition of 2-, 3-, or 4-chlorobenzoate or 3,4-dichlorobenzoate ions in sewage, for example, bacteria acting on these compounds multiply, and the increase in cell numbers parallels the destruction of the molecules that serve as their source of carbon.
- Similarly, bacteria capable of metabolizing 4-nitrophenol proliferate in sewage samples as the compound disappears from the water phase.
- Bacteria using 2,4-D similarly increase in numbers as the microbial community in soil destroys this herbicide.
- Pure cultures grow as they utilize synthetic chemicals, for example, during the decomposition of the herbicide IPC by *Arthrobacter sp*.
- The herbicide Endothal is converted to typical constituents of microbial cells as the compound is used as carbon and energy source.

3.2.1.1
Assimilation of Carbon

Many measurements have been made of the % carbon in the organic substrate which is converted into the cells that are carrying out the biodegradation. Such measurements are simple and straightforward in liquid media with water-soluble substrates since the biomass is particulate and thus can be readily distinguished from carbon in solution [47, 56, 84–86]. In contrast, the measurements in soils, wastewater, sewage, or sediments are complicated because: (1) other organic particulate matter is present in addition to the microbial cells and (2) complex water-insoluble products are often formed which must be distinguished from the cell material. In samples of such environments, carbon assimilation is estimated as:

$$C_{\text{assimilated}} = C_{\text{substrate}} - C_{\text{mineralized}} \tag{6}$$

The assimilated carbon is further mineralized, as the cells which metabolized the original substrate are themselves decomposed or consumed by protozoa or other predators. The values from the measurements in pure cultures of microorganisms are often expressed as growth yield, which can be calculated on a weight basis:

$$\text{Growth Yield} = \left(\frac{\text{Weight of Biomass Formed}}{\text{Weight of Substrate Metabolized}} \right) \tag{7}$$

or as a molar growth yield:

$$\text{Molar Growth Yield} = \left(\frac{\text{Weight of Biomass Formed}}{\text{Moles of Substrate Metabolized}} \right) \tag{8}$$

The estimated values for the efficiency of biomass production vary appreciably, for both aerobes and anaerobes. Some species are efficient in capturing the energy in the organic substrate and converting the carbon to cells, but others are notably inefficient.

Under certain conditions in fresh and wastewaters, essentially all the carbon is mineralized, and little or none accumulates in the biomass. This is unexplained because mineralization generates energy, and the metabolic pathways leading to the formation of CO_2 are assumed to involve biochemical sequences that result in carbon assimilation. The following are some examples reported by several workers [40, 47, 57, 87]:

- 93–98% of benzoate ion, benzylamine, aniline, phenol, and 2,4-D added to samples of lake water or sewage at levels below 300 µg/l was converted to CO_2, and direct measurements revealed no carbon assimilation during the mineralization of 24 ng/l to 250 µg/l of benzylamine.
- Only 1.2% of the carbon of 2,4-D added to stream water was converted to organic particulate matter, the solids fraction in water containing the microbial cells. This lack of significant carbon assimilation may be a result of the inability of the microorganisms to obtain carbon and energy for biosynthetic purposes at these low concentrations, the immediate use of the carbon for respiration in order for the cells to maintain their viability (i.e., for maintenance energy), or the rapid decomposition and mineralization of the cells and their constituents.

In contrast, a high percentage of the carbon in other compounds in different environments is incorporated and accumulates in the biomass, even at low substrate concentrations. With some bacteria moreover, the efficiency of incorporation of substrate-C into cells is essentially the same from 43 ng to 100 mg of glucose-C/L [43, 47, 50, 58]. This constancy is especially surprising at substrate concentrations so low that presumably all the carbon is being diverted to respiration by the microorganisms to maintain their viability (maintenance metabolism), although it is possible that bacteria use other organic molecules in their environment for maintenance and not the compound whose biodegradation is being determined [41, 43, 51, 88].

The cells in natural communities which grow on the compound of interest are themselves decomposed or grazed upon by other species and the carbon respired as CO_2 by the predators. Hence, the percentage of substrate-C incorporated into the biomass of natural communities declines and the percentage mineralized increases with time, at least in the presence of O_2. The values initially reflect the populations acting on the organic compound but with time reflect the activities of the community of microorganisms [47, 49, 50]. Thus, patterns of mineralization have a characteristic initial phase which to a significant degree represents the species acting on the parent molecule. Thereafter, a slower phase of mineralization is evident as the original cells, as well as their excretions, are destroyed and converted to CO_2 and other products. In soil and sediment environments, a small or a large part of the substrate-C is also converted to high-molecular-weight complexes which are resistant to rapid biodegradation. Such humic substances (see Chap. 2) may contain much of the carbon originally added to that environment, and this organic matter is only very slowly converted to CO_2 [41, 43, 51, 89].

3.2.1.2
Assimilation of Other Elements

Synthetic organic molecules may be used as sources of required elements other than carbon. Microorganisms need N, P, and S, and hence these nutrient requirements may be satisfied as the responsible species degrade the compound of interest. It is common for the element in the organic compound to be converted to the inorganic form before it becomes utilized for cell components. The following are some examples reported by several authors [43, 47, 49–51, 90–92]:

- *Klebsiella pneumoniae* uses Bromoxynil as a nitrogen source, but it does so only after converting the nitrite to NH_3, which is then assimilated.
- A strain of *Pseudomonas sp.* uses 2,6-dinitrophenol as a nitrogen source for growth by first cleaving the nitro groups to free nitrite which, presumably after reduction to NH_3, sustains replication of the bacteria.
- Bacteria are also able to use a large number of organophosphorus insecticides, alkyl phosphates, phosphonates, and the herbicide Glyphosate as phosphorus sources.
- Sulfur may also be extracted from organic molecules and then support replication, as indicated by utilization of *O,O*-diethylphosphorothioate and *O,O*-diethylphosphoro-dithioate as sulfur sources by *Pseudomonas acidovorans*.

For heterotrophic microorganisms in most natural ecosystems, the limiting element is generally C, and usually sufficient N, P, and S are present to satisfy the microbial demand. Because carbon is limiting and because it is the element for which there is intense competition, a species with the unique ability to grow on synthetic molecules has a selective advantage. No such selective advantage exists for a microorganism using an organic compound as the source of an element that is already available in abundant supply. Hence, it is unlikely that microorganisms obtaining other nutrient elements from synthetic molecules are selectively enhanced in such environments. Nevertheless, as the microorganisms use the molecules as carbon or energy sources, the biodegradative process will usually still lead to the mineralization of the other elements in the compound.

3.2.2
Acclimation

Prior to the degradation of many organic compounds, a period is noted in which no destruction of the compound is evident. This time interval is designated as an *acclimation* period or, sometimes, an *adaptation* or *lag* period [93–98]. It may be defined as *the length of time between the addition or entry of the compound into an environment and evidence of its detectable loss.* During this interval, no change in concentration is noted, but then the disappearance becomes evident and the rate of destruction often becomes rapid.

This acclimation phase may be of considerable public health or ecological significance because the compound is not destroyed. Hence, the period of exposure of humans, animals, or plants is prolonged, and the possibility of an undesirable effect increased. Furthermore, if the pollutant is present in flowing waters above or below ground, it may be widely disseminated laterally or vertically because of the lag of detectable biodegradation. In the case of toxicants, such increased dispersal may result in the exposure of susceptible species at distant sites before the harmful substance is destroyed.

Acclimation periods have been reported for many compounds that are introduced into soil, fresh water, sediment, and sewage. The following compounds exhibit an acclimation period, either aerobically or anaerobically [12–15, 21, 32, 57, 75, 99–108]:

- Herbicides: 2,4-D, MCPA, Mecoprop, 4-(2,4-DB), TCA, Amitrole, Dalapon, Monuron, Chlorpropham, Endothal, Pyrazon, and DNOC.
- Insecticides: methyl parathion and Azinphosmethyl.
- Quaternary ammonium compounds: dodecyltrimethylammonium chloride.
- Polycyclic aromatic hydrocarbons: naphthalene and anthracene.
- Others: phenol, 4-chlorophenol, 4-nitrophenol, chlorobenzene, 1,2- and 1,4-dichlorobenzene, 3,5-dichlorobenzoic acid, PCP, diphenylmethane, and NTA.

The length of the acclimation period varies enormously, from less than 1 h to many months. The duration varies among chemicals and environments, and it also depends on the concentration of the compound and a number of environmental conditions. The time period can be especially long in anaerobic environments for some compounds, such as chlorinated molecules [14, 15, 21, 57, 109–111].

The acclimation phase is considered to end at the onset of detectable biodegradation. After acclimation, the rate of metabolism of the compound may be slow or rapid, but if a second addition of the chemical is made during this time of active metabolism, the loss of the second increment characteristically occurs with little or no acclimation. The disappearance or marked reduction in the acclimation period has been noted in solid particles amended with 2,4-D, DNOC, Amitrole, Methomyl, 4-(2,4-DB), river water supplemented with 4-nitrophenol, and marine waters containing 4-chlorophenol [62, 112–117]. It is generally assumed that biodegradation is detected immediately following the second introduction of the compound because the microorganisms responsible for transformation enumerated as they utilized the organic compound following its first introduction.

The rate of biodegradation of the second addition may be the same as the final rate evident during the active phase of breakdown of the first addition [106, 107, 118–120]. However, it is far more common to have a greater rate of biodegradation, which is usually measured as the loss of parent compound or the formation of $^{14}CO_2$ from labeled compound, following the second rather than the first application. The rate is further enhanced with still more additions. This enhancement of rate upon repeated additions of chemical substrate has been reported frequently for several pesticides and surfactants [14, 15, 99, 100, 113, 123–126] as follows:

- The rate of Parathion loss and its conversion to CO_2 rises as soil receives additional monthly treatments with the insecticide.
- The degradation of Iprodione and Vinclozolin similarly becomes more rapid as a result of prior additions of these fungicides to soil.
- In soil with applied EPTC or Butylate, the rate of mineralization increases as a result of prior treatments with these herbicides.
- Greater rates of disappearance of the nematocides Enthoprop and Diphenamid are evident following the second than after the first introduction into soil.
- Dodecyltrimethylammonium chloride is rapidly mineralized in fresh water after an acclimation period, and the rate is faster following the second rather than the first addition of the quaternary ammonium compound.

Once the indigenous community of microorganisms has become acclimated to the degradation of a chemical at an interface and the activity becomes marked, the community may retain its active state for some time. Too little information is presently available to permit generalizations to be made among compounds regarding the duration of the beneficial influence of prior additions of the compound. It is presently not clear why a microbial community which has acclimated to a particular substrate loses that activity. This could be a result of the decline in numbers or biomass of the responsible microorganisms or a loss of the metabolic activity in the absence of the specific compound.

3.2.2.1
Factors Affecting Acclimation

Acclimation of a microbial community to one substrate frequently results in the simultaneous acclimation to some, but not all, structurally related molecules. Because individual species often act on several structurally similar substrates, the species favored by the first addition may then quickly destroy the analogues [100, 107, 108, 110, 111, 127].

The duration of acclimation is affected by several environmental factors, such as temperature, pH, aeration status, and nutrients. The concentration of the compound that is being metabolized greatly affects the length of time before a decline in its concentration is detectable. The rate of biodegradation of trace compounds increases with concentration, but because compound loss is usually determined and not CO_2 or product formation, the low precision of analysis leads to data indicating a longer acclimation at higher concentration [104, 106, 113, 128].

There appear to be concentration thresholds for some compounds below which no acclimation occurs. A typical case is 4-nitrophenol, which is destroyed at concentrations above but not below 10 mg/l in samples containing sediments and natural water [113, 121, 122]. On the other hand, microorganisms in fresh or marine waters may acclimate to destroy compounds at levels below which they can use single compounds as sole carbon sources for growth (i. e., below the threshold).

3.2.2.2
Explanations

Many explanations have been proposed for the acclimation of microbial communities to the biodegradation of organic compounds, especially at aqueous-solid phase interfaces. Many of these were proposed based on early studies of pure cultures of bacteria growing in media containing single organic substrates, often at cell densities far higher than is common for individual species of bacteria in nature. Some were based on investigations of the biochemistry or genetics of individual species acting in pure culture on very high concentrations of sugars, amino acids, or other natural products that can readily be metabolized by a diverse array of microbial species. Few of the explanations, however, were derived from studies of natural microbial communities acting on synthetic compounds at environmentally relevant concentrations, and hence the original emphasis placed on some of these hypotheses must be considered with skepticism.

On the other hand, more recent studies have been designed to evaluate these hypotheses as they relate to: (1) natural communities as contrasted to pure cultures, (2) cell densities more characteristic of natural ecosystems than those bacterial densities commonly used in tests of pure cultures, (3) synthetic compounds acted on by only a few rather than a diversity of microbial genera or species, and (4) compound concentrations which are characteristic of environmental pollutants rather than organic nutrients included in culture media. In general, all these explanations are related mainly to: (1) proliferation of small populations, (2) presence of toxins, (3) predation by protozoa, and (4) appearance of new genotypes [101, 104, 106–108, 110, 111].

3.2.3
Detoxification

The most important role of microorganisms in the transformation of pollutants at aqueous-solid phase interfaces is their ability to bring about *detoxification* (i. e., the change in a molecule that renders it less harmful to one or more susceptible species). Detoxification results in inactivation, with the toxicologically active substance being converted to an inactive product. Because toxicological activity is associated with many chemical entities, substituents, and modes of action, detoxifications similarly include a large array of different types of reactions. A simple way of demonstrating detoxification is to measure the effect of environmental samples on the behavior, growth, or viability of susceptible species.

Bioassays are especially useful inasmuch as they reflect the loss of biological activity of a molecule, but they are often replaced by chemical analysis showing the loss of the parent compound or the formation of products [128–135].

Detoxification is advantageous to the microorganisms carrying out transformations at interfaces if the concentration of the chemical is in a range which suppresses these species. Several processes may result in detoxification, such as hydrolysis, hydroxylation, dehalogenation, dealkylation, methylation, nitro reduction, deamination, ether cleavage, nitrile conversion to an amide, and conjugation. The following is a brief summary of these processes [108, 136–168].

3.2.3.1 Hydrolysis

Microorganisms can inactive toxicants by cleavage of a bond by the addition of water. Such reactions may involve a simple hydrolysis of an ester bond, as with the insecticide Malathion by a carboxyesterase enzyme:

$$[\text{R-COO-R}'] + [\text{H}_2\text{O}] \rightarrow [\text{R-COOH}] + [\text{R}'\text{-OH}] \tag{9}$$

3.2.3.2 Hydroxylation

The addition of an OH to an aromatic or aliphatic molecule often makes it less harmful. Thus, simple replacement of H by OH inactivates the herbicide 2,4-D, as follows:

$$[\text{R-NH-CO-CH}_2\text{O-R}'] + [\text{H}_2\text{O}] \rightarrow [\text{R-NH}_2] + [\text{HOOC-CH}_2\text{-O-R}'] \tag{10}$$

The hydroxylation of the ring moiety of 2,4-D similarly converts the parent herbicide to a non-toxic product. Microorganisms may bring about such a detoxification when they hydroxylate the ring in the 4-position, a process that leads to a migration of the chlorine to give 2,5-dichloro-4-hydroxyphenoxyacetic acid.

3.2.3.3 Dehalogenation

Many pesticides contain chlorine or other halogens, and removal of the halogen often converts the toxicant to an innocuous product. The enzymes are designated dehalogenases. These dehalogenations may involve replacement of the halogen by H (i.e., reductive dehalogenation, reaction 11), by OH (i.e., hydrolytic dehalogenation, reaction 12) or it may result in removal of the halogen and an adjacent H (i.e., dehydrodehalogenation, reaction 13):

$$[\text{R-Cl}] \rightarrow [\text{RH}] \tag{11}$$

$$[\text{R-Cl}] \rightarrow [\text{R-OH}] \tag{12}$$

$$[\text{R-CH}_2\text{-CHCl-R}'] \rightarrow [\text{R-CH=CH-R}'] \tag{13}$$

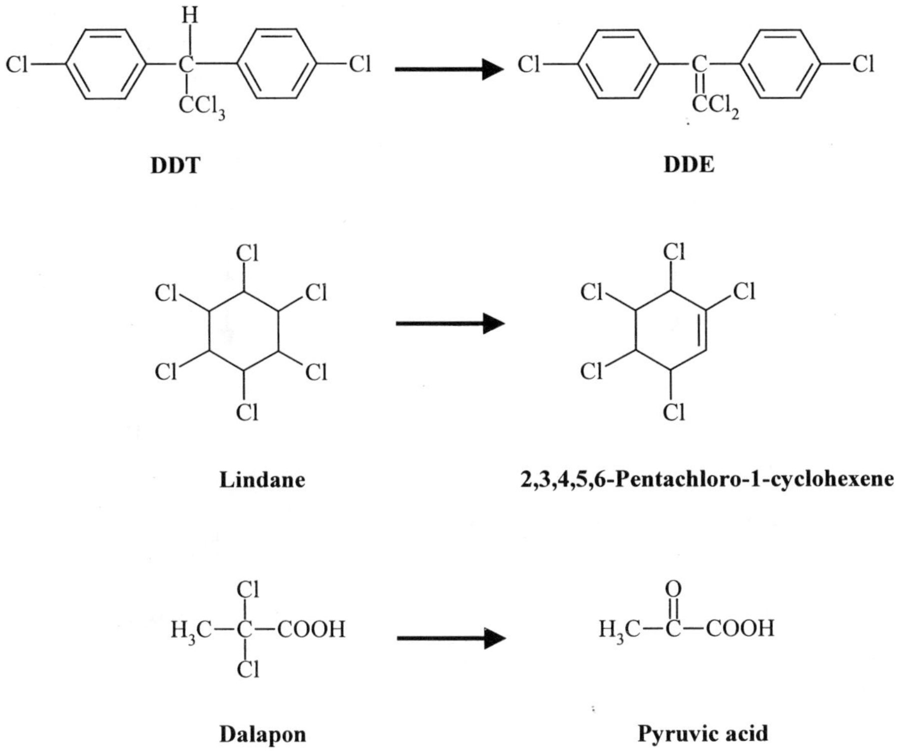

Fig. 6. Detoxification of DDT, Lindane, and Dalapon by dehalogenation

Some examples of dehalogenation of pesticides are shown in Fig. 6, indicating the microbial conversion of DDT, Lindane, and Dalapon to non-toxic products such as DDE, 2,3,4,5,6-penta-chloro-1-cyclohexene, and pyruvic acid, respectively.

3.2.3.4
Dealkylation

Pesticides containing methyl or other alkyl substituents may be linked to N or O (i.e., *N*- or *O*-alkyl substitution). An *N*- or *O*-dealkylation catalyzed by microorganisms frequently results in loss of the pesticide activity. Phenylurea (see Chap. 1) becomes less active when microorganisms *N*-demethylate the molecules (e.g., the conversion of Diuron to the normethyl derivative, Fig. 7). The subsequent removal of the second *N*-methyl group renders the molecule fully non-toxic [169]. On the other hand, the microbial *O*-demethylation of Chloroneb creates the non-toxic product 2,5-dichloro-4-methoxyphenol (Fig. 7).

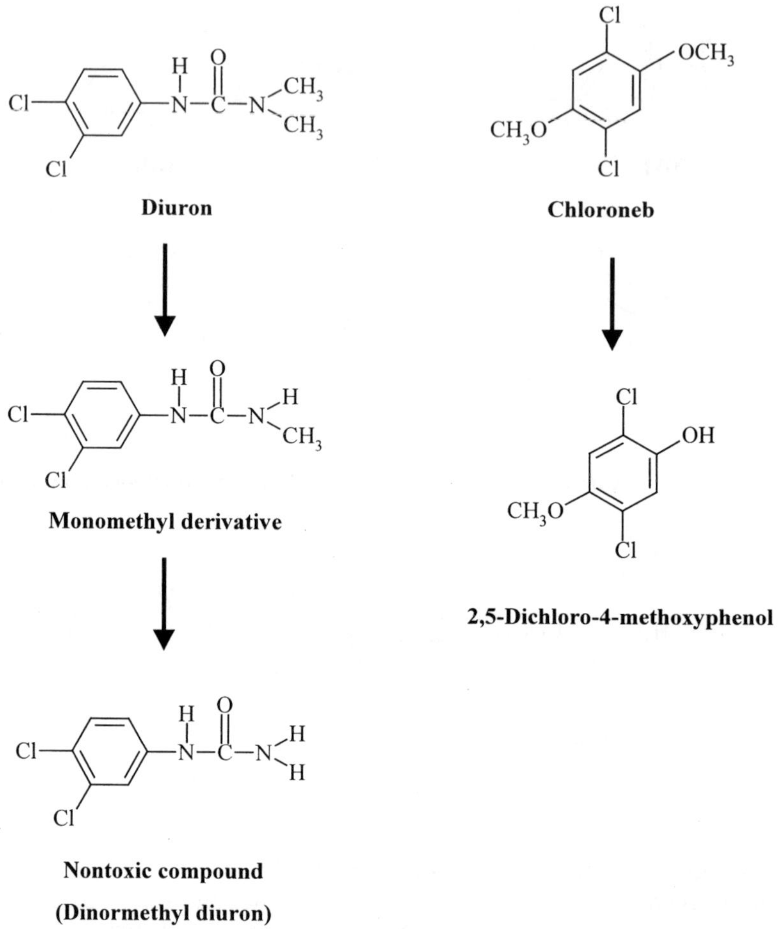

Fig. 7. Detoxification of Diuron and Chloroneb by dealkylation

3.2.3.5
Methylation

The addition of a methyl group may inactivate toxic phenols. Thus, penta- and tetrachlorophenols, which are fungicides with the former in especially wide use, can be detoxified microbiologically by addition of a methyl group in a reaction representing an *O*-methylation:

$$[\text{R-OH}] \rightarrow [\text{R-O-CH}_3] \quad (14)$$

3.2.3.6
Nitro Reduction

Nitro compounds are harmful to many types of microorganisms. They may be rendered less toxic by reduction of the nitro to an amino group:

$$[R\text{-}NO_2] \rightarrow [R\text{-}NH_2] \quad (15)$$

Such reductions may result in loss or diminution of the harmful effects as microorganisms convert the broad-spectrum poison 2,4-dinitrophenol to 2-amino-4- and 4-amino-2-nitrophenol, the fungicide pentachloronitrobenzene to pentachloroaniline, and the insecticide Parathion to aminoparathion.

3.2.3.7
Deamination

The herbicide known as Metamitron (Fig. 8) can be transformed microbiologically to yield a deaminated product which is non-toxic [169].

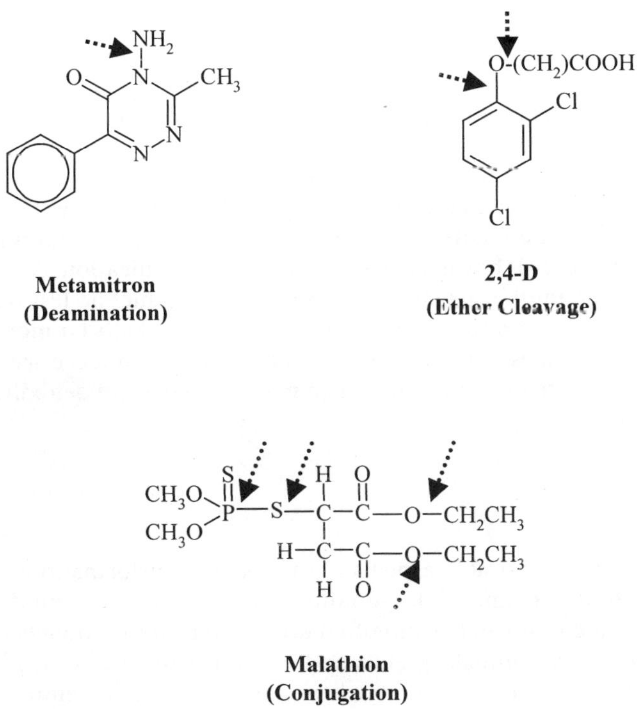

Fig. 8. Initial detoxification of Metamitron, 2,4-D, and Malathion by deamination, ether cleavage, and conjugation, respectively

3.2.3.8
Ether Cleavage

Phenoxy herbicides (see Chap. 1) contain ether linkages (-C-O-C-), and the cleavage of these linkages destroys the phytotoxicity of the molecule. This is illustrated by the cleavage of the ether bond in 2,4-D (Fig. 8). This microbial conversion is somewhat surprising because of the bond energy between carbon and O, which is 85.5 kcal/mole [170], and thus the need of the microorganism to provide the energy to cleave the bond.

3.2.3.9
Conversion of Nitrile to Amide

A herbicide such as 2,6-dichlorobenzonitrile (Dichlobenil), is converted to 2,6-dichlorobenzamide, the molecule is rendered inactive in solid phases.

$$[\text{R-C}\equiv\text{N}] \rightarrow [\text{R-CH}_2\text{-NH}_2] \qquad (16)$$

3.2.3.10
Conjugation

Conjugation involves a reaction between a common intermediate in some natural metabolic pathway with a synthetic molecule. Products of the combination of a normal metabolite with a toxicant frequently are harmless. Malathion conjugation is shown in Fig. 8.

In general, a particular microorganism or a microbial community may detoxify a single toxicant in multiple ways/pathways. Such pathways are initiated by entirely different enzymes. The previously mentioned reaction types given here (reactions 9–16), however, are not always detoxification. A contaminant altered by one or another mechanism may yield a product no less toxic than its precursor. Indeed, several such reactions may yield products far more toxic than the original substrates. Furthermore, a reaction or a sequence which yields a product non-toxic to one microorganism may not represent detoxification for a second species.

3.2.4
Activation

One of the most undesirable aspects of microbial transformations in nature is the formation of toxicants. A large number of organic compounds which are themselves innocuous can be, and often are, converted to products that may be harmful to humans, animals, plants, and microorganisms. By such means, the environment may create a pollutant where none was present before. The process of forming toxic products from innocuous precursors is known as *activation* [171–177].

Activation is a major reason for studying the pathways and products from the breakdown of organic molecules in natural ecosystems and waste disposal systems which lead to environmental discharges [9, 23, 93–97, 127, 146, 147, 149,

150, 161, 164, 165, 178–193]. Activation occurs at aqueous-solid phase interfaces where microorganisms are active and the products thus created may have a short residence time or persist for long periods (Fig. 9). The harmful product may be an intermediate in mineralization, yet it may persist long enough to create a pollution problem. Moreover, the mobility of the activation product is sometimes different from that of its precursor, so that the product may be transported to distant sites to a greater or to a smaller extent than the contaminant molecule from which it was formed [43, 73, 194, 195].

Many different pathways, mechanisms, and enzymes are associated with activation. These include dehalogenation, *N*-nitrosation of secondary amines, epoxidation, conversion of phosphothionates to phosphate, metabolism of phenoxyalkanoic acids, oxidation of thioethers, hydrolysis of esters and peroxides. The following is a summary.

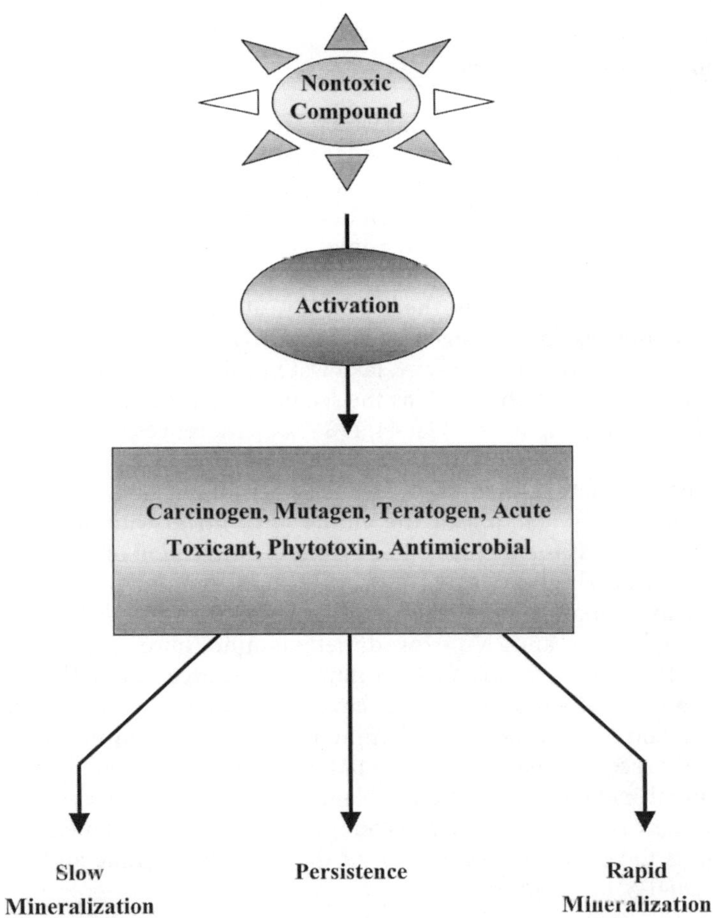

Fig. 9. Processes associated with an activation process

3.2.4.1
Dehalogenation

Significant activation occurs during the microbial metabolism of trichloroethylene (TCE). This compound was once widely used and now represents a major contaminant of many aquifers. Because TCE is metabolized by many bacteria, its elimination by bioremediation is being actively pursued. However, a major product frequently encountered is vinyl chloride, a potent carcinogen:

$$[Cl_2\text{-}C{=}CH\text{-}Cl] \rightarrow [Cl\text{-}HC{=}CH_2] \quad (17)$$

The same carcinogen can also be formed during the anaerobic metabolism of 1,1- and *trans*-1,2-dichlorethylene [196]. TCE can also be converted in cultures of methanotrophs to 2,2,2-trichloroacetaldehyde [197]:

$$[Cl_2\text{-}C{=}CH\text{-}Cl] \rightarrow [Cl_3\text{-}C\text{-}CHO] \quad (18)$$

3.2.4.2
N-Nitrosation of Secondary Amines

Many activations involve compounds which are used as pesticides. In the case of *N*-nitrosation, the precursors are secondary amines and nitrate. The former are common synthetic compounds and the latter is an anion found in nearly all solid and aqueous phases. The *N*-nitrosation of a secondary amine [R-NH-R′] occurs in the presence of nitrite formed microbiologically from nitrate. The product is an *N*-nitroso compound (i.e., a nitrosamine [RR′-N-N=O]). The reason for concern with nitrosamines is their potency, at low concentrations, as carcinogens, teratogens, and mutagens.

Nitrosations can occur in sewage, lake water, soil, and wastewater by microbial activity from precursors such as the secondary amines dimethylamine and diethanolamine, among others [23, 44, 198–200]. Moreover, microbial enzymes can also *N*-nitrosate several amines in the presence of nitrite [201]. Nevertheless, the actual nitrosation step may be nonenzymatic and may result from a spontaneous reaction of the amine and nitrite with some metabolic product or cell constituents [200]. The extent of conversion of the amine to the nitrosamine is nearly always small at the pH values common in nature, although the yield can be high in artificially acidified solutions [98, 202]. However, *N*-nitrosodiethylamine and *N*-nitrosodimethylamine have been reported in aqueous solid phase systems such as municipal sludge [203]. These two carcinogens and *N*-nitrosomorpholine have been found in sewage-treatment operations, and *N*-nitrosodiethanolamine has been detected in the outlet of a cutting-fluid recovery plant [204]. The latter nitrosamine probably is a result of the microbial metabolism of diethanolamine [199], which is a common constituent of cutting fluids and many other products. Hence, the microbial role in activation is the enzymatic formation of the secondary amine and nitrite and not the actual *N*-nitrosation.

3.2.4.3
Epoxidation

Microorganisms are able to form epoxides from several compounds with double bonds:

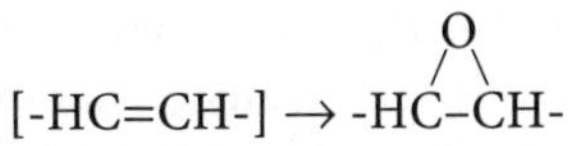

(19)

In the case of insecticides, this oxidation converts the precursor to a product which is more toxic (e.g., the conversion of Heptachlor and Aldrin to epoxides).

3.2.4.4
Conversion of Phosphothionates to Phosphate

Phosphothionate molecules, a group of insecticides, have little toxicity, but when they are converted to the corresponding phosphates they become potent insecticides which are highly toxic to humans and other mammals (Fig. 10).

A typical example is the conversion of Parathion to its oxygen analog (i.e., Paraoxon) in soil and microbial cultures Fig. 11).

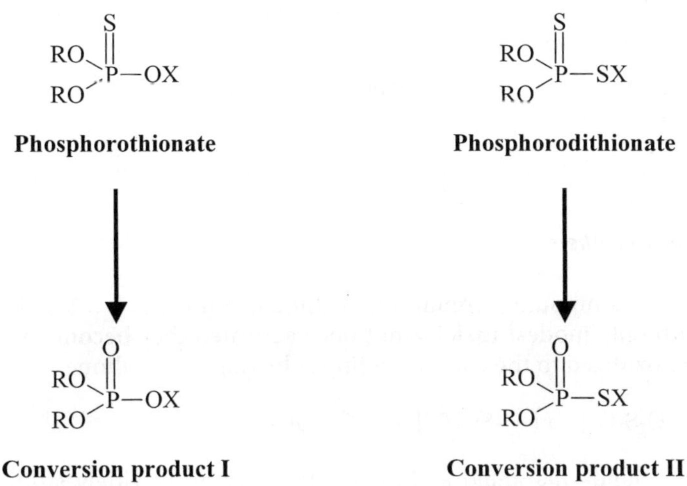

Fig. 10. Phosphothionate molecules and their conversion products

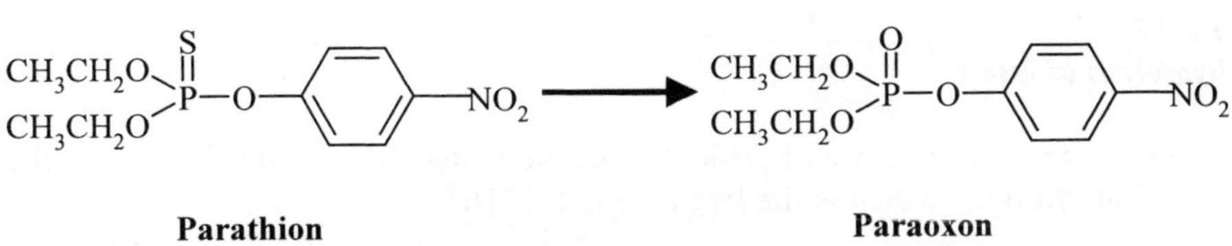

Fig. 11. Conversion of Parathion to its oxygen analog

3.2.4.5
Metabolism of Phenoxyalkanoic Acids

The herbicide 2,4-D is itself a potent phytotoxin. However, a number of structurally related but inactive compounds may be converted by plants to 2,4-D following the activation process and thus act as herbicides. These phenoxyalkanoic acids are ω-(2,4-dichlorophenoxy) alkanoic acids. The transformation may be viewed as shown in Fig. 12 as 6-(2,4-dichlorophenoxy)hexanoic acid as the parent compound. The sequence is called β-oxidation because the steps in which two carbons are removed initially involve the oxidation of the β-carbon to the aliphatic acid moiety.

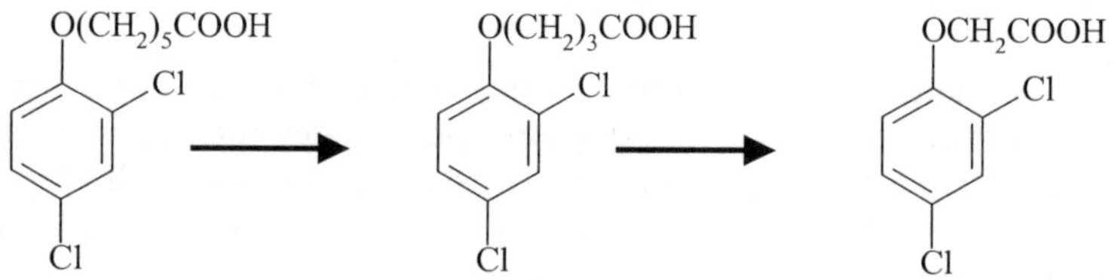

6-(2,4-Dichlorophenoxy)-hexanoic acid **4-(2,4-DB)** **2,4-D**

Fig. 12. Transformation of 6-(2,4-dichlorophenoxy)hexanoic acid to 4-(2,4-DB) and finally to the actual phytotoxin (2,4-D)

3.2.4.6
Oxidation of Thioethers

A number of compounds containing a thioether linkage [-C-S-C-] are insecticides with only modest toxicity, but once activated they become more potent as they are oxidized to the corresponding sulfoxides and sulfones:

$$[\text{-C-S-C-}] \rightarrow [\text{-C-SO-C-}] \rightarrow [\text{-C-SOO-C}] \qquad (20)$$

Three compounds marketed as insecticides have been widely studied in this regard, namely Aldicarb, Phorate, and Disulfoton [93, 164, 190].

3.2.4.7
Hydrolysis of Esters

Several esters marketed as herbicides are activated by hydrolysis to give the actual phytotoxin, which is the free acid [94, 127]:

$$[\text{R-COO-R}'] \rightarrow [\text{R-COOH}] \qquad (21)$$

This reaction occurs in soils amended with Flamprop-methyl [205], Benzoylprop-ethyl, and Diclofop-methyl [206]. As the names of these pesticides indicate, R′ is CH_3 or CH_2CH_3, respectively. The second product of the conversion is presumably the non-toxic alcohol [R′-OH].

3.2.4.8
Peroxidase

Chlorinated dibenzo-*p*-dioxins and dibenzofurans are among the most toxic substances known, especially 2,3,7,8-tetrachloro-*p*-dibenzodioxin (TCDD). These extremely hazardous compounds can be produced from 3,4,5- and 2,4,5-trichlorophenols by peroxidases [207]. However, the biological formation of such toxicants in nature or by microorganisms has not been described.

Many chlorophenols are harmful and persistent. It is possible that these may be produced microbiologically in nature in view of the finding that a fungal chloroperoxidase halogenates phenol to yield monochlorophenols and the latter to give dichlorophenols. The sequence continues with producing trichlorophenols, tetrachlorophenols, and even pentachlorophenol [208].

Fungal peroxidases may also dimerize 3,4-dichloroaniline to 3,4,3′,4′-tetrachloro-azobenzene, a compound similar in toxicity to TCDD [209, 210].

3.2.5
Defusing

A compound that is potentially activated may pose a health or environmental hazard if it undergoes that type of reaction. However, if the microorganisms convert that substrate to a different metabolite which is both harmless and not subject to activation, the potential problem posed by the initial substrate does not arise [185, 186]. Thus, compound A is converted to carbon rather than to compound B (Fig. 13).

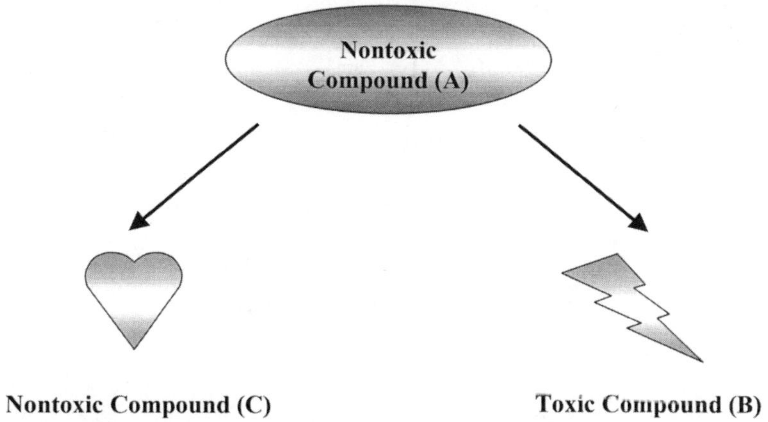

Fig. 13. Mechanism of defusing

Defusing is best illustrated by an example that undergoes activation. Among the phenoxy herbicides, 4-(2,4-DB) is activated when it undergoes oxidation to yield 2,4-D. Hence, bacteria which cleave the molecule in culture by removing butyric acid from the side chain to release 2,4-dichlorophenol are defusing the molecule (Fig. 14). Defusing has also been reported for a number of other insecticides that are activated by the conversion of [-P=S] to [-P=O] as, for example, when Parathion or Malathion are cleaved in part or completely in bacterial cultures or Dimethoate is cleaved in soil before the compound is activated [96, 101, 111, 184, 211–219].

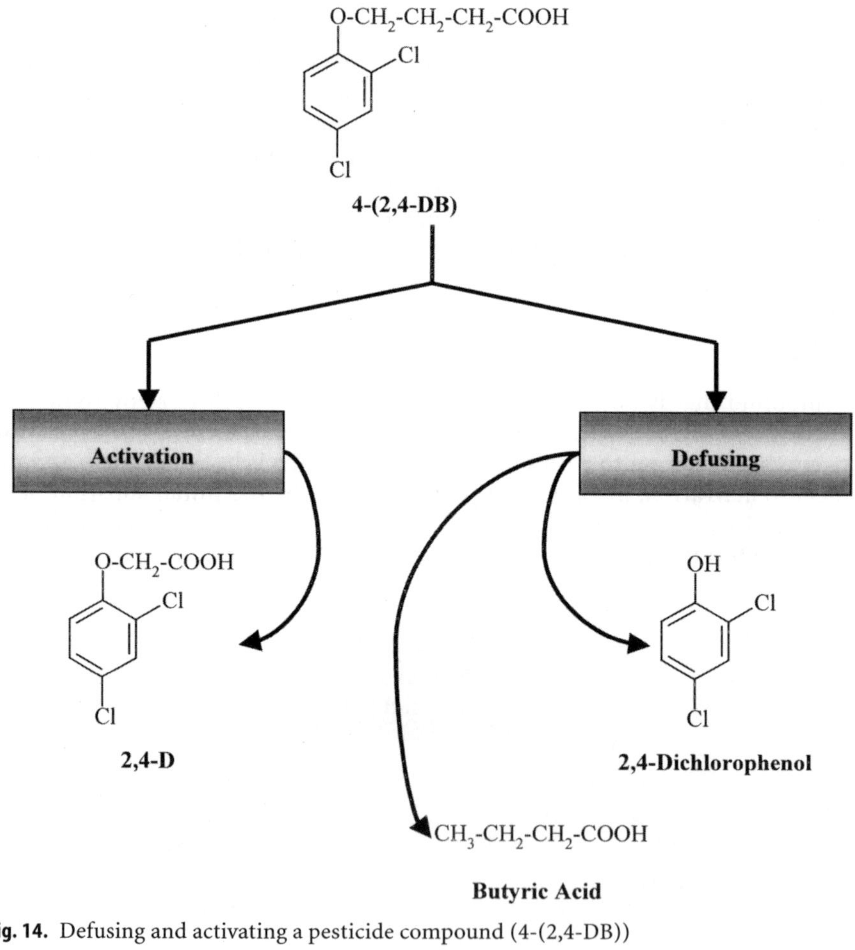

Fig. 14. Defusing and activating a pesticide compound (4-(2,4-DB))

3.2.6
Threshold

Many organic pollutants at aqueous-solid phase interfaces are present at low concentrations. Even at these trace levels, they may be of concern because of the following:

- Risk analyses suggest that many of the chronic toxicants can be injurious to a small portion of the human population consuming waters or foods containing them. Chronic toxicants include a diversity of carcinogens, mutagens, and teratogens.
- Some of the compounds at these low concentrations (e.g., microgram per liter levels) are acutely toxic to aquatic microorganisms.
- Several low level pollutants are subject to bioconcentration within tissues of microorganisms in natural food chains and ultimately reach levels that are injurious to species at higher trophic levels in these food chains.
- Regulatory agencies of national or local governments have established concentration levels for many organic compounds which are deemed to be safe, especially for public health, and the concentrations given by these regulatory guidelines or standards are often quite low.

The public health and ecological concerns with low chemical concentrations have fostered interest in the biodegradative processes affecting trace concentrations of organic compounds. In the past, microbiologists have not paid attention to the problem because it was deemed far easier to grow microorganisms at high substrate concentrations which would yield large cell numbers. However, as interest grew, previously unanticipated phenomena became apparent. One such phenomenon is the existence of a threshold, i.e., a concentration of a nutrient source below which microorganisms cannot grow.

To maintain its viability, every microorganism must expend energy. The amount of energy to permit the microorganism to remain alive is designated "*maintenance energy*" [220, 221]. For heterotrophs, this energy is derived from the oxidation of organic compounds. When the concentration of the carbon source for growth is high, diffusion of the substrate from solution to the cell surface and the subsequent transfer of the molecule across the surface into the cell provide enough of the substrate to satisfy the needs for maintenance energy and for processes that lead to increases in cell size, growth, and multiplication. The same is not the case at low substrate levels. Considering only diffusion of the molecule from the liquid to the cell surface, as a low substrate concentration is reduced to a still lower level, the energy for maintenance represents an ever-higher percentage of substrate-C which reaches the microorganism by diffusion, and an ever-smaller percentage is used for growth and replication. At some lower value, all the energy in the form of carbon which reaches and/or enters the cell is used simply to keep the cell alive, and none is used for growth. At this concentration, although the substrate is being metabolized, the cells are not growing and the population size and biomass are not increasing. This concentration represents the threshold [222, 223].

Moreover, if the population size initially is small so that biodegradation is inconsequential and/or undetectable, then no replication is reflected by the absence of significant or detectable biodegradation, even though the microorganisms are metabolizing part of the substrate pool to maintain themselves. The threshold is the lowest concentration that sustains growth. It represents the level below which a species that needs to proliferate to cause a detectable change brings about little or no chemical destruction.

3.2.6.1
Explanations

The possible existence of a threshold was first postulated because of the presence of relatively constant levels of dissolved organic carbon (DOC) in the oceans. This C, presumably because of its low concentration, was not available to support microbial proliferation and hence mineralization of the carbon [224]. The level of such DOC is approximately 1 mg/l in marine waters and is commonly less than 5 mg/l in oligotrophic fresh waters. Moreover, if significant decomposition of this organic matter were occurring, the concentration should fall at increasing distances away from the water's surface, where the organic matter is being generated photosynthetically by the phytoplankton [94, 225–230]. Because no such marked decline is evident with depth, it was hypothesized that biodegradation must be slow. However, this line of evidence in support of the existence of a threshold for growth is weak because: (1) much of the organic matter, when concentrated, is intrinsically resistant to microbial degradation, and (2) the concentration of some aquatic constituents may represent a steady state, that is, a balance between continuous formation and continuous mineralization.

More convincing evidence has come from studies of biodegradable synthetic compounds in waters and soils. Because these compounds are not formed biologically, their presence at reasonably constant levels or their persistence at low levels indicates that the biodegradation one might expect is not occurring. These studies indicate that no biodegradation occurs in the test period below a certain concentration or the rate is less than what might be expected from the rates observed at higher levels.

Analogous observations have been made when wastewaters are passed through solid particles as a means of destroying a harmful chemical by microbial action. In experimental trials, the concentrations of many compounds decreased to undetectable levels as solutions containing them passed through soil columns. However, a minor percentage of the 1,2-, 1,3-, and 1,4-dichlorobenzenes and diisobutyl phthalate in the influent water was still present in the effluent, and a readily biodegradable molecule like di-(2-ethylhexyl) phthalate at 70 ng/l did not disappear at all as a result of passage through soil [231]. Benzophenone and diethyl and dibutyl phthalate have also been reported to persist when passed at low concentrations through soil columns set up to simulate the rapid infiltration of contaminated waters through soil [90, 127, 232–238].

Investigations of pure cultures of bacteria clearly show the existence of a threshold concentration for the carbon source below which replication does not occur. This value is about 18 μg/l for *Escherichia coli* and *Pseudomonas sp.* growing on glucose, 180 μg/l for *Aeromonas hydrophila* growing on starch,

210 µg/l for a *Coryne bacterium* using glucose, approximately 300 µg/l for a strain of *Pseudomonas* growing at the expense of 2,4-dichlorophenol, about 5 µg/l for *Salmonella typhimurium* provided with glucose, and 2 µg/l for a bacterium mineralizing quinoline [129, 214, 239–241]. Such information, as well as individual studies of a variety of marine bacteria for which threshold concentrations of 0.15mg/l to greater than 100 mg/l were found [224], demonstrate that the threshold concentrations below which individual bacterial species are unable to multiply vary enormously.

A threshold may also exist for the acclimation of microbial communities. Thus, a freshwater microbial community became acclimated to the mineralization of 4-nitrophenol at levels above but not below 10 µg/l [113]. This acclimation probably is merely an indication of the time for the cells to become sufficiently numerous to cause a detectable loss of the compound, and thus the threshold may only reflect growth. On the other hand, the induction of metabolic activity in bacterial cells may have a threshold even in the absence of growth, as for example the reported induction of 3- and 4-methylchlorobenzoate ion degradation by *Acinetobacter calcoaceticus* at concentrations above 160 µg/l but not below [242]. The threshold phenomenon may not be restricted to carbon sources, and growth may not take place at concentrations of other nutrients below some threshold value [243]. At this time, however, the occurrence of thresholds for other nutrients and their significance for biodegradation has scarcely been explored. The fact that the biodegradation of some compounds, both in pure culture and in nature, does not occur below some measurable concentration does not mean that thresholds always exist or at least at concentrations measurable by currently available methods.

Many environments contain levels of organic carbon in excess of that needed to support growth, or the levels may be regenerated constantly by excretions of other microorganisms (e.g., phytoplankton) or by new additions. Under these conditions, the energy needs for maintenance and growth of the populations degrading the compounds of interest may be met by use of the other organic molecules. Microorganisms may metabolize two, or sometimes more, organic substrates simultaneously provided that their concentrations are not excessively high. The compound sustaining the growth which is present at levels above the threshold has been called the primary substrate, and the compound that is below the threshold but is still catabolized has been designated the secondary substrate [244–247].

The apparent existence in natural waters and wastewaters of traces of potentially degradable organic pollutants may thus be attributable to the thresholds below which growth does not occur. A microorganism whose sole selective advantage in these environments is its ability to grow by using particular novel substrates therefore may not increase in abundance, and the substrate may then not disappear. Moreover, the fact that thresholds exist points to the danger of drawing conclusions about what will happen at low chemical concentrations in nature based on laboratory tests with solutions containing much higher concentrations of the substrate. Nevertheless, it is not presently possible to predict which biodegradable compounds will persist in what environments because of the threshold, and which will be destroyed because of the ability of the responsible populations to function at still lower levels of the substrate.

3.2.7 *Co-Metabolism*

The transformation of an organic compound by a microorganism that is unable to use the substrate or one of its constituent elements as a source of energy is termed *co-metabolism*. The active microbial populations thus derive no nutritional benefit from the substrates they co-metabolize. The energy sufficient to sustain growth fully is not acquired even if the conversion is an oxidation and releases energy, and the C, N, S, or P that may be in the molecule is not used as a source of these elements for biosynthetic purposes [93–95, 185, 188–190, 202].

In co-metabolism, a partial oxidation of the substrate occurs, but the energy derived from the oxidation is not used to support growth of new microbial cells [96–98]. This phenomenon arises when microorganisms possess enzymes that

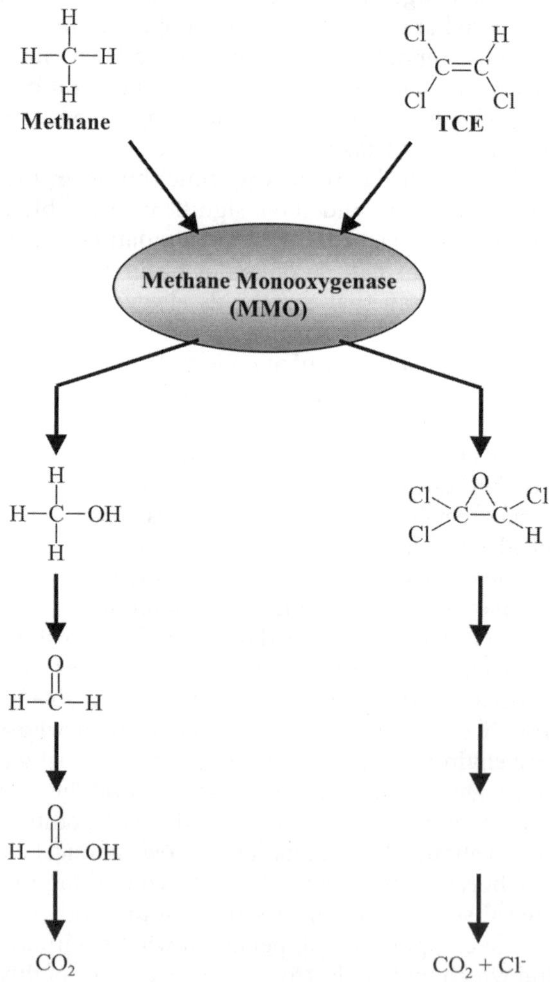

Fig. 15. Co-metabolism of TCE

coincidentally degrade a particular pollutant; that is, their enzymes are nonspecific. Co-metabolism can occur not only during periods of active growth, but also during periods in which resting (non-growing) microbial cells interact with an organic compound. Although difficult to measure in the environment, co-metabolism has been demonstrated for some environmental pollutants. For example, the industrial solvent trichloroethene (TCE) can be oxidized co-metabolically by methanotrophic bacteria, whose sole carbon substrate is methane. TCE is currently of great interest because it is one of the most frequently reported contaminants at hazardous waste sites, a suspected carcinogen, and generally resistant to biodegradation.

As shown in Fig. 15, the first step in the oxidation of methane is catalyzed by methane monooxygenase, the enzyme produced by methanotrophic bacteria. This enzyme is so nonspecific that it can also co-metabolically catalyze the first step in the oxidation of TCE when both methane and TCE are present. The bacteria receive no energy benefit from this co-metabolic degradative step of the reaction. The subsequent degradation steps shown in Fig. 15 may be catalyzed spontaneously, by other bacteria, or in some cases by the methanotrophs themselves. This co-metabolic reaction may have great significance in remediation. Currently, research is underway to investigate the application of these methanotrophs to TCE-contaminated sites. Other co-metabolizing microorganisms that grow on toluene, propane, and even ammonia are also being evaluated for use in bioremediation.

3.3 Factors Affecting Biodegradation

It is often difficult to predict the fate of a pollutant in an interfacial microenvironment because the interactions between the microbial, chemical, and physical components of the environment are still not well understood. The total microbial activity at aqueous-solid phase interfaces depends on a variety of factors, such as numbers of microbes, available nutrients, environmental conditions, and pollutant chemical structure. The impact of some of the most important factors affecting microbial activity, with the implicit understanding that microbial activity can be inhibited by any one of these factors, will be discussed in the present sections.

The interfacial microenvironment around a microbial community, that is the sum of the physical, chemical, and biological parameters which affect a microorganism, determines whether a particular microorganism will survive and/or metabolize. The occurrence and abundance of microorganisms in an environment are determined by nutrient availability, and various physicochemical factors such as pH, redox potential, temperature, and solid phase texture and moisture. Because a limitation imposed by any one of these factors can inhibit biodegradation, the cause of the persistence of a pollutant is sometimes difficult to pinpoint. The summary follows [39, 92, 94, 97, 109, 110, 172, 173, 176, 189, 190, 195, 248–252, 256–300].

3.3.1
Oxygen

Oxygen is very important in determining the extent and rate of pollutant biodegradation. In general, biodegradation is much faster under aerobic (i.e., oxygen is present) conditions than under anaerobic (i.e., no oxygen is present) conditions. Also, some pollutants that are degraded aerobically are not degradable anaerobically. Thus, the saturated aliphatic hydrocarbons found in petroleum are readily degraded aerobically; but, unless an oxygen atom is present in the initial compounds, they are quite stable under anaerobic conditions (Fig. 16). Hydrocarbons with no oxygen, such as hexane, are only degraded aerobically, while the presence of a single oxygen atom (hexanol) results in both aerobic and anaerobic degradation. This anaerobic stability explains why underground petroleum reservoirs, which contain no oxygen, have remained intact for millions of years, even though microorganisms may be present. In contrast, highly chlorinated organic compounds are more stable under aerobic conditions. That is, increasing chlorine content favors anaerobic dehalogenation (removal of chlorine) over aerobic dehalogenation.

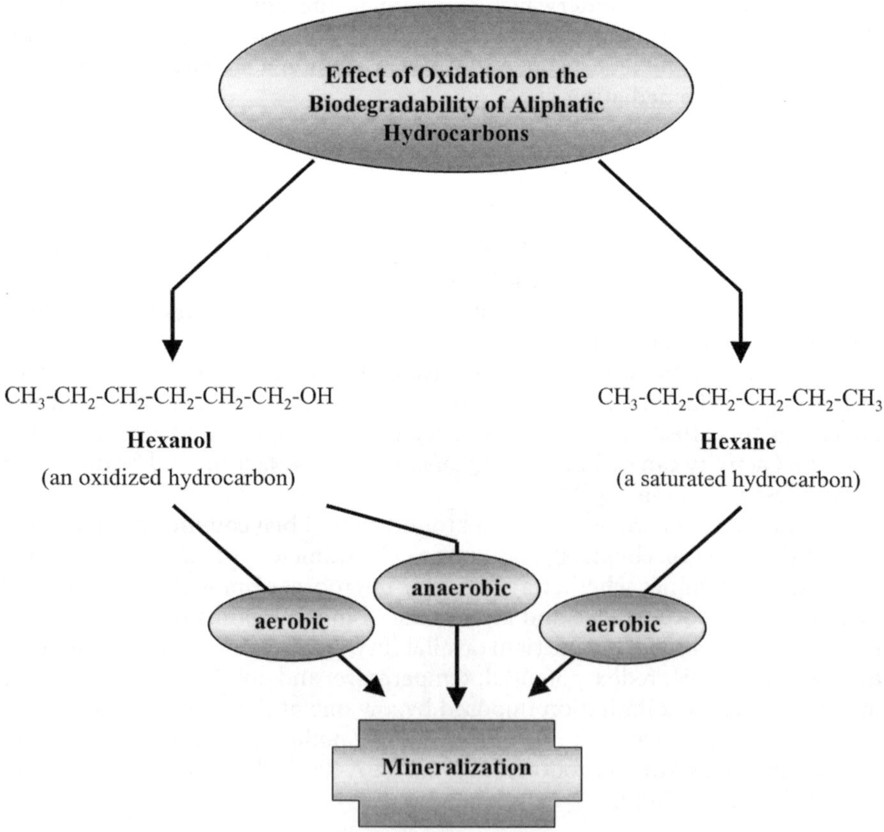

Fig. 16. The effects of oxidation on the biodegradability of aliphatic compounds

In terms of oxygen availability, surficial bottom sediments of oxic aquatic environments, surface soils, and the vadose zone (i.e., the water-unsaturated and generally unweathered material between groundwater and the land surface) are similar, being primarily aerobic regions. Thus, these regions tend to favor aerobic degradation of pollutants. However, these regions may contain pockets of anaerobic activity generated by localized conditions (e.g., high biodegradative activity) which reduce oxygen levels. In contrast, the oxygen concentrations in groundwaters or saturated regions are low. The only oxygen that exists in these regions are dissolved oxygen, and the oxygen levels are low because it is not very water soluble. Therefore, if significant microbial activity occurs, the limited supply of oxygen is rapidly used up, causing anaerobic conditions to develop. Addition of air or oxygen can often improve biodegradation rates, particularly in subsurface areas with a high clay content.

3.3.2
Organic Matter Content

Solid particles of bottom sediments and surface soils have large numbers of microorganisms ranging from 10^6 to 10^9 cells per gram of solid phase. Fungal numbers are somewhat lower, 10^4 to 10^6 per gram of solid phase. In contrast, microbial populations in deeper regions, such as the vadose zone and groundwater region, are often lower by two orders of magnitude or more. This large decrease in microbial numbers with depth is primarily due to differences in organic matter content. Whereas bottom sediments and soil surfaces may be rich in organic matter, both the vadose zone and the groundwater region often have low amounts of organic matter. One consequence of low total numbers of microorganisms is that the population of pollutant degraders initially present is also low. Thus, biodegradation of a particular pollutant may be slow until a sufficient biodegrading population has been built up. A second reason for slow biodegradation in the vadose zone and groundwater region is that the microorganisms in this region are often dormant owing to the low amount of organic matter present. If microorganisms are dormant, their response to an added carbon source is slow, especially if the carbon source is a pollutant molecule to which they have not been exposed.

Given these two main factors (i.e., oxygen availability and organic matter content), several generalizations can be made about solid phase surfaces, the vadose zone, and the groundwater region as follows:

- Biodegradation at aqueous-solid phase interfaces is primarily aerobic and rapid.
- Biodegradation in the vadose zone is also primarily aerobic, but significant acclimation times may be necessary for sufficient biodegrading populations to build up.
- Biodegradation in the groundwater region is initially slow owing to low microbe numbers, and can rapidly become anaerobic due to lack of available oxygen.

3.3.3
Nitrogen

Nitrogen is another macronutrient that often limits microbial activity because it is an essential part of many key microbial metabolites and building blocks, including proteins and amino acids. It is also subject to removal from the aqueous-solid phase interface by various processes such as leaching or denitrification. Many xenobiotics are carbon-rich and nitrogen-poor, and thus nitrogen limitation can inhibit their biodegradation while the simple addition of nitrogen-rich compounds can often improve it. For example, in the case of petroleum spills, where nitrogen shortages can be acute, biodegradation can be significantly accelerated by adding nitrogen fertilizers. In general, microbes have an average C:N ratio within their biomass of about 5:1 to 10:1, depending on the type of microorganism, so the C:N ratio of the material to be biodegraded must be 20:1 or less. The difference in the ratios is due to the fact that approximately 50% of the carbon metabolized is released as carbon dioxide, whereas almost all of the nitrogen metabolized is incorporated into the microbial biomass.

3.3.4
Pollutant Structure

The rate at which a pollutant molecule is degraded in the environment depends largely on its chemical structure. If the molecule is not normally found in the environment or if its structure does not resemble that of a molecule usually found in the environment, a suitable biodegrading microorganism may not be present. In this case, chances are slim for biodegradation to occur. The bioavailability of the pollutant is also extremely important in determining the rate of biodegradation. If the water solubility of the pollutant is extremely low, then it has a low bioavailability (see Chap. 4). Many pollutant molecules which are persistent in the environment share the property of low water solubility. Examples include DDT, PCBs, and petroleum hydrocarbons (see Chap. 1). Both PCBs and petroleum hydrocarbons are liquids at room temperature and actually form a hydrophobic phase which is separate from the aqueous phase. Although microorganisms are not excluded from this phase, active metabolism seems to occur only in the aqueous phase or at the oil-water interface. The second factor that reduces bioavailability is sorption of the pollutant by soil. Organic compounds that have a low water solubility are also prone to sorption by solid phase surfaces (see Chap. 2).

Many pollutants have extensive branching or functional groups which block or sterically hinder the pollutant carbon skeleton at the reactive site, that is, the site at which the substrate and enzyme come into contact during a transformation step. This can best be explained and illustrated by comparing the differences between linear (i.e., readily biodegradable) and non-linear (i.e., slowly biodegradable) alkylbenzenesulfonates (ABSs), i.e., the branching between them (Fig. 17).

As a result of our increasing knowledge of the effect of pollutant structure on biodegradation in the environment, efforts are being focused on developing and

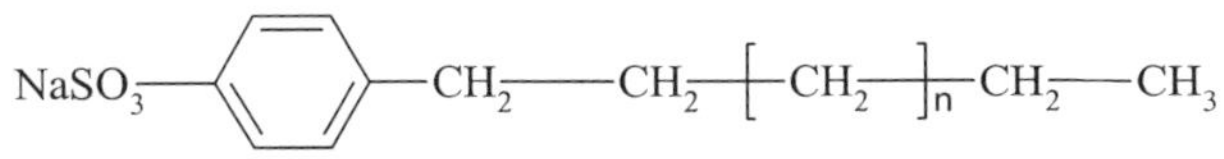

(a) Linear alkylbenzenesulfonate (n=8)

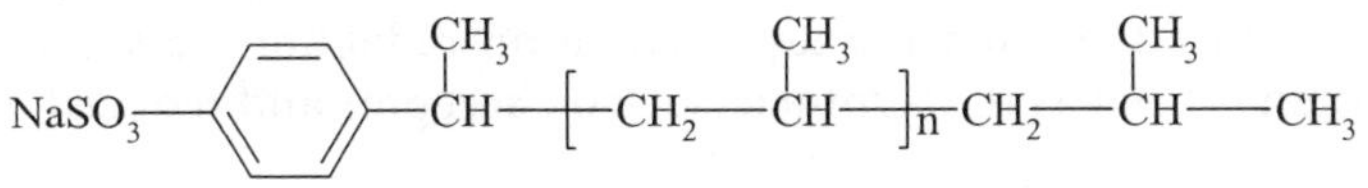

(b) Branched alkylbenzenesulfonate (n=8)

Fig. 17. **a** Linear (readily biodegradable) alkylbenzenesulfonates (ABS). **b** Branched (slowly biodegradable) alkylbenzenesulfonates (ABS)

utilizing "environmentally friendly" compounds. For example, slowly biodegradable pesticides are being replaced by rapidly biodegradable compounds, which are used in conjunction with integrated pest-management approaches. This approach means that pesticides are not used on a yearly basis but, rather, are rotated. Thus, insects do not become fully acclimated to these easily degraded pesticides, and soil microorganisms degrade them so rapidly that they are active only during the intended time frame.

3.4 Biodegradation Pathways

The vast majority of the organic carbon available to microorganisms at aqueous-solid phase interface microenvironments is material which was fixed photosynthetically. Anthropogenic activity has resulted in the addition of many industrial and agricultural organic compounds, including petroleum products, chlorinated solvents, and pesticides (see Chap. 1). Many of these molecules are readily degraded because of their similarity to photosynthetically produced organic matter. This allows microorganisms to utilize preexisting biodegradation pathways. However, some chemical structures are unique, or have unique components, which result in slow or little biodegradation (e.g., high molecular weight PAHs). To understand and predict biodegradation of organic pollutants in this interfacial microenvironment, these contaminants can be classified into one of three basic structural groups: the aliphatics, the alicyclics, and the aromatics. Constituents of each of these groups can be found in all three physical states: gaseous, solid, and liquid. The general degradation pathways for each of these structural classes are delineated below. These pathways differ for aerobic and anaerobic conditions and can be affected by structural modifications of the contaminant.

3.4.1
Aerobic Conditions

In the presence of oxygen, many heterotrophic microorganisms rapidly mineralize organic compounds. During degradation some of the carbon is completely oxidized to CO_2 to provide energy for growth, and some carbon is used as structural material in the formation of new microbial cells (Fig. 3). Energy used for growth is produced through a series of oxidation-reduction (redox) reactions in which oxygen is used as the terminal electron acceptor and reduced to water.

3.4.1.1
Aliphatic Hydrocarbons

Aliphatic hydrocarbons are straight chain and branched-chain structures (see Chap. 1). Most aliphatic hydrocarbons introduced into the environment come from industrial solvent waste, the petroleum industry, and vehicular traffic. Liquid aliphatic hydrocarbons readily degrade under aerobic conditions, especially when the number of carbons is between 8 and 16. Longer chain aliphatic compounds are usually waxy substances. Biodegradation of these longer carbon chains is slower due to limited water solubility, while biodegradation of shorter chains may be impeded by the toxic effects of the short-chain aliphatic compounds on microorganisms. In addition several common structural modifications can result in severely reduced biodegradation, such as:

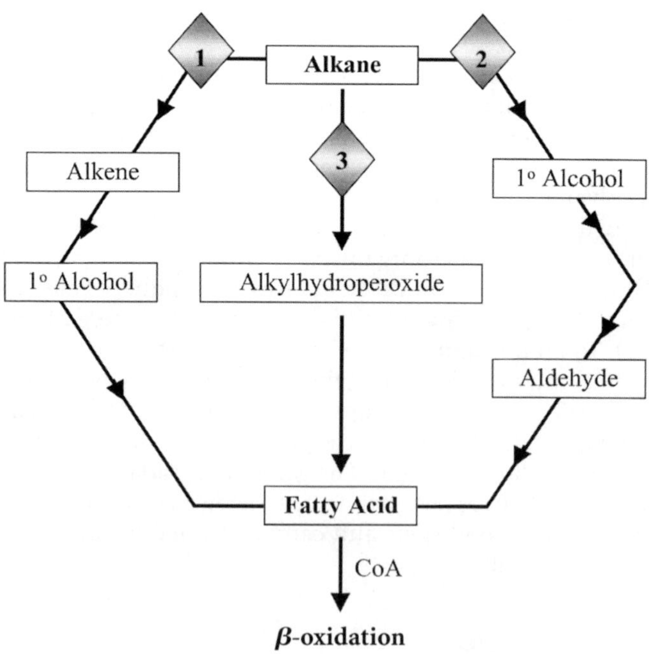

Fig. 18. Aerobic biodegradation pathways for aliphatic hydrocarbons

(1) extensive branching in the hydrocarbon chain found in petroleum, where branched hydrocarbons comprise one of the slowest degradable fractions, and (2) halogen substitution, as represented by the compound TCE. Chlorinated solvents such as TCE have become a serious environmental pollution problem. Although such severely branched and highly chlorinated hydrocarbons degrade slowly, these solvents pose a more severe problem because of their toxicity. Thus, both the rate of biodegradation and toxicity must be considered in evaluating the potential hazard of such pollutants at aqueous-solid phase environments.

Biodegradation of aliphatic compounds generally occurs by one of the three pathways as shown in Fig. 18. The most common is a direct enzymatic incorporation of molecular oxygen (pathway 1). All three pathways result in the

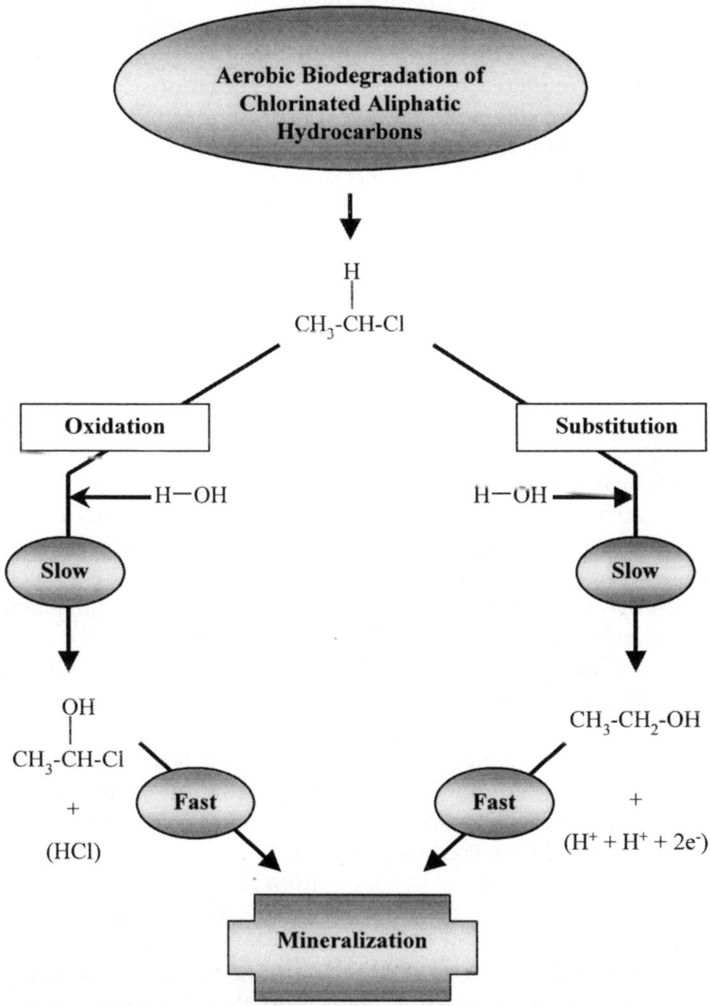

Fig. 19. Aerobic biodegradation of chlorinated aliphatic compounds

formation of a primary fatty acid. The fatty acid formed in degradation of an alkane is subject to normal cellular fatty acid metabolism and includes β-oxidation which cleaves consecutive two-carbon fragments. Each two-carbon fragment is removed by coenzyme A (CoA) as acetyl-CoA and shunted to the tricarboxylic acid (TCA) cycle for complete degradation to CO_2 and H_2O. If the alkane has an even number of carbons, acetyl-CoA is the last residue. If the alkane has an odd number of carbons, propionyl-CoA is the last residue, which is also shunted to the TCA cycle after conversion to succinyl-CoA.

Both branching and halogenation can slow biodegradation. In the former case, extensive branching causes interference between the degrading enzyme and the enzyme-binding site. In the latter case, the bonds and the reactions involved play a major role. For halogenated compounds, the relative strength of the carbon-halogen bond requires two things: (1) an enzyme that can act on the bond, and (2) a large input of energy to break the bond. In general, monochlorinated alkanes are considered degradable; however, increasing halogen substitution results in increased inhibition of degradation. Halogenated aliphatic compounds can be degraded by two types of reactions that occur under aerobic conditions. The first is substitution, which is a nucleophilic reaction (i.e., the reacting species donates an electron pair) in which the halogen is substituted by a hydroxyl group. The second is an oxidation reaction, which requires an external electron acceptor. These two reactions are shown in Fig. 19. Although increasing halogenation generally slows degradation, aerobic oxidation of highly chlorinated aliphatic hydrocarbons can occur co-metabolically (see Sect. 3.2.7).

3.4.1.2
Alicyclic Hydrocarbons

Alicyclic hydrocarbons are saturated carbon chains that form ring structures. Naturally occurring alicyclic hydrocarbons are common (Chap. 1). For example, alicyclic hydrocarbons are a major component of crude oil, comprising 20–67 vol.%. Other examples of complex, naturally occurring alicyclic hydrocarbons include camphor (a plant terpene) and cyclohexyl fatty acids (components of microbial lipids). Anthropogenic sources of alicyclic hydrocarbons to the environment include fossil-fuel processing and oil spills, as well as the use of such agrochemicals as the pyrethrin insecticides (Chap. 1, and references therein).

It is very difficult to isolate pure cultures of bacteria which can degrade alicyclic hydrocarbons. For this reason, biodegradation of an alicyclic hydrocarbon is thought to take place as a result of teamwork among mixed microbial populations, commonly referred to as a microbial consortium. For example, in the degradation of cyclohexane, one population in the consortium performs the first two degradation steps, cyclohexane to cyclohexanone via cyclohexanol, but is unable to lactonize and open the ring. Subsequently, a second population in the consortium, which cannot oxidize cyclohexane to cyclohexanone, performs the lactonization and ring-opening steps, and then degrades the compound completely (Fig. 20).

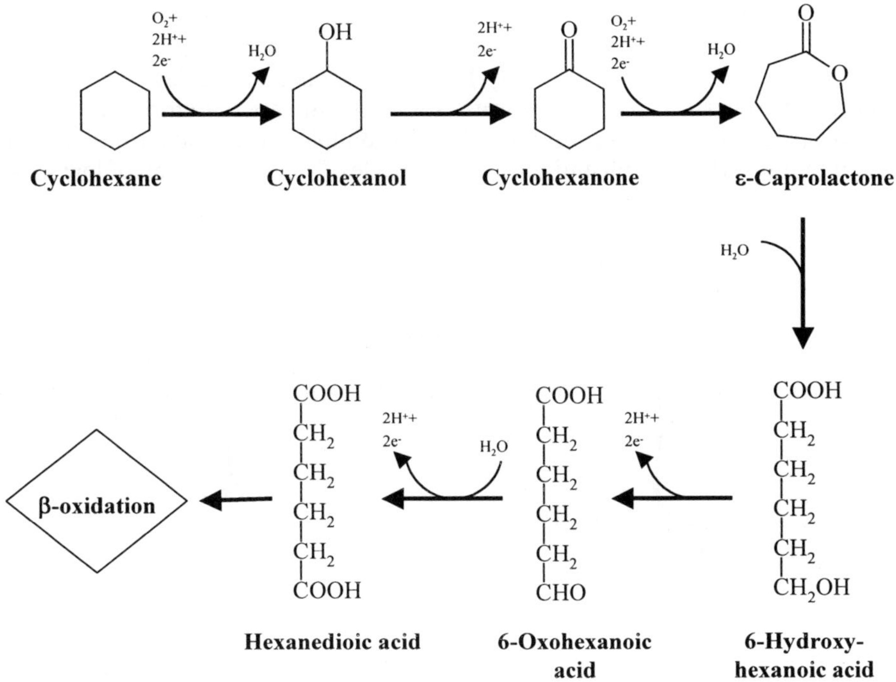

Fig. 20. Aerobic biodegradation of cyclohexane by a microbial consortium

Interestingly, cyclopentane and cyclohexane derivatives, which contain one or two hydroxyl, carbonyl, or carboxyl groups, degrade more readily in the environment than do their parent compounds. In fact, microorganisms capable of degrading of cycloalkanols and cycloalkanones are ubiquitous in environmental samples.

3.4.1.3
Aromatic Hydrocarbons

The aromatic hydrocarbons contain at least one unsaturated ring system with the general structure C_6R_6, where R is any functional group (see Chap. 1). The parent hydrocarbon of this class of compounds is benzene (C_6H_6), which exhibits the resonance, or delocalization of electrons, typical of unsaturated cyclic structures. Owing to its resonance energy, benzene is remarkably inert.

Aromatic compounds, excluding polycyclic aromatic hydrocarbons (PAHs), which contain one or more benzene rings, are synthesized naturally by plants. For example, they are a major component of the common plant polymer, lignin. Release of aromatic compounds into the environment occurs as a result of natural processes such as forest and grass fires, which also generate PAHs from these aromatic precursors. The major anthropogenic sources of aromatic compounds are fossil fuel processing and utilization (burning). For example, benzene is one component of gasoline that is often released into the environment; it is of particular concern because it is a carcinogen.

Aromatic compounds, especially PAHs and higher molecular weight compounds, are characterized by low water solubility and are therefore very hydrophobic (see Chap. 4). As is common with hydrophobic compounds, aromatics are often found sorbed to soil and sediment particles. The combination of low solu-

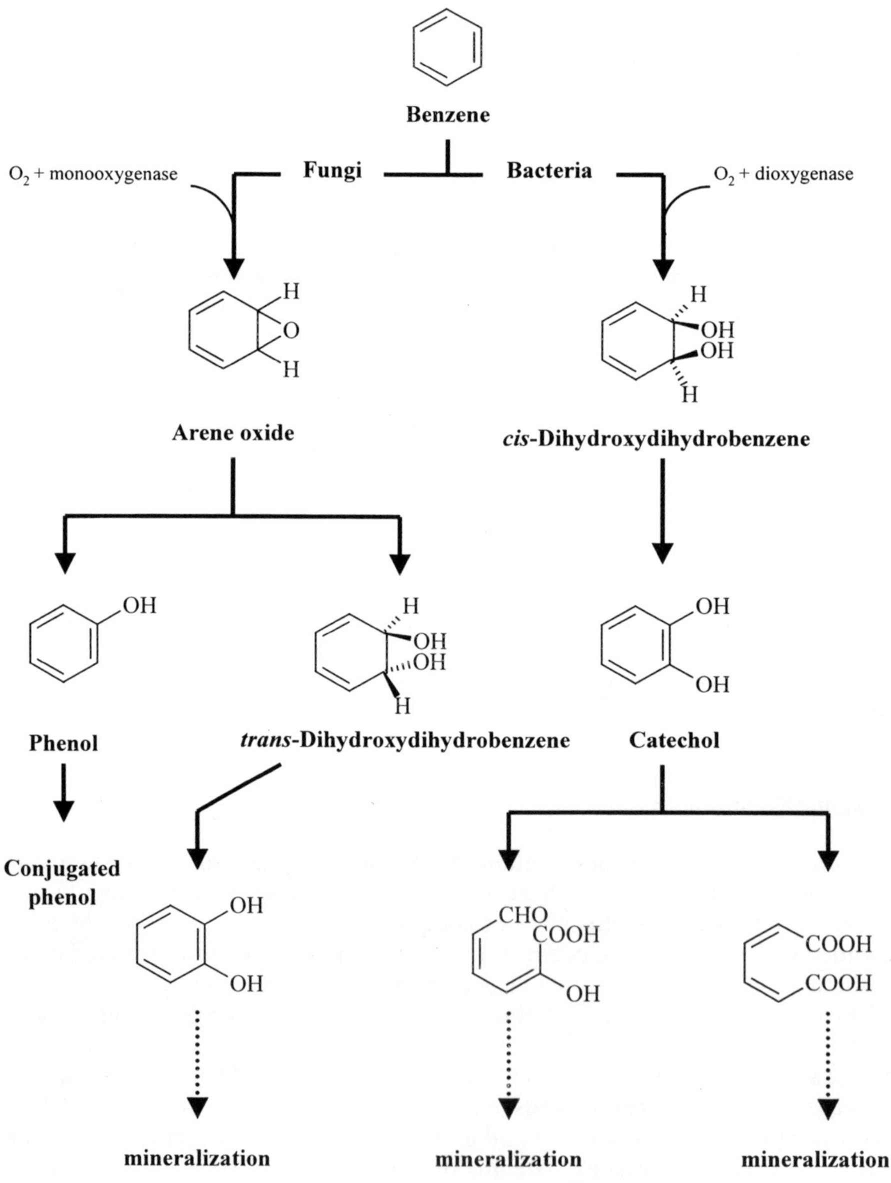

Fig. 21. Aerobic biodegradation pathways of aromatic compounds by bacteria and fungi

bility and high sorption results in low substrate bioavailability and slow biodegradation rates. This is particularly true for PAHs having three or more rings because water solubility decreases as the number of rings increase.

A wide variety of bacteria and fungi can degrade aromatic compounds. Under aerobic conditions, the most common initial transformation is hydroxylation, which involves the incorporation of molecular oxygen. The enzymes involved in these initial transformations fall into two groups: (1) *dioxygenases*, which incorporate both atoms of molecular oxygen into the PAH, and (2) *monooxygenases*, which incorporate only one atom of molecular oxygen. In general, bacteria transform PAHs by an initial dioxygenase attack to form a *cis*-dihydrodiol, which is subsequently rearomatized to afford a dihydroxylated intermediate (phenol) called catechol. The ring is then cleaved by a second dioxygenase, as shown in Fig. 21, using either an *ortho*- or a *meta*-pathway, and then further degraded.

Fungi transform PAHs by an initial monooxygenase attack. This enzyme incorporates one atom of molecular oxygen into the PAH and reduces the second oxygen to water. The result is the formation of an arene oxide, followed by the enzymatic addition of water to yield a *trans*-dihydrodiol (Fig. 21). Alternatively, the arene oxide can be isomerized to form phenols, which can be conjugated with sulfate, glucuronic acid, and glutathione. These conjugates are similar to those formed in higher microorganisms and aid in detoxification and elimination of PAHs. In general, PAHs having two or three condensed rings are transformed rapidly, often mineralizing completely, whereas PAHs with four or more condensed rings are transformed much more slowly, often as a result of cometabolic attack.

3.4.2
Anaerobic Conditions

Anaerobic conditions are not uncommon in the environment and can develop in water or saturated sediment/soil environments. However, even in well-aerated solid phase systems, there are interfacial microenvironments with little or no oxygen. In all of these environments, *anaerobiosis* occurs when the rate of oxygen consumption by microorganisms is greater than the rate of oxygen diffusion through either air or water. In the absence of oxygen, organic compounds can be mineralized through anaerobic respiration, in which an electron acceptor other than oxygen is used. The series of alternative electron acceptors in the environment includes iron, nitrate, manganese, sulfate, and carbonate, which are listed in order from most oxidizing to most reducing. This progression means they are usually utilized in this order because the amount of energy generated for growth depends on the oxidation potential of the electron acceptor. Since none of these electron acceptors are as oxidizing as oxygen, growth under anaerobic conditions is never as efficient as growth under aerobic conditions (Fig. 22). As shown in Fig. 22, aerobic conditions refer to specific potential facultative and obligate anaerobes which metabolize over a spectrum of redox potentials.

Anaerobic degradation pathways have not been as extensively studied as aerobic degradation of organic compounds. Interestingly, many compounds that

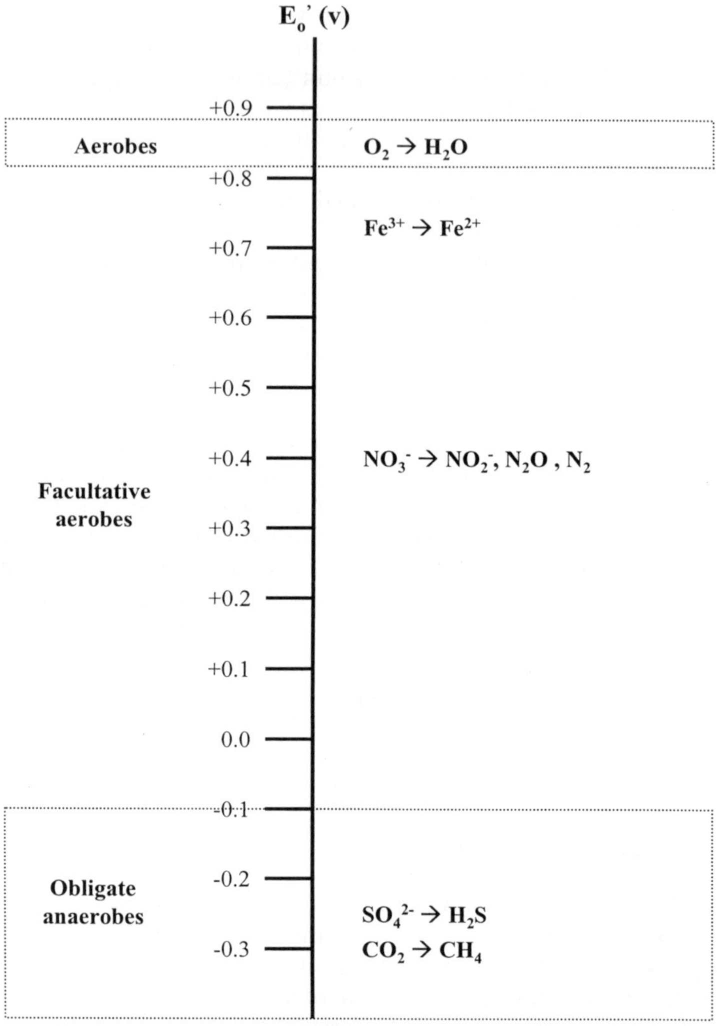

Fig. 22. The range of redox potentials found in environments commonly inhabited by actively metabolizing microorganisms

are easily degraded aerobically, such as saturated hydrocarbons, are far more difficult to degrade anaerobically. However, in at least one group of compounds, those that are highly halogenated, the halogen substituents are removed more rapidly under anaerobic conditions. However, once dehalogenation has occurred, the remaining molecule behaves more typically; that is, it is generally degraded more rapidly and extensively under aerobic conditions. As a consequence of this sequential process, bioremediation technologies have been developed that utilize sequential anaerobic-aerobic treatments to optimize degradation of highly halogenated compounds.

3.4.2.1 Aliphatic Hydrocarbons

Saturated aliphatic hydrocarbons are degraded slowly, if at all, under anaerobic conditions. Evidence of this slow to non-existent degradation can be seen in nature. For example, hydrocarbons in natural underground reservoirs of oil (which are under anaerobic conditions) are not degraded, despite the presence of microorganisms. However, both unsaturated aliphatics and oxygen-containing aliphatics (alkenes, alcohols, and ketones) are readily biodegraded anaerobically. The suggested pathway of biodegradation for unsaturated hydrocarbons is the hydration of the double bond to an alcohol, with further oxidation to a ketone or aldehyde, followed finally by formation of a fatty acid (Fig. 23).

Halogenated aliphatics can be partially or completely degraded under anaerobic conditions through a transformation reaction called *reductive dehalogenation*. Often a co-metabolic degradation step, reductive dehalogenation

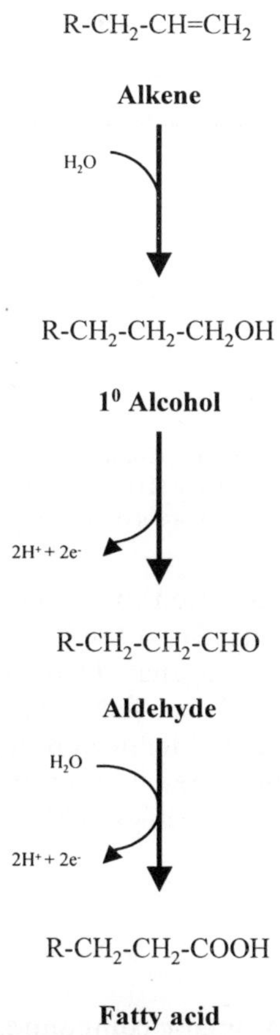

Fig. 23. General anaerobic biodegradation pathway for an alkene

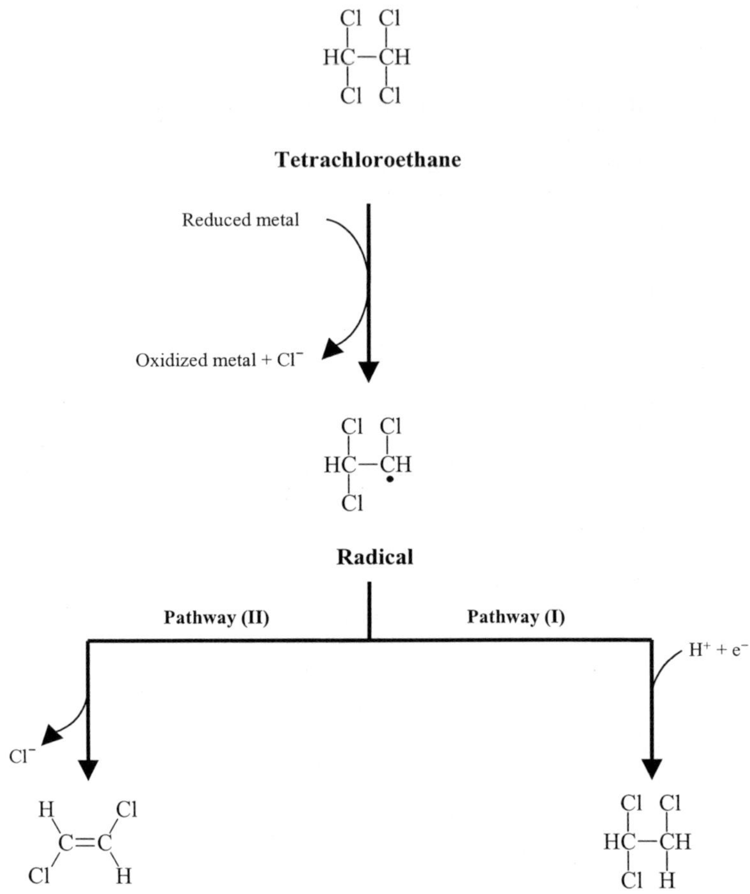

Fig. 24. Reductive dehalogenation of a chlorinated hydrocarbon in the presence of a metal forming an alkyl radical, showing: (Pathway (I)) the alkyl radical scavenging a hydrogen atom, and (Pathway (II)) the alkyl radical losing a second halogen to form an alkene

may be mediated by reduced transition-metal/metal complexes. The steps in this transformation are shown in Fig. 24. In the first step, electrons are transferred from the reduced metal to the halogenated aliphatic compound, resulting in an alkyl radical and free halogen. Then, the alkyl radical can either scavenge a hydrogen atom (I), or lose a second halogen to form an alkene (II). In general, anaerobic conditions favor the degradation of highly halogenated compounds, while aerobic conditions favor the degradation of mono- and disubstituted halogenated compounds.

3.4.2.2
Aromatic Hydrocarbons

Like aliphatic hydrocarbons, aromatic compounds can be completely degraded under anaerobic conditions if the aromatic is oxygenated. Recent evidence also indicates that even unsubstantiated aromatics are degraded slowly under anaerobic conditions. Anaerobic mineralization of aromatics often requires a mixed microbial community whose populations work in consortia under different redox potentials. For example, mineralization of benzoate ion can be achieved by growing an anaerobic benzoate degrader in co-culture with an aerobic methanogen or sulfate reducer. In this consortium, benzoate ion is transformed by one or more anaerobes to yield aromatic acids, which in turn are transformed to methanogenic precursors such as acetate, carbon dioxide, or formate. These small molecules can then be utilized by methanogens (Fig. 25). This process can be described as an anaerobic food chain because the microorganisms higher in the food chain cannot utilize acetate or other methanogenic precursors, while the methanogens cannot utilize larger molecules such as benzoate ion. Methanogens utilize carbon dioxide as a terminal electron acceptor, thereby forming methane. Methanogens should not be confused with methanotrophic bacteria, which aerobically oxidize methane to carbon dioxide.

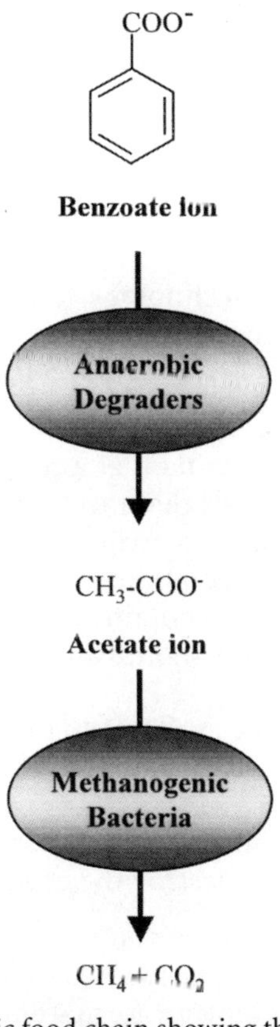

Fig. 25. An example of an anaerobic food chain showing the formation of simple compounds from benzoate ion by a population of anaerobic bacteria and the subsequent utilization of the newly available substrate by a second anaerobic population (the methanogenic bacteria)

4 Field Applications

The controlled, practical use of microorganisms for the destruction, biodegradation, and biotransformation of organic contaminants at aqueous-solid phase environments has recently become widely used [8, 20, 22, 24, 71, 146, 147, 193, 301]. These various technologies rely on the biodegradation capability of microorganisms, and focus mainly on enhancing slow biodegradation processes in nature and/or engineering technologies which bring organic compounds into contact with microorganisms in some types of reactors allowing their rapid transformation [187, 202].

Bioremediation of contaminated sites is a new field of endeavor and many new or altered technologies are now appearing. These processes are being used mainly for the destruction of organic compound mixtures and restoration of impacted environments. The goal of bioremediation is to degrade organic contaminants to concentrations that are either undetectable or, if detectable, to concentrations below the limits established as safe or acceptable by regulatory agencies [93, 97, 179, 180, 189, 232]. Bioremediation is being applied in soils, bottom sediments, wastewaters, ground waters, and industrial waste systems. The list of compounds that may be subject to biological destruction by one or another bioremediation engineering system is long. However, because they are widespread and are susceptible to microbial detoxification, most interest has been directed to oil and oil products, gasoline and its constituents, PAHs, chlorinated aliphatic compounds, and polychlorinated biphenyls [93, 97, 179, 180, 189, 232].

In order for bioremediation technology to be considered seriously, the following criteria must be met and none of the key points mentioned below can be disregarded [150, 165, 174, 176, 191, 192, 195, 200, 302]:

- Microorganisms must exist, which have the needed catabolic activity and ability to transform the contaminant at a reasonable rate, reducing the contaminant concentration to levels that meet regulatory standards.
- Microorganisms must not generate toxic products at the concentrations likely to be achieved during the remediation.
- Contaminated sites must not contain concentrations or combinations of organic compounds which are significantly inhibiting to the biodegrading species.
- The target compound must be available to the microorganisms.
- Conditions at the site or in a bioreactor must be made conducive to microbial growth or activity.
- The cost of the technology must be less or at least no more expensive than other technologies which can destroy the chemical.

In the next few sections, several case studies will be discussed to illustrate the different types of xenobiotics at various contaminated sites, the ways of enhancing their biodegradation/biotransformation techniques, and the verification of the results.

4.1 Case Studies

This includes bioremediation cases of contaminated sites with several toxic and carcinogenic pollutants, such as petroleum hydrocarbons, PAHs, dichlorobenzene, chlorinated hydrocarbons, carbon tetrachloride, Dicamba, methyl bromide, trinitrotoluene, silicon-based organic compounds, dioxins, alkylphenol polyethoxylates, nonylphenol ethoxylates, and polychlorinated biphenyls. The following is a brief summary of each case.

4.1.1 *Petroleum Hydrocarbons*

Due to widespread use, petroleum hydrocarbons are ubiquitous groundwater contaminants. They enter the subsurface environment via leakage from underground storage tanks, industrial discharge, improper disposal techniques, and accidental spills (see Chap. 1). Approximately 15% of regular gasoline is comprised of benzene, toluene, ethylbenzene, and *m*-, *p*-, and *o*-xylene (BTEX), relatively water-soluble monoaromatic hydrocarbons which are toxic and confirmed or suspected carcinogens [303]. Many cleanup efforts have focused on bioremediation and in particular on in situ or intrinsic biodegradation, taking advantage of the ability of indigenous microbial populations to degrade hydrocarbons [58, 304].

The ability of microorganisms to biodegrade BTEX compounds under aerobic conditions is well documented in the literature [103, 305, 306]. However, aerobic processes are limited by the slow rate at which oxygen can be supplied to the contaminated zone [307, 308]. While less is known about anaerobic BTEX biodegradation, laboratory results have shown that some BTEX compounds can be degraded under denitrifying [27, 140], iron(III)-reducing [39, 277], sulfate-reducing [225, 257, 259, 275, 309], and methanogenic [258, 310] environmental conditions. However, these results have not been widely verified in the field [310–313]. Although the rate of microbial transformation of BTEX slows down in the absence of molecular oxygen, anaerobic biodegradation nonetheless can provide significant remediation potential at many contaminated sites.

It is currently not possible to predict complete anaerobic transformation pathways for BTEX because the factors that promote or inhibit the process are not completely understood. Furthermore, it is uncertain whether rates measured in the laboratory can be fully applied to the field. Laboratory studies have shown that anaerobic degradation rates can be sensitive to the presence of readily degradable co-substrates and geochemical factors. For instance, when multiple BTEX compounds are present simultaneously, anaerobic biotransformation was found to be sequential with toluene being the most readily degraded compound followed by *p*- and *o*-xylene [220, 259, 314]. The place of ethylbenzene in the sequence depends on the geochemical conditions, e.g., it is high under nitrate-reducing conditions but low under sulfate-reducing conditions. Benzene is generally the most persistent compound. Hydrogen sulfide inhibits degradation of BTEX compounds under sulfate-reducing conditions [85, 86, 259]. Ferric or ferrous iron may aid in initiating or accelerating BTEX transformation by removing free hydrogen sulfide from solution, thereby preventing sulfide toxicity [85, 86]. Field studies have demonstrated that nitrate can enhance BTEX transformation at sites contaminated by hydrocarbons [212, 315, 316]. Thierrin et al. [133] observed toluene, *p*-xylene, and naphthalene transformation in a sulfate-reducing aquifer contaminated by gasoline. Reinhard et al. [318] investigated the in situ anaerobic biotransformation of BTEX under enhanced nitrate- and sulfate-reducing conditions.

There are a wide range of bioremediation technologies either in use or proposed for use on oil/gasoline-contaminated land [301, 319], and these can be divided into two broad groups. *In situ* techniques treat the contamination at the site of the pollution event, whereas *ex situ* techniques remove the contamination from the ground and transfer it to another location for treatment. The use of in situ treatment is often preferable in terms of financial considerations, due to the cost of moving large quantities of soil [20]. Some novel approaches to the problem of hydrocarbon contamination of contaminated aqueous-solid phase environments is the use of: (1) gas-liquid foams to enhance in situ bioremediation, and (2) biostimulation, as follows.

4.1.1.1
Foaming

Foams are dispersing systems containing at least two distinct phases. A continuous liquid phase surrounds bubbles of air and may enclose droplets of a secondary liquid phase or particles of a solid phase. Surfactants are essential for the generation and stabilization of foams, accumulating as a viscoelastic layer at various interfaces to maintain the structural integrity of the foam [299]. This has an important stabilizing effect by altering surface properties at the interfaces, particularly by lowering the surface tension.

The most widespread large-scale application of gas-liquid foams is in fire fighting, where air is excluded from the combustible material by a thick blanket of foam [320]. These fire-fighting foams are supplied as liquid concentrates, which can be diluted on-site to the required strength. The foam is formed from this premixture by an aerating device. Several studies have been undertaken to investigate the suitability of foams for bioremediation applications, as follows:

- Li et al. [321–323] investigated the degradation of oil using a solid alginate foam carrier inoculated with a marine oil-degrading yeast and nutrients. The foam carrier was prepared from chicken egg and bovine serums. They observed that a floating alginate carrier could both adsorb and hold the oil and that the immobilized nutrients contained in the serum were of use to the microbial population. Using this system, they found that 61% of the model oil, *n*-tetradecane, was degraded in 14 days.
- Stabnikova et al. [295] showed enhanced degradation of crude oil in soil columns using a foamed preparation referred to as Lestan. This contained a hydrocarbon-degrading microbial component, a biological surfactant, and a carrier. Eighty-nine percent of the oil was degraded after 35 days of treatment with foamed Lestan (applied at one-week intervals), which was 43% higher than the untreated controls.

- A number of different systems have been investigated to treat nonaqueous-phase liquids (NAPLs) in subsurface soils [287, 324]. In these studies, synthetic surfactants were injected directly into soil to mobilize hydrocarbons.
- The potential use of foams has also been demonstrated for the decontamination of nerve agents [80]. In these applications, the detoxification of the nerve agent was carried out by immobilizing the enzyme organophosphorous acid anhydrase within either a fire fighting or blast-containment foam carrier.

Just recently, Ripley et al. [325] described the development of a "*protein-based foam*" formulation and subsequent investigations into its suitability for enhancing the degradation of *n*-hexadecane using a novel bench-scale soil microcosm. High-density protein-based foam concentrates, developed by the fire-fighting industry, were selected for experimental investigation. Using crude protein hydrolysate as a starting material, a foam formulation was developed with properties suitable for bioremediation studies. This formulation incorporated eight species of hydrocarbon-degrading bacteria (i. e., seven individual *Acinetobacter* species and a *Pseudomonas* species) which were selected for their ability to degrade *n*-hexadecane. In addition to their ability to utilize *n*-hexadecane, the bacteria were tested for compatibility with the foam formulation and each other. The use of this "bioactive foam" led to an enhanced *n*-hexadecane degradation when compared to controls without foam.

4.1.1.2
Biostimulation

Without appropriate cleanup measures, BTEX often persist in subsurface environments, endangering groundwater resources and public health. Bioremediation, in conjunction with free product recovery, is one of the most cost-effective approaches to clean up BTEX-contaminated sites [326]. However, while all BTEX compounds are biodegradable, there are several factors that can limit the success of BTEX bioremediation, such as pollutant concentration, active biomass concentration, temperature, pH, presence of other substrates or toxicants, availability of nutrients and electron acceptors, mass transfer limitations, and microbial adaptation. These factors have been recognized in various attempts to optimize clean-up operations. Yet, limited attention has been given to the exploitation of favorable substrate interactions to enhance in situ BTEX biodegradation.

BTEX bioremediation projects often focus on overcoming limitations to natural degradative processes associated with the insufficient supply of inorganic nutrients and electron acceptors. However, other limitations associated with the presence and expression of appropriate microbial catabolic capacities may also hinder the effectiveness of bioremediation. Thus, while subsurface addition of oxygen or nitrate has proven sufficient to remove BTEX below detection levels [134, 145, 292, 315, 316], it has been only marginally effective at some sites [6]. Sometimes, the concentration of a target BTEX compound fails to decrease below a threshold level even after years of continuous addition of nutrients and electron acceptors [317]. This phenomenon has also been observed for many other xenobiotic and natural substrates under various experimental conditions [327–332].

Residual concentrations of carcinogenic compounds such as benzene could exceed applicable clean-up standards and remain a threat to public health. Possible reasons for residual BTEX concentrations include mass transfer and diffusion limitations [333], the requirement for a minimum substrate concentration to satisfy the maintenance energy demand and sustain a sufficient concentration of BTEX degraders [334] and the existence of a threshold substrate concentration below which induction of the necessary catabolic enzymes does not occur [109, 128, 273]. Therefore, overcoming limitations associated with the presence and expression of appropriate catabolic capacities might be required in some BTEX bioremediation projects. Hypothetically, this might be accomplished by the addition of supplemental substrates which increase the concentration of desirable phenotypes without repressing the required catabolic enzymes.

Biostimulation through substrate addition is commonly practiced to support co-metabolic biodegradation processes [30, 268, 272]. Addition of stimulatory substrates to enhance bacterial growth and metabolic activity has also been used in bio-augmentation experiments involving both environmental clean-up [183] and agricultural applications [73]. This approach, however, has not yet been used to enhance BTEX bioremediation because BTEX are often present in hydrocarbon plumes at sufficiently high concentrations to induce and sustain their degradation. In addition, there are concerns about potential effects and exacerbation of the oxygen demand when additional substrates are added. Nevertheless, controlled addition of stimulatory substrates to groundwater contaminated with traces of BTEX could help achieve lower residual BTEX concentrations.

Biostimulation through substrate addition may be even more valuable to support a new technological area quickly developing in the remediation field (i.e., in situ reactive zones). This novel remediation approach is based on the creation of a subsurface zone where migrating contaminants are intercepted and immobilized or degraded [335]. This is different from reactive walls or funnel and gate systems where the groundwater flow pattern is also controlled. In situ reactive zones allow groundwater to continue to flow naturally and are particularly attractive in that they conserve energy and water and, through long-term low operating and maintenance costs, have the potential to be considerably less costly than conventional clean-up methods [60]. Thus, injecting a non-toxic stimulatory substrate downgradient of a BTEX plume might be a cost effective approach to enhance the growth and viability of BTEX degraders before the arrival of the plume. This would attenuate BTEX migration and protect down-gradient groundwater resources.

Benzoic acid, a common food preservative, may be a suitable substrate to achieve biostimulation. It is a relatively inexpensive, harmless aromatic compound that has been previously used in "analogue enrichment" schemes to enhance biodegradation of the aromatic herbicide, 2,3,6-trichlorobenzoic acid (2,3,6-TBA) [336]. Benzoate ion is also an intermediate in the toluene pathway and it can induce related enzymes involved in the degradation of toluene and *m*- and *p*-xylenes [336]. In addition, the anionic nature of benzoic acid would minimize its retardation and facilitate its distribution when injected into an aquifer. Thus, the addition of benzoate ion to establish and sustain a reactive zone would depend mainly on its ability to acclimate the indigenous microbial consortium and enhance the growth and viability of BTEX degraders, even in the absence of BTEX.

Accordingly, Alvarez et al. [28] used flow-through aquifer columns to evaluate the efficacy of using benzoate ion (from sodium benzoate) as a biostimulatory substrate to enhance the aerobic biodegradation of benzene, toluene, and *o*-xylene (i.e., BTX), fed continuously at low concentrations. They reported the following key points:

- When used as a co-substrate, benzoate addition enhanced BTX degradation kinetics and attenuated BTX breakthrough relative to acetate-amended or unamended control columns.
- The benzoate-amended column experienced an increase in the predominance of pseudomonad species capable of degrading BTX.
- The feasibility of injecting benzoate to enhance the growth of BTX degraders and establish a buffer zone downgradient of a BTX plume was also investigated.
- Using pristine aquifer material without previous exposure to BTX, aquifer columns were fed benzoate, acetate, or mineral medium without supplemental substrates during a two-day acclimation stage. All columns were subsequently fed BTX alone, and their breakthrough was monitored.
- Previous exposure to benzoate, but not to acetate, shortened the acclimation period to BTX degradation and enhanced the short-term bio-attenuation potential of the indigenous consortium, suggesting that benzoate could potentially be used to establish and sustain in situ reactive zones to attenuate BTX migration and protect downgradient groundwater resources.

4.1.2
Polycyclic Aromatic Hydrocarbons

Polycyclic aromatic hydrocarbons (PAHs) are common pollutants in contaminated bottom sediments and soils and usually occur as a complex mixture of low- to high-molecular weight (HMW) compounds (see Chap. 1). These compounds are of concern due to their acute toxicity, mutagenicity, or carcinogenicity. Prior laboratory and larger scale work on the biodegradation of PAHs has indicated that the removal of compounds with four or more rings (defined here as high-molecular weight PAHs) is often less extensive than the removal of lower molecular weight compounds [153, 337].

Hydrocarbon-degrading microorganisms are ubiquitous in most ecosystems [32, 117]; however, it is often very difficult to prove that transformation, degradation, and mineralization actually occur in the environment because it is difficult to distinguish contributions from biotic and abiotic processes under uncontrolled conditions in the natural environment [338]. In contrast, laboratory assays can provide definitive evidence for microbial degradation, and sterilized samples can be used to determine abiotic losses. Thus, contributions from microbial degradation can be differentiated from abiotic loss by a mass balance study performed in a sealed vessel. Such studies have led to a better understanding of biodegradation of organic compounds [48, 57, 100, 107, 117, 118, 127, 306]. Results obtained from laboratory studies have also been applied to in situ bioremediation of gasoline-contaminated aquifers and soils with oil spills [32, 117, 138, 153, 339–341]. Several workers have shown that microorganisms enriched from seawater and sediment samples are capable of utilizing PAHs such as phenanthrene and pyrene [322, 342]. The microbial degradation of pyrene was further confirmed by the production of metabolites and $^{14}CO_2$ from ^{14}C-labeled pyrene [322].

Three approaches have been recommended to obtain evidence for in situ biodegradation [71, 343, 344], including: (1) quantitative determination of the pollutant of interest in samples collected at different times to show a decrease in its concentration over time, (2) laboratory-based microbial degradation studies under conditions that mimic the environment to show the potential of biodegradation in the field, and (3) searching for a particular metabolite of biodegradation in samples collected from the field. Thus, without knowing the amount and nature of PAH inputs, it is impossible to estimate any biotic loss of PAHs.

Because weathering and other abiotic processes simultaneously occur and contribute to changes in the concentrations of PAHs in the field, laboratory microbial degradation and the determination of a target transformation metabolite appear to be useful to evaluate the possibility of microbial transformation in any contaminated environment. Such case studies follow:

- Li et al. [323] studied the bacterial transformation of pyrene in an estuarine environment (Kitimat Arm, British Columbia, Canada), where they separated a metabolite (i.e., *cis*-4,5-dihydroxy-4,5-dihydropyrene) from the sediment and pore waters. The presence of this key metabolite from the dioxygenase-mediated transformation of pyrene [100, 186, 342], along with previous pyrene degradation studies using cultures isolated from the same sediment samples, suggested a possible in situ bacterial transformation of pyrene in the Kitimat Arm environment.
- Wilson and Madsen [152] used the metabolic pathway for bacterial naphthalene oxidation as a guide for selecting 1,2-dihydroxy-1,2-dihydronaphthalene as a unique transient intermediary metabolite whose presence in samples from a contaminated field site would indicate active in situ naphthalene biodegradation (Fig. 26). Naphthalene is a component of a variety of pollutant mixtures. It is the major constituent of coal tar [345], the pure compound was commonly used as a moth repellant and insecticide [345], and it is a predominant constituent of the fraction of crude oil used to produce diesel and jet fuels [346]. Prior studies at a coal tar-contaminated field site have focused upon contaminant transport [10, 347], the presence of naphthalene catabolic genes [348, 349], and non-metabolite-based in situ contaminant biodegradation [343].

It should be mentioned that bioremediations of PAH contaminated sites are mainly affected by the degree of bioavailability (Sect. 4.1.2.1) of the PAHs as well as ways which enhance the rates and modes (Sect. 4.1.2.2). The following is a summary.

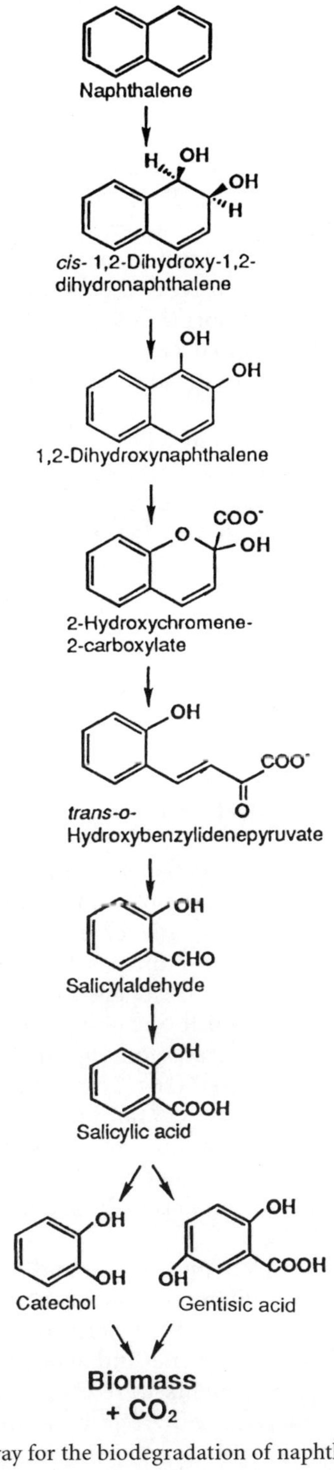

Fig. 26. The metabolic pathway for the biodegradation of naphthalene

4.1.2.1
Bioavailability

Studies of the fate of PAHs have shown that their microbial mineralization, especially PAHs with four or more benzene rings, decreases with increasing contaminant residence time in soils [150, 230, 350]. Decreased microbial mineralization is often attributed to PAHs association with the soil organic matrix (SOM) [282,230]. Proposed interactions between PAHs and SOM include adsorption and absorption, chemisorption, partitioning, and covalent binding to the soil matrix (see Chap. 2). Sorptive and partitioning processes reduce PAH mineralization by slowing PAH desorption from SOM into soil aqueous phases where biodegradation is believed to occur [25, 40, 217, 351]. Non-sorptive interactions may inhibit complete PAH degradation by hindering desorption of PAH transformation products.

The type of PAH-SOM interaction will significantly affect long-term contaminant fate and bioavailability [137]. Irreversible binding of pesticide residues in soil, a result of either biological or abiotic oxidative coupling reactions, has been proposed to limit residue desorption and transport [352, 353]. Recent evidence suggests that a significant fraction of bound pesticide residues may not irreversibly bind to soil but may sorb to soil via cation and hydrophobic interactions which do not necessarily limit residue mobility [351]. Both covalent and non-covalent interactions can contribute to non-linear, non-equilibrium distributions of contaminants in aqueous and solid phases of soils [223, 279, 354]. For nonionic, recalcitrant compounds such as DDT or higher molecular weight PAHs, adsorption and partitioning within SOM or soil micropores is considered a primary mechanism for association with SOM [64, 137, 355]. These associations involve mainly non-covalent interactions between pollutant and SOM [284].

If sorption and partitioning mechanisms dominate the fate of PAHs in soils, then the PAHs remaining in SOM should be primarily parent compounds which are sorbed to organic surfaces. Slow rates of desorption become the primary limitation for biodegradation; however, the presence of adapted PAH-mineralizing communities in contaminated soils suggests that PAH desorption occurs at sufficient rates over time to establish and maintain adapted microbial communities [36, 264, 356]. PAH biodegradation appears to proceed, albeit at much slower rates than predicted or desired [264, 278, 279].

Previous research has shown that contaminant biodegradation by specific microorganisms can alter desorption rates of contaminants from sorbing surfaces [226, 357–359]. For pesticides, biodegradation has been shown to contribute to significant residue accumulation in soil at rates much greater than surface sorptive interactions [352].

The primary focus of the study by Guthrie and Pfaender [360] was to assess how biological activity influenced interactions of pyrene and pyrene derivatives with soil organic matter, by determining how microbial activity influenced associations between pyrene and particular SOM fractions over extended periods of time. Experiments were then conducted to determine if pyrene-SOM associations altered the pyrene bioavailability, and designed to follow the fate of pyrene in a consistent soil matrix with and without microbial activity. Pyrene degradation and association with SOM were quantified by systematic removal and analysis of gas phase traps and soil subsamples from aerated soil chambers. The soil matrix was extensively fractionated to separate soluble SOM (lipids, carbohydrates, and humic/fulvic acids) and insoluble SOM (humin). SOM extracts were analyzed by HPLC and liquid scintillation counting (LSC) to determine residual pyrene concentrations and the formation of intermediate products. The ^{14}C activity in soils and SOM fractions was assayed after 270 days for bioavailability by incubating soils or soil fractions with a microbial community shown to mineralize pyrene in static microcosms and measuring the amount of evolved $^{14}CO_2$ over time. Comparisons were made between soils with and without microbial activity to determine the extent of biological influence on pyrene-SOM interactions and pyrene biodegradation with time.

4.1.2.2
Enhancement

One of the strategies applied to enhance the degradation of specific PAH is to offer bacteria one or more known inducers to stimulate both selective growth of PAH degraders and induction of PAH metabolism [38, 73, 132, 148, 182, 194]. However, little has been reported on the regulation of PAH metabolism by bacteria for compounds other than naphthalene. The transformation of benz[*a*]-anthracene was found to be inducible by salicylate [188] in a strain that has recently been identified as *Sphingomonas yanoikuyae* [172], but little else is known about the regulation of the metabolism for HMW PAH.

Many bacteria with PAH-transforming capabilities have a relatively broad substrate range [23, 98, 149, 164, 178, 186, 361], and pre-exposure of an individual species or a microbial community to one PAH can result in enhanced degradation of other PAHs [127, 177]. Such observations suggest that these microorganisms might possess one or more broad-specificity enzymes for PAH metabolism. Naphthalene dioxygenase, the enzyme responsible for the initial oxidation of naphthalene, has a wider substrate specificity which permits the *cis*-dihydroxylation of several aromatic compounds [175, 184] and consequently has been referred to as PAH dioxygenase [44]. Molecular evidence also indicates that an individual bacterium species may transform multiple PAHs through a common pathway found in several bacterial strains [94, 96, 161, 173, 190].

If PAH-degrading microorganisms use broad-specificity enzymes or common pathways to transform multiple PAHs, then inducers for the metabolism of one PAH substrate might co-induce the transformation of a range of PAHs. Preliminary evidence indicated that the transformation of naphthalene, phenanthrene, fluoranthene, and pyrene by *Pseudomonas saccharophila* P15 was stimulated by salicylate [132], a known inducer of naphthalene metabolism in pseudomonads [43]. However, Chen and Aitken [181] reported in more detail the inducing effects of salicylate on the transformation of various HMW PAHs by *Pseudomonas saccharophila* P15 isolated from contaminated soil, including initial rates of transformation and the mineralization of benz[*a*]anthracene, chrysene, and benzo[*a*]pyrene. They reported the following:

- Strain P15 was grown on phenanthrene by a known pathway in which salicylate is an intermediate. Pre-incubation with phenanthrene and downstream intermediates through salicylate stimulated PAH dioxygenase activity and initial rates of phenanthrene removal, suggesting that salicylate was the inducer of this activity.
- Salicylate also greatly enhanced initial rates of removal of fluoranthene, pyrene, benz[*a*]anthracene, chrysene, and benzo[*a*]pyrene, HMW PAH substrates which strain P15 did not use for growth.
- The specific rate of removal of benzo[*a*]pyrene was at least two orders of magnitude lower than that of the four-ring compounds and nearly five orders of magnitude lower than that of phenanthrene.
- The mineralization of phenanthrene, benz[*a*]anthracene, chrysene, and benzo[*a*]pyrene was stimulated by pre-incubation with phenanthrene or salicylate, although significant mineralization of phenanthrene, benz[*a*]-anthracene, and chrysene occurred in un-induced cultures.
- Further experiments with chrysene indicated that it did not induce its own mineralization.
- In general, the study suggested that *Pseudomonas saccharophila* P15 expressed a low level of constitutive PAH metabolism which was inducible to much higher levels and that HMW PAH metabolism by this microorganism was induced by the low-molecular weight substrates phenanthrene and salicylate.

4.1.3
Dichlorobenzidine

Several million kilograms of 3,3′-dichlorobenzidine (DCB) and benzidine were produced in the United States up to 1977 for the production of dyes and pigments [363]. Recognition of the carcinogenic nature of DCB and its lesser-chlorinated congeners including benzidine resulted in a reduction of their use [362, 363]). Benzidine has been found to be carcinogenic in the human bladder and in oral passages in animals. DCB has likewise been acknowledged to induce cancer in animals and is considered a potential carcinogen in humans [363]. The carcinogenicity of DCB toward humans is believed to be attributable to dehalogenation in the digestive system, resulting in benzidine formation [364]. The U.S. Environmental Protection Agency [365] established water quality criteria for DCB and benzidine of 10 ng/l and 0.12 ng/l, respectively. Ambient water containing DCB and benzidine at these concentrations was estimated to result in an incremental increase of human cancer risk of 10^{-6} over the lifetime of an exposed population.

Several factors govern the transport and fate of hydrophobic organic chemicals in sediment/water environments; microbially mediated reactions and sorption are major processes affecting the fate of these compounds in aquatic systems [166, 366–368]. Aryl halides have been shown to undergo microbially-mediated dehalogenation under anaerobic conditions [38, 52, 68, 105, 116,

369–371]. For example, chloroanilines and polychlorinated biphenyl congeners have been shown to alter by microbially-mediated reductive dehalogenation in sediment/water systems, yielding less chlorinated congeners [38, 48, 52, 68, 105, 116, 119, 369–371].

To elucidate the fate of these compounds at sediment-water interfaces, sediment/water mixtures (Lake Macatawa, Holland, MI) were spiked with DCB and incubated at 20 °C for 12 months under anaerobic conditions [72]. Dehalogenation of DCB to benzidine appeared to take place through a transient intermediate, 3-monochlorobenzidine (Fig. 27), which was observed in time-course analyses of the sediment/water mixtures. No metabolites were observed in autoclaved samples, suggesting that dehalogenation of DCB in anaerobic sediment/water systems was mediated by microbial activity. The product of dehalogenation (benzidine, Fig. 27) is more toxic to humans than the parent compound, DCB. From sediment/water distribution experiments, DCB showed greater affinity for the sediment phase than its non-chlorinated derivative,

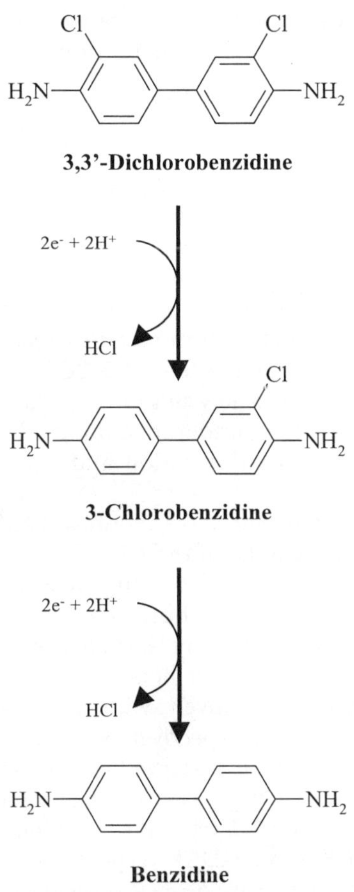

Fig. 27. Reductive dehalogenation of dichlorobenzidine

benzidine. Therefore, progressive dehalogenation of DCB in anaerobic lake sediments was expected to yield a greater total concentration of benzidine in the solution phase, a shift to a more toxic form, and greater potential for transport in the environment.

4.1.4
Chlorinated Hydrocarbons

1,1,2,2-Tetrachloroethane (TeCA) was the first chlorinated hydrocarbon solvent produced in large quantities before World War I [371]. It was used as a solvent for cellulose acetate, fat, waxes, greases, rubber, and sulfur. In a few cases, TeCA is used as a carrier or reaction solvent in manufacturing processes for other chemicals and as an analytical reagent for polymers [371]. TeCA was largely replaced by less toxic solvents after 1945. TeCA release in the United States varied from 44,000 pounds in 1988 to 66,000 pounds in 1991 [372].

Little information about TeCA transformation in groundwater is available; however, reductive dechlorination, dehydrochlorination, and dichloroelimination are three possible reactions for TeCA transformations. The following is a brief summary of biotransformation research conducted for TeCA:

- Reductive dechlorination or reductive hydrogenolysis is a common transformation of 1- and 2-carbon chlorinated aliphatics under methanogenic conditions [373, 374]. 1,1,1-Trichloroethane (1,1,1-TCA), for example, is converted to 1,1-dichloroethane (1,1-DCA) [375], and Perchloroethylene (PCE) is successively converted to TCE, cDCE, VC, and ethane [274]. Each reductive dechlorination is a two-electron transfer reaction.
- Dehydrochlorination has been observed, for example, in the abiotic conversion of pentachloroethane to PCE [376] and 1,1,1-TCA conversion to 1,1-DCE [375]. Dehydrochlorination is not a redox reaction.
- Bouwer and McCarty [374] suggested 1,1,2-TCA is an intermediate of TeCA transformation in continuous-flow column experiments and TCE as an intermediate in a batch experiment under methanogenic conditions. Those transformations are reductive dechlorination and dehydrochlorination, respectively.
- Dihaloelimination is a two-electron transfer reaction. Thompson et al. [377] reported reductive dichloroelimination of 1,1,2-TCA and TeCA by hepatic microsomes from rat liver, with VC and both tDCE and cDCE as metabolites. Reductive dichloroelimination from hexa- and pentachloroethane by microsomal cytochrome P450 was studied by Nastainczyk et al. [378]. The main products of the in vitro metabolism of hexa- and pentachloroethane were PCE (99.5%) and TCE (96%), respectively, with minor amounts of pentachloroethane (0.5%) and TeCA (4%), respectively, via reductive dechlorination.
- Dihaloelimination has also been observed under partially aerobic conditions [274]. With cytochrome P-450_{CAM} as a primary catalyst, dichloroelimination from hexa-, penta-, and 1,1,1,2-tetrachloroethane were catalyzed, and the products were PCE, TCE, and 1,1-DCE, respectively; no reaction was observed with TeCA. Significant rates were observed for these reactions at 5% oxygen concentration.

- Schanke and Wackett [379] reported TeCA degradation by transition-metal coenzymes. cDCE (53%), tDCE (29%), VC (14%), ethylene (1%), and traces of 1,1,2-TCA were the products from this abiotic transformation with vitamin B_{12} and titanium(III) citrate. Both dechlorination and dichloroelimination had occurred; the major route of degradation was reductive dihaloelimination.
- Chen et al. [380] studied the abiotic and biotic transformations of 1,1,2,2-TeCA under methanogenic conditions. They reported that TeCA degradation started without lag with municipal digester sludge. 1,1,2-TCA, tDCE, and cDCE were products of biotic transformation, while TCE resulted from abiotic degradation. TCE was further transformed to cDCE, VC, and ethene. Ethene, VC, and tDCE were the persistent products of TeCA transformations. With the same municipal digester sludge culture, 1,1,2-TCA was removed and converted to 1,2-DCA and VC. 1,2-DCA partially degraded, resulting in chloroethane and ethene formation. Reductive dechlorination, dichloroelimination, and dehydrochlorination simultaneously took place during the degradation of TeCA. Dichloroelimination and dehydrochlorination played important roles in the removal of TeCA and 1,1,2-TCA under methanogenic conditions.

4.1.5
Carbon Tetrachloride

Carbon tetrachloride (CT) is a significant pollutant at hazardous waste sites, and its biodegradation has been reported extensively. While aerobic transformation is not favorable, CT can be degraded under denitrifying conditions [294, 374, 381, 382], sulfate reducing conditions [383–385], methanogenic conditions [69, 70, 238, 386, 387] and fermentation conditions [374, 381]. The transformation products reported include less chlorinated methanes, i.e., chloroform (CF), dichloromethane (DCM), and chloromethane (CM), with CO, CO_2, and CS_2, suggesting both reductive and substitutive pathways for CT transformation [383]. Although some bacteria can use such chlorinated methanes as CM and DCM to support growth [388, 389], no growth has been shown with CT, suggesting that microbial CT transformation is a co-metabolic process.

Several authors indicated that the microbial transformation of CT is considered to be closely related to the presence of microbial cofactors, such as porphinoids (cofactor F_{430}) and corrinoids (vitamin B_{12}) [60–70, 238, 386]. In vitro, abiotic degradation of CT, mediated by these cofactors under reducing conditions, has been widely reported. Such cofactors can serve as electron carriers passing electrons from a donor to reduce CT, as follows:

- In the presence of a strong reductant such as titanium citrate, dithiothreitol, or sulfide, cofactor F_{430}, and vitamin B_{12} can dechlorinate CT to either less chlorinated products (CF, DCM, and CM) or to completely non-chlorinated products as CO, CO_2, and formic acid at relatively high rates [262, 390].
- Hashsham et al. [230] reported a tenfold increase in the CT degradation rate when 2 µmol/l of vitamin B_{12} was added to their culture grown anaerobically on DCM.

- Workman et al. [155] studied CT dechlorination by an iron-reducing microbial culture amended with vitamin B_{12}. They found that the culture reduced cobalt(III) in vitamin B_{12} to cobalt(II), and that the reduced vitamin B_{12} carried out CT dechlorination. While vitamin B_{12} addition can significantly enhance microbial CT degradation under reducing conditions, some microorganisms such as methanogens and some acetogens contain elevated levels of this cofactor and have shown CT degradation capability [69, 70, 238, 383, 386, 387]. However, the relationship between the cellular vitamin B_{12} content and CT degradation performance has not been well defined.
- In addition to methanogens and some acetogens, bacteria capable of 1,2-propanediol fermentation have been reported to produce vitamin B_{12} [286]. The transformation of 1,2-propanediol (propylene glycol) to propionaldehyde requires vitamin B_{12}. Further fermentation of propionaldehyde to *n*-propanol and propionic acid does not require vitamin B_{12} but yields energy for growth.
- There are no reports regarding CT degradation by these bacteria containing vitamin B_{12}. Although vitamin B_{12} is only produced under fermentation conditions, the bacteria can grow aerobically [286]. Therefore, an anaerobic/aerobic operating sequence with anaerobic propanediol feeding might be advantageous for CT degradation by selecting for vitamin B_{12}-producing bacteria and then maximizing their biomass production through the oxidation of fermentation products under aerobic conditions. The aerobic step inhibits methanogenic activity so that fermentation is the main reaction in the anaerobic step.
- Zou et al. [167] evaluated the CT degradation kinetics for the different cultures, investigated the relationship between intracellular vitamin B_{12} content and CT degradation, and determined the effect of the presence of growth substrate on CT degradation. The effect of the aerobic step in the anaerobic/aerobic operating sequence on biomass growth and CT degradation was also evaluated.

4.1.6
Dicamba

Dicamba (3,6-dichloro-2-methoxybenzoic acid) is primarily used as a post-emergence broadleaf herbicide, which interferes with normal plant auxin function, subsequently causing uncontrolled growth and the inhibition of the phototropic and geotropic function. Cumulative response results in plant death. The success of auxinic analogues such as Dicamba and 2,4-dichlorophenoxyacetic acid in weed control has led to widespread manufacturing and use. Estimated U.S. production for Dicamba was 5 million kg in 1990 [391].

The possibility for transport of Dicamba in subsurface soils, resulting in subsequent groundwater pollution, is potentially high. Both Dicamba and its initial transformation product 3,6-dichlorosalicylic acid have pK_a values of 1.95 [392]. The high solubility of these weak acids at neutral to high pH makes it feasible for them to be mobile in lime treated or neutral pH soils. In the field, Dicamba: (1) has been found to leach to a depth of 1 m over a 2-month period following application in a Missouri clay pan soil [296], (2) was discovered in approximately

2% of pesticide monitoring wells tested in Iowa [393], (3) was one of six pesticides found in the shallow aquifers on the Delmarva Peninsula in Maryland [235], (4) was detected in 21% of the groundwater samples taken during a field study on pesticide leaching from historically sprayed agricultural plots [283], and (5) was also found in over 4% of 45 wells tested in 1992 by the U.S. Geological Survey [234].

The occurrence of Dicamba in groundwater at sites of herbicide application and drainage is an impetus for studying the fate of this compound in anoxic environments. Anaerobic microbial respiration in aquatic bottom sediments and aquifers can take place via a variety electron acceptors [309, 394–396], and microbial degradation of herbicides and substituted aromatic compounds (herbicide metabolites) can be influenced by the type of electron acceptors present [55, 115, 397, 398]. In general, the electron acceptors are utilized in order of their relative energy potential following the sequence O_2, NO_3^-, Mn(IV), Fe(III), SO_4^{-2}, and the redox zones in anoxic aquifers and sediments can become stratified [276].

Biodegradation of Dicamba in the presence of oxygen, through the *O*-demethylated product 3,6-dichlorosalicylic acid, is well documented [160, 261, 399, 400]. The metabolite 2,5-dihydroxy-3,6-dichlorosalicylic acid has been reported as an intermediate after *O*-demethylation of Dicamba [251], but the pathway for degradation of 3,6-dichlorosalicylic acid has not been investigated in detail. Under anaerobic, methanogenic conditions, transformation of Dicamba through *O*-demethylation and subsequent reductive dehalogenation of 3,6-dichlorosalicylic acid to 6-chlorosalicylic acid has been observed [401]. The metabolite 6-chlorosalicylic acid was resistant to further degradation and anaerobic mineralization has not been demonstrated. Furthermore, the biodegradability of Dicamba under different anaerobic conditions has not been investigated.

Milligan and Häggblom [65] examined the anaerobic biodegradability and transformation of Dicamba under denitrifying, iron Fe(III) reducing, sulfate reducing, and methanogenic conditions. Anaerobic microcosms were established with Dicamba treated agricultural soil and stream bottom sediments receiving golf course drainage, which were each spiked with Dicamba as a sole carbon source. In general, the study revealed that:

- The predominant electron-accepting process can affect the rate and extent of Dicamba degradation in anaerobic environments.
- The degradation activity depended on the anaerobic conditions and ranged between complete inhibition of biotransformation and mineralization of the herbicide. The degradation and transformation pathways that were observed under different reducing conditions are summarized in Fig. 28.
- Mineralization of Dicamba was demonstrated under methanogenic conditions and the degradation pathway elucidated.
- Methanogenic enrichments resulted in *O*-demethylation of Dicamba to 3,6-dichlorosalicylic acid which was reductively dechlorinated to 6-chlorosalicylic acid and to salicylic acid, which was in turn further degraded to CH_4 and CO_2.

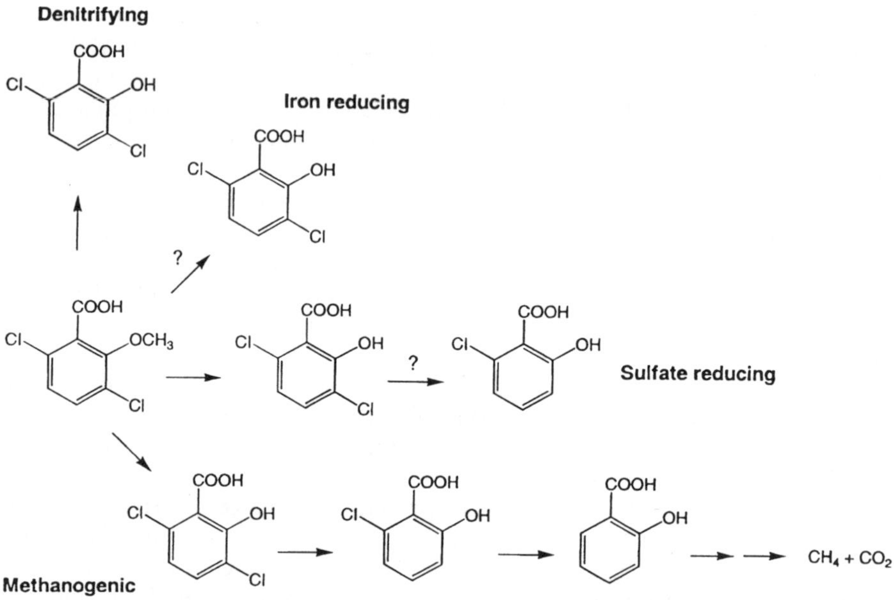

Fig. 28. Transformation and degradation of Dicamba under different reducing conditions [65]

- Transformation of Dicamba under sulfate reducing conditions did occur, but the extent of dehalogenation after *O*-demethylation remained unclear.
- Anaerobic *O*-demethylation was a prerequisite in all the cultures before Dicamba could be degraded, and reductive dehalogenation of the dichloroanisic acid prior to *O*-demethylation was not observed.
- The data provided clear evidence that anaerobic respiratory conditions must be taken into consideration when performing degradation feasibility studies and determining herbicide application practices in the future.

The finding that nitrate can inhibit the anaerobic transformation of Dicamba may have environmental implications, especially in agricultural areas where Dicamba is used extensively and where nitrogen from nitrate often exceeds the EPA maximum contaminant level of 10 mg/l in the groundwater. This suggests that applications of Dicamba where nitrate levels in groundwater are high may risk prolonging the anaerobic half-life of the herbicide in the aquifer. The observed accumulation of 6-chlorosalicylic acid by Milligan and Häggblom [65] may indicate that assessment of the toxicity and the recalcitrance of this chlorinated aromatic compound in anaerobic environments may be of more relevance than that of Dicamba.

4.1.7
Methyl Bromide

Methyl bromide is presently the most important preplanting soil fumigant commercially available [402]. This compound is used extensively in the United States in the production of many economically important crops for the management of plant-pathogenic nematodes, soil-borne fungi and bacteria, and weeds [403]. This compound is also used as a space fumigant for commodities, for structural pest control, and for quarantine and regulatory purposes. For preplanting soil fumigation, methyl bromide is generally applied under a sheet of polyethylene plastic, which may remain in place until the crop cycle is completed.

Due to its gaseous nature it may have an effect on the stratospheric ozone layer [281, 402, 404]. After injection into soil for fumigation, methyl bromide rapidly diffuses through the soil pore space to the soil surface and then into the atmosphere [159, 162, 163, 405, 406]. Since a plastic sheet typically covers the soil surface, the rate of emission into the atmosphere depends upon the thickness and density of the plastic, if other conditions are the same [159, 406]. Other routes of disappearance from soil include chemical hydrolysis, methylation to soil organic matter through free radical reactions, and microbial degradation [136, 159, 405, 407]. Several reports appeared on the study of the microbial transformations of methyl bromide, summarized as follows:

- Yagi et al. [159] reported that up to 70% of the injected methyl bromide was degraded in soil.
- Shorter et al. [405] suggested that bacteria were responsible for the biological degradation of methyl bromide in several soil samples studied for biotransformation.
- Microorganisms, capable of utilizing short-chained halogenated hydrocarbons (e.g., methyl bromide) as a sole source of carbon and energy for growth, degraded these compounds through co-metabolic processes [33, 45, 154, 266, 408–410].
- Rasche et al. [410] reported that some terrestrial and marine nitrifiers had the capacity to oxidize methyl bromide to formaldehyde and bromide ion. They concluded that ammonia monooxygenase produced by the nitrifiers, which catalyzes the oxidation of ammonia to hydroxylamine, was responsible for the oxidation of methyl bromide to formaldehyde.
- Oremland et al. [136] subsequently demonstrated that methane-oxidizing bacteria also had the capacity to co-oxidize methyl bromide by methane monooxygenase produced during the oxidation of methane to methanol. They also showed that methanotrophic soils that had a high capacity to oxidize methane degraded ^{14}C-labeled methyl bromide to $^{14}CO_2$.
- Ou et al. [74] reported the enhancement of the degradation of methyl bromide in soil pretreated with an ammonia-based nitrogen fertilizer (i.e., $(NH_4)_2SO_4$) and stimulation of methyl bromide degradation in soil inoculated with a nitrifier, *Nitrosomonas europaea*.

4.1.8
Trinitrotoluene

The relevant authorities and remediation companies of many industrialized countries have made numerous efforts to develop and establish efficient and reasonable techniques for the cleanup of contaminated sites with explosives. 2,4,6-Trinitrotoluene (TNT) was the most widely produced and used explosive in both World Wars [215]. The remediation of soils and groundwater contaminated with TNT is of particular concern, since this compound and its reduced metabolites (i.e., aminodinitrotoluenes and diaminonitrotoluenes) are toxic to a variety of biota and show a broad spectrum of toxicological behavior ranging from mutagenic to carcinogenic activity [256, 411–413].

Various soil remediation techniques such as incineration, soil washing, or biological soil treatment were applied in the past, but the microbiological degradation of TNT-contaminated soils is considered to be the most favorable technique as far as costs are concerned [414]. The following is a summary of these TNT remediation technologies:

- Several workers recommended a promising strategy by boosting the bioremediation of contaminated soil with cheap biomass products such as alfalfa, sawdust, chopped potato waste, apple pomace, cow and chicken manure, straw, or molasses in compost systems [215, 415–417]. These applications have led to transformations of TNT of more than 95% [414, 415, 417] and were often accompanied by detoxification effects [414, 418].
- Rieger and Knackmuss [419] and Lenke et al. [420] tested and evaluated an anaerobic/aerobic bioremediation process in a technical scale volume of up to 18 m^3 of contaminated soil. They also recommended the use of such anaerobic/aerobic processes for contaminated sites with various levels of TNT.
- Drzyzga et al. [411] conducted experiments to evaluate the levels of incorporation and transformation of TNT and metabolites into the organic soil matrix of anaerobic and sequential anaerobic-aerobic treated soil/molasses mixtures. They proposed a two-step treatment process (i.e., anaerobic-aerobic bioremediation process) with some special procedures during the anaerobic and the aerobic treatment phases. The transformation of TNT at the end of the experiments was above 95% and 97% after anaerobic and sequential anaerobic-aerobic treatment, respectively. This technique is considered the most promising method for effective, economic, and ecologically acceptable disposal of TNT from contaminated soils by means of immobilization (e.g., humification) of this xenobiotic.

4.1.9 *Silicon-Based Organic Compounds*

Tetraalkoxysilanes, a group of silicon-based compounds such as tetrabutoxysilane [i.e., TBOS, $(CH_3CH_2CH_2CH_2O)_4Si$] and tetrakis(2-ethylbutoxy)silane [TKEBS, $(CH_3CH_2CH{-}(CH_3CH_2)CH_2O)_4Si$], contain four oxygen bridges from the central silicon atom to the corresponding organic (alkoxy) groups. These compounds are widely used as heat-exchange fluids, sealants, and lubricants because of their excellent thermal properties [24, 421–423].

At Lawrence Livermore National Laboratory site 300, these compounds along with trichloroethylene (TCE) were used in heat-exchanger pipes at their materials testing facility [421–423]. Subsurface contamination by these compounds resulted from leaking heat-exchanger pipes. TBOS and TKEBS were present as light non-aqueous phase liquids whereas TCE was present as a dense non-aqueous phase liquid in the subsurface and was also dissolved in both pure phase alkoxysilanes and groundwater.

Available literature on TBOS and TKEBS mainly focuses on their thermal properties [24]. Specific research work related to the transformation of these compounds under environmental conditions is limited, and biological degradation of these compounds has not been investigated [423]. However, numerous hydrolysis studies have been conducted on the lower homologues of the tetraalkoxysilanes such as tetramethoxysilane and tetraethoxysilane [229, 423]. These compounds hydrolyze abiotically to give the corresponding alcohols and silicic acid [424].

TCE is the other major contaminant at the site and is a common groundwater contaminant in aquifers throughout the United States [425]. Since TCE is a suspected carcinogen, the fate and transport of TCE in the environment and its microbial degradation have been extensively studied [25, 63, 95, 268, 426, 427]. Reductive dechlorination under anaerobic conditions and aerobic co-metabolic processes are the predominant pathways for TCE transformation. In aerobic co-metabolic processes, oxidation of TCE is catalyzed by the enzymes induced and expressed for the initial oxidation of the growth substrates [25, 63, 268, 426]. Several growth substrates such as methane, propane, butane, phenol, and toluene have been shown to induce oxygenase enzymes which co-metabolize TCE [428].

Vancheeswaran et al. [421–423] investigated the attenuation of silicon-based organic compounds (i.e., tetraalkoxysilanes) along with TCE as subsurface contaminants by abiotic hydrolysis and biological mineralization at Lawrence Livermore National Laboratory site 300. They reported, under abiotic conditions, the hydrolysis of the alkoxysilanes such as TBOS and TKEBS to 1-butanol and 2-ethylbutanol, respectively, and silicic acid. An aerobic microbial culture from the local wastewater treatment plant that could grow and mineralize the alkoxysilanes was enriched. The enriched culture was reported to hydrolyze rapidly TBOS and TKEBS and grow on the hydrolysis products. The microorganisms grown on TBOS co-metabolized TCE and *cis*-1,2-dichloroethene (cDCE). TCE and cDCE degradation was inhibited by acetylene, indicating that a monooxygenase was involved in the co-metabolism process.

4.1.10
Dioxins

The environmental burden of waterways with polychlorinated dibenzo-*p*-dioxins (PCDD) has been at the forefront of public and regulatory concern, because of the toxicity associated with particularly the 2,3,7,8-(laterally) substituted congeners, which have a tendency to bioaccumulate throughout the trophic food chain. Contamination of aquatic sediments by dioxins includes both non-point (e.g., atmospheric deposition) and point sources (e.g., industrial effluents, combined sewage overflows), and is generally characterized by a dominance of hepta- and octa-CDD, with minor contributions of hexa- to tetra-CDD [429]. Elevated concentrations of the 2,3,7,8-TCDD isomer tend to be associated with direct discharge from sources such as 2,4,5-trichlorophenol production [54, 430].

Despite the environmental stability of these compounds, a number of reports have indicated that under reducing conditions prevailing in sediments dioxins may undergo transformation reactions, including dechlorination. The potential importance of reductive dechlorination, and perhaps one of the reasons for the emphasis on this transformation process, is illustrated by recent evidence that 2,3,7,8-TCDD may be in a state of flux, resulting from dechlorination of octa- and hepta-CDD and being further dechlorinated to 2-mono-CDD [54]. Beside dechlorination reactions in sediments [4], dioxin dechlorination reactions have been demonstrated in the presence of microorganisms ([5, 12, 13, 431–433], dihydroxylated monoaromatic compounds [433], vitamin B_{12}, and zero valent metals [3].

Based on information by Fu et al. [219], other reactions include trans-chlorinations (migration of chlorine from PCDD to organic matter) and polymerizations, which have not been quantified. In spiked sediments from the Hudson and Passaic Rivers and sediment microorganisms, lesser-chlorinated products accounted for 10–15% of the decrease in octa-CDD. Hepta-, tetra-, tri-, and 2-mono-CDD congeners tend to dominate the de-chlorination pattern [432, 433]. The microbial dechlorination sequence of octa-CDD is provided in Fig. 29 [432], which distinguishes a pathway via 2,3,7,8-TCDD (*peri*-dechlorination), from one, which does not produce this tetra-CDD isomer (*peri*-lateral dechlorination). The relative contribution of each pathway (i.e., the ratio of 2,3,7,8- to other tetra-substituted congeners) observed is dependent on the system tested, whereby the presence of phenolic compounds appears to shift the pathway toward *peri*-dechlorination and enhances the total yield of lesser chlorinated products [433].

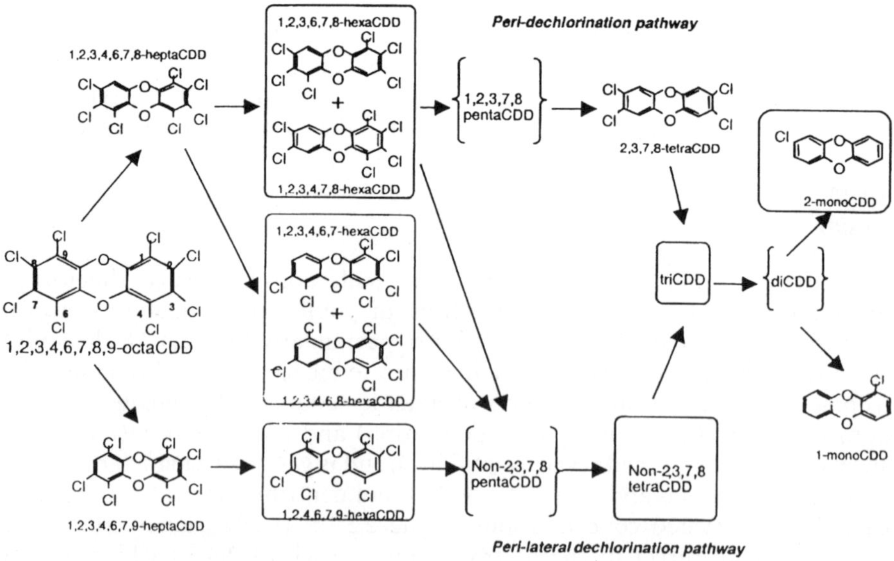

Fig. 29. Dechlorination pathway of octa-CDD in the presence of microbial cells [432]

Fu et al. [219] investigated the susceptibility of dioxins to dissolved organic carbon (DOC)-mediated dechlorination reactions by using 1,2,3,4,6,7,9-heptachlorodibenzo-*p*-dioxin (HpCDD), Aldrich humic acid (AHA), and polymaleic acid (PMA) as model compounds. The dechlorination yields were on the order of 4–20% which, when normalized to phenolic acidity, was comparable to yields observed in the presence of the humic constituents catechol and resorcinol. Based on the ratio of dechlorination yields as a function of phenolic acidity and electron transfer capacity, differences in electron transfer efficiency to dioxins were likely combined effects of specific interactions with the functional groups and nonspecific hydrophobic interactions. Hexa- and penta-CDD homologs were dominant in all incubations, and di-CDD constituted the final product of dechlorination. The rates of appearance of lesser chlorinated products were similar to those observed in sediment systems and followed thermodynamic considerations as they decreased with a lower levels of chlorination. Generally, both absolute and phenolic acidity-normalized rate constants for AHA-mediated reactions were up to twofold higher than those effected by PMA. These results indicated that the electron shuttling capacity of sediment DOC might significantly affect the fate of dioxins, in part through dechlorination reactions.

4.1.11
Alkylphenol Polyethoxylates

Alkylphenol polyethoxylates (APEO) are a major class of nonionic surfactants; over 230 million kg were sold in the United States in 1990 [434, 435]. They are most important in industrial applications but are also used in institutional and household cleaners and personal care products [436]. In recent years, APEs have received widespread attention in the United States and abroad because of their incomplete elimination during sewage treatment and the detection of their biodegradation intermediates in secondary effluents [17, 18, 21, 106, 115] and rivers receiving such effluents [17, 18, 115, 434, 437]. Reported concentrations in rivers range from less than 1 µg/l to greater than 100 µg/l for the various metabolites [436, Aboul-Kassim and Simoneit, unpublished report). The most common residues detected are those with shortened ethoxy chains so that just one or two ethoxy groups remain (AP1EO and AP2EO), the alkylphenol ethoxycarboxylates (APEC, or more specifically, AP*n*EC, where *n* is the number of ethoxy units plus a terminal acetic acid unit), and the alkylphenols (AP). The general structures of these compounds are presented in Fig. 30.

Because of their hydrophobicity, AP and APEO are often removed from the water phase by sorption onto sewage sludge [17, 18] and sediments [434]. In contrast, the hydrophilic APECs are especially difficult to remove from the aqueous phase; in most studies where APEC concentrations in effluents or receiving waters have been reported, they are the dominant APEO residues measured. Di Corcia et al. [254] reported APEC concentrations up to 145 µg/l in biologically treated wastewaters. APECs can persist in treated wastewater at low microgram per liter levels even after granular activated carbon contact or reverse osmosis [216] and have also been detected in finished drinking water [82, 438]. Reports of the environmental persistence of APEO residues are of concern because they

R = nonyl or octyl, and branched

m = usually 1-30

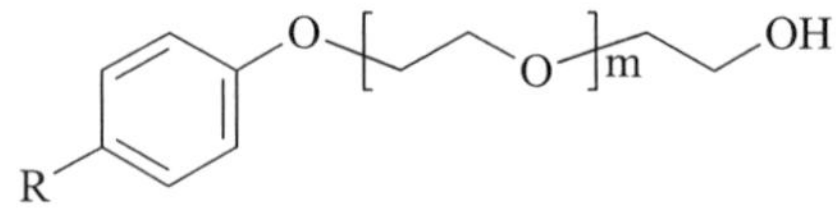

Alkylphenol Polyethoxylate (APEO)

R' = nonyl or octyl, and branched

n = usually 0-2

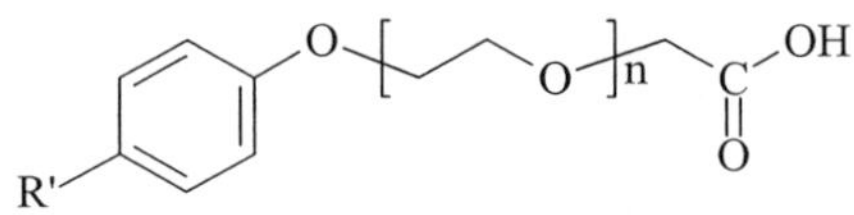

Alkylphenol Ethoxycarboxylate (APEC)

Fig. 30. Structures of APEO and major residues detected after biological treatment

can be toxic to aquatic life [249, 435, 437], and AP, AP1EO, and AP2EO have been shown to bioaccumulate in aquatic microorganisms [19, 439]. Consequently, the use of nonylphenol-based surfactants is being phased out in the European Community, and APEs in general are increasingly being replaced by the more easily degraded alcohol ethoxylates. Several reports have indicated that APE metabolites are estrogen mimics for both mammals and fish [247, 435, 440, 441]. Jobling and Sumpter [441] suggested that levels as low as 10 µg/l of NP, NP1EO, and NP2EO, within the ranges reported for polluted rivers, could affect fish reproduction.

The biodegradation of APEO has been studied by many researchers and has been the subject of extensive reviews [17, 18, 35, 237, 435, 436, 442, 443]. However, few studies have gone beyond the report of primary biodegradation or the removal of the parent compound. Those that have looked in detail at the progression of APEO degradation generally report the formation of the recalcitrant AP and short-chain APEO and APEC residues [17, 18, 237], but few details have been reported about the ultimate fate of the aromatic ring and alkyl side chain. Ding et al. [444] recently presented evidence for the carboxylation of alkyl side chains in APECs detected in tertiary treated wastewater effluents, but the exact structure of the carboxylated side chains could not be determined from the data.

The results of a detailed study of the further degradation of one isomer of APEC and its brominated analog by groundwater microorganisms are presented next, with the identification of persistent novel metabolites [444]. Bromination of the aromatic ring of APEO metabolites can occur during chlorine disinfection in the presence of bromide ion [21], and both APECs and brominated APECs (BrAPECs) have previously been detected at microgram per liter levels in reclaimed water produced at Water Factory 21 (WF21) in Orange County, California [216, 445]. APECs were also detected in groundwater from an aquifer recharged by direct injection of the WF21 effluents, but the BrAPECs were not [216].

Degradation of mixtures of brominated and non-brominated octylphenol ethoxycarboxylates (BrOPEC and OPEC, respectively) by sewage microorganisms was previously studied by Ball et al. [221]. They observed that OPEC could be transformed aerobically or anaerobically, but BrOPEC with fewer than three ethoxy units were recalcitrant under aerobic conditions, and only brominated octylphenoxyacetic acid (BrOP1EC) and possibly BrOP5EC were transformed anaerobically. However, because the studies were performed with mixtures of similar structures, only removal of the parent compounds could be reported.

Fujita and Reinhard [111] studied the aerobic biological transformation of octylphenoxyacetic acid (OP1EC) and its brominated analog (BrOP1EC) by groundwater enrichment cultures, and several persistent metabolites were identified by GC-MS. OP1EC is a representative of the class of alkylphenol ethoxycarboxylates (APEC), formed from alkylphenol polyethoxylate nonionic surfactants during sewage treatment. BrOP1EC is a byproduct formed during chlorine disinfection in the presence of bromide. The metabolite 2,4,4-trimethyl-2-pentanol was detected in stoichiometric quantities in OP1EC-metabolizing enrichment cultures, representing the intact alkyl side chain as a tertiary alcohol. BrOP1EC was transformed by the OP1EC-utilizing cultures only if OP1EC was simultaneously metabolized, suggesting a co-metabolic mechanism of transformation. Brominated intermediates were also detected, such as brominated octylphenol and a compound tentatively identified as 2-aminomethoxy-3-bromo-5-(1,1,3,3-tetramethylbutyl)phenol.

4.1.12
Nonylphenol Ethoxylates

Nonylphenol ethoxylates (NPEOs) are extensively used as surfactants in industrial products (see Chap. 1). NPEOs are a mixture of polyethoxylated mono-alkylphenols, predominantly para-substituted, and are used in the manufacturing of paints, detergents, inks, and pesticides [435, 446]. Surfactants are common water pollutants because of their use in aqueous solutions, which are discharged into the environment in the form of wastewater from treatment plants or sludge stored in landfills. Degradation products of alkylphenol polyethoxylates, i.e., nonylphenol (NP), have the potential to be bioaccumulated, thereby becoming toxic to aquatic [447] and soil microorganisms [435, 448].

The partial degradation of NPEOs can proceed both aerobically and anaerobically, and although the metabolic pathways are not completely understood, it is believed that biotransformation commences at the hydrophilic part of the molecule and that C-2 units (ethylene glycol) are removed one at a time [435, 443], giving rise to nonylphenol mono- and diethoxylates (NPEO1–2). Complete degradation of NPEO1–2 may occur under aerobic conditions [17, 114, 439, 449], but they have been reported to be more persistent in anaerobic environments [62, 103]. Under aerobic conditions carboxylated metabolites may be formed [18, 62, 103, 114]. Furthermore, because NPEOs with one or two ethoxy groups are less hydrophilic than polyethoxylated NPEOs, they are subjected to non-biological elimination by sorption to hydrophobic sludge constituents and organic matter, among other materials.

Ejlertsson et al. [450] investigated qualitatively the anaerobic biotransformation and degradation of nonylphenol mono- and diethoxylates by microorganisms derived from: (1) an anaerobic sludge digester treating wastewater from a pulp plant and industrial wastewater containing NPEOs as a pollutant, (2) a landfill site where the very same sludge is deposited, and (3) a municipal waste landfill, the latter acting as a reference source. NPEO1 and NPEO2 (i.e., NPEO1–2) used in a mixture were chosen as model compounds. Anaerobic experimental bottles were amended with 100% digester sludge at three different concentrations of NPEO1–2. Unlabeled [U-^{14}C]-NPEO1–2 was used to detect any possible decomposition of the aromatic moiety of the NPEO1–2. All inoculates used degraded NPEO1–2, with nonylphenol (NP) forming the ultimate degradation product. The NP formed was not further degraded, and the incubations with labeled NPEO showed that the aromatic structure remained intact. Both landfill inoculates also transformed NPEO1–2.

4.1.13
Polychlorinated Biphenyls

Polychlorinated biphenyls (PCBs) were manufactured by catalytic chlorination of biphenyl to produce complex mixtures, each containing 60–90 different PCB molecular species or congeners (see Chaps. 1 and 4). In the United States, PCB mixtures were manufactured by Monsanto under the trade name Aroclor and were widely used as dielectric fluids in capacitors and transformers from 1929 to 1978. PCBs are widespread contaminants of aquatic sediments and continue to be a focus of environmental concern because they tend to accumulate in biota and are potentially toxic. The following sections show the most effective bioremediation techniques applied to various PCB contaminated environments:

4.1.13.1
Aerobic Degradation

Aerobic degradation in a mesophilic temperature range of 18–35 °C of river sediments contaminated by PCBs was reported by several authors [77, 79, 151]. The PCB-degrading bacterium *Alcaligenes eutrophus* H850 was isolated by Bedard and co-workers from PCB-contaminated dredge materials of the upper Hudson River [77, 79]. This bacterium has a particularly broad PCB congener specificity as compared to many other PCB-degrading bacterial isolates from the upper Hudson River and other sites [77, 79, 300]. In addition to PCB degradation by bacterial isolates, in situ microbial aerobic PCB degradation was demonstrated at a PCB-contaminated site in the upper Hudson River [104]. Fish and Principe [101] and Fish [102] also described microbial aerobic PCB degradation and anaerobic dechlorination of Aroclor 1242 in test tube microcosms of PCB-contaminated upper Hudson River sediment.

In contrast, Williams and May [151] investigated the low-temperature (e.g., 4 °C) microbial aerobic PCB degradation of PCB-contaminated upper Hudson River sediments. They reported the depletion of specific di- and trichlorobiphenyls in the surface layer of PCB-contaminated sediments. The loss of specific PCB congeners from the sediment was indicative of microbial aerobic PCB degradation and demonstrated that this degradation can occur in sediment samples at low temperatures.

4.1.13.2
Reductive Dechlorination

The discovery that microbial dechlorination of PCBs was occurring in many aquatic sediments brought the hope that this process would provide a natural means of remediation [371, 451]. Dechlorination decreases the bioaccumulation potential of PCBs by making them more degradable and is expected to decrease the potential toxicity of PCBs [2, 34, 105, 371, 451–453]. Extensive microbial dechlorination of PCBs has occurred in some aquatic sediments including those of the Hudson River (NY) and Silver Lake (Pittsfield, MA) [371].

Reductive dechlorination of PCBs is important because it reduces their potential toxicity and persistence. In situ dechlorination of PCBs attributed to microorganisms in the anaerobic sediments has been documented in the Hudson River, Silver Lake (MA), the St. Lawrence River (NY), and New Bedford Harbor (MA) [76, 371, 451, 454–456].

Reductive dechlorination is the only biodegradation mechanism known for highly chlorinated PCB congeners, such as the majority of PCB congeners found in Aroclors 1254 and 1260 [34]. It has been well established that microbial dechlorination of Aroclor 1260 can take place both in the environment and under laboratory conditions [14, 15, 34, 245, 371, 451, 456]. Some studies have also quantified the extent of dechlorination of octa- and nona-chlorobiphenyls in Aroclor 1260 [14, 15, 75, 108]. However, because of the complexity of both the starting Aroclor 1260 mixture and the product mixture formed, it is impossible to identify characteristic products and degradation pathways for specific congeners in the general mixture.

Several studies have investigated the dechlorination of single PCB congeners under anaerobic conditions [298, 457–460], but these have all focused on PCBs with six or fewer chlorine substituents. The dehalogenation of decachlorobiphenyl over time has also been reported [228]; however, the products were only tentatively identified and not quantified.

Bedard and May [452] used congener-specific GC with electron capture detection and mass spectrometric detection to determine the PCBs in sediments of Woods Pond (Lennox, MA). The congener distributions of all samples showed the hexa-, hepta-, and octachlorobiphenyls characteristic of Aroclor 1260, but key hexa- and hepta-CBs had decreased by as much as 45% relative to Aroclor 1260, and the tri-, tetra-, and penta-CBs had increased. GC-MS analysis revealed unusual tetra-, penta-, and hexa-CBs, many containing 2,4- and 2,4,6-chlorophenyl rings, which are uncommon in higher Aroclors, and provided strong

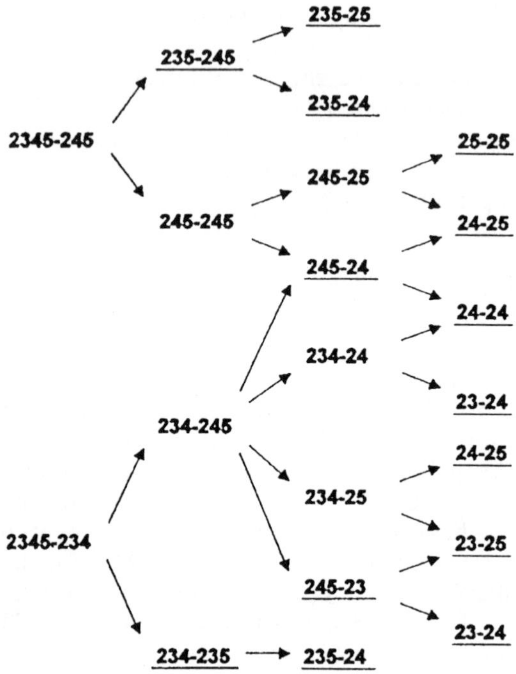

Fig. 31. Proposed routes of dechlorination for major hexa- and heptachlorobiphenyl components of Aroclor 1260 (after Bedard and May [452], with permission)

evidence of dechlorination. The proposed routes of dechlorination for major hexa- and hepta-CB components of Aroclor 1260 are shown in Fig. 31, indicating the following [453]:

- Congeners that were elevated in the sediment samples relative to Aroclor 1260 are underlined.
- One putative dechlorination product, 235–24-CB, co-migrates with an isomer, 245–25-CB, which was proposed for further dechlorination.
- The total PCBs showed little change relative to Aroclor 1260, but the composition changed from mainly 245–2′5′-CB to mainly 235–2′4′-CB.

Kuipers et al. [461] investigated the anaerobic dechlorination of four octachlorobiphenyls [i.e., 2,3,4,5,6,2′,3′,4′-octachlorobiphenyl (23456–2′3′4′-octaCB, Fig. 32), 2345–2′3′4′6′-octaCB (Fig. 33), 2345–2′3′5′6′-octaCB, and 23456–2′4′5′-octaCB (Fig. 34)] and three nonachlorobiphenyls [i.e., 23456–2′3′4′5′-nonaCB, 23456–2′3′4′6′-nonaCB, and 23456–2′3′5′6′-nonaCB]. They reported that all seven congeners were reductively dechlorinated; with dechlorination predominance patterns showing *meta*-dechlorination of doubly flanked *m*-chlorines followed by *meta*-dechlorination of singly flanked *m*-chlorines. Some *ortho*- and *para*-dechlorination was also observed. Figures 32–34 illustrate the dechlorination of several congeners, with relative amounts of products.

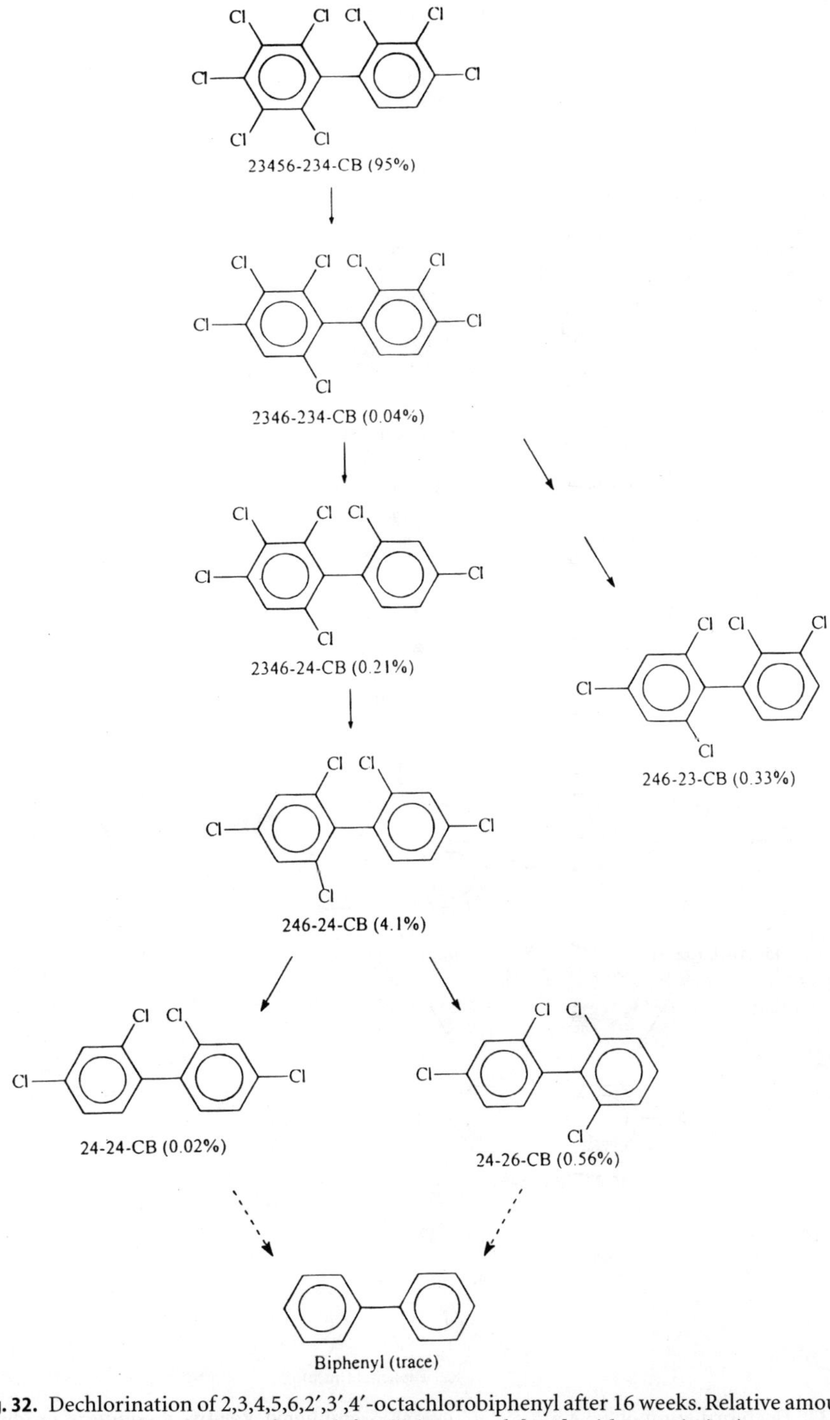

Fig. 32. Dechlorination of 2,3,4,5,6,2′,3′,4′-octachlorobiphenyl after 16 weeks. Relative amounts of products are given in parentheses (after Kuipers et al. [461], with permission)

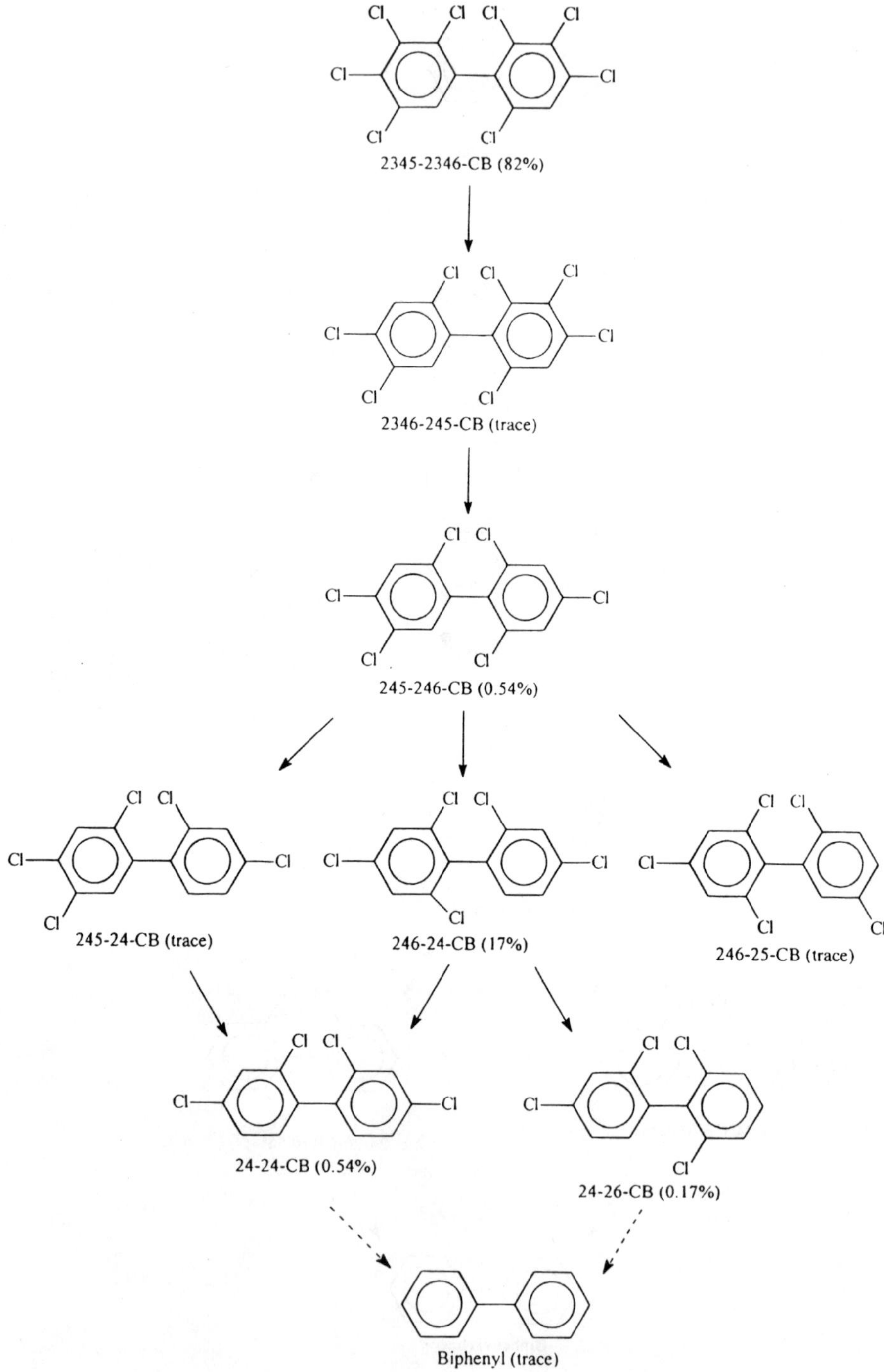

Fig. 33. Dechlorination of 2,3,4,5,2′,3′,4′,6′-octachlorobiphenyl. Relative amounts of products are given in parentheses (after Kuipers et al. [461], with permission)

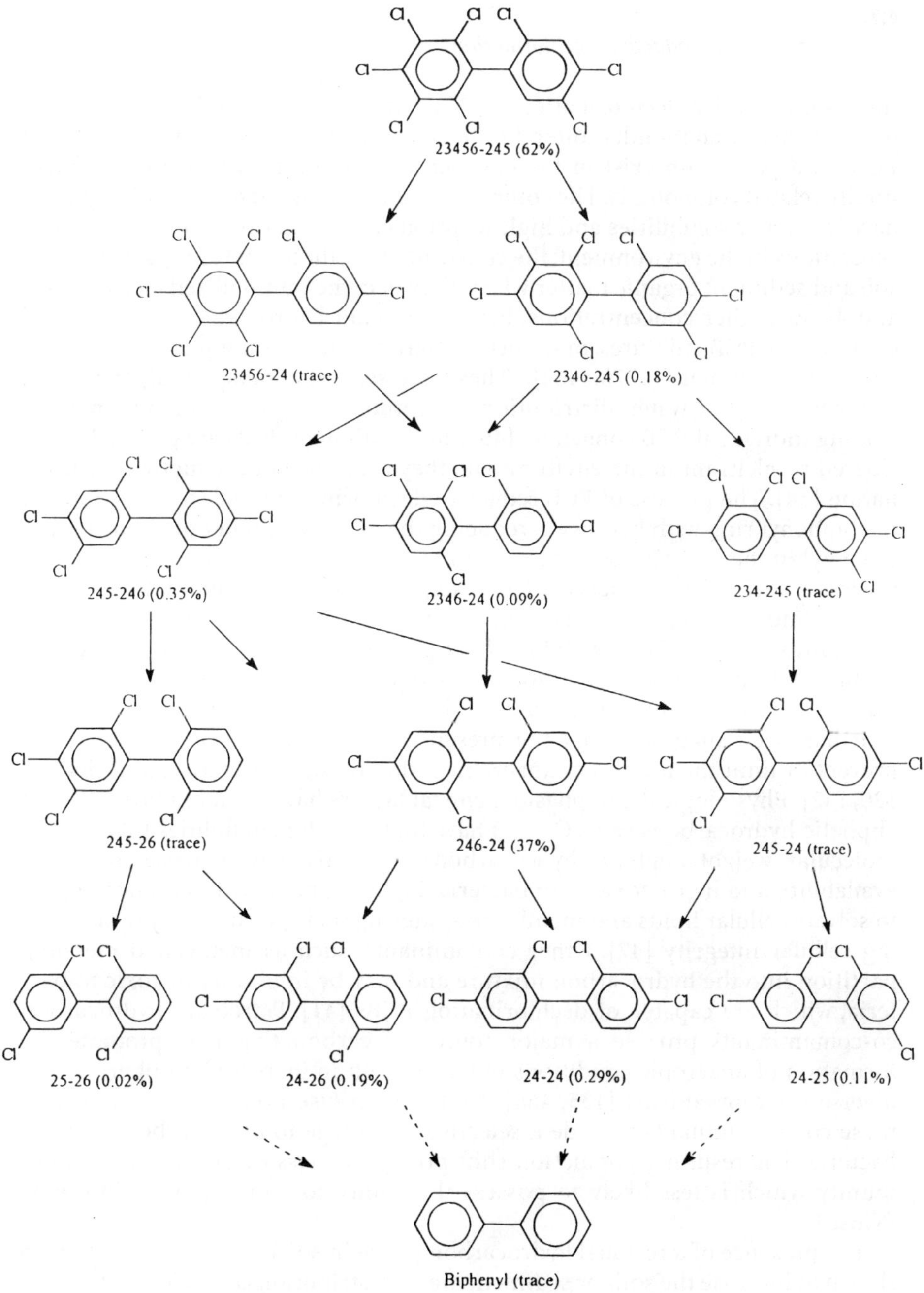

Fig. 34. Dechlorination of 2,3,4,5,6,2′,4′,5′-octachlorobiphenyl. Relative amounts of products are given in parentheses (after Kuipers et al. [461], with permission)

4.1.13.3
Bioavailability and Reductive Dechlorination

Intrinsic reductive dechlorination of PCBs at sites, where residual petroleum products are often found, is often limited or nonexistent [34]. Both commercial PCBs and petroleum exist in the environment as complex mixtures of structurally related compounds. The compounds comprising these mixtures typically have low water solubilities and high sorption coefficients. At relatively low concentrations in the environment, the constituents of these mixtures partition into soil and sediment organic matter where they become immobilized (see Chaps. 2 and 4). At higher concentrations, both petroleum hydrocarbon mixtures and commercial PCB mixtures may form separate stable non-aqueous phases in soils and sediments [270, 462]. These phases may substantially alter the sediment- or soil-water distribution of nonionic organic contaminants, including individual PCB congeners [463, 464]. Although PCBs are generally considered recalcitrant in the environment, they are subject to reductive dechlorination [34]. The process of PCB reductive dechlorination replaces chlorines on the biphenyl ring with hydrogen, reducing the average number of chlorines per biphenyl in the resulting product mixture. This reduction is important because the less chlorinated products are less toxic, have lower bioaccumulation factors, and are more susceptible to aerobic metabolism, including ring opening and mineralization [22, 79, 139, 465]. Although the intrinsic anaerobic reductive dechlorination of PCBs is well documented, the extent and rate of dechlorination varies considerably among sites [34].

It has been suggested that the presence of petroleum hydrocarbons may prevent or limit the process of anaerobic reductive dechlorination of PCBs [34, 120, 142]. Physiological and physiochemical factors have been implicated. Light aliphatic hydrocarbons (e.g., C_3-C_8) have higher water solubilities than higher molecular weight aliphatic hydrocarbons, and this may increase their bioavailability and hence toxicity to bacteria. Light aliphatic hydrocarbons appear to solvate cellular lipids and membranes, altering their permeability or destroying cellular integrity [47]. Other contaminants, such as methylated mercury, partition into the hydrocarbon mixture and may be inhibitory or toxic to bacteria, which are capable of dechlorinating PCBs [41]. Petroleum hydrocarbon co-contaminants provide a major source of carbon that may promote the formation of anaerobic conditions but also result in increased numbers of less diverse microorganisms [135, 466]. Under otherwise nonlimiting conditions, these co-contaminants provide a selective advantage to hydrocarbon-utilizing bacteria. The resulting population shift produces a less diverse bacterial community which is less likely to possess the ability to reductively dechlorinate PCBs.

The presence of a residual hydrocarbon phase in soils or sediments has been shown to increase the soil- or sediment-water distribution coefficients of poorly water-soluble organic contaminants [463, 464]. Such petroleum-hydrocarbon-based phases have been shown to function as effective partition media for PCB congeners [467]. In general, sorption of contaminants by soils and sediments reduces their bioavailability to microorganisms [468, 469]. In this fashion, the presence of an additional partition phase (e.g., a residual petroleum hydrocarbon phase) may reduce bioavailability, thereby limiting the rate and/or extent of PCB dechlorination. This finding was recently supported by the investigation carried out by Zwiernik et al. [168], who determined whether and to what extent petroleum hydrocarbons can inhibit the reductive dechlorination of PCBs in anaerobic contaminated sediment slurries from Silver Lake, MA. They reported the following:

- Sediments of Silver Lake, which contain ~6.2% petroleum hydrocarbons, did not support PCB dechlorination in laboratory assays.
- Removal of petroleum components from lake sediments by solvent extraction did not alter their inability to support dechlorination.
- When other sediments known to support PCB dechlorination were inoculated with PCB-dechlorinating microorganisms and amended with incremental increases of pure petroleum hydrocarbons (0–4 wt%) or 6.2% petroleum hydrocarbons extracted from Silver Lake sediments, a reduction in both the rate and extent of PCB dechlorination occurred.
- The maximal rate of dechlorination observed in these assays depended singularly on the aqueous-phase PCB concentrations.
- Petroleum components in sediments provided a sorptive phase which lowered the solution concentrations of PCBs, thus diminishing the bioavailability of PCBs and rate of dechlorination.

4.1.13.4
Priming and Reductive Dechlorination

The persistence of PCBs in river and harbor sediments and contaminated soils worldwide has become a focus for environmental regulation because PCBs accumulate in fauna and flora and are potentially toxic to wildlife and humans. The discovery that microbial PCB dechlorination was occurring in situ in freshwater and estuarine sediments [371, 451, 454] raised hopes for natural restoration because dechlorination is expected to detoxify the PCBs and at the same time make them more degradable and less persistent [34, 59]. Microbial dechlorination of PCBs has had a major impact at some locations such as the Hudson River [2, 371, 451], but it has had a much smaller impact at other locations such as the Housatonic River (Pittsfield, MA) [453]. An effective method for stimulating or "priming" the activity of indigenous PCB-dechlorinating microbes would have great potential for accelerating natural restoration at the latter sites.

The following section discusses the priming of PCB-dechlorinating microorganisms with various compounds, such as: (1) chlorobiphenyls, (2) bromobiphenyls, and (3) benzoate ions.

4.1.13.4.1 Chlorobiphenyls

Reductive dehalogenation of chloroaromatic compounds can lead to energy conservation [67, 141, 469–471]. For example, it was recently demonstrated that the molar growth yields from the reductive dehalogenation of 3-chloro-4-hydroxyphenylacetate yielded energy for growth equivalent to that obtained from the reduction of nitrate, sulfite, or fumarate [471]. Several studies have proposed that PCB-dechlorinating microorganisms also derive energy by transferring electrons to PCBs [105, 371, 451]. It has also been proposed that high concentrations of halogenated biphenyls (e.g., 2,3,4,6-tetrachlorobiphenyl [2346-CB], 23456-CB, and 2,6-dibromobiphenyl [26-BB]) prime PCB dehalogenation because they support the growth of PCB-dechlorinating microorganisms [59, 156, 158, 245, 246].

The different PCB dechlorination patterns reported by several authors suggested that specific chlorobiphenyls prime dechlorination by enriching distinct microbial populations that exhibit unique PCB dechlorination specificities. The following is a summary of this research:

- Priming was reported by several authors [105, 245, 371] to stimulate selectively the growth of PCB-dechlorinating microorganisms by providing them with a high concentration of a preferred dehalogenation substrate which can act as an electron acceptor in an environment where electron acceptors are limiting. Thus, it should be possible to enrich selectively PCB dechlorinators by sequential transfers with a PCB congener used as a primer. This would most likely enhance the effectiveness of the dechlorination and could lead to a way to stimulate microbial dechlorination of PCBs by adding an inoculum which is highly enriched in PCB-dechlorinating microorganisms.
- Bedard et al. [245] reported that PCB dechlorination was stimulated by adding 2,5,3′,4′-tetrachlorobiphenyl (25–3′4′-CB) to slurries (incubated under methanogenic conditions) of sediments contaminated with Aroclor 1260 from Woods Pond (MA). The 25–3′4′-CB was converted stoichiometrically to 25–3′-CB and stimulated a selective *para*-dechlorination which decreased the penta- through heptachlorobiphenyls containing 234-, 245-, or 2345-chlorophenyl groups by up to 83% in 12 weeks.
- Bedard et al. [14, 15] reported that enrichment with 2,3,4,5,6-pentachlorobiphenyl (23456-CB) greatly enhanced the broad specificity *meta*-dechlorination activity known as Process N and fostered a new *para*-dechlorination activity, Process LP.
- DeWeerd and Bedard [253] investigated novel approaches for enhancing microbial PCB dechlorination in aquatic sediments of the Housatonic River. They reported that PCB dechlorinating microorganisms were activated (i.e., primed) to dechlorinate the PCBs that have persisted for years in these sediments. Several PCB congeners, especially 2,3,4,5,6-pentachlorobiphenyl (23456-CB), 2,3,4,6-tetrachlorobiphenyl (2346-CB), and 2,3,6-trichlorobiphenyl (236-CB), were shown to prime and sustain *meta*-dechlorination of Aroclor 1260 in the river sediment, whereas the PCB congener 245-CB primed *para*-dechlorination of PCBs in the same sediments.

4.1.13.4.2
Bromobiphenyls

The success of priming PCB dechlorination with specific chlorobiphenyl congeners prompted further investigation of PCB priming using individual bromobiphenyl congeners. All of the tested bromobiphenyl congeners were completely dehalogenated to biphenyl, and most required a relatively short acclimation period of 1–2 weeks, which was considerably less time than the corresponding chlorobiphenyl isomers [246]. In addition, specific bromobiphenyl congeners such as 2,6-dibromobiphenyl (26-BB), 2,5,3′-tribromobiphenyl (25–3′-BB), 25–4′-BB, and 245-BB primed more extensive PCB dechlorination in Woods Pond sediments than was observed with the best results from 23456-CB [78]. The complete dehalogenation of the bromobiphenyl primers, the shorter lag times, and the more extensive PCB dechlorination showed great promise for the use of compounds other than chlorobiphenyls to stimulate PCB dechlorination in situ.

The reason for the effectiveness of bromobiphenyls in priming PCB dechlorination is due to the enrichment of bacteria which can use halobiphenyls as electron acceptors. Reductive dehalogenation reactions have been calculated to be energy-yielding reactions for a variety of halogenated aromatic compounds including PBBs, and there is an evidence that they can supply sufficient energy as a respiratory process to support the growth of halorespiring bacteria [472–474]. Although no microorganisms capable of dechlorinating PCBs have been isolated, two studies have shown by most probable number (MPN) methods that the number of PCB dechlorinating microorganisms increased 200–1000-fold as a result of either PCB or bromobiphenyl dehalogenation [157, 271]. These results suggested that halogenated substrates could potentially be used to increase the population size of dehalogenating microorganisms and influence the bioremediation of habitats contaminated with chlorinated compounds.

The ability of enriched microbial populations using halogenated substrates as nutrients also to dehalogenate specific targeted contaminants is considered a type of *cross-acclimation*. Cross-acclimation of dehalogenation activity has been observed previously in sediments and soils and with isolated microorganisms [59, 66, 68, 78, 143, 144, 245, 297, 370, 469]. In some cases, specific halogenated and nonhalogenated compounds that were not transformed still induced the dehalogenation of structural analogs [68]. In addition, some chlorinated substrates co-induced the dechlorination of compounds that were not structural analogs. For example, perchloroethylene (PCE) and trichloroethylene (TCE) were dechlorinated by *Desulfomonile tiedjei* after induction with 3-chlorobenzoate ion [297, 469]. These results suggested that some microorganisms may possess enzymes or cofactors that have a broader specificity for halogenated substrates and perform fortuitous dehalogenation of a variety of halogenated compounds.

Wu et al. [157] applied the most probable number (MPN) method to test the hypothesis that 2,6-dibromobiphenyl (26-BB) primes PCB dechlorination by stimulating the growth of microorganisms which dehalogenate 26-BB and PCBs. The experiments were conducted in anaerobic microcosms of Aroclor 1260-contaminated sediment from Woods Pond (Lenox, MA). They reported that: (1) the number of microorganisms capable of dehalogenating 26-BB and PCBs increased approximately 1000-fold after priming for 48 days with 26-BB in the presence of maleate at 22 °C, and (2) debromination of 26-BB dehalogenated Aroclor 1260 even at high dilutions. In general, these results demonstrated that halogenated biphenyls primed PCB dechlorination by stimulating the growth of PCB-dechlorinating microorganisms. 26-BB primed exclusively *meta*-dechlorination of the PCBs, which effected extensive decreases in the hexa- through nonachlorobiphenyls in only 5–8 months at temperatures as low as 8 °C.

4.1.13.4.3
Benzoic Acids

DeWeerd and Bedard [253] tested the ability of halogenated benzoic acids and other halogenated aromatic compounds to prime PCB dechlorination in contaminated bottom sediments. They found that none of the fluorinated or chlorinated benzoic acids primed PCB dechlorination, but several brominated (e.g., 4-bromobenzoic acid; 2,5-dibromobenzoic acid) and iodinated (e.g., 4-iodobenzoic acid) benzoic acids initiated this activity and primed the most extensive PCB dechlorination, decreasing the hexa- through nonachlorobiphenyl fraction of Aroclor 1260 by 40–70%, 10–50%, and 10–50%, respectively. None of the halogenated benzoic acids were as effective in priming PCB dechlorination as 2,6-dibromobiphenyl, which primed a 60–80% decrease of the hexa- through nonachlorobiphenyl fraction of Aroclor 1260. Several other brominated aromatic compounds were also tested for their ability to prime PCB dechlorination. Monobrominated isomers of acetophenone, phenol, or toluene did not prime PCB dechlorination, but all monobrominated isomers of benzonitrile, 2-bromo-, 4-bromo-, and 2,5-dibromonitrobenzene, 4-bromobenzamide, 4-bromobenzophenone, 4-bromobenzoic hydrazide, methyl 4-bromobenzoate, and 2,5-dibromobenzene sulfonate primed PCB dechlorination in Housatonic River sediments. All of the compounds primed PCB-dechlorination Process N (primarily flanked *meta*-dechlorination) except 4-bromonitrobenzene, which primed dechlorination Process P (flanked *para*-dechlorination). These results indicated that halogenated aromatic compounds that are not structural analogs to PCBs can prime PCB dechlorination.

4.2
Bioremediation Enhancement

There is tremendous interest in using in situ bioremediation for the cleanup of contaminated soil/sediment sites and ground waters. However, biodegradation/biotransformation rates, especially in the subsurface aqueous-solid phase environment, are often constrained by a limited oxygen supply and by factors related to bioavailability (e.g., solubility, dissolution rate, sorption) [150, 191, 200]. Recent research has examined the possibility of enhancing the bioavailability of low solubility and highly sorptive organic compounds by adding a "solubilization" agent (see Chap. 2), such as a surfactant, to the contaminated aqueous/solid phase system [97, 165, 180, 185–187, 192, 195, 202].

On the other hand, Wang and Brusseau [147]) and Wang et al. [193] have been investigating cyclodextrins as an alternative and powerful agent for enhancing solubilization of organic contaminants. Cyclodextrins are cyclic, non-reducing maltooligosaccharides produced from the enzymatic degradation of starch and related compounds by certain bacteria which contain the cyclodextrin glycosyltransferases [93]. The most pertinent property of cyclodextrins is that they have a hydrophilic shell and a hydrophobic cavity. Thus, cyclodextrins have the ability to form water-soluble inclusion complexes by incorporating suitably sized low-polarity molecules in their cavities. Through research on these applications, it has been shown that cyclodextrins can aid the microbial transformation of water-soluble compounds, such as vanillin [232], and low solubility compounds, such as cholesterol [171] and other steroids [176]. Recently, cyclodextrins have been used in environmental applications to improve the remediation of contaminated soil and groundwater by:

- Increasing the apparent water solubilities of low-polarity organic compounds such as trichloroethene, naphthalene, anthracene, chlorobenzene, and DDT [147].
- Reducing the sorption and facilitating the transport of these compounds through soil [179].
- Removing significant amounts of multicomponent, immiscible-organic liquid contamination from an aquifer [189].
- Decreasing β-cyclodextrin to the microbial toxicity of some pesticides and aromatics for wastewater treatment and bioreactor applications [174, 302].
- Investigating the bioavailability and biodegradation of various organic compounds in the presence of cyclodextrins for in situ environmental applications.
- Evaluating the effect of hydroxypropyl cyclodextrine (HPCD) on phenanthrene solubilization and biodegradation, showing HPCD significantly increased the apparent solubility (i.e., the bioavailability) of phenanthrene, having a major impact on the biodegradation rate of phenanthrene [193].

4.3 Verification of Intrinsic Bioremediation

Techniques traditionally used to verify the occurrence of intrinsic bioremediation at contaminated field sites include monitoring indirect indicators of biological activity such as depletion of contaminant and electron acceptors and production of dissolved inorganic carbon (DIC) [310] or methane (CH_4) [475], and the enumeration of BTEX-degrading microorganisms [134]. Although determining the geochemical and microbiological characteristics at a specific contaminated location is essential to any remediation protocol, this approach alone will not provide irrefutable proof of intrinsic bioremediation. This stems from: (1) the difficulty in obtaining accurate mass balances of contaminant, (2) the electron acceptors and end products in heterogeneous soil and groundwater systems, (3) the inability to distinguish between biodegradation and contaminant concentration decreases due to physical processes (e.g., sorption, dissolution, volatilization), and (4) the inability to extrapolate laboratory-based microbiological assays and microcosm studies to intrinsic biodegradation in the field [7, 343].

Stable carbon isotopes (see Chap. 1) provide a promising new method of validating intrinsic bioremediation. Carbon has two stable isotopes, with ^{12}C comprising 98.89% and ^{13}C comprising 1.11% of the total natural abundance [476]. Because of the magnitude of this abundance gap, the ratios of ^{13}C to ^{12}C in carbon-bearing compounds are expressed as per mil (‰) differences relative to a standard (i.e., $\delta^{13}C$ vs PDB, see Chap. 1).

Isotopically distinct molecules will participate in reactions at slightly dissimilar rates. This is known as the *kinetic isotopic effect* and occurs as the result of differences in activation energies of the isotopic forms caused by differences in mass [477]. In particular, biologically mediated reactions tend to favor the lighter isotope. For the stable carbon isotopes (^{13}C and ^{12}C), this typically results in the residual substrate becoming more enriched in ^{13}C (i.e., a less negative $\delta^{13}C$ value) as the reaction proceeds. This phenomenon has been observed in a variety of microbial processes; for example, large isotopic shifts have been recorded during the bacterial oxidation of methane [248, 478] and during the biodegradation of chlorinated hydrocarbons [211, 479–481].

In the past few years, the development of continuous flow compound-specific isotope analysis (CSIA, see Chap. 1) has made it possible to perform rapid isotopic analyses of organic contaminants present as dissolved constituents in groundwater at very low concentrations, providing the potential to use CSIA as a means of validating bioremediation at BTEX-contaminated sites [252, 255, 269, 481]. Within the accuracy and reproducibility typically associated with CSIA (±0.5‰), recent studies have demonstrated that dissolution [481], sorption [227], and volatilization [227, 481] do not significantly alter the isotopic signature of aromatic hydrocarbons. Until recently, however, less was known about carbon isotope fractionation produced during biodegradation. Sherwood Lollar et al. [479] found no significant change in the isotopic composition of the residual toluene during aerobic biodegradation of toluene carried out in laboratory experiments using two mixed microbial consortia cultured from different field sites. Only a small isotopic fractionation effect (i.e., ≅2‰ isotopic enrichment in residual contaminant) was observed during aerobic biodegradation of benzene by a mixed microbial culture [482].

In contrast, Meckenstock et al. [280] reported larger isotopic enrichments in residual toluene, 3–6‰ and up to 10‰ during anaerobic and aerobic biodegradation experiments, respectively. These results indicated that isotopic fractionation effects may be different for different compounds, terminal electron-accepting processes (TEAP), degradative metabolic pathways, or microbial populations.

Significantly, Hall et al. [265] found that two different species of bacteria capable of aerobically degrading phenol produced distinctive fractionation signals in the respired CO_2. More detailed characterization of the magnitude of the carbon isotope fractionation associated with these different parameters must be carried out before the potential of using CSIA as a tool for monitoring biodegradation of aromatic hydrocarbons can be fully assessed. The goal of this experiment was to characterize carbon isotope fractionation associated with anaerobic biodegradation of toluene under two different TEAP.

Ahad et al. [8] evaluated carbon isotope fractionation produced by anaerobic biodegradation of toluene in laboratory experiments under both methanogenic and sulfate-reducing conditions. A small (≅2‰) but highly reproducible ^{13}C-enrichment in the residual toluene at advanced stages of microbial transformation was observed in both cultures. The maximum isotopic enrichment observed in the residual toluene was +2.0‰ and +2.4‰ for the methanogenic and sulfate-reducing cultures, respectively, corresponding to isotopic enrichment factors of −0.5 and −0.8. Because the accuracy and reproducibility associated with gas chromatography combustion-isotope ratio mass spectrometry (GC-C-IRMS or CSIA) is ±0.5‰, delineating which of these two terminal electron-accepting processes (TEAP) is responsible for the biodegradation of toluene at field sites is not possible. However, the potential does exist to use CSIA, in conjunction with other methodologies, as a means of validating advanced stages of intrinsic bioremediation in anaerobic systems. It is important that caution be urged because relating this small (≅2‰) fractionation to biodegradation at complex contaminated field sites will prove a challenge.

5 Conclusion

Of the many biogeochemical processes catalyzed by microorganisms in field sites, one of particular relevance to contemporary society, is the biodegradation of environmental contaminants. Microorganisms can carry out biodegradation in different environmental compartments. Of particular relevance for several organic pollutants is the aqueous-solid phase environment. Aerobic and anaerobic microbial processes are extremely important for the destruction of synthetic organic compounds. Solid phase particulate media such as soils and sediments, receive countless synthetic organic molecules from atmospheric fallout, farming operations, industrial wastes, accidental land and marine spills, or sludge disposal.

Just recently, the disposal of industrial and domestic wastes on or below land surfaces (landfills) became widespread before evidence of groundwater pollution became prominent. However, the soils adjacent to these waste disposal locals contain microbial communities, which should destroy many of the organic compounds, if they are not directly affected by the toxicity of the wastes. Ground water adjacent to these waste-disposal sites, and waters in lakes, rivers, estuaries, and oceans, which receive inadvertent or deliberate discharges of organic chemicals similarly contain highly diverse and often highly active microbial communities (e.g., bacteria, fungi, protozoa). They metabolize numerous natural products as well as various synthetic organic compounds. In addition, a variety of pollutants is retained by the bottom sediments in fresh water or marine environments, and these sediments also contain large and metabolically active communities of heterotrophic microorganisms.

Natural communities of microorganisms in these various habitats have amazing physiological versatilities, where they can metabolize and often mineralize a large number of organic molecules. Probably every natural product, regardless of its complexity, is degraded by one species or another in some particular environment. If not, such organic compounds would have accumulated in enormous amounts. The lack of significant accumulations of natural products in oxic ecosystems is an indication that the indigenous microorganisms utilize these products. A particular species may metabolize only a small number of compounds from this array, but another species in the same habitat may be able to make up for the deficiencies of its neighbor. Although certain bacteria and fungi act on a broad range of organic compounds, no microorganism known to date is sufficiently omnivorous to utilize a large percentage of the natural product compounds that are biosynthesized by plants, animals, and other microorganisms.

Communities of bacteria and fungi can metabolize a multitude of synthetic organic compounds. It is not known how many of the diverse organic molecules synthesized in the laboratory or made industrially can be modified in these ways. However, of the list of compounds presently regarded as pollutants, many can be modified and often are biodegraded by actions of these natural communities. Because few of the known organic compounds have been tested, however, it is not yet certain to what degree the impressive microbial versatility applies to all organic compounds. However, at least this versatility has been amply demonstrated with regard to many of the environmental pollutants of current concern.

In this regard, several conditions must be satisfied for microbial degradation to take place in aqueous-solid phase interfacial environments. These include the following key points:

- *A microorganism with the necessary enzymes must exist to bring about the biodegradation.* The mere existence of a microorganism with the appropriate catabolic potential is necessary; however, it is not sufficient for biodegradation to occur at an interface.
- *A microorganism must be present at the aqueous-solid phase microenvironment containing the organic compound.* Although some microorganisms are present in essentially every environment near the earth's surface, a particular aqueous-solid phase environment may not harbor a microorganism with the necessary enzymes.
- *The organic chemical compound must be accessible to the microorganism having the requisite enzymes.* Many synthetic compounds persist even at interfaces containing the biodegrading species, because the microorganism does not have access to the organic compound which it would otherwise metabolize. Inaccessibility may result from the substrate being in a different microenvironment from the microorganism.
- *If the initial enzyme bringing about the degradation is extracellular, the bonds acted upon by that enzyme must be exposed to the catalyst in order to proceed.* This is not always the case because of sorption of many organic molecules.
- *Should the enzymes catalyzing the initial degradation be intracellular, the molecule to be degraded must penetrate the microbial cell wall to the internal sites where the enzyme acts. Alternatively, the products of an extracellular reaction must penetrate the cell wall for the transformation to proceed further.*

Because biodegradation processes have the potential to eliminate the toxicity of organic contaminants in aqueous-solid phase environments, it is important to know if and when the reactions are actually progressing in real time. In this regard, the methodological limitations of environmental microbiology have major practical implications for safe guarding human health and environmental quality. Within the last few years, significant conceptual and technological improvements in environmental microbiology have been made, which have advanced our understanding of how to demonstrate microbial biodegradation activity in the field. Detection of unique intermediate metabolites in field samples is perhaps the most elegant of the variety of criteria that have recently become accepted as evidence for the occurrence of in situ contaminant biodegradation and bioremediation. This detection is governed by a combination of both the understanding of the biochemistry of the metabolic process and the degree to which sample handling methods have precluded artifacts. Metabolite intermediates are unstable, both chemically and physiologically, so that their detection is best explained by the metabolic process being actively in progress in situ at the time and place of sample removal from the field site. But, given the propensity for microorganisms to respond to environmental changes implicit in field site sample removal, the utmost care must be taken in preventing metabolic change in the microbial community during the interim between sample removal and metabolite analysis.

It is important to indicate that bioremediation technology is not only useful but also should be without risk. Every new technology has a risk which may be large or small, but it does exist. Recognizing the issues or factors coupled to the risk is a first step in reducing or avoiding the risk itself. These issues are not insubstantial, and by learning more about the dangers associated with microbial metabolites, it should be possible to enhance the establishment of various approaches to avoid their occurrence and/or reduce their concentrations. The biologically active metabolite formed from a toxicant may not always be toxic, sometimes it can also be stimulative.

References

1. Aboul-Kassim TAT (1998) PhD Dissertation, Department of Civil, Construction and Environmental Engineering, College of Engineering, Oregon State University, Corvallis, Oregon, USA
2. Abramowicz DA (1994) Res Microbiol 145:42
3. Adriaens P, Chang PR, Barkovskii AL (1996) Chemosphere 32:43
4. Adriaens P, Fu Q, Grbic'-Galic D (1995) Environ Sci Technol 29:2252
5. Adriaens P, Grbic'-Galic D (1994) Chemosphere 29:2253
6. Aelion CM, Bradley PM (1991) Appl Environ Microbiol 57:57
7. Aggarwal PK, Fuller ME, Gurgas MM, Manning JF, Dillon MA (1997) Environ Sci Technol 31:590
8. Ahad JME, Lollar BS, Edwards EA, Slater GF, Sleep BE (2000) Environ Sci Technol 34:892
9. Babcock RW Jr, Stenstrom MK (1993) Water Environ Res 65:26
10. Backhus DA, Ryan JN, Groher DM, Macfarlane JK, Gschwend PM (1993) Ground Water 31:466
11. Bailey RE, Gonsior SJ, Rhinehart WL (1983) Environ Sci Technol 17:617

12. Ballapragada BS, Stensel DH, Puhakka JA, Ferguson JF (1997) Environ Sci Technol 31:1728
13. Ballerstedt H, Kraus A, Lechner U (1997) Environ Sci Technol 31:1749
14. Bedard DL, Van Dort HM, May RJ, Smullen LA (1997) Environ Sci Technol 31:3308
15. Bedard DL, Van Dort HM, May RJ, Smullen LA (1997) Environ Sci Technol 31:3300
16. Ahel M, Conrad T, Giger W (1987) Environ Sci Technol 21:697
17. Ahel M, Giger W, Koch M (1994) Water Res 28:1131
18. Ahel M, Hrsal D, Giger W (1994) Arch Environ Contam Toxicol 26:540
19. Ahel M, McEvoy J, Giger W (1993) Environ Pollut 79:243
20. Anderson WC (ed)(1995) Bioremediation: innovative site remediation technology, vol 1. Springer, Berlin Heidelberg New York
21. Ball HA, Reinhard M (1984) In: Jolley RL, Bull RJ, Davis WP, Katz S, Roberts MH Jr, Jacobs VA (eds) Water chlorination chemistry, environmental impact and health effects. Lewis Publishers, Chelsea, MI, vol 5, pp 1505
22. Bedard DL (1990) In: Kamely D, Chakrabarty A, Omenn GS (eds) Biotechnology and biodegradation. Portfolio Publishing, The Woodlands, Houston, TX, 369
23. Aitken MD, Stringfellow WT, Nagel RD, Kazunga C, Chen S-H (1998) Can J Microbiol 44:743
24. Alagar A, Krishnasamy V (1988) Chem Technol Biotechnol 36:577
25. Al-Bashir B, Cseh T, Leduc R, Samson R (1990) Appl Microbiol Biotechnol 34:414
26. Allard AS, Remberger M, Neilson AH (1985) Appl Environ Microbiol 49:279
27. Altenschmidt U, Fuchs G (1991) Arch Microbiol 156:152
28. Alvarez PJJ, Cronkhite LA, Hunt CS (1998) Environ Sci Technol 32:509
29. Alvarez-Cohen L, McCarty PL (1993) Appl Environ Microbiol 57:228
30. Alvarez-Cohen L, McCarty PL (1991) Appl Environ Microbiol 57:1031
31. Pitter P, Chudoba J (1990) Biodegradability of organic substances in the aquatic environment. CRC Press, Boca Raton, Florida
32. Pritchard PH, Mueller JG, Rogers JC, Kremer FV, Glaser JA (1992) Biodegradation 3:315
33. Arciero DM, Vannelli T, Logan M, Hooper AB (1989) Biochem Biophys Res Commun 159:640
34. Bedard DL, Quensen JF (1995) In: Young LY, Cerniglia C (eds) Microbial transformation and degradation of toxic organic chemicals, Wiley, New York, p 127
35. Cain RB (1981) In: Leisinger T, Hütter R, Cook AM, Nüesch J (eds) Microbial degradation of xenobiotics and recalcitrant compounds. Academic Press, New York, p 326
36. Caldini G, Cenci G, Manenti R, Morozzi G (1995) Appl Microbiol Biotechnol 44:225
37. Cole JR, Fathepure BZ, Tiedje JM (1995) Biodegradation 6:167
38. Kuhn EP, Suflita JM (1989) In: Brown S (ed) Reaction and movement of chemicals in soils. Soil Science Society of America, Madison, WI, p 111
39. Lovely DR, Woodward JC, Chapelle FH (1994) Nature 370:128
40. Alexander M (1995) Environ Sci Technol 29:2713
41. Atlas RM, Bartha R (1993) Microbial ecology, 3rd edn. Benjamin/Cummings Publishing Company, New York
42. Chapelle FH (1993) Ground water microbiology and geochemistry. Wiley, New York
43. Schell MA (1990) In: Chakrabarty AM, Iglewski B, Kaplan S, Silver S (eds) Pseudomonas: biotransformation pathogenesis and evolving biotechnology. American Society for Microbiology, Washington, DC, p 165
44. Takizawa N, Kaida N, Torigoe S, Moritani T, Sawada T, Satoh S, Kiyohara HJ (1994) Bacteriol 176:2444
45. Wackett LP (1995) In: Young LY, Cerniglia CE (eds) Microbial transformation and degradation of toxic organic chemicals. Wiley-Liss, New York, p 217
46. Atlas RM (1984) Petroleum microbiology, MacMillan: New York
47. Britton LH (1984) In: Gibson DT (ed) Microbial degradation of organic compounds, Marcel Dekker, New York, p 89
48. Gibson DT, Subramian V (1984) Microbial degradation of organic compounds. Marcel Dekker, New York

49. Harder W, Dijkhuisen L, Veldkamp H (1984) In: Kelly DP, Carr NG (eds) The microbe. Cambridge Univ Press, Cambridge, UK, part 11, p 51
50. Hobbie JE, Ford TE (1993) In: Ford TE (ed) Aquatic microbiology. Blackwell Scientific Publishers, Boston, MA, p 1
51. Lappin-Scott HM, Greaves MP, Slater JG (1986) In: JensenV, Sorensen LH (eds) Microbial communities in soil. Elsevier Applied Science, London, p 211
52. Schwarzenbach RP, Gschwend PM, Imboden DM (1993) Environmental organic chemistry. Wiley, New York, p 410
53. Young LY, Cerniglia CE (eds) (1995) Microbial transformation and degradation of toxic organic chemicals. Wiley, New York
54. Albrecht A, Barkovskii AL, Adriaens P (1999) Environ Sci Technol 33:737
55. Colberg PJS (1991) Geomicrobiology J 8:147
56. Aldridge WN, Brown AW (1988) In: Craig PJ, Glockling F (eds) The biological alkylation of heavy elements. Royal Society of Chemistry, London, p 147
57. Alexander M (1981) Science 211:132
58. Bowlen GF, Kosson DS (1995) In: Young LY, Cerniglia CE (eds) Microbial transformation and degradation of toxic organic chemicals, Wiley-Liss, New York, p 77
59. Bedard DL, Van Dort HM (1997) In: Sayler GS, Sanseverino J, Davis K (eds) Biotechnology in the sustainable environment, Plenum, New York, p 65
60. Blowes DW, Ptacek CJ, Cherry JA, Gillham RW, Robertson WD (1995) Geoenvironment 2000. American Society of Civil Engineers, New York, p 1588
61. Major EM, Hodgins EW, Butler BJ (1991) In: Hinchee RE, Olfenbuttel RF (eds) On-site bioreclamation. Butterworth-Heinemann, Stoneham, MA, p 147
62. Marcomini A, Capel PD, Lichtensteiger T, Brunner PH, Giger W (1989) J Environ Qual 18:523
63. Mars AE, Houwing J, Dolfing J, Janssen DB (1996) Appl Environ Microbiol 62:886
64. Maruya KA, Risebrough RW, Horne AJ (1996) Environ Sci Technol 30:2942
65. Milligan PW, Häggblom MM (1999) Environ Sci Technol 33:1224
66. Mohn WW, Kennedy KJ (1992) Appl Environ Microbiol 58:1367
67. Mohn WW, Tiedje JM (1991) Arch Microbiol 157:1
68. Mohn WW, Tiedje JM (1992) Microbiol Rev 56:482
69. Novak PJ, Daniels L, Parkin GF (1998) Environ Sci Technol 32:1438
70. Novak PJ, Daniels L, Parkin GF (1998)Environ Sci Technol 32:3132
71. NRC (1993) National Research Council in situ bioremediation: when does it work? National Academy Press. Washington, DC
72. Nyman MC, Nyman AK, Lee LS, Nies LF, Blatchley ER (1997) Environ Sci Technol 31:1068
73. Ogunseitan OA, Delgado IL, Tsai Y-L, Olson BH (1991) Appl Environ Microbiol 57:2873
74. Ou L-T, Joy FJ, Thomas JE, Hornsby AG (1997) Environ Sci Technol 31:717
75. Quensen JF, Boyd SA, Tiedje JM (1990) Appl Environ Microbiol 56:2360
76. Lake JL, Pruell RJ, Osterman FA (1992) Mar Environ Res 33:31
77. Bedard DL, Unterman R, Bopp LH, Brennan MJ, Haberl ML, Johnson C (1986) Appl Environ Microbiol 51:761
78. Bedard DL, Van Dort HM, DeWeerd KA (1998) Appl Environ Microbiol 64:1786
79. Bedard DL, Haberl ML, May RJ, Brennan M (1987) J Appl Environ Microbiol 53:1103
80. Cheng T-C, Calomiris J (1996) J Enzyme Microb Technol 18:597
81. Chesney RH, Sollitti P, Rubin HE (1985) App Environ Microbiol 49:15
82. Clark LB, Rosen RT, Hartman TG, Louis JB, Suffet IH, Lippincott RL, Rosen JD (1992) Int J Environ Anal Chem 47:167
83. Kunc F, Rybarova (1983) J Bio Biochem 15:141
84. Beller HR, Ding W-H, Reinhard M (1995) Environ Sci Technol 29:2864
85. Beller HR, Grbic'-Galic D, Reinhard M (1992) Appl Environ Microbiol 58:786
86. Beller HR, Reinhard M (1995) Microb Ecol 30:105
87. Subba-Rao RV, Rubin HE, Alexander M (1982) Appl Environ Microbiol 43:1139
88. Seto M, Alexander M (1985) App Environ Microbiol 50:1132
89. Stott DE, Martin JP, Focht DD, Haider K (1983) Scisoc Am J 47:66

90. Balthazor TM, Hallas LE (1986) Appr Environ Microbiol 51:432
91. Bruhn C, Lenke H, Knackmuss H-J (1987) Appl Environ Microbiol 53:208
92. McBride KE, Kenny JW, Stalker DM (1986) Appl Environ Microbiol 52:325
93. Bender H (1986) Adv Biotechnol Proc 6:31
94. Denome SA, Stanley DC, Olson ES, Young KD (1993) J Bacteriol 175:6890
95. Gibson SA, Sewell GW (1992) Appl Environ Microbiol 58:1392
96. Goyal AK, Zylstra G (1996) J Appl Environ Microbiol 62:230
97. Grimberg SJ, Stringfellow WT, Aitken MD (1996) Appl Environ Microbiol 62:2387
98. Mueller JG, Chapman PJ, Blattmann BO, Pritchard PH (1990) Appl Environ Microbiol 56:1079
99. Brown JF Jr (1987) Science 236:709
100. Cerniglia CE (1992) Biodegradation 3:351
101. Fish KM, Principe JM (1994) Appl Environ Microbiol 60:4289
102. Fish KM (1996) Appl Environ Microbiol 62:3014
103. Giger G, Brunner PH, Schaffner C (1984) Science 225:623
104. Harkness MR, McDermott JB, Abramowicz DA, Salvo JJ, Flanagan WP, Stephens ML, Mondello FJ, May RJ, Lobos JH, Carroll KM, Brennan MJ, Bracco AA, Fish KM, Warner GL, Wilson PR, Dietrich DK, Lin DT, Morgan CB, Gately WL (1993) Science 259:503
105. Quensen JF III, Tiedje JM, Boyd SA (1988) Science 242:752
106. Schröder HFJ (1993) Chromatogr 647:219
107. Simkins S, Alexander M (1984) Appl Environ Microbiol 47:1299
108. Wu Q, Sowers KR, May HD (1998) Appl Environ Microbiol 64:1052
109. Linkfield TG, Suflita JM, Tiedje JM (1989) Appl Environ Microbio 55:2713
110. Foght JM, Fedorak PM, Westlake DWS (1990) Can J Microbiol 36:169
111. Fujita Y, Reinhard M (1997) Environ Sci Technol 31:1518
112. Kuiper J, Hanstveit AO (1984) Ecotoxicol Environ Saf 8:15
113. Spain JC, Van Veld PA (1983) Appl Environ Microbial 45:428
114. Jones FW, Westmoreland DJ (1998) Environ Sci Technol 32:2623
115. Kubeck E, Naylor CG (1990) J Am Oil Chem Soc 67:400
116. Kuhn EP, Townsend TG, Suflita JM (1990) Appl Environ Microbiol 56:2630
117. Prince RC (1993) Crit Rev Microbiol 19:217
118. Pritchard PH, Bourquin AW (1984) Adv Microbiol Ecol 7:133
119. Quensen JF, Tiedje JM, Boyd SA (1988) Science 242:752
120. Quensen JFI, Zwiernik MJ, Boyd SA, Tiedje JM (1993) In: Abramowicz D, Hamilton S (eds) Twelfth Progress Report for the Research and Development Program for the Destruction of PCBs. General Electric Company Corporate Research and Development, Schenectady, NY, p 117
121. Chapman RA, Harris CR (1986) J Environ Sci Health B21:125
122. Chapman RA, Harris CR, Moy P, Henning K (1986) J Environ Sci Health B21:269
123. de Andrea MM, Lord KA, Bromilow RH, Ruegg EF (1982) Environ Pollut Ser A 27:167
124. Kaufman DD, Katan Y, Edwards DF, Jordan EG (1985) In: Hilton JL (ed) Agricultural chemicals of the future. Rowman and Allenheld, Totowa, p 437
125. Obrigawitch T, Wilson RG, Martin AR, Roeth FW (1982) Weed Sci 30:175
126. Obrigawitch T, Martin AR, Roeth FW (1983) Weed Sci 31:187
127. Bauer EJ, Capone DG (1988) App Environ Microbiol 54:1649
128. Hoover DG, Borgonovi GE, Jones SH, Alexander M (1986) Appl Environ Microbiol 51:226
129. Schmidt SK, Alexander M (1985) Appl Environ Microbiol 49:822
130. Schmidt SK, Alexander M, Schuler ML (1985) J Theor Biol 114:1
131. Schmidt SK, Gier MK (1989) Microb Ecol 18:285
132. Stringfellow WT, Chen SH, Aitken MD (1995) In: Hinchee RE, Vogel CM, Brockman FJ (eds) Microbial processes for bioremediation. Battelle Press, Columbus, OH, p 83
133. Thierrin J, Davis GB, Barber C (1995) Ground Water 33:469
134. Thomas JM, Gordy VR, Fiorenza S, Ward CH (1990) Water Sci Technol 6:53
135. Thomas JM, Ward CH (1992) J Hazard Mater 32:179

136. Oremland RS, Miller LG, Culbertson CW, Connell TL, Jahnke L (1994) Appl Environ Microbiol 60:3640
137. Pignatello JJ, Xing B (1996) Environ Sci Technol 30:1
138. Pritchard PH, Costa CF (1991) Environ Sci Technol 25:372
139. Quensen JF III, Mousa MA, Boyd SA, Sanderson JT, Froese KL, Giesy JP (1998) Environ Toxicol Chem 17:806
140. Rabus R, Widdel F (1996) Appl Environ Microbiol 62:1238
141. Sanford RA, Cole JR, Löffler FE, Tiedje JM (1996) Appl Environ Microbiol 62:3800
142. Unterman R (1996) In: Crawford R, Crawford D (eds) Bioremediation: principles and applications. Cambridge University Press, New York, p 209
143. Van Dort HM, Smullen LA, May RJ, Bedard DL (1997) Environ Sci Technol 31:330
144. Van Dort HM, Smullen LA, May RJ, Bedard DL (1997) Environ Sci Technol 31:321
145. Verheul JH, van den Berg R, Eikelboom DH (1988) In: Wolf K (ed) Contaminated soil. Kluwer Academic Publishers, Norwell, MA, p 705
146. Wang X, Brusseau ML (1993) Environ Sci Technol 27:2821
147. Wang X, Brusseau ML (1995) Environ Sci Technol 29:2346
148. Wang X, Yu X, Bartha R (1990) Environ Sci Technol 24:1086
149. Weissenfels WD, Beyer M, Klein J (1990) Appl Microbiol Biotechnol 32:479
150. Weissenfels WD, Klewer HJ, Langhoff J (1992) Appl Microbiol Biotechnol 36:689
151. Williams WA, May RJ (1997) Environ Sci Technol 31:3491
152. Wilson MS, Madsen EL (1996) Environ Sci Technol 30:2099
153. Wilson SC, Jones KC (1993) Environ Pollut 81:229
154. Winter RB, Yen K-M, Ensley BD (1989) Biotechnology 7:282
155. Workman DJ, Woods SL, Gorby YA, Fredrickson JK, Truex MJ (1997) Environ Sci Technol 31:2292
156. Wu Q, Bedard DL, Wiegel J (1997) Appl Environ Microbiol 63:4818
157. Wu Q, Bedard DL, Wiegel J (1999) Environ Sci Technol 33:595
158. Wu Q, Wiegel J (1997) Appl Environ Microbiol 63:4826
159. Yagi K, Williams J, Wang N-Y, Cicerone RJ (1995) Science 267:1979
160. Yang J, Wang X, Hage DS, Herman PL, Weeks DP (1993) Anals Biochem 219:37
161. Yang Y, Chen RF, Shiaris MP (1994) J Bacteriol 176:2158
162. Yates SR, Ernst FF, Gan J, Gao F, Yates MV (1996) J Environ Qual 25:192
163. Yates SR, Gan J, Ernst FF, Mutziger A, Yates MV (1996) J Environ Qual 25:184
164. Ye D, Siddiqi MA, Maccubbin AE, Kumar S, Sikka HC (1996) Environ Sci Technol 30:136
165. Zhang Y, Maier WJ, Miller RM (1997) Environ Sci Technol 31:2211
166. Zierath DL, Hassett JJ, Banwart WL (1980) Soil Sci 129:277
167. Zou S, Stensel SD, Ferguson JF (2000) Environ Sci Technol 34:1751
168. Zwiernik MJ, Quensen JF III, Boyd SA (1999) Environ Sci Technol 33:3574
169. Halhaway DW (1986) Appl Environ Microbiol 29:463
170. Speier LK (1984) Appl Environ Microbiol 51:683
171. Hesselink PGM, van Vliet S, de Vries H, Witholt B (1989) Enzyme Microb Technol 11:398
172. Khan AA, Wang R-F, Cao W-W, Franklin W, Cerniglia CE (1996) Int J Syst Bacteriol 46:466
173. Kiyohara H, Torigoe S, Kaida N, Asaki T, Iida T, Hayashi H, Takizawa N (1994) J Bacteriol 176:2439
174. Schwartz A, Bar R (1995) Appl Environ Microbiol 61:2727
175. Selifonov SA, Grifoll M, Eaton RW, Chapman PJ (1996) Appl Environ Microbiol 62:507
176. Singer Y, Shity H, Bar R (1991) Appl Microbiol Biotechnol 35:731
177. Stringfellow WT, Aitken MD (1994) Can J Microbiol 40:432
178. Boldrin B, Tiehm A, Fritzsche C (1993) Appl Environ Microbiol 59:1927
179. Brusseau ML, Wang X, Hu Q (1994) Environ Sci Technol 28:8
180. Bury SJ, Miller CA (1993) Environ Sci Technol 27:104
181. Chen S-H, Aitken MD (1999) Environ Sci Technol 33:435
182. Colbert FF, Hendson M, Ferri M, Schroth M (1993) Appl Environ Microbiol 59:2071

183. Colbert SF, Isakeit T, Ferri M, Weinhold AR, Hendson M, Schroth MN (1993) Appl Environ Microbiol 59:2056
184. Gibson DT, Resnick SM, Lee K, Brand JM, Torok DS, Wackett LP, Schocken MJ, Haigler BE (1995) J Bacteriol 177:2615
185. Guha S, Jaffé PR (1996) Environ Sci Technol 30:605
186. Guha S, Jaffé PR (1996) Environ Sci Technol 30:1382
187. Laha S, Luthy RG (1991) Environ Sci Technol 25:1920
188. Mahaffey WR, Gibson DT, Cerniglia CE (1988) Appl Environ Microbiol 54:2415
189. McCray JE, Brusseau ML (1998) Environ Sci Technol 32:1285
190. Menn F-M, Applegate BM, Sayler GS (1993) Appl Environ Microbiol 59:1938
191. Thomas JM, Yordy JR, Amador JA, Alexander M (1986) Appl Environ Microbiol 52:290
192. Tiehm A (1994) Appl Environ Microbiol 60:258
193. Wang J-M, Marlowe EM, Miller-Maier RM, Brusseau ML (1998) Environ Sci Technol 32:1907
194. Ogunseitan OA, Olson BH (1993) Appl Microbiol Biotechnol 38:799
195. Rouse JD, Sabatini DA, Suflita JM, Harwell JH (1994) Environ Sci Technol 24:325
196. Wilson BH, Smith GB, Rees JF (1986) Environ Sci Technol 25:1997
197. Newman LM, Wackett LP (l99l)Appl Environ Microbi l 57:2399
198. Greene S, Alexander M, Leggett D (1981) J Environ Qual 10:416
199. Yordy JR, Alexander M (1981) J Environ Qual 10:266
200. Stucki G, Alexander M (1987) Appl Environ Microbiol 53:292
201. Ayanaba A, Alexander M (1973) Appl Microbiol 25:862
202. Guerin WF, Jones GE (1988) Appl Environ Microbiol 54:937
203. Brewer WS, Draper AC, Wey SS (1980) Environ Pollut Ser B 1:37
204. Richardson ML, Webb KS, Gough TA (1980) Ecototicol Environ Saf 4:207
205. Roberts TR, Standen ME (1978) Pestic Biochem Physiol 9:322
206. Gaynor JD (1984) Can J Soil Sci 64:283
207. Svenson A, Kjeller L-O, Rappe C (1989) Environ Sci Technol 23:901
208. Wannstedt C, Rotella D, Siuda JF (1990)Environ Contam Toxicol 44:282
209. Pieper DH, Winkler R, Sandermann H Jr (1992)Angew Chem Int Ed Engl 31:68
210. Rosenberg A, Alexander M (1979) Appl Environ Microbiol 37:886
211. Hunkeler D, Aravena R, Butler BJ (1999) Environ Sci Technol 33:2733
212. Hutchins SR, Downs WC, Wilson JT, Smith GB, Kovacs DA, Dine DD, Douglass RH, Hendrix DJ (1991) Ground Water 29:571
213. Hutchins SR, Tomson MB, Ward CH (1983) Environ Toxicol Chem 2:195
214. Goldstein RM, Mallory LM, Alexander M (1985) Appl Environ Microbiol 50:977
215. Gorontzy T, Drzyzga O, Kahl MW, Bruns-Nagel D, Breitung J, von Löw E, Blotevogel K-H (1994) Crit Rev Microbiol 20:265
216. Fujita Y, Ding W-H, Reinhard M (1996) Water Environ Res 68:867
217. Fu G, Kan AT, Tomson MB (1994) Environ Toxicol Chem 13:1559
218. Fu MH, Alexander M (1992) Environ Sci Technol 26:1540
219. Fu QS, Barkovskii AL, Adriaens P (1999) Environ Sci Technol 33:3837
220. Ball HA, Reinhard M (1996) Environ Toxicol Chem 15:114
221. Ball HA, Reinhard M, McCarty PL (1989) Environ Sci Technol 23:951
222. Jones S, Alexander M (1988) Appl Environ Microbiol 54:3177
223. Kan AT, Fu G, Tomson MB (1994) Environ Sci Technol 28:859
224. Jannasch HW (1967) Limnol Oceanogr 12:264
225. Beller HR, Spormann AM, Sharma PK, Cole JR, Reinhard M (1996) Appl Environ Microbiol 62:1188
226. Harms H, Zehnder A (1995) Appl Environ Microbiol 61:27
227. Harrington RR, Poulson SR, Drever JI, Colberg PJS, Kelly EF (1999) Org Geochem 30:765
228. Hartkamp-Commandeur LCM, Gerritse J, Govers HAJ, Parsons JR (1996) Chemosphere 32:1275
229. Hasegawa H, Sakka SJ (1988) Non-Cryst Solids 100:201
230. Hashsham SA, Scholze R, Freedman DL (1995) Environ Sci Technol 29:2856

231. Bouwer EJ, McCarty PL, Lance JC (1981) Water Res 15:151
232. Bar R (1989) Appl Microbiol Biotechnol 31:25
233. Bartha R (1986) Microbiol Ecol 12:155
234. Kolpin DW, Goolsby DA, Thurman MEJ (1995) Environ Qual 24:1125
235. Koterba MT, Banks WSL, Shedlock RJ (1993) J Environ Qual 22:500
236. Kravetz L (1990) In: Glass JE, Swift G (eds) Agricultural and synthetic polymers: biodegradation and utilization. American Chemical Society, Washington, DC, p 96
237. Kravetz L, Salanitro JP, Dorn PB, Guin KF (1991) J Am Oil Chem Soc 68:610
238. Krone UE, Laufer K, Thauer RK, Hogenkamp HP (1989) Biochemistry 28:10,061
239. Boethling RS, Alexander M (1979) Appl Environ Microbiol 37:1211
240. Boethling RS, Alexander M (1979) Environ Sci Technol 13:989
241. Brockman FJ, Denovan BA, Hicks RJ, Fredrickson JF (1989) Appl Environ Microbiol 55:1029
242. Button DK (1985) Microbiol Rev 49:270
243. Button DK, Robertson BR (1985) Mar Ecol Prog Ser 26:187
244. Rittmann BE (1985) Sci Total Environ 47:99
245. Bedard DL, Bunnell SC, Smullen LA (1996) Environ Sci Technol 30:687
246. Bedard DL, Van Dort HM (1998) Appl Environ Microbiol 64:940
247. Jobling S, Sheahan D, Osborne JA, Matthiessen P, Sumpter JP (1996) Environ Toxicol Chem 15:194
248. Coleman DD, Risatti JB, Schoell M (1981) Geochim Cosmochim Acta 45:1033
249. Comber MHI, Williams TD, Stewart KM (1993) Water Res 27:273
250. Corcia AD, Samperi R, Marcomini A (1994) Environ Sci Technol 28:850
251. Cork DJ, Krueger JP (1991) Adv Appl Microbiol 36:1
252. Dempster HS, Sherwood Lollar B, Feenstra S (1997) Environ Sci Technol 31:3193
253. DeWeerd KA, Bedard DL (1999) Environ Sci Technol 33:2057
254. Di Corcia A, Samperi R, Marcomini A (1994) Environ Sci Technol 28:850
255. Dias RF, Freeman KH (1997) Anal Chem 69:944
256. Dilley JV, Tyson CA, Spanggord RJ, Sasmore DP, Newell GW, Dacre JCJ (1982) Toxicol Environ Health 9:565
257. Edwards EA, Grbic'-Galic D (1992) Appl Environ Microbiol 58:2663
258. Edwards EA, Grbi-Gali D (1994) Appl Environ Microbiol 60:313
259. Edwards EA, Wills LE, Reinhard M, Grbic'-Galic D (1992) Appl Environ Microbiol 58:794\
260. Flanagan WP, May RJ (1993) Environ Sci Technol 27:2207
261. Fogarty AM, Tuovinen OH (1995) J Ind Microbiol 14:365
262. Gantzer CG, Wackett LP (1991) Environ Sci Technol 25:715
263. George SE, Whitehouse DA, Claxton LD (1992) Environ Toxicol Chem 11:733
264. Ghiorse WC, Herrick JB, Sandoli RL, Madsen EL (1995) Environ Health Perspect 3:107
265. Hall JA, Kalin RM, Larkin MJ, Allen CCR, Harper DB (1999) Org Geochem 30:801
266. Harker AR, Kim Y (1990) Appl Environ Microbiol 56:1179
267. Harkness MR (1993) Science 259:503
268. Hopkins GD, Semprini L, McCarty PL (1993) Appl Environ Microbiol 59:2277
269. Kelley CA, Hammer BT, Coffin RB (1997) Environ Sci Technol 31:2469
270. Killidromitou D, Bonazountas M (1993) In: Calabrese EJ, Kostecki P (eds) Principles and practices for petroleum contaminated soils. Lewis Publishers, Chelsa, MI, p 111
271. Kim J, Rhee GY (1997) Appl Environ Microbiol 63:1771
272. Kohler H-P, Kohler-Staub D, Focht DD (1988) Appl Environ Microbiol 54:1940
273. Linkfield TG, Suflita JM, Tiedje JM (1989) Appl Environ Microbiol 55:2773
274. Logan MSP, Newman LM, Schanke CA, Wackett LP (1993) Biodegradation 4:39
275. Lovely DR, Coates JD, Woodward JC, Philips EJP (1995) Appl Environ Microbiol 61:953
276. Lovley DR (1991) Microbiol Rev 55:259
277. Lovley DR, Lonergan DJ (1990) Appl Environ Microbiol 56:1858
278. McGroddy SE, Farringtion JW, Gschwend PM (1996) Environ Sci Technol 30:172
279. Means JC, Wood SG, Hassett JJ, Banwar WL (1980) Environ Sci Technol 14:1524

280. Meckenstock RU, Morasch B, Warthmann R, Schink B, Annweiler E, Michaelis W, Richnow HH (1999) Environ Microbiol 10:409
281. Mellouki A, Talukdar R, Schmoltmer AM, Gierczak T, Mills MJ, Solomon S, Ravishankara A (1992) Geophys Res Lett 19:2059
282. Mihelcic JR, Luthy RG (1991) Environ Sci Technol 25:169
283. Miller JJ, Foroud NB, Hill D, Linwall CW (1994) Can J Soil Sci 75:145
284. Nanny MA, Bortiatynski JM, Hatcher PG (1997) Environ Sci Technol 31:530
285. Roszak DB, Colwell RR (1987) Microbiological Reviews 51:365
286. Roth JR, Lawrence JG, Bobik TA (1996) Annu Rev Microbiol 50:137
287. Rothmel RK, Peters RW, Martin E, Deflaun MF (1998) Environ Sci Technol 32:1667
288. Rouchaud J, Roucourt P, Vanachter A, Benoit F, Ceustermans N (1988) Reu Agric 41:889
289. Rubin HE, Alexander M (1983) Environ Sci Technol 17:104
290. Rubin HE, Subba-Rao RV, Alexander M (1982) Appl Environ Microbio 43:1133
291. Shannon MJR, Unterman R (1993) Annu Rev Microbiol 47:715
292. Sheehan PJ, Schneiter RW, Mohr TKG, Gersberg RM (1988) Second National Outdoor Action Conference on Aquifer Restoration Groundwater Monitoring and Geophysical Methods, Las Vegas, NV
293. Singh G, Kathpal TS, Spencer WF, Dhankar IS (1991) Environ Pollut 70:219
294. Skeen RS, Petersen JN (1995) Biotech Bioeng 45:279
295. Stabnikova EV, Selezneva MV, Dulgerov AN, Ivanov VN (1996) Appl Biochem Microbiol 32:219
296. Tindall JA, Vencill WK (1995) J Hydrol 166:37
297. Townsend GT, Suflita JM (1997) Appl Environ Microbiol 63:3594
298. Van Dort HM, Bedard DL (1991) Appl Environ Microbiol 57:1576
299. Walstra P, De Roos, AL Food (1993) Rev Int 9:503
300. Williams WA, Lobos JH, Cheetham WE (1997) Intl J Syst Bacteriol 47:207
301. Brown AB, Hinchee RE, Norris RD, Wilson JT (1996) Remediation 95:109
302. Oláh J, Cserháti T, Szejtli J (1988) Water Res 22:1345
303. Johnson PC, Kemblowski MW, Colhart JD (1990) Ground Water 28:413
304. Barker JF, Patrick GC, Major D (1987) Ground Water Monitor Rev 7:64
305. Ridgway HF, Safarik J, Phipps D, Carl P, Clark D (1990) Appl Environ Microbiol 56:3565
306. Smith MR (1990) Biodegradation 11:191
307. Norris RD, Hinchee RE, Brown R, McCarty PL, Semprini L, Wilson JT, Kampell DH, Reinhard M, Bouwer EJ, Borden RC, Vogel TM, Thomas JM, Ward CH (1994) The handbook of bioremediation. Lewis Publishers, Boca Raton, FL
308. Morgan P, Watkinson RJ (1992) Water Res 26:73
309. Harms G, Zengler K, Rabus R, Aeckersberg F, Minz D, Rossello-Mora R, Widdel F (1999) Appl Environ Microbiol 65:999
310. Chapelle FH, Bradley PM, Lovley DR, Vroblesky DA (1996) Ground Water 34:691
311. Schmitt R, Langguth HR, Puttmann W, Rohns HP, Eckert P, Schubert J (1996) Org Geochem 25:41
312. Berry DF, Francis AJ, Bollag J-M (1987) Microbiol Rev 51:43
313. Krumholz LR, Caldwell ME, Suflita JM (1996) In Crawford RL, Crawford DL (eds) Bioremediation: principles and applications. Cambridge University Press, Cambridge, UK, p 61
314. Haag F, Reinhard M, McCarty PL (1991) Environ Toxicol Chem 10:1379
315. Werner P (1985) Water Supply 3:41
316. Werner P (1985) Water Supply 3:53
317. Barbaro JR, Barker JF, Lemon LA, Gillham RW, Mayfield CI (1990) Canadian Petroleum Products Institute, Ottawa, Canada, Rep 91
318. Reinhard M, Shang S, Kitanidis PK, Orwin E, Hopkins GD, LeBron CA (1997) Environ Sci Technol 31:28
319. Sturman PJ, Stewart PS, Cunningham AB, Bouwer EJ, Wolfram JH (1995) J Contam Hydrol 19:17

320. Fitch F (1989) In: Wilson AJ (ed) Foams: physics, chemistry, structure. Springer, Berlin Heidelberg New York, p 207
321. Li Z, Obika H, Fukuoka S, Kakita H, Kobayashi Y, Higashihara T (1994) Biotechnology 72:363
322. Li X-F, Cullen WR, Reimer KJ, Le X-C (1995) Sci Total Environ 177:17
323. Li X-F, Le X-C, Simpson CD, Cullen WR, Reimer KJ (1996) Environ Sci Technol 30:1115
324. Chaphalikar PG, Valsaraj KT, Roy D, Constant WD, Lee P (1997) In Tedder DW, Pohland FG (eds) Emerging technologies in hazardous waste management, Plenum Press, New York, p 113
325. Ripley MB, Harrison AB, Betts WB, Dart RK, Wilson AJ (2000) Environ Sci Technol 34:489
326. NAS (1994) Alternatives for groundwater cleanup. Report of the National Academy of Science Committee on Groundwater Cleanup Alternatives. National Academy Press, Washington DC
327. Bosma TNP, Ballemans EMW, Hoekstra NK (1996) Ground Water 34:49
328. Jetten MSM, Stams AJM, Zehnder AB (1990) Microbiol Ecol 73:339
329. LaPat-Polasko LT, McCarty PL, Zehnder AJB (1984) Appl Environ Microbiol 47:825
330. Rittmann BE, McCarty PL (1980) Biotechnol Bioeng 22:2359
331. Tros ME, Schraa G, Zehnder AJB (1996) Appl Environ Microbiol 62:437
332. van der Meer JR, Roelofsen W, Schraa G, Zehnder AJB (1987) Microbiol Ecol 45:333
333. Bosma TNP, Middeldorp PJM, Schraa G, Zehnder AJB (1997) Environ Sci Technol 31:248
334. Bouwer EJ, McCarty PL (1984) Ground Water 45:433
335. Nyer EK, Suthersan S (1996) Ground Water Monit Rem 16:70
336. Horvath AMJ (1988) Bacteriol 170:3742
337. Shiaris MP (1989) Appl Environ Microbiol 55:1391
338. Madsen EL, Sinclair JL, Ghiorse WC (1991) Science 52:830
339. Bragg JR, Prince RC, Harner EJ, Atlas RM (1994) Nature 368:413
340. Rittmann BE, Johnson NM (1989) Water Sci Technol 21:209
341. Bourquin AW (1989) Hazard Mater Control 2:16
342. Cullen WR, Li X-F, Reimer KJ (1994) Sci Total Environ 156:27
343. Madsen EL (1991) Environ Sci Technol 25:1662
344. MacDonald JA, Rittmann BE (1993) Environ Sci Technol 27:1974
345. The Merck Index (1989) 11th edn, Merck, Rahway, NJ
346. Speight JG (1991) The chemistry and technology of petroleum, 2nd edn. Marcel Dekker, New York
347. Murarka I (1992) Hazard Mater 32:245
348. Herrick JB, Madsen EL, Batt CA, Ghiorse WC (1993) Appl Environ Microbiol 59:687
349. Moré MI, Herrick JB, Silva MC, Ghiorse WC, Madsen EL (1994) Appl Environ Microbiol 60:1572
350. Erickson DC, Loehr RC, Neuhauser EF (1993) Water Res 27:911
351. Lerch RN, Thurman EM, Kruger EL (1997) Environ Sci Technol 31:1539
352. Verstraete W, Devliegher W (1996) Biodegradation 7:471
353. Bollag J-M (1992) Environ Sci Technol 26:1876
354. McCarthy JF, Jimenez BD (1985) Environ Sci Technol 19:1072
355. Weber WJ, Huang W (1996) Environ Sci Technol 30:881
356. Carmichael LM, Pfaender FK (1997) Environ Toxicol Chem 16:666
357. Guerin WF, Boyd SA (1992) Appl Environ Microbiol 58:1142
358. Bollag J-M (1983) In: Christman RF, Gjessing ET (eds) Aquatic and terrestrial humic materials. Ann Arbor Science, Ann Arbor, MI
359. Burgos WD, Novak JT, Berry DF (1996) Environ Sci Technol 30:1205
360. Guthrie EA, Pfaender FK (1998) Environ Sci Technol 32:501
361. Foght JM, Westlake DWS (1988) Can J Microbiol 34:1135
362. EPA (1979) Survey of the manufacture import and uses for benzidine related substances and related dyes and pigments. US Environmental Protection Agency, Office of Toxic Substances, Washington DC, EPA-560/13-79-005

363. EPA (1979) TSCA Chemical assessment series preliminary risk assessment phase i: benzidine. it's congeners and their derivative dyes and pigments. US Environmental Protection Agency, Washington DC, EPA-560/11–80–019
364. EPA (1975) Review of the environmental fate of selected chemicals. US Environmental Protection Agency, Office of Toxic Substances, Washington, EPA-560/5–75–001
365. EPA (1986) Quality criteria for water. US Environmental Protection Agency, Office of Water Regulations and Standards, Washington DC, EPA 440/5–86–001
366. EPA (1978) Fate of 3,3′-dichlorobenzidine in aquatic environments. US Environmental Protection Agency, Environmental Research Laboratory, Athens, EPA–600/3–78–068
367. Binkley BJ (1993) MS Thesis, Purdue University
368. Wu S-C, Gschwend PM (1986) Environ Sci Technol 20:717
369. Kuhn EP, Suflita JM (1989) Environ Sci Technol 23:842
370. Struijs J, Rogers JE (1989) Appl Environ Microbiol 55:2527
371. NIOSH (1976) Occupational exposure to 1122-tetrachloroethane. National Institute for Occupational Safety and Health.Division of Criteria Documentation and Standards Development,US Government Printing Office, Washington,DC
372. EPA (1991) Toxics release inventory-public data release. US Environmental Protection Agency, Office of Pollution Prevention and Toxics, Washington DC
373. Vogel TM, Criddle CS, McCarty PL (1987) Environ Sci Technol 21:722
374. Bouwer EJ, McCarty PL (1983) Appl Environ Microbiol 45:1295
375. Vogel TM, McCarty PL (1991) Environ Sci Technol 21:76
376. Roberts AL, Gschwend PM (1987) Environ Sci Technol 25:1208
377. Thompson JA, Ho B, Mastovich SL (1985) Anal Biochem 145:376
378. Nastainczyk W, Ahr H, Ulrich V (1982) Biochem Pharmacol 3:391
379. Schanke CA, Wackett LP (1992) Environ Sci Technol 26:830
380. Chen C, Puhakka JA, Ferguson JF (1996) Environ Sci Technol 30:542
381. Criddle CS, Dewitt JT, Grbic'-Galic D, McCarty PL (1990) Appl Environ Microbiol 56:3240
382. Stensel HD, DeJong L (1994) Bioremediation of chlorinated and polycyclic aromatic hydrocarbon compounds. Lewis Publishers, Boca Raton, FL
383. Egli C, Leisinger T (1988) Appl Environ Microbiol 54:2819
384. Bouwer EJ, Wright JP (1988) J Contam Hydrol 2:155
385. Cobb GD, Bouwer EJ (1991) Environ Sci Technol 25:1068
386. Van Eekert MHA, Schroder TJ, Stams AJM, Schraa G, Field JA (1998) Appl Environ Microbiol 64:2350
387. Komisar SJ (1993) PhD Thesis, Department of Civil and Environmental Engineering, University of Washington, WA
388. Mägli A, Wendt M, Leisinger T (1996) Arch Microbiol 166:101
389. Freedman DL, Gossett JM (1991) Appl Environ Microbiol 57:2847
390. Lewis TA, Morra MJ, Brown PD (1996) Environ Sci Technol 30:292
391. Gianessi LP, Puffer C (1991) Quality of the Environment Division Resources for the Future. 1616 P Street NW Washington DC, 20036
392. Murray MR, Hall JK (1989) J Environ Qual 18:51
393. Hallberg GR (1989) Agri Ecosyst Environ 26:299
394. Canfield DE, Thamdrup B, Hansen JW (1993) Geochim Cosmochim Acta 57:3867
395. Blackburn HT, Blackburn ND (1992) Microbiol Lett 100:517
396. Myers CR, Nealson KH (1988) Science 240:1319
397. Kazumi J, Häggblom MM, Young LY (1995) Appl Microbiol Biotechnol 43:929
398. Häggblom MM, Rivera MD, Young LY (1993) Appl Environ Microbiol 59:1162
399. Ferrer MR, del Moral A, Ruiz-Berraquero F, Ramos-Cormenzana A (1985) Chemosphere 14:1645
400. Krueger JP, Butz RG, Atallah YH, Cork DJ (1989) J Agric Food Chem 37:534
401. Taraban RH, Berry DF, Berry DA, Walker HL (1993) Appl Environ Microbiol 59:2332

402. Noling JW, Becker JO (1994) Suppl J Nematol 26 : 73
403. Fergusion W, Padula A (1994) Economic effects of banning methyl bromide for soil fumigation. USDA Economic Research Service Agricultural Economic Report 677, USDA, Beltville, MD
404. Watson RT, Albritton DL, Anderson SO, Lee-Bapty S (1992) Methyl bromide: its atmospheric science technology and economics. United Nations Environmental Programme, United Nations Headquarters, Nairobi, Kenya
405. Shorter JH, Kolb CE, Crill PM, Kerwin RA, Talbot RW, Himes ME, Harris RC (1995) Nature 377 : 717
406. Yagi K, Williams J, Wang N-Y, Cicerone RJ (1993) Proc Natl Acad Sci USA 90 : 8420
407. Gan J, Yates SR, Anderson MA, Spenser WF, Ernst FF, Yates MV (1994) Chemosphere 29 : 2685
408. Fox BG, Borneman JG, Wackett LP, Lipscomb JD (1990) Biochemistry 29 : 6419
409. Oldenhuis R, Oedze JY, van der Warrde JJ, Janssen DB (1991) Appl Environ Microbiol 55 : 2819
410. Rasche ME, Hicks RE, Hyman MR, Arp DJ (1990) J Bacteriol 171 : 5368
411. Drzyzga O, Bruns-Nagel D, Gorontzy T, Blotevogel K-H, Gemsa D, von Löw E (1998) Environ Sci Technol 32 : 3529
412. Kaplan DL, Kaplan AM (1982) Appl Environ Microbiol 44 : 757
413. Palazzo AJ, Leggett DC (1986) J Environ Qual 15 : 49
414. Griest WH, Tyndall RL, Stewart AJ, Caton JE, Vass AA, Ho C-H, Caldwell WM (1995) Environ Toxicol Chem 14 : 51
415. Isbister JD, Anspach GL, Kitchens JF, Doyle RC (1984) Microbiologica 7 : 47
416. Kästner M, Lotter S, Heerenklage J, Breuer-Jammali M, Stegmann R, Mahro B (1995) Appl Microbiol Biotechnol 43 : 1128
417. Caton JE, Ho C-H, Williams RT, Griest WH (1994) J Environ Sci Health 29 : 659
418. Bruns-Nagel D, Breitung J, von Löw E, Steinbach K, Gorontzy T, Kahl M, Blotevogel K-H, Gemsa D (1996) Appl Environ Microbiol 62 : 2651
419. Rieger P-G, Knackmuss H-J (1995) In: Spain JC (ed) Biodegradation of nitroaromatic compounds. Plenum Press, New York
420. Lenke H, Warrelmann J, Daun G, Walter U, Sieglen U, Knackmuss H-J (1997) In: Alleman BC, Leeson A (eds) In situ and on-site bioremediation. Batelle Press, Columbus, OH, vol 2, p 1
421. Vancheeswaran S, Semprini L, Williamson KJ, Ingle JD Jr (1998) Final Report to the Lawrence Livermore National Laboratory Site-300. Project: Intrinsic Transformation of Alkoxysilanes and Chlorinated Ethenes at Site-300
422. Vancheeswaran S, Semprini L, Williamson KJ, Ingle JD Jr, Daley P (1998) Proceedings of the 1st International Conference on Remediation of Chlorinated and Recalcitrant Compounds, Monterey, CA
423. Vancheeswaran S, Halden RU, Williamson KJ, Ingle JD Jr, Semprini L (1999) Environ Sci Technol 33 : 1077
424. Arkles BA (1995) Silicon, germanium, tin and lead compounds: metal alkoxides, diketonates and carboxylates: a survey of properties and chemistry. Gelest, Tullytown PA
425. Westrick JJ, Mello JW, Thomas RF (1984) J Am Water Works Assoc 76 : 52
426. Hamamura N, Page C, Long T, Semprini L, Arp D (1997) J Appl Environ Microbiol 63 : 3607
427. Fennell DE, Gossett JM, Zinder SH (1997) Environ Sci Technol 31 : 918
428. Semprini L (1997) Curr Opin Biotechnol 8 : 296
429. Fiedler H, Lau C, Kjeller LO, Rappe C (1996) Chemosphere 32 : 421
430. Bopp RF, Gross ML, Tong H, Simpson HJ, Monson SJ, Deck BL, Moser FC (1991) Environ Sci Technol 25 : 951
431. Beurskens JEM, Toussaint M, de Wolf J, van der Steen JMD, Slot PC, Commandeur LCM, Parsons JH (1995) Environ Toxicol Chem 14 : 939
432. Barkovskii AL, Adriaens P (1996) Appl Environ Microbiol 62 : 4556
433. Barkovskii AL, Adriaens P (1998) Environ Toxicol Chem 17 : 1013

434. Naylor CG, Mieure JP, Adams WJ, Weeks JA, Castaldi FJ, Ogle LD, Romano RR (1992) J Am Oil Chem Soc 69:695
435. Aboul-Kassim TAT, Simoneit BRT (1993) CRC-Crit Rev Environ Sci Technol 23:325
436. Talmage S (1994) Environmental and human safety of major surfactants: alcohol ethoxylates and alkylphenol ethoxylates. Lewis Publishers, Ann Arbor, MI
437. Yoshimura K (1986) J Am Oil Chem Soc 63:1590
438. Ventura F, Figueras A, Caixach J, Espadaler I, Romero J, Guardiola J, Rivera J (1988) Water Res 22:1211
439. Ekelund R, Bergman Å, Granmo Å, Berggren M (1990) Environ Pollut 64:107
440. Soto AM, Justicia H, Wray JW, Sonnenschein C (1991) Environ Health Perspect 92:167
441. Jobling S, Sumpter JP (1993) Aquat Toxicol 27:361
442. Maki H, Masuda N, Fujiwara Y, Ike M, Fujita M (1994) Appl Environ Microbiol 60:2265
443. Swisher RD (1987) Surfactant biodegradation. 2nd edn. Marcel Dekker, New York
444. Ding W-H, Fujita Y, Aeschimann R, Reinhard M (1996) Fresenius J Anal Chem 354:48
445. Reinhard M, Goodman N, Mortelmans KE (1982) Environ Sci Technol 16:351
446. Weinberger P, Greenhalg R (1984) In: Garner WY, Harvey J (eds) Chemical and biological controls in forestry. ACS Symposium Series 238, American Chemical Society, Washington, DC, p 351
447. Lewis MA (1991) Wat Res 25:101
448. Trocmé M, Tarradellas J, Védy J-C (1988) Biol Fertil Soils 5:299
449. Tanghe T, Devriese G, Verstraete W (1998) Water Res 32:2889
450. Ejlertsson J, Nilsson M-L, Kylin H, Bergman Å, Karlson L, Öquist M, Svensson BH (1999) Environ Sci Technol 33:301
451. Brown JF Jr, Wagner RE, Feng H, Bedard DL, Brennan MJ, Carnahan JC, May R (1987) J Environ Toxicol Chem 6:579
452. Bedard DL, May RJ (1995) Environ Sci Technol 30:237
453. Brown JF (1994) Environ Sci Technol 28:2295
454. Brown JF Jr, Wagner RE (1990) Environ Toxicol Chem 9:1215
455. Brown JF Jr, Wagner RE, Bedard DL, Brennan MJ, Carnahan JC, May RJ, Tofflemire TJ (1984) Northeast Environ Sci 3:167
456. Sokol RC, Kwon O-S, Bethoney CM, Rhee G-Y(1994) Environ Sci Technol 28:2054
457. Rhee G-Y, Sokol RC, Bethoney CM, Bush B (1993) Environ Sci Technol 27:1190
458. Williams WA (1994) Environ Sci Technol 28:630
459. Berkaw M, Sowers KR, May HD (1996) Appl Environ Microbiol 62:2534
460. Cutter L, Sowers KR, May HD (1998) Appl Environ Microbiol 64:2966
461. Kuipers B, Cullen WR, Mohn WW (1999) Environ Sci Technol 33:3579
462. Mackay D (1982) Studies on the use of polychlorinated biphenyls. United States Environmental Protection Agency
463. Boyd SA, Sun S (1989) Environ Sci Technol 24:142
464. Luthy RG, Dzombak DA, Shannon MJR, Unterman R, Smith JR (1997) Water Res 31:561
465. Mousa MA, Quensen JF III, Chou K, Boyd SA (1996) Environ Sci Technol 30:2087
466. Song H, Bartha RA (1990) Environ Microbiol 56:646
467. Sun S, Boyd SA (1991) J Environ Qual 20:557
468. Volkering F, Breure AM, Van Andel JG (1993) Appl Microbiol Biotechnol 40:535
469. Cole JR, Cascarelli AL, Mohn WW, Tiedje JM (1994) Appl Environ Microbiol 60:3536
470. Dolfing J (1990) Arch Microbiol 153:264
471. Mackiewicz M, Wiegel J (1998) Appl Environ Microbiol 64:352
472. Dolfing J, Harrison BK (1992) Environ Sci Technol 26:2213
473. Holmes DA, Harrison BK, Dolfing J (1993) Environ Sci Technol 27:725
474. El Fantroussi S, Naveau H, Agathos SN (1998) Biotechnol Prog 14:167

475. Chapelle FH, McMahon PB, Dubrovsky NM, Fujii RF, Oaksford ET, Vroblesky DA (1995) Water Resour Res 31:359
476. Faure G (1986) Principles of isotope geology. Wiley, New York
477. Galimov EM (1985) The biological fractionation of isotopes. Academic Press, Orlando
478. Barker JF, Fritz P (1981) Nature 273:289
479. Sherwood Lollar B, Slater G, Ahad J, Sleep B, Spivack J, Brennan M, MacKenzie P (1999) Org Geochem 30:813
480. Heraty LJ, Fuller ME, Huang L, Abrajano T Jr, Sturchio NC (1999) Org Geochem 30:793
481. Slater GF, Dempster HS, Sherwood Lollar B, Ahad J (1999) Environ Sci Technol 33:190
482. Stehmeier LG, Francis MM, Jack TR, Diegor E, Winsor L, Abrajano TA Jr (1999) Org Geochem 30:821

Subject Index

Printing (Computer to Film): Saladruck Berlin
Binding: Stürtz AG, Würzburg